ALLE ZEIT WACH 1842

V. Ittekkot S. Kempe
W. Michaelis A. Spitzy (Eds.)

# Facets of Modern Biogeochemistry

## Festschrift for E.T. Degens

With 161 Figures

Springer-Verlag Berlin Heidelberg New York
London Paris Tokyo Hong Kong

Dr. VENUGOPALAN ITTEKKOT
Dr. STEPHAN KEMPE
Dr. WALTER MICHAELIS
Dr. ALEJANDRO SPITZY

Institut für Biogeochemie und Meereschemie
der Universität Hamburg
Bundesstraße 55
2000 Hamburg 13, FRG

ISBN-13:978-3-642-73980-4 e-ISBN-13:978-3-642-73978-1
DOI: 10.1007/978-3-642-73978-1

Library of Congress Cataloging-in-Publication Data. Facets of modern biogeochemistry / V. Ittekkot ... [et al.] eds. p. cm. ISBN-13:978-3-642-73980-4 (U.S.) 1. Geochemistry. 2. Marine sediments. 3. Biogeochemistry. I. Ittekkot, V. (Venugopalan), 1945-. QE515.F23 1988 551.9 – dc20 89-21742


Softcover reprint of the hardcover 1st edition 1990



Typesetting: International Typesetters Inc., Makati, Philippines

2132/3145-543210 – Printed on acid-free paper

# Contents

## *Chapter 5 Inorganic Geochemistry*



## *Chapter 6 Organic Geochemistry*

# Contributors

Botz, R., Geologisch-Paläontologisches Institut der Christian-Albrechts-Universität zu Kiel, Olshausenstraße 40, D-2300 Kiel 1, FRG

Brockmann, Uwe H., Institut für Biogeochemie und Meereschemie der Universität Hamburg, Martin-Luther-King-Platz 6, D-2000 Hamburg 13, FRG

Calvert, S. E., Department of Oceranography, University of British Columbia, Vancouver, British Columbia, V6T 1W5, Canada

Carey, Robert, U.S. Department of Commerce, National Environmental Satellite, Data and Information Services, Washington, DC 20233, USA

Clüsener Godt, M., General Ecology Group, Biology/Chemistry Department of the University, Barbarastraße 11, D-4500 Osnabrück, FRG

Dickman, M. D., Department of Biological Sciences, Brock University, St. Catharines, Ontario L2S 3A1, Canada

Eisma, Doeke, Netherlands Institute for Sea Research, P.O. Box 59, NL-1790 AB Den Burg, Texel, The Netherlands

Emeis, Kay-Christian, Geologisch-Paläontologisches Institut der Christian-Albrecht-Universität Kiel, Olshausenerstraße 40, D-2300 Kiel 1, FRG

Emery, K. O., Woods Hole Oceanographic Institution, Woods Hole, MA 02543, USA

Esser, G., General Ecology Group, Biology/Chemistry Department of the University, Barbarastraße 11, D-4500 Osnabrück, FRG

Faber, E., Bundesanstalt für Geowissenschaften und Rohstoffe, Stilleweg 2, Alfred-Bentz-Haus, D-3000 Hannover, FRG

Farrington, J.W., Environmental Sciences Programme, University of Massachusetts-Boston, Boston, MA 02125-3393, USA

Feng, Xuan-Zeng, Central Laboratory of Petroleum Geology, P.O. Box 916, Wuxi, Jiangsu, PR China

Fyfe, W.S., Dean of Science, University of Western Ontario, London, Ontario N6A 5B7, Canada

Haake, B., Institut für Biogeochemie und Meereschemie der Universität Hamburg, Bundesstraße 55, D-2000 Hamburg 13, FRG

Haddad, R.I., NASA, Ames Research Center, Moffett Field, CA 94035, USA

Honjo, Susumu, Woods Hole, Oceanographic Institution Woods Hole, MA, 02543, USA

Irion, G., Forschungsinstitut Senckenberg, Abt. Meeresgeologie und Meeresbiologie, D-2940 Wilhelmshaven, FRG

Ittekkot, V., Institut für Biogeochemie und Meereschemie der Universität Hamburg, Bundesstraße 55, D-2000 Hamburg 13, FRG

Jenisch, Angela, Institut für Biogeochemie und Meereschemie der Universität Hamburg, Bundesstraße 55, D-2000 Hamburg 13, FRG

Kaplan, I.R., Institute of Geophysics and Planetary Physics, University of California, Los Angeles, CA 90024, USA

Kaźmierczak, Josef, Institute of Paleobiology, Polish Academy of Sciences, Zwirki i Wigury 93, PL-02089 Warszawa, Poland

Kempe, Stephan, Institut für Biogeochemie und Meereschemie der Universität Hamburg, Bundesstraße 55, D-2000 Hamburg 13, FRG

Konta, Jiri, Department of Petrology, Charles University, Albertrov 6, 12843 Prague 2, CSSR

Krumbein, Wolfgang E., Geomicrobiology Division, University of Oldenburg, P.O. Box 2503, D-2900 Oldenburg, FRG

Laane, R.W.P.M., Tidal Waters Division, P.O. Box 20904, NL-2500 EX The Hague, The Netherlands

Lieth, H., General Ecology Group, Biology/Chemistry Department of the University, Barbarastraße 11, D-4500 Osnabrück, FRG

McCaffrey, M. A., Chemistry Department, Woods Hole Oceanography Institution, Woods Hole, MA 02543, USA

McGinnis, David, Land Sciences Branch, U.S. Department of Commerce, National Environmental Satellite, Washington, DC 20233, USA

Michaelis, Walter, Institut für Biogeochemie und Meereschemie der Universität Hamburg, Bundesstraße 55, D-2000 Hamburg 13, FRG

Milliman, John D., Woods Hole Oceanographic Institution, Woods Hole, MA 02543, USA

Mook, W. G., Centre of Isotope Research, Westersingel 34, NL-9718 CM Groningen, The Netherlands

Müller, German, Institut für Sedimentforschung der Universität Heidelberg, D-6900 Heidelberg, FRG

Mycke, Bernd, LABOFINA, Centre de Recherches du Groupe Petrofine, Chaussèe de Vilvarde 100, B-1120 Bruxelles, Belgium

Neev, D., Geological Survey of Israel, Jerusalem, Israel

Nissenbaum, Arie, Weizmann Institute of Science, Rehovot 76100, Israel

ODP Leg 107 Scientific Party, Ocean Drilling Program, Texas A&M, College Station, TX 77843, USA

Onken, Birte, Institut für Biochemie und Lebensmittelchemie der Universität Hamburg, Martin-Luther-King-Platz 6, D-2000 Hamburg 13, FRG

Postma, H., Netherlands Institute of Sea Research, P.O. Box 59, 1790 AB Den Burg, Texel, The Netherlands

Richnow, Hans Hermann, Institut für Biogeochemie und Meereschemie der Universität Hamburg, Bundesstraße 55, D-2000 Hamburg 13, FRG

Rohrbach, B. G., Institute of Geophysics and Planetory Physics, University of California, Los Angeles, CA 90024, USA

Romankevich, E. A., Shirshov Institute of Oceanology, USSR Academy of Sciences, Moscow, USSR

Ross, David A., Department of Geology and Geophysics, Woods Hole Oceanographic Institution, Woods Hole, MA 02543, USA

Schulze, Thomas, Institut für Biogeochemie und Meereschemie der Universität Hamburg, Bundesstraße 55, D-2000 Hamburg 13, FRG

Schellnhuber, H.-J., Theoretical Physics, University of Oldenburg, P.O. Box 2503, D-2900 Oldenburg, FRG

Serban, Andrei, Weizmann Institute of Science, Rehovot 76100, Israel

Simoneit, Bernd R. T., Petroleum Research Group, College of Oceanography, Oregon State University, Corvallis, OR 97331, USA

Spitzy, Alejandro, Institut für Biogeochemie und Meereschemie der Universität Hamburg, Bundesstraße 55, D-2000 Hamburg 13, FRG

Stahl, W. J., Bundesanstalt für Geowissenschaften und Rohstoffe, Stilleweg 2, D-3000 Hannover, FRG

Stoffers, P., Geologisch-Paläontologisches Institut der Christian-Albrechts-Universität Kiel, Olshausenstraße 40, D-2300 Kiel 1, FRG

Sulanowski, J., Earth Sciences and Geography Department, Bridgewater State College, Bridgewater, MA 02324, USA

Szekielda, Karl-Heinz, United Nations, Department for Technical Cooperations, New York, NY 10017, USA

Thode, H. G., Department of Chemistry, McMaster University, Hamilton, Ontario L8S 4K1, Canada

Turkstra, E., Fresh Water Division, P.O. Box 510, NL-3300 AM Dordrecht, The Netherlands

Uchupi, Elazar, Department of Geology and Geophysics, Woods Hole Oceanographic Institution, Woods Hole, MA, 02543, USA

Whiticar, M.J., Bundesanstalt für Geowissenschaften und Rohstoffe, Stilleweg 2, D-3000 Hannover, FRG

Wong, How Kin, Institut für Biogeochemie und Meereschemie der Universität Hamburg, Bundesstraße 55, D-2000 Hamburg 13, FRG

Zhang, Yi-Gang, Central Laboratory of Petroleum Geology, P.O. Box 916, Wuxi, Jiangsu, PR China

**Chapter 1**

# Introduction

Scientists who have had the opportunity of being associated with Professor Egon T. Degens, to whom this Festschrift is devoted, have been influenced by his ideas on subjects as varied as: extraterrestrial organic matter, origin of life, evolution of organisms, isotope biogeochemistry down to more imminent ones such as the carbon cycle and its implications on climate. This variety is also reflected in the papers in the present volume contributed by colleagues who have known Egon or have worked with him.

Egon Theodor Degens was born on April 16, 1928 at Inden, Germany and had his education in Bonn and Würzburg. After a stint at the Pennsylvania State University he returned to Würzburg to help set up one of the first organic geochemistry laboratories in the world. This laboratory was the breeding ground for some of the eminent organic geochemists at work today. Later, he joined the California Institute of Technology and began his work on stable carbon isotopes, and later on biogeochemical compounds in natural waters. From California he moved on to the east coast, which led to yet another productive phase at the Woods Hole Oceanographic Institution. He was instrumental in the pioneering work carried out by the Woods Hole scientists in the Black Sea which is the largest anoxic basin in the world, and in the Red Sea where the first hydrothermal ore deposits on the seafloor were discovered. These were followed by expeditions to the East African Rift Lakes where his team discovered hydrothermal activity in the Lake Kivu. His major research interest remained, however, natural organic matter, its composition and cycling in the oceans and sediments. Of special note is his work on stable carbon isotopes in marine plankton and its fractionation, and biomineralization, and the significance of the associated organic matter to decipher evolutionary patterns observed in certain marine organisms.

He became Professor at the University of Hamburg in 1973 where he established a research group dealing with the biogeochemical cycling of elements and its perturbation by man. Over the years he has initiated major research efforts within the framework of SCOPE (Scientific Committee on Problems of the Environment) and UNEP (United Nations Environmental Programme) projects on Biogeochemical Cycling of Elements in rivers, lakes, estuaries and oceans. Within the framework of these projects he stimulated and encouraged colleagues from a wide range of disciplines to sit together and discuss global issues of environmental change.

During his last years in Hamburg he also conducted extensive research on the state of the North Sea, on sedimentation processes in the open ocean, on theoretical questions concerning biomineralization and on the biogeochemistry of regional and intercontinental seas. His exceptionally widely oriented interests led to his collaboration on more than 300 articles and the publication of 19 books. He succumbed to a prolonged illness on February 19, 1989. In his book *Perspectives on Biogeochemistry*, which was completed last year and is being published by Springer-Verlag, he provides a personal view of the knowledge acquired in more than 30 years of research and teaching.

Egon T. Degens was a scientist of immense productivity and creativity and was endowed with a powerful integrative spirit. With his passing away, we have lost a challenging, supportive colleague, an understanding friend and an outstanding scientist.

Chapter 2

# General Concepts in Biogeochemistry

# Geophysiology of Carbonates as a Function of Bioplanets

W.E. Krumbein and H.-J. Schellnhuber

## 1 Introduction

The interaction between life and carbon is expressed by the simple relation

Carbon Dioxide – Organic Carbon – Carbonate

in all living systems (Fig. 1). It is also more or less accepted that practically all carbonates and organic carbon material formed at present and in the geological past are biogenic and produced by sun energy input from outside the planetary system (see Fig. 3). Organic carbon storage takes place in thousands of compounds that undergo later alterations into more complex organic polymers and into simpler substances such as methane. The latter may migrate upwards to the hydrosphere and atmosphere where it is biologically oxidized to carbon dioxide under anaerobic and aerobic conditions. Some biogenic carbon compounds in sedimentary and metamorphic rocks are:

Carbohydrates, Hydrocarbons, Steroids, Humic substances, Methane, Graphite, and Diamonds

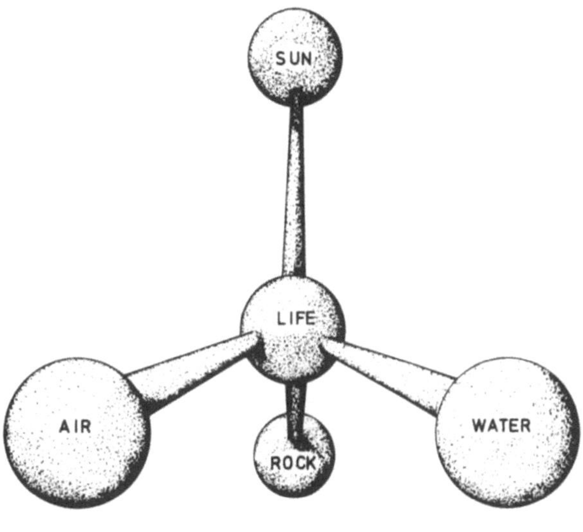


**Fig. 1.** In one of his books E.T. Degens combined the essentials of life in a form which had a spark of fire in the center. In this interpretation, life is the spark of fire, while sun is the constant source of energy, which keeps the bioid in its living equilibrium

Inorganic biogenic carbonates are stored mainly in the form of Calcium and Magnesium carbonates including some minor amounts of Fe, Mn, Sr, and other element carbonates.

The major enzymes catalyzing the biotransfer of carbon from and into the sedimentary reservoirs are summarized in Table 1. Vernadsky (1929, 1932) and Degens (1976) mentioned much of the biochemical potential in the carbon, phosphorus, and silica transfer via enzymatic reactions. The sedimentary reservoir of sun energy and biologically-derived reduced compounds, produced and stored by bygone biospheres is comparable to a memory effect. It memorizes former catalytic processes and the products of the latter can be reactivated when they are stored in a reduced, i.e., energetisized stage. The extent of storage and ways to analyze how much was stored was discussed by Degens and Epstein (1962) at a quite early stage. Degens himself (1976), however, warned against any "deus ex machina" explanations such as "explosive evolution" of calcified tissue animals at the onset of the Cambrian. In this paper, we will not discuss the ways in which calcification was introduced, and we will not commit the teleological error of discussing "why" it was introduced. The important point is that most organisms are capable of calcium pumping and other biochemical reactions in order to favor the dissolution and/or precipitation of calcium carbonate by very few enzymes.

There are many individual physical and chemical factors which are involved in the biotransformation of carbon into organic carbon compounds and into carbonates. Practically all of them are under biological control on this planet. They are summarized in Table 2.

In order to understand the presentation in Table 2, it has to be mentioned that the oxygen content of the atmosphere, its water content, and the solutes in waters are products of biological processes. Oxygen, water, and solutes, however, regulate the solubility and actual dissolution/precipitation equilibria of carbonates almost exclusively. Furthermore atmospheric pressure itself is biologically modified (flight acceleration and retardation of gases). And (!) finally in terms of geophysiology and geo-energetics, most of the reduced organic and inorganic compounds in the thin layer of earth we are used to calling Crust are a huge reservoir of biologically trapped energy, upheaved and redistributed by complex geophysiological processes regulated through Vernadski's "living matter" (Vernadsky 1938).

**Table 1.** Enzymes involved in carbon transfer

| |
|---|
| Ribulose bisphosphate carboxylase (Rubpcase) |
| Carbonic anhydrase |
| Decarboxylases |
| Carboxylases |
| Ca-ATPase |
| Calcium pumps |
| Phosphatases |
| and other pumps |

**Table 2.** Carbonate precipitation and dissolution

| | Biological processes or products | Abiogenic processes or products |
|---|---|---|
| pH | + | – |
| Protons | + | – |
| Redox-potential | + | – |
| Photosynthesis | + | – |
| Respiration | + | – |
| Matrix calcification | + | – |
| Enzyme calcification | + | – |
| Oxygen pressure | + | – |
| Carbon dioxide press | + | – |
| Solution partner: | | |
| Ammonia | + | – |
| Sulfate | + | – |
| Nitrate | + | – |
| Ferric iron | + | – |
| Phosphate | + | – |
| Ca ion | + | +/– |
| Mg ion | + | +/– |
| and others | | |
| Temperature | + | +/– |
| Pressure | indirect | + |

## 2 Geophysiology of the Bioplanet

We have pointed out that microbial pathways control the major biogeochemical and other cycles (Krumbein 1983). Furthermore Vernadsky (1929) and others have hinted at the fact that biogeochemical global activities of biota exist and that their mode of action may span millions of years. The stability of the ratio of organic and inorganic carbon reservoirs in sedimentary rocks over large sections of Earth history, as well as the astonishing speed of biologically induced weathering of silicate minerals on this planet, allow us to deduce that more than just atmospheric composition, temperature, and its regime of atmosphere and oceans are under biological control on this planet.

Therefore we have recently forwarded the hypothesis of a global living system that is self-equilibrated by present and past biospheres and largely influences the sedimentary distribution of the elements and their redox states as well as some parts of global dynamics of the crust (speed of weathering, sedimentation and, to some extent crust deformation). Probably this is the only way of explaining why life has such a tradition on this planet. This hypothesis is forwarded here under the name of "bioid" or "bioplanet" hypothesis (Krumbein 1969, 1971, 1981; Krumbein and Dyer 1985; Krumbein 1986, 1987, 1988).

The term *bioid* designates an almost spherical planet or planetoid or a mass of planetary dimensions that is controlled by life. Bioid can be synonymously used with *bioplanet*. This definition includes the presence of an external energy source such as a sun. The term is derived from the term geoid which is an idealized planetary body. Anderson (1984) named it a *geoid with memory effect*. Vernadsky (1938) 50 years ago called it: *A special geometry of space inside living matter*!

Geophysiology rather than biogeochemistry is needed in order to understand the functioning of the system. Internal and external dynamics of bioplanets are operated by a huge number of biological or biogenic feedback circuits. All of them contain different goemetries and assumptions or appreciations of time. The feed-back circuits may be direct ones, such as photosynthesis which directly controls the atmospheric gas composition, albedo and temperature of the earth. Indirect biogenic feedback circuits are, e.g., the speed of cycling of carbonates, organic carbon and heat from the mantle and crust to the surface and from the surface down to deeper parts of the crust over time lapses between a few hundred thousand to millions of years. The fast circuits concern sun, atmosphere and ocean as they interact with biota. Slower circuits are bioweathering, bioerosion, biosedimentation and biotransfer or storage of sun energy in sediments and sedimentary rocks. The slowest circuits are global tectonics and feedback exchange between remanent (radioactivity of crust and upper mantle) energy and "eternal" energy (i.e. sunlight trapped mainly as organic carbon, iron sulfide, and other reduced compounds "eternally" cycled). Vernadsky (1938) called all geometrical and time relations eternal that are related to the durability of the sun system, while he conceded that radioactivity as such is defined by its constant decrease. We know, that the moon lost its activity about 3 Ga ago and little if any activity is taking place on Mars. It seemed impossible for living natural bodies to "breed" new radioactivity. On the other hand the Gabun events 2 Ga ago seem to have been natural reactors through the upheaval of U-235 – which was then much commoner – in a restricted seam by bacterial activities. Anderson (1984) further stated that biological limestone production maybe the reason for plate dynamics. In viewing the planet as whole and its interesting biological activities, we feel that some kind of a physiology study of the planet as a living entity is urgently needed. We define geophysiology as follows:

Physiology is the science of the normal functions and phenomena of living natural bodies. Up to the end of the 18th Century physiology embraced the study and description of the functions and phenomena of all natural bodies. Hitherto – and because there is no rule against defining an animal or a plant with its up to hundreds of symbiotic microorganisms as one single species – we define geophysiology as the science of the normal functions and phenomena of this planet as a living natural body. Consequently the species name *Terra sempervirens* may be suggested.

Geophysiology embraces:

1. the biochemical or microbial processes which regulate the budget of materials and gases of the atmosphere, hydrosphere, pedosphere and lithosphere and hereby also the temperature budget of the planet at least in the near-surface layers (uppermost mantle and crust);
2. the biotransfer and bioaccumulation of energy carriers and reservoirs and minerals, many of which mankind is presently exploiting;

3. the biochemically, microbially or plant-related biotransfer processes of biodeterioration, bioerosion, and decay of rocks and minerals resulting in nutrient liberation for new life and in the formation of soil and the solute content of the seas;
4. and perhaps the biotransport of chemical species that cannot be transported as fast by diffusion (chemical or eddy) as by plant fueled animals which carry materials actively (living locomotion) or passively (falling of dead bodies or fecal pellets is faster than diffusion).

Geophysiology thus is the sum of all biotransfer, biotransformation and biotransport activities of micro- and macroorganisms that are living presently and have lived in the past in bygone biospheres (Lapo 1979, 1982, 1987). The geophysiological transfer and transport processes, however, reach dimensions of physical, chemical and geological nature which e.g. keep the atmosphere equilibrated in such a way that life stays possible. This seems only possible when continents are slowly transferred into motion and endogenic forces are created that had not existed on this planet prior to the establishment of life or were dying out with decreasing radioactive primary energy.

If we accept geophysiology as the embracing term for the natural history of a planet we will have fewer problems with eliminating unfortunate terms such as ecology, megaecosystem or global ecosystem. Also formerly widely accepted terms as ethology (St. Hilaire and Dollo) would remain obsolete. Both terms (ecology and ethology) can be summarized under the term of physiology of one single organism, namely Earth, as was the view for quite some time in the past (see Albertus Magnus, Giordano Bruno, Leibnitz). All this, however is only possible when the functions and phenomena of the bioplanet can be ascertained as being driven by one single cause, namely the unifying concept of one single bioplanet filled formerly and now with one type of genetically interconnected (standing genetic diversity) life which organizes itself by biological transformation of sunlight or other "eternal" energy sources into chemically bound energy.

The fact that oxygen, carbon, sulfur, hydrogen, iron, and calcium are intimately related by planetary geophysiology and organized into carbonate, organic carbon compounds, gypsum, pyrite, and iron oxides has to be accepted as one of the major long-range feedback circuits existing within the bioplanet. The amounts of pyrite, which were not originally sedimentary are negligible as well as is the sulfate in the ocean that is not cycled biologically. It is not very well understood yet how water and silicate are interlocked with the elements depicted in Fig. 3 and their cycles. Much research and calculation is still needed before we obtain a clear picture of the functions and phenomena of the bioid. Biotransfer of minerals and bioweathering speed control make it possible, however, that silica, aluminium and sodium and potassium are also deeply involved in geophysiology. It is obvious from observations on planets without life (which have weathering and erosion rates more than 10 000 times slower than earth) that weathering is under strong geophysiological control. This includes biogeneration and biodestruction of clay minerals. These multiply interfere with biota and rocks – as soil structuring, mobility enhancing, ion exchanging, water storing minerals.

The energy content of organic carbon differs considerably from the energy content of carbonates. The trapping of sun energy and the transfer of sun energy into the crust and mantle are of the same order of magnitude as the transfer upwards of remanent energy – radioactively produced heat. At the earth's surface itself, the energy derived from sun power accounts for 99.98% (Hubbert 1974). It can even be assumed that without the trapping of sun energy for a period of now more than 2 Ga (2 000 000 000 years) the planet may have already completely lost its internal dynamics. Such a statement has also recently been implied by geophysicists (Anderson 1984). The contribution of evaporites to plate tectonics is also a case of geophysiology for the evaporitic production of carbonate, sulfate and other salts as related to the directly or indirectly metabolized salts (biogenic minerals) is temperatures related. The temperature, however, on bioplanets is regulated biologically as we know from global crisis scenarios of changes of the composition of atmospheric gases by anthropogenic influences and their mutual influences on global temperatures (e.g., Lovelock 1979).

Evolution and bolide impacts (as an example of abiological crises) cannot be immediately accepted as arguments against the geophysiology of Earth. Firstly the most important group of organisms presently living on earth are microbes and these stabilized themselves at least 3000 million years ago in a system of natural genetic engineering. Catastrophes or, better, events (excursions from "normal", sometimes through species extinction) are not necessarily extraterrestrial. Many of them seem to be related to geophysiology as cataclysmic readaptations which are needed for the functions of the bioid. There seems to be a certain pattern in the event discussion which relates plate tectonics to extinction and extinction to microbial mat explosions. These in turn reduce biotransport and increase biotransfer rates functionally by redistribution between microorganism and macroorganism activity over several hundred thousand years. There is also almost an agreement about the possibilities that not only accumulations of carbonates but also of uranium, gold and – iridium – are processed by microbial mats (Krumbein 1983a). In addition, we may assume that man also is such a regulatory cataclystic excursion from normal and therefore environmental problems on a global scale as they are at present are at an unprecedented level (although even Pliny the Elder wrote long passages about global pollution and environmental anthropogenic problems) may be within the borderlines and buffer capacity of the bioplanet.

In order to understand the interplay of the bioid with carbonates one has to go back to the onset of carbonate deposition, i.e., approx. 3.5 Ga. If plate tectonics really have something to do with the carbonate cycle as suggested here, then plate tectonics should not have existed prior to massive turnover of carbonates in the earth crust? There are hints that as a matter of fact plate tectonics were not operating in the early Precambrium.

In earlier approaches (Baas Becking 1931, 1934; Lovelock 1979) the term Gaia was used to describe the influence of life on the environment and on the shaping of the atmosphere and on the geochemical cycles. Both, however, remained in the frame that the environment selects the "dormant possibilities" in a classical Darwinian evolutionary sense.

It is proposed here that the biota and the environment shape themselves reciprocally and continuously thus maintaining a synergistic system powered by varying sunlight input. The variations of energy input, in turn need to be overcome by changes in the operational mode. From the sum of the latter considerations and assumptions the model of a self-organizing and self-controlling global living system (bioid instead of geoid earth) is derived and it is largely influenced by the organization of organic carbon compounds and carbonates as they interfere with calcium sulfate, and via sulfate reduction with iron sulfides and oxides. The latter cycle in turn is controlled by oxygenic, photosynthetic, and respiratory oxygen equilibria and interferes in many ways with the biologically controlled cycling of silicates and the water they contain. There is no doubt that weathering of silicates and clay formation is considerably accelerated and controlled by microbial activities (Krumbein 1983).

The bioid could perhaps be compared to a laser system. It (or the bioplanet) produces its own physical, geochemical and biological environment, which favors the expansion or at least the stabilisation of life. We state the case of a self-enhancing process in which (as in the laser) coherent self-enhancement at a sufficient pumping rate (solar energy input) and appropriate construction (inorganic and organic reservoirs turned over by innumerable microbial and macroorganismic systems) is focused. The laser is, however, an insufficient comparison for life itself controls its conditions of existence to a much greater extent. Energy from outside is the key function in all steady state equilibria which are far off thermodynamic balance. Therefore the geological factors of energetization of the planet will be discussed first.

## 3 Energetic Evolution of the Earth and Carbonates

Geologists tend to state that the interior of the Earth is a gigantic machine fueled by radioactivity. The assume that with insufficient radioactivity continents might not have evolved to their present form and dynamics and volcanoes might not have spilled out all the water and gases and liquid materials necessary for the formation of the balanced rock crust and life processes. In the case of too much radioactivity existing in the core, the planet might have had tremendous volcanic activity with no stabilizing crust and a dense suffocating atmosphere.

The sun is a major energy source for all planetary processes on Earth. Astrophysicists tend to believe and have forwarded data in proof of the assumption that while radioactivity decreases after an initial buildup to a maximum on our planet, solar luminosity went through a minimum and has increased slowly over the past 4 billion years (Figs. 2 and 3). The present-day geological textbook view is anthropocentric in one sense and not anthropic (human oriented or seen through the filter of human life) in another sense. In geology it is widely assumed that after several formational events which can be summarized into (1) a radioactivity heat scenario, (2) a gravitational heat scenario (iron sinking down to the core and thus releasing large amounts of heat, magnetic and reoriented gravitational forces) earth somehow stabilized into a surface temperature situation suitable for the establishment of life, after some differentiation had taken place (Press and Siever 1974). This apparently had to proceed relatively fast (approx. 1000 million years or 1 to 1.5

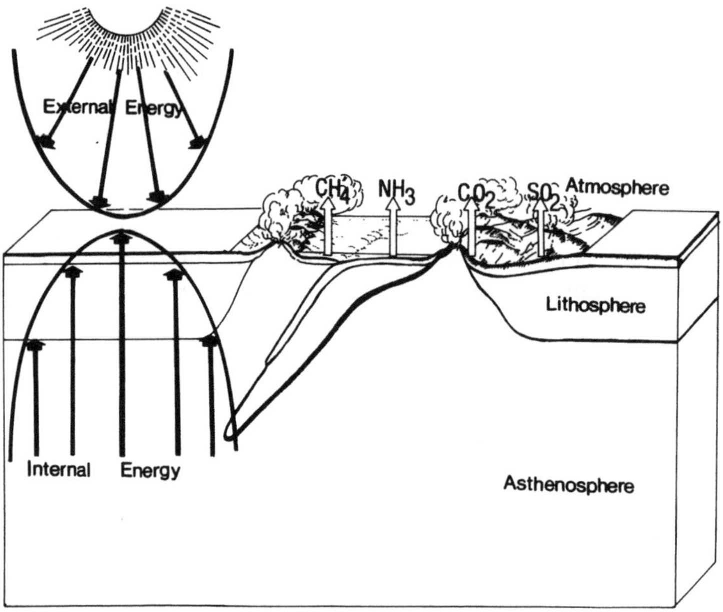


**Fig. 2.** Possible relations between external energy input (sun luminosity, adsorption, reflection) and internal energy input (radioactivity and other processes) into the crust area of an early Earth without life and with strong degassing

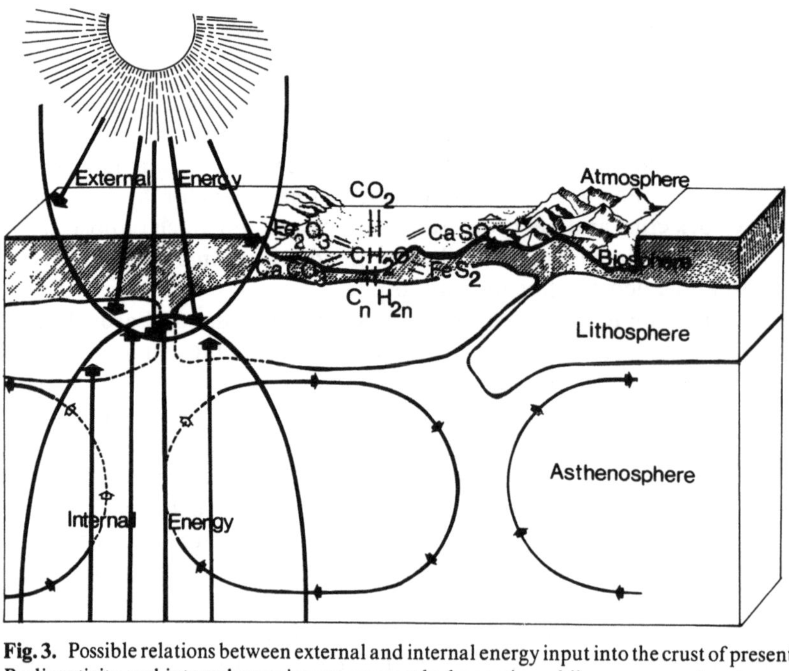


**Fig. 3.** Possible relations between external and internal energy input into the crust of present-day Earth. Radioactivity and internal energies are constantly decreasing while sun energy input according to the bioid hypothesis is regulated by the bioid's energy demand and the control function of bygone biospheres

Ga). With a then molten core and at least large parts of the lower mantle molten, another process could take over – the dynamics of differentiation and convection flows. The materials that became segregated were few. The distribution pattern of elements in the universe, inner planetary core of the Earth and the Earth's crust thus became very different from each other at different times and, in addition, patchy on and in the Earth, at least for the crust and upper mantle. Some authors additionally believe that this universal fractionation, in which earth represents a strange sink of several strange isotopes, may also be linked to life (nickel 78 theory and the final anthropic principle of Barrow and Tipler 1986). Also the early work of Kant (on the true estimation of living forces) where he stated that there is a direct connection between the inverse square of gravitation and the existence of exactly three spatial dimensions, may have led to the assumption that atomic matter and its odd distribution is only possible in a three dimensional space universe (Gurevich and Mostepanenko 1971). The most abundant elements are given in Table 3.

After this major segregation occurred, it appears that some kind of a separation of the heavy elements which do not easily react to form oxides and silicates from those that easily do so (including radioactive elements) had changed the composition of the crust in such a way that it differed considerably from the total composition of Earth as does the total composition of Earth from the absolute distribution of elements in the universe. It seems that with the separation and segregation over, some planetary bodies such as The Moon, Mars, and Mercury, after some time of activity, have now lost any considerable internal dynamics. Jupiter, on the other hand, emits three times the energy it receives from the sun and sometimes is regarded as the second sun of a double-star system. After the first Ga of development on the Earth with no solid crust, geologists conceive another system, namely a system of rock formation and weathering by rain water and gravity and

**Table 3.** Abundance of elements in %

| | Universe atom % | Whole Earth % | Earth's crust % |
|---|---|---|---|
| Hydrogen | 93.5400 | [a] | 0.14 |
| Iron | 0.0031 | 35 | 5.0 |
| Helium | 6.3200 | [a] | [a] |
| Oxygen | 0.0640 | 30 | 46.6 |
| Silicon | 0.0040 | 15 | 27.7 |
| Magnesium | 0.0040 | 13 | 2.09 |
| Nickel | 0 | 2.4 | 0.008 |
| Sulfur | 0.0020 | 1.9 | 0.520 |
| Calcium | 0 | 1.1 | 3.63 |
| Aluminum | 0 | 1.1 | 8.13 |
| Potassium | 0 | [a] | 2.59 |
| Sodium | 0 | [a] | 2.83 |
| Carbon | 0.0390 | [a] | 0.32 |
| Nitrogen | 0.0080 | [a] | 0.0046 |
| Phosphorus | [a] | [a] | 0.118 |
| Manganese | [a] | [a] | 0.100 |

[a] Not indicated.

some other physical factors such as Coriolis forces, tides of the moon and the sun etc., and especially the miraculously initiated convectional cell movement, led to apparently relative equilibrium for another 4 Ga. This relative equilibrium is, however, difficult to explain unless life is considered as a driving force.

At present and dating from approx. 300 million years ago, about a dozen plates seem to exist and they move at a speed of about 10 cm per annum. Material seems to flow down and to come up at most of the meeting edges of the plates. Where plates move apart, material flows upwards. The energizing of this machinery is considered as being manufactured by convectional forces and slightly influenced by lunar tides. This view is a very abbreviated version of the present-day geotectonic ideas. It is true – since the time of Alexander von Humboldt (1844), who revived ancient myths from the Upanishads of India (possibly influenced by the activity of the nearby Himalayas) – that geologists have increasingly painted a picture of a dynamic Earth. Humboldt in his book *Kosmos* phrased it the following way:

"With the latter phenomenon of the rotation or gliding of strips of land, in which one piece of land takes the place of the other, we can state a translatory movement or mixing of layers" . . . "The expansion of the circles of earthquakes can perhaps be imagined via the upwelling of large masses into the cleavages formed by earthquakes. It is thus an additional phenomenon which may explain the expansion of the radius of earthquakes."

And finally:

"The followers of Sakhyamuni claim that the real cause of earthquakes is a gigantic steel wheel that rotates underneath the solid masses. I regard this mechanistic explanation not as sillier than some of our now outdated geological and magnetic (mesmeristic) ideas."

In modern times and after Wegener had elaborated the theory of continental drift, Press and Siever (1986) state the following:

"Most now believe that the Earth's lithosphere is broken into about a dozen plates, which for reasons not fully understood move over the interior."

These twelve plates and their movement of 10 cm annually originated only about 0.3 billion years ago. Solid sedimentary rock crust, on the other hand is known to have existed for 3.7 to 3.8 billion years and life for about the same time period. Pangäa – a large supercontinent of former times – was stable for an extended period. The reasons why the plates move today across the surface of the Earth is not fully understood. The interaction of (1) narrow belts, (2) growth zones along the mid-oceanic ridges, (3) material and mineral producing systems and traps along the margins of the continents, (4) filling of large troughs with sedimentary material, (5) weight increase by this accumulation of sedimentary rocks and several other factors have to be considered in this puzzle of plates and their motion (mobilization). Continental plates have a mean thicknesses of 35 km, while the Earth measures 6378.245 or 6378.077 km in mean equatorial radius. The total weight of earth has been calculated as $5.9763 \times 10^{27}$ g. The mass of the total lithosphere is less than $2.4 \times 10^{25}$ g. Thus the part of earth that is considered by us to be under influence of present and bygone biospheres is more or less 0.4% of the total. Therefore subduction, transform faults, "terrane" transformations, and other "substantials" of plate tectonics are such a small factor in planetary evolution that sunpowered biological processes may well have had their influence and impact on continental motion and geomorphology.

The constant build up and destruction of geomorphological relief, however, is much more important for life than plate tectonics by itself. It seems, however, that plate tectonics – be they powered endogenic radioactively or exogenic biologically – are the most important factor in the creation of a permanent exogenic geomorphological biotransfer, bioweathering, and bioerosion cycle which is the most efficient process of supplying organisms with nutrients to build cell mass and store energy. Oceans and continents have average ages of about 200 and 500 million years, which is about 4% to 10% of the age of the Earth. This brings some of us to start thinking of the possibilities of a connection between life and earth dynamics. Anderson (1984) stated with some reluctance:

"It has often been suggested that life originated on earth because of a coincidence between the narrow temperature interval over which water is liquid and the temperature extremes that actually occur on the earth. The earth apparently is also exceptional in having plate tectonics. If the carbon in the atmosphere of Venus could turn into limestone, the surface temperatures and those of the upper mantle would drop. The basalt eclogite phase change would migrate to shallow depths, causing the lower part of the crust to become unstable. Thus there is the interesting possibility that plate tectonics may exist on earth because limestone-generating life evolved here."

He calls Earth a "Geoid" with a memory. We call it a "Bioid" with a memory and its innate geophysiology which Lovelock (1984) also found appropriate in a book review: "An interdisciplinary approach to earth science is more in need of contributions from geophysiology than from biogeochemistry." When Lovelock (1979) and Anderson (1984) promoted these ideas they were probably not aware of Vernadsky (1929, 1930) nor of the Vernadski laudation by Lapo (1979). Stepping forward from the ideas laid out by Vernadski (1929) and Lovelock (1979) we have derived the idea of Earth as a "bioid with its memory effects" (Krumbein and Dyer 1985; Krumbein 1986) instead of the "geoid with memory effects" of Anderson (1984).

Krumbein (1983b) stated: "The sum of life on earth (or in the universe) will tend to keep itself constant" – Masses and concentrations of matter can be regarded as transformable into energy and vice versa, and they have their relative and often biologically catalyzed velocities. Thus accelerations and retardations of mass and energy biotransfer and biotransport may explain much of the dynamics of Earth". This as well as the general statement for some basic lines of thought in geophysiology (Krumbein 1984) were then vigorously rejected. The book was re-published in an altered form under a different cover. Some of these thoughts, however, have – mostly independently – found acceptance at other places. The astrophysicists Barrow and Tipler (1986) state for example: "Paradoxically this appears to be possible only in a closed Universe with a very special global causal structure, and thus the requirement that life never becomes extinct – which we define precisely by a new 'final anthropic principle' – leads to definite testable predictions about the global structure of the Universe" and "Final Anthropic Principle (FAP): Intelligent information processing must come into existence in the Universe, and, once it comes into existence, it will never die out."

In talking mainly about man, the authors of the "Final Anthropic Principle" describe the physical causation of the "Bioid paradox" as we use to call it: "An intelligent being – or more generally, any living creature – is fundamentally" (comparable to) "a kind of computer, and is thus subject to the limitations imposed on computers by the laws of physics. However, the really important part of a computer is not the particular hardware, but the program; we may even say that a human being is a program designed to run on a specific hardware, called a human body, coding its data in very special types of data storage devices called DNA molecules and nerve cells. . . In fact defining the soul to be a type of program has much in common with Aristotle and Aquinas' definition of the soul as 'the form of activity of the body'. . . An intelligent (or functional) "program can in principle be run on many types of hardware if there is sufficient energy in the environment to run the programs."

Thus one may tend to define the bioid as a program running on a specific hardware system of planetary dimensions: the geoid. DNA and organic and inorganic compounds stored and enriched in the sedimentary and metamorphic rocks are the primary and secondary memory effects. The estimated $5 \times 10^{22}$ organisms (Fischer 1984; Seilacher 1985) finally act as the geophysiological agents, namely as transfer and transport cells, nerve cells and data storage devices. The energy for the bioid's operation may initially have been remanent energy from radioactive decay and iron segregation. We assume, however, that the bioid now uses external (sun) energy as a program-controlled energy source for stabilizing the system.

In this context, the memory effect can be further elaborated. More than 98% of the $5 \times 10^{22}$ living organisms are bacteria, with an average lifetime of let us say 30 minutes. From this we may estimate the total number of organisms that have previously lived with the bioid to be approximately $2.5 \times 10^{36}$. Practically all of them were microorganisms. If we assume that 1% of the total were macroorganisms with 1 g weight and if we further assume that microorganisms weigh 1 $\mu$g we will reach a total weight of all organisms that have ever lived of between $2.5 \times 10^{26}$ and $2.5 \times 10^{33}$g. This would be the equivalent of between 10% to $10 \times 10^5$ equivalents of the Earth's mass. Considering the relatively large amounts of energy and matter that are channelled through the living organism throughout its lifetime there is serious reason to come to the conclusion that the bioid is biologically energetisized and that biotransfer and biotransformation of elements is a major driving force of planetary dynamics. The initial energy input for the gradual change from internal to external energitization of the bioid is either derived from chemolithoautotrophy or photolithoautotrophy. Photolithoautotrophy seems to have been the most important process for more than 2.5 billion years. These calculations are not unrealistic like the calculation that within two days, (energy and nutrients given) one single yeast cell could give rise to a cell mass of the dimension of the earth. The calculations above state nothing but the high probability that the geochemical cycling of most elements, as a matter of fact, is a biogeochemical type of dynamics which we assume to be of geophysiological dimensions.

With another approach we can calculate the amount of carbonates and organic carbon compounds presently existing in the bioid's crust. Kempe (1979) derived values of $60 \times 10^{21}$ g carbonate carbon and $12 \times 10^{21}$ g organic carbon compounds with a ratio of 1:5. This ratio may have been 1:4 through the late Precambrian. This would, however, only modify the general line of thought. The turnover time of carbonate and carbon has been calculated for the different compartments of the bioid as being 11 years in the biosphere, 4–7 years in the atmosphere and about 342 million years in the lithosphere.

It seems more important, however, to compare the energetic values of carbon turnover and the energies fixed and stored in the lithosphere via photosynthesis through geological time. Also we would need to calculate the individual energy content of organic and iron sulfide rich geological eras as compared to poor systems. The free energy of an average organic molecule (sugar) is approximately 3000 kJ. If we assume that all organic matter in the present day rocks contains more or less the same amounts of energy (with a confidence of one order of magnitude) we will arrive at values between $15 \times 10^{23}$ to $3.6 \times 10^{24}$ kJ energy fixed in organic carbon compounds within the crust at present. The annual capture of energy via photosynthesis and its storage within sedimentary rocks by itself is of the same order of magnitude as the theoretically calculated heat flow from the "interior" as stated before. Furthermore we can calculate the potentially maximal energy content of of the present day crust of the earth being transformed into organic carbon compounds by photosynthesis using sunlight. This would, however, drastically change all scenarios inasmuch as no carbon would be fixed in carbonate rocks.

Most of the data presented in Tables 4 and 5 are unreliable. The error factors involved are between 1 and 2 orders of magnitude. Furthermore, we can assume that metal sulfides and in particular iron sulfides make up at least 10% of the

**Table 4.** Reservoirs of carbon

| Reservoir | $\times 10^{15}$ g C |
|---|---|
| Atmosphere (before 1850) | 560– 610 |
| Atmosphere 1978 | 692 |
| Oceans inorganic | 35 000 |
| Dissolved organic | 1 000 |
| Living | 3 |
| Land biota | 600– 900 |
| Land organic (humus) | 1 000–3000 |
| | |
| Sediments total | 10 000 000 |
| Carbonates in sediments | 8 000 000 |
| Organic carbon in sediments | 2 000 000 |
| Fossil fuels (mineable) | 5 000 |

The ratio of 1:4 for organic carbon versus carbonate may have changed between Precambrian to Phanerozoic from 1:5 to 1:4. This, however, is difficult to calculate.

**Table 5.** Fluxes of carbon

| Flux | $\times 10^{15}$ g a |
|---|---|
| Atmosphere/ocean | 100 |
| Atmosphere land | 50– 80 |
| Ocean biota | 15–126 |
| Land to ocean transfer inorganic | 0.4 |
| Land to ocean transfer organic | 0.1 |
| Deposition in ocean | 1.–10 |
| Fossil fuel burning | 5 |

mentioned energy content stored in sedimentary rocks via sulfate reduction in which a energy rich compound is biogenically produced. All other biogenic or possibly biogenic energy rich sedimentary reservoirs are negligibly small.

Another scenario of importance for the theory of global "dynamisation" by biological transfer of matter and minerals is lubrication. Everybody today assumes an automobile to be an ingenious vehicle for transportation. This, however, would be very difficult to deduce from a fossil car, without its lubricating parts and substances such as oil, fat, surface active substances etc. Microorganisms produce many substances known to be important in the dynamics of tectonical movements. These substances are: Gases, liquids, organic matter, clays, salts, graphite. For those who do not know how clays are produced by microorganisms – it has been demonstrated quite well by field and laboratory studies (see Berthelin 1983). Therefore the mobility of plates and parts of plates may be biologically modified if not controlled. Anderson (1984) assume buoyancy and basin-filling capacity to be largely modified by biogenic carbonate production.

All of us know by now that the atmosphere is largely controlled through life processes and undergoes annual equilibrated cycles of global dimensions. It is also a fascinating thought, that microbial life has ruled the composition and distribution of the atmosphere for about 3 to 3.5 Ga including a period without free oxygen that lasted for about 0.5 to 1.5 Ga. Oxygenated oceans and atmosphere with all the consequences of oxidized rock surfaces on solid land may have already existed for 2.5 Ga. Thus the modern era with trees on land and many fairly motile animals around, accounts only for about 10% of the life history of the Earth.

Looking at such a scenario, we may conceive several changes in the mode of geochemical cycles as influenced by life on Earth. The following biologically (and mostly microbially) controlled geophysiological factors and processes can thus be listed:

In summary it seems that we have alternative ways of discussing global tectonics as a part of the bioid's geophysiology. These dynamics, when calculated and analyzed properly are gigantic and span the total mass of the crust and its heat potential. The idea that all this is actually fueled by sun energy and functionally controlled by organismic memory effects is somewhat strange. It is clear, however, that many geophysicists and geologists admit that, so far, no other explanation is satisfactory for plate tectonics and the distribution and cycling of oxidized and reduced compounds in the crust. The idea, however, that geophysiological control really exists on the sun energy pumping rates, is very close since hardly anybody denies the influence of biologically cycled oxygen, carbon dioxide, methane, dimethyl sulfoxide and other compounds on the albedo and on surface temperature. This sun energy pumping has already been compared with the pumping rates and energetization of coherent laser systems.

The best physical elaboration of the geological facts and their new interpretation through the geophysiological calculations presented here may be derived from a series of statements developed by us. These are some of the main outlines or cornerstones of the bioid hypothesis and of geophysiology. We will use these statements as a kind of summarizing conclusion of the global carbonate scenario we have tried to depict in the previous pages.

# 4 Outlook for Geophysiology

The Earth is not singled out among the planets because it offers the physicochemical conditions for life. On the contrary: the Earth exhibits very special conditions far away from astrophysical equilibrium because this planet is alive (Krumbein and Dyer 1985).

The Earth is a "bioplanet" or a "bioid" whose surficial layers (upper mantle, crust + hydrosphere + atmosphere) represent a gigantic dynamical system made up of mutually selfstabilizing components

(a) inorganic, organic, and energy reservoirs
(b) living and fossil biospheres
(c) biologically controlled energy input and output (sun, internal heat, radioactivity)

coordinated in time, mass and energy by geophysiological processes in order to maintain equilibrium for life.

The reactions of bioplanets can be described in terms of geophysiology, i.e. the functions and phenomena of the Earth as a living natural body. The bioid or bioplanet as a dynamical system can choose between a number of global operational modes, i.e. stationary dynamical states which should be deduceable in principle from basic physical and chemical laws. The states are metastable and respond elastically to minor disturbances. Major impacts (collisions with bolides etc.) or tectonical events may cause a transition to another operational mode. Most operational modes can be classified as various different procedures of gaining, storing and controlling energy on a global scale.

A description could be given in terms of synergetic models, which show the existence of ordered structures born out of chaos by collective interaction. These fine structures supported by the primary operational mode tend to evolve more or less rapidly in time to higher and higher degrees of hierarchical organization embracing generations of secondary, tertiary etc. submodes. It is known from the theory of dissipative systems that the degree of structural organization is directly related to the complexity of the driving energy flux distribution.

Constant energy losses of the initial formational radioactive materials will be compensated for by increased energy trapping through the bioid and storage by

**Table 6.** Important geophysiological operations of the bioplanet

1. Cycling of reduced carbon compounds
2. Equilibrated production of carbonates by interaction with organic carbon compounds
3. Oxygenic photosynthesis and its impact on the distribution of water, oxygen, oxides, sulfides, carbonates and on bioweathering
4. Respiratory microbial and macroorganism activities including aerobic and anaerobic respiration. The oxygen usually assumed as main electron acceptor can be replaced by sulfate, nitrate, oxidized metals, carbon dioxide. Respiration and – to some extent – fermentation rule thus the distribution of oxidized and reduced carbon and sulfur reservoirs. The relevances of the individual respiratory pathways are ruled by the fact that the total carbon reservoir as compared to the total sulfur reservoir in the crust is in the order of 10:1 as previously stated
5. Weathering and solution rates (fractionation of particles, solution of chemical compounds, precipitation of chemical compounds, rate and form of clay mineral formation) regulated mostly by (3) and (4)
6. Absolute control of phosphorus compounds and their cycling
7. Almost total control of the fractionation of reduced and oxidized iron, manganese compounds, and other heavy elements such as uranium, gold, silver, lead including the still puzzling phenomenon of natural biogenic chain reactions.
8. Partial control of the mechanical and chemical cycling of silica
9. Total control of the inflowing sun energy and its storage and release assumed over geological time via albedo control
10. Accelerated mass transfer via macroorganisms (horizontally and vertically)
11. Control of the so-called convection cells and plate tectonics by sun energy storage, calcium carbonate, iron sulfide, calcium sulfate, and iron oxide reservoir balances, biologically controlled weathering rates and biogenic production of the lubricants clays, graphite, organics including oil and gases.

bygone and recent biospheres within the crust and mantle. In the course of this process bioenergy (i.e. "memorized" sunlight) penetrates the crust and contributes to global morphogenesis and plate tectonics.

There is no competition between morphologies but rather between stationary processes. The traditional evolutionary model – continuous development through stochastic mutations in response to permanent selective pressure as a consequence of stochastic changes of the environment – does not offer a satisfactory explanation of the dynamics of the bioid or for the significant lack of dynamics in many groups of organisms (e.g. procaryotes). Properly deduced from the bioid hypothesis we infer that the bioid acts by creating environments and these environments reorganize the informational level (i.e. micro- and macroorganisms). In this scenario microorganisms act as the ground level or ground mass in which different operational modes and different morphologies are nested as a reaction to external or internal condition changes.

Homogeneous-stochastic "challenge-and-response" models imply the unrestrained evolution of organisms into omnipotent superorganisms where time represents the exclusive dynamical parameter. The study of bygone biospheres demonstrates, however, that global evolution no longer took place at a high rate after the establishment of the procaryotic operational mode including the fact, that practically all the enzymatic pathways were firmly established about 2 Ga ago. The standing genetic diversity of procaryotes was also established through symbiosis in eucaryotes so that in addition to chemotrophy and autotrophy we may also infer symbiotrophy as the operational mode for mixed bacteria-eucaryote systems. The genetic diversity represents a huge reservoir of reaction mechanisms to (often self-produced) environmental changes that are constantly buffered. Evolution in the sense of higher organization and more and more differentiation cannot be deduced from a look at the *whole* history of life on Earth.

Internal energies (heat, radioactivity, crystal energy) have decreasing tendencies. During the history of the bioplanet these were, however, balanced by bio-accumulation of energy in organic carbon compounds and reduced metal compounds produced and cycled faster to compensate for the losses of original energy (memory-effect, new regional and internal driving forces).

The "higher" a life form is located on a scale of increasing complexity in the hierarchy of driving and driven processes of the bioplanet the more sensitive it is to strong external disturbances and also to larger intrinsic fluctuations (excursions). Thus changes in macroorganisms are expected to be much faster and "drastic" than with microorganisms. Changes in the latter, however, would have the greater impact on geophysiology as a whole.

## Post-Script

This article was dedicated to my elder brother in science, Egon T. Degens on whose advice I used to build. At the occasion of his 60th birthday I wanted to use the following quotations pointing downwards to him through the growing science of geophysiology. At that time I did not know, how fast I would no longer be able to stand on his shoulders (Degens 1989, Krumbein 1984).

*Lucretius* (50 B.C.): Corporibus caecis igitur natura gerit res. . .

*Albertus Magnus* (1250): Quocirca propter hoc quod unius generatio est alterius corruptio, et e converso unius est alterius generatio, est generatio transmutatio quae nunquam secundum naturam quiescit.

*Linnaeus* (1760–1761): Petrificata montium calcariorum non filii sed parentes sunt, cum omnis calx oriatur ab animalibus.

*Immanuel Kant*: (1763): Burnet thought the mountains and oceanic abysses of Earth to be useless and a part of divine punishment of mankind, a thought in which Burnet was doubtlessly wrong. Whether the spherical form of Earth or its deviations from it are more important, needs, however, further consideration (in his physicotheology).

*Wladimir I. Vernadskij* (1938): Living matter is the sum of all organisms. Living matter, soil and rock are peculiar natural bodies. The fundamental property of biogeochemical energy is the growth of free energy of present and bygone biospheres with the progress of geological time.

*Wolfgang E. Krumbein* (1988): The continuous horizontal and vertical movement of the Earth crust is a consequence and not a causation of the manifold expressions of life on this planet (bioid). Die horizontale und vertikale Bewegung der Erdkruste ist Folge und nicht Ursache der vielfältigen Äußerungen des Lebens auf dem Bioplaneten Erde.

*Egon T. Degens* (1989): I picked up the pen again and again to bring to a close this biogeochemical Odyssy. . .

# References

Anderson DL (1984) The earth as a planet: paradigms and paradoxes. Science 223:347–354
Baas Becking LGM (1931) Gaia of Leven en Aarde
Baas Becking LGM (1934) Geobiologie of Inleiding tot de Milieurkunde. van Stockum and Zoon, Den Haag
Barrow JD, Tipler FJ (1986) The anthropic cosmological principle. Clarendon, London
Berthelin J (1983) Microbial weathering processes: In: Krumbein WE (ed) Microbial Geochemistry, 223–262 Blackwell (Oxford) 330 p
Degens ET, Epstein S (1962) Relationship between $O^{18}/O^{16}$ ratios in coexisting carbonates, cherts and diatomites. AAPG Bull 46:534–542
Degens ET (1976) Molecular mechanisms on carbonate, phosphate, and Silica Deposition in the Living Cell. Topics in Current Chemistry 64:1–112
Degens ET, Izdar E, Honjo S (eds) (1987) Particle flux in the ocean. Geol Pal Inst, Hamburg
Degens ET (1989) Perspectives on Biogeochemistry Springer, Berlin Heidelberg New York
Fischer AG (1984) Biological innovations and the sedimentary record. In: Holland HD, Trendall AF (eds) Patterns of change in Earth evolution. Springer, Berlin Heidelberg New York, p 145
Gurevich L, Mostepanenko V (1971) On the existence of atoms in n-dimensional space. Phys Lett 35A:201–202
Holland HD (1984) The chemical evolution of the atmosphere and oceans. Princeton University Press, Princeton
Hubbert MK (1974) The energy resources of the Earth. In: Press F, Siever R (eds) Planet Earth Readings. Sci Am, p 5
Humboldt Av (1844) Kosmos – Entwurf einer physischen Weltbeschreibung. Cotta, Stuttgart
Kempe S (1979) Carbon in the rock cycle. In: Bolin B, Degens ET, Kempe S, Ketner P (eds) The global carbon cycle scope 13. Wiley New York, p 343
Krumbein WE (1969) Über den Einfluß der Mikroflora auf die exogene Dynamik (Verwitterung und Krustenbildung). Geol. Rdsch. 58:333–363
Krumbein WE (1971) Sedimentmikrobiologie und ihre goelogischen Aspekte. Geol. Rdsch. 60:438–471
Krumbein WE (1983a) Stromatolites – the challenge of a term in space and time. Precamb Res 20:493–531
Krumbein WE (1983b) Introduction. In: Krumbein WE (ed) Microbial geochemistry. Blackwell, Oxford, p 1
Krumbein WE (1984) Auf den Schultern des Riesen – Vom Zeitgeist in der Erforschung geomikrobiologischer Zusammenhänge. In: Degens ET, Krumbein WE, Prashnowksy AA (eds) Ein Nord-Süd Profil: Zentraleuropa – Mittelmeerraum – Afrika Festband Georg Knetsch. Mitt Hbg Geol Staatsint Geol-Paläontol Inst, Hamburg 56:435
Krumbein WE (1986) Biotransfer of minerals by microbes and microbial mats. In: Leadbeater BSC, Riding R (eds) Biomineralization in lower plants and animals. Oxford University Press, Oxford, p 55
Krumbein WE (1987) Geomikrobiologie an der Universität Oldenburg. Forum Mikrobiol. 10:16–17
Krumbein WE (1988) Biotransfer in monuments – a sociobiological study. Durability of building materials 5:359–382
Krumbein WE, Dyer BD (1985) This planet is alive – weathering and biology, a multifaceted Problem. In: Drever JI (ed) The chemistry of weathering, Vol 149. Reidel, Dordrecht, p 143
Krumbein WE, Lasserre P, Nixon SW (1981) Biological processes and ecology. In: UNESCO IABO (ed) Coastal lagoon research, present and future. UNESCO technical papers in marine Science Vol. 32:51–79, Paris 95 p
Lapo AV (1982) Traces of bygone biospheres. Mir, Moscow (mss. 1979, eucl. Frurl. 1982 revised ed. 87)
Lovelock JE (1979) Gaia a new look at life on Earth. Oxford University Press, oxford
Lovelock JE (1984) Holland's Earth. Nature 312:571
Press F, Siever R (1986) Planet Earth. Freeman, San Francisco
Seilacher A (1985) Discussion of Precambrian metazoans. Philos Trans Roy Soc Lond B Biol Sci 311:47–48
Vernadsky WJ (1929) La biosphere. Alcan, Paris
Vernadsky WJ (1930) Geochemie in ausgewählten Kapiteln. Akademische Verlagsgesellschaft, Leipzig
Vernadsky WI (1938, 1944) Problems of biogeochemistry II, The fundamental matter-energy difference between the living and the inert natural bodies of the biosphere. Trans Conn Acad Arts Aci 35:487–517

# Biomimetic Geochemistry – A Speculation

A. NISSENBAUM and A. SERBAN

## 1 Rationale

The following article is based on laboratory studies of the involvement of natural polymers and their possible laboratory analogues in reactions simulating processes, which in the environment, are usually ascribed to biological activity.

On an extremely simplified level, the chemical interactions between the biosphere, geosphere and hydrosphere can be classified into two major divisions. One is the mineralization of dead organisms into inorganic moieties such as ammonia, phosphate, carbon dioxide etc. and the reincorporation of those molecules into fresh organic matter and the other, the activity of living organisms, and in particular microorganisms, in the transformation of organic and inorganic compounds. To cite a few examples – oxygen consumption, fermentation, methanogenesis, sulfate reduction etc. In many of those processes the role of biochemical reactions is in mediating electron transfer fluxes, or in other words oxidation and reduction.

Yet, occasionally a situation exists in which a direct biological involvement does not seem to be a dominant factor. Certain oceanic environments and in particular the deep water are very oligotrophic. However, a "biological" type of activity seems to be going on. An example would be the "benthic metabolism" invoked by Craig (1970) to explain certain aspects of the $CO_2$ cycle of the deep water masses. Even in surface waters it is not always possible to unequivocally relate certain organic compounds to biological activity, and frequently all that can be said is that life processes must be involved in the reactions without specifying a mechanistic relationship.

On the other hand, virtually all natural waters and sediments do contain dissolved or solid organic matter of high molecular weight, which is often called "polymer" even though no proof has been forwarded as to its being composed of repeating units, and which is usually considered to be the "kitchen sink" of geochemistry. Although the nomenclature of those polymers is a matter of debate, they will be refered to in this article, for simplicity sake, as humic substances. It is frequently assumed that these compounds play at best a passive role in the environment. They are capable of chelating or absorbing chemical species or produce various compounds by diagenetic heating or cracking.

However, some evidence exists to show that humic-type polymers can be involved in reactions which mimic reactions occurring in living cells. We propose the use of the term "biomimetic geochemistry" to describe such non-biological reactions.

## 1.1 Humic Substances in the Environment – a (very) Brief Review

Humic substances can be defined as a general class of dissolved, colloidal or solid refractory organic matter that is obiquitous, occurring in all terrestrial and aquatic environments (Aiken et al. 1985). According to Hedges (1988) some of the definitive characteristics of humic substances are: acidity, molecular weight of 500 to > 250,000 Daltons, the fact that usually less than 20% of the polymers can be accounted as recognizable biochemicals, a significant nitrogen content, color which ranges from yellowish to black and their ubiquitous distribution. As to general nomenclature perhaps the term – heteropolycondensates – introduced by Degens (1970) is an appropriate, even if somewhat cumbersome, being devoid of any source connotations.

Using those very broad definitions, which can only be slightly refined by a three-fold division into humic acids (base soluble, acid-insoluble fraction), fulvic acid (base-soluble, acid-soluble fraction) and humin (insoluble fraction), it can be shown that this class of compounds comprises one of the largest, if not the largest reservoir of organic carbon in the recent environment, comprising 20 to 60% of the total organic carbon.

The elemental composition of the humic polymers from various sources gives average values of C = 50–55%, H = 4–5%, N = 2.5%, S = 1–4%, O = 35–40% (Rashid 1985). The intractibility of those polymers to classical methods of organic analysis led to two main experimental approaches which have been used in order to decipher the composition of chemical and structural units of those polymers. The methods are degradation (by oxidation or reduction) and non-destructive spectroscopic techniques such as IR and $^{13}C$ and $^{1}H$ NMR. Although such techniques have been widely practiced, no general agreement exists as even to the question whether the humic substances represents a class of closely related polymers or whether they represent a wide spectrum of compounds with similar physical properties. Recent studies, for example, indicate that paraffinic structures may be a much more important component of humic polymers than previously thought (Hatcher and Spiker 1988). It should be borne in mind that the question of whether aromatic or paraffinic moieties are more important in humic acid structure is a basic problem which theoretically should have been resolved long ago. The fact that there is still no general agreement on the subject testifies to the complexity of the fundamental problem of elucidating the structure of humic substances.

Five major pathways have been proposed to account for the source of humic substances: formation from lignin derivatives (e.g. Waksman 1983), non-biological condensation of polyphenols (Martin and Haider 1971), condensation of amino containing groups with aldehydes (Nissenbaum and Kaplan 1972), autooxidative crosslinking of unsaturated lipids (Harvey et al. 1983), or by degradative alteration of an insoluble biopolymer (Hatcher et al. 1981).

## 1.2 Model Compounds

The intractability of humic polymers to structural studies as well as the ambiguity of its sources and pathways of formation, led to the widespread use of model compounds synthesized in the laboratory. Bracewell et al. (1988) say "Models first arose from a need to represent the then known chemistry in terms of structures which are highly hypothetical but which are based on chemically perceived experience."

The three most common models are: the reactions involving quinones which are easily produced from polyphenols by non-biological oxidation (Flaig 1988), the production of brown polymers by in vivo enzymatic oxidation of phenolics (Martin and Haider 1971), and the condensation of aldehydes such as carbohydrates with compounds containing amino groups such as amino acids or proteins to form melanoidins (Maillard 1913; Nissenbaum and Kaplan 1972).

Each of these model approaches has its advantages and disadvantages. None of the products has all the properties of natural polymers and therefore they must be used very gingerly in applying experimental results obtained on them to the natural environment. Yet, several of their properties, may have resemblance to natural polymers. This assumption is used throughout the present report.

## 2 Biomimetic Reactions

### 2.1 "Respiration"

Aerobic respiration is the process through which organisms oxidize organic matter to $CO_2$ while reducing oxygen to $H_2O$. In the natural environment such processes frequently lead to anoxification of the environment and enrichment in inorganic carbon.

Reactions which can be formally analogous occur during the synthesis of melanoidins. Serban and Nissenbaum (1981) showed that during the reaction of glucose with either glycine or lysine, the oxygen concentration in the gas phase above the reaction mixture decreases (Fig. 1). The amounts of oxygen consumed is

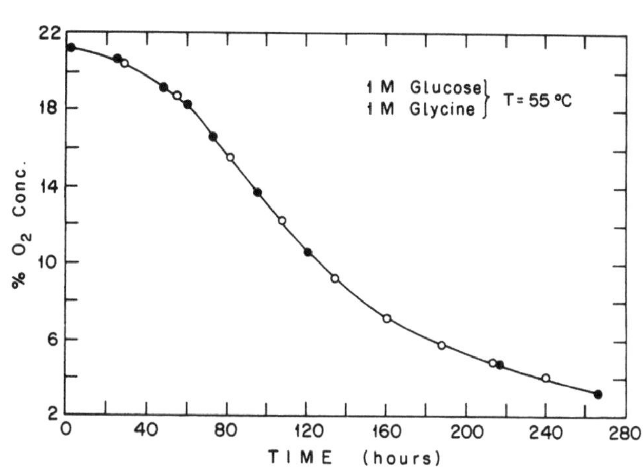


**Fig. 1.** Oxygen consumption during melanoidin formation

strongly dependent on temperature and on the nature of the amino acids. In this respect it is similar to the formation of the melanoidin polymer which is dependent on the same two factors (Hodge 1953).

The mechanism of oxygen removal probably involves several parallel reactions with one or more intermediates, resulting in an S-shaped curve. Each reaction possibly follows different kinetics. The nature of the intermediate(s) is not known, but we have observed the appearance of an intermediate which is UV absorbing at 295 nm which appears when the oxygen consumption rate is highest and disappear towards the end of the reaction.

During the melanoidin reaction, large amounts of $CO_2$ are evolved (Fig. 2). It is frequently assumed that the $CO_2$ originates from the Strecker degradation of α-amino acids by β-dicarbonyl intermediates (Stadman et al. 1952; Cole 1967).

In order to investigate the source of the evolved $CO_2$, we used, in a set of experiments, uniformly $C_1$ and $C_6$ $^{14}C$ labelled D-glucose and $C_1$ and $C_2$ labelled

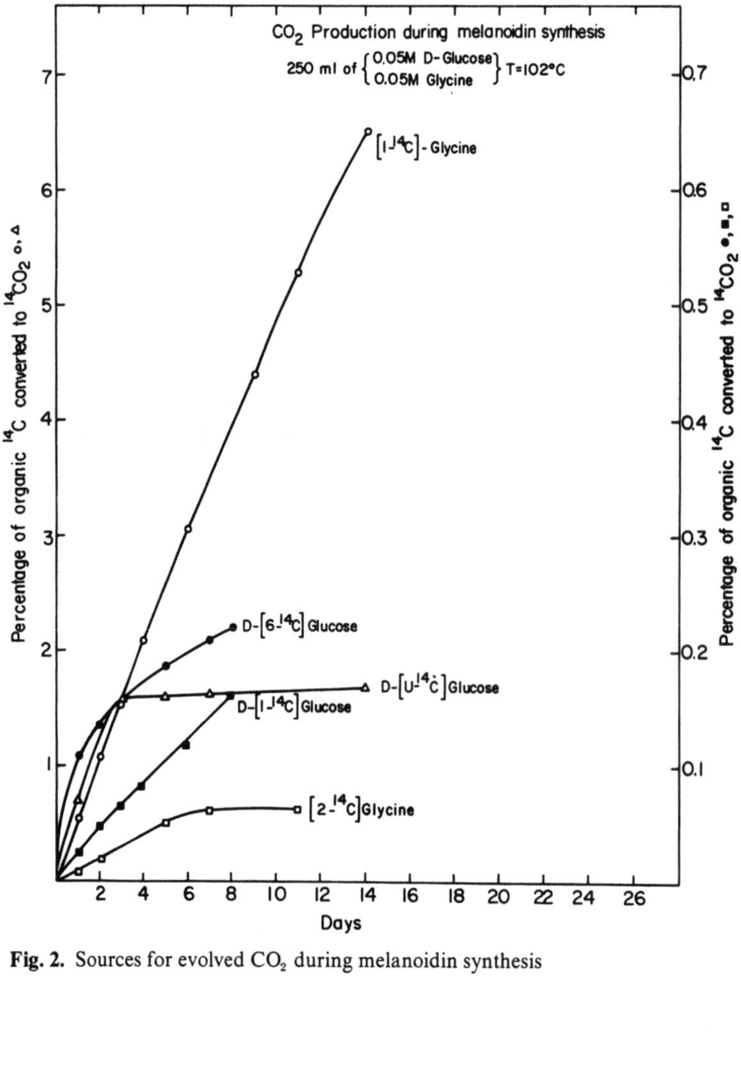


**Fig. 2.** Sources for evolved $CO_2$ during melanoidin synthesis

glycine. We found (Fig. 2) that 79% of the $CO_2$ was produced from the carboxyl group of glycine and 21% from the glucose. Thus, Strecker degradation seems to be indeed a major contributor of $CO_2$, although the role of carbohydrates should not be discounted. It is not known if such reactions occur in nature. A pertinent observation was made by Huc and Durand (1977) who noted the progressive loss of $CO_2$ during the diagenesis of humic substances to immature kerogens.

The mechanism of $O_2$ removal is not known. In order to check the possibility of production of $CO_2$ by total oxidation of part of the organic matter, we introduced into the reaction vessel $O_2$ gas labelled with 97.8% $^{18}O$. We found very little labelled oxygen in the $CO_2$, but much heavier label in the solid melanoidin, suggesting the incorporation of molecular oxygen into the polymer.

The analogy between biological respiration and melanoidin "respiration" is only expressed in the removal of oxygen and production of $CO_2$. They can be decoupled in the abiological reaction, since $CO_2$ production does not depend on presence of oxygen, but not so in aerobic biological respiration.

However, in general terms, if non-biological "respiration" occurs in nature, its environmental impact will be similar to that of biological, true respiration.

## 2.2 "Enzymatic Activity"

Serban and Nissenbaum (1985) reported that humic substances, as well as synthetic melanoidins, exhibit catalytic activity in reactions usually associated with enzymes such as peroxidase, diaphorase, NADH dehydrogenase and esterase, albeit with much reduced efficiency. The analogy between humic substances, melanoidins and genuine enzymes was particularly conspicuous in oxidation-reduction reactions. This is probably due to the occurrence of a large number of free radicals in humic substances (Nissenbaum and Kaplan 1972) and melanoidins (Aizenshtat et al. 1987) and of the formation of ε-donor-acceptor complexes in humic systems (Ziechman 1988).

In contrast to enzymes which are characterized by very high substrate specificity, the polymers are much less restricted. However, since humic substances originate from many sources of materials, as well as under different environments, it is conceivable that in the multitude of structural regimes of the humic polymers, which are probably highly unordered, a region can be found to catalyze a particular reaction. The fact that only a tiny part of the polymer will have the right configuration, will be one of the explanations to the low activity of the polymers as compared with custom-tailored enzymes.

Another mode of "geo-enzymatic" activity is through the association of cell-free enzymes with humic substances. This subject has been extensively discussed in the soil science literature (Skujins 1976; Burns 1978). Nissenbaum and Serban (1987) proposed that the catalase and peroxidase activity detected in marine sediments as old as the Upper Miocene (approximately 8,000,000 years B.P.) are due to scavenging of the enzymes, which are liberated by lysis of cells, by humic substances. Laboratory studies, in which free enzymes were interacted with humic substances (Serban and Nissenbaum 1986) showed that the resulting complexes retain the enzymatic activity although at a much reduced rate. The complexes were much more resistant to environmental conditions such as pH, temperature and enzymic degradation than the free enzyme. Such formation of "immobilized enzymes" may explain the preservation of the enzymes over a time span and under such conditions where the free enzymes would have been long denatured. It has to be remembered, though, that the presence of potential "geo-enzymes" does not prove that they are catalytically active in the environment. However, if they are, then they can be included in the concept of biomimetic geochemistry.

### 2.3 "Photoprocesses"

Humic substances (and model compounds) absorb virtually the whole spectra of sunlight. It has been shown (Slawinksa et al. 1975; Serban, unpublished data) that visible light enhances the ESR signal of free radicals in the humic substances. Recently, evidence emerged (Zepp 1988) that singlet oxygen can form by irradiation of humic acid solutions. This excited oxygen species can act as a powerful oxidant. Other reactive species can be photoformed as well such as hydrogen peroxide, and organoperoxyl radicals.

Thus, humic substances can act as powerful environmental agents in reactions involving electron transfer (Zika and Cooper 1987). The reactions may involve inorganic species, as is the case in reducing metal cations, such as $Fe^{+3}$, or in the alteration of various organic compounds. It has been shown, for example, that pollutants may be degraded by photoreactions of humic material (Zepp 1988).

Thus, electron or charge transfer photochemical processes may be responsible for environmental reactions, which are frequently assumed to be biological.

Photoprocesses are relevant not only for degradation processes but also in formative reaction. It has recently been shown that photochemical reactions of dissolved organic matter can produce a large variety of low molecular weight organic compounds. For example, formaldehyde and acetaldehyde (Zafiriou 1977) aldehydes and ketones (Mopper and Stahovec 1986) and glyoxilic and pyruvic acids (Kieber and Mopper 1987). Although such compounds are usually regarded as biochemically produced metabolites, these studies provide evidence that they may be produced in a nonbiological process mimicking biological activity.

## 3 In Times Past

There can be no doubt that if such biomimetic reactions as described above are active in the environment, their role is very minor as compared to that of active biological community. It is interesting however to speculate, that perhaps in the era before development of life, natural organic polymers may have been involved in various reactions. We have no direct evidence that such polymers did exist in this era of pre-biological evolution. However, polymerized organic matter is known from the most ancient sediments from the Isua group (3.8 Ga ago) and from meteorites.

Nissenbaum et al. (1975) speculated that non-biological formation of amino acids and sugars would lead to rapid formation of melanoidins and that this material would be very rapidly removed into the sediments. Therefore, under non-steady state conditions of precursor formation, the amount of soluble organic matter in the primitive oceans will be very small, moving the locus of certain reactions to the solid-liquid interface. Nissenbaum et al. (1975) further speculated that such polymers could play a role in mimicking enzymatic redox reactions, with the polymers acting as co-enzymes. For example, the formation of volatile heterocyclic nitrogen compounds (Koehler et al. 1969) in the melanoidin reaction, and pyrolysis of melanoidins which produces heterocyclic nitrogen compounds (Rubinsztain et al. 1984) suggest the existence of structural units in the polymer which resembles the isoalloxazine moiety of flavin co-enzymes, which are involved in redox reactions.

Another possible role is through the photochemical production of low molecular weight compounds such as glyoxilic and pyruvic acids (Kieber and Mopper 1987) which can potentially serve in primitive biochemistry and later be incorporated into biochemical cycles as we know them today.

Of course, once life has risen, the role of those polymers becomes miniscule due to the competitive edge of living systems. Perhaps we have here a very early case of the Darwinian principle of "survival of the fittest".

## References

Aiken GR, McKnight DM, Wershaw RL, MacCarthy P (1985) An Introduction to the Humic Substances in Soil, Sediment and Water. Ch. 1. In: Humic Substances in Soil, Sediment and Water (eds Aiken GR, McKnight DM, Wershaw RL, MacCarthy P), Wiley, New York, pp 1–12

Aizenshtat Z, Rubinszstain Y, Ioselis P, Miroslavski I, Ikan R (1987) Long Living Free Radicals Study of Stepwise Pyrolyzed Melanoidins and Humic Subtances. Org Geochem 11:65–72

Bracewell JM, Abbt-Braun G, de Leeuu JW, Hayes MHB, Nimz DL, Norwood EM, Perdue EM, Schnitzer M, Visser SA, Wilson MA, Ziech Mann W (1988) The Characterization and Validity of Structural Hypothesis: Group Report 3. In: Frimmel FH, Christman RF (eds) Humic Substances and their Role in the Environment. Dahlem Workshop, Wiley, New York, pp 151–164

Burns RG (1978) Soil Enzymes. Academic Press, London

Cole SJ (1967) The Maillard Reaction in Food Production and $CO_2$ Production. J Food Sci 32:245–250

Craig H (1970) Deep metabolism. Oxygen Consumption in Abyssal Ocean Water. J Geophys Res 76:5078–5086

Degens ET (1970) Molecular Nature of Nitrogenous Compounds in Sea Water. In: Organic Matter in Natural Waters. Hood DW (ed), Inst Mar Sci, Univ of Alaska, pp 77–106

Flaig W (1988) Generation of Model Chemical Precursors. In: Frimmel FH, Christman RF (eds) Humic Substances and their Role in the Environment. Dahlem Workshop, Wiley, New York, pp 75–92

Harvey GR, Boran DA, Chesal LA, Tokar JM (1983) The Structure of Marine Humic and Fulvic Acids. Mar Chem 12:119–132

Hatcher PG, Spiker EC (1988) Selective Degradation of Plant Biomolecules. In: Frimmel FH, Christman RF (eds) Humic Substances and their Role in the Environment. Dahlem Workshop, Wiley, New York, pp 59–74

Hatcher PG, Spiker EC, Szeverenyi NM, Maciel GE (1981) Selective Preservation and Origin of Petroleum Forming Aquatic Kerogen. Nature (Lond) 305:498–501

Hedges JI (1988) Polymerization of Humic Substances in Natural Environments. In: Frimmel FH, Christman RF (eds) Humic Substances and their Role in the Environment. Dahlem Workshop, Wiley, New York, pp 45–58

Hodge JE (1953) The Browning Reaction. J Agric Food Chem 1:928–943
Huc AY, Durand BM (1977) Occurrence and Significance of Humic Acids in Ancient Sediments. Fuel 56:73–80
Kieber DJ, Mopper K (1987) Photochemical Formation of Glyoxylic and Pyruvic Acids in Seawater. Mar Chem 21:135–150
Koehler PE, Mason ME, Newell JA (1969) Formation of Pyrazine Compounds in Sugar-Amino Acid Model Systems. J Agric Food Chem 17:393–396
Maillard LC (1913) Formation des Matieres Humiques par Action de Polypeptide sur les Sucres. CR Acad Fr 156:1159–1160
Martin JP, Haider K (1971) Microbial Activity in Relation to Soil Humus Formation. Soil Sci 111:54–63
Mopper K, Stahovec WL (1986) Sources and Sinks of Low Molecular Weight Organic Carbonyl Compounds in Seawater. Mar Chem 19:305–322
Nissenbaum A, Kaplan IR (1972) Chemical and Isotopic Evidence for the *in situ* Origin of Marine Humic Substances. Limnol Oceanogr 17:570–582
Nissenbaum A, Serban A (1987) Enzymatic (?) Activity Associated with Humic Substances in Deep Sediments for the Cariaco Trench and the Walvis Ridge. Geochim Cosmochim Acta 51:377–378
Nissenbaum A, Kenyon DH, Oro J (1975) On the Possible Role of Organic Melanoidin Polymers as Matrices for Prebiotic Activity. J Mol Evol 6:253–270
Rashid MA (1985) Geochemistry of Marine Humic Compounds. Springer, Berlin Heidelberg New York Tokyo, 300 pp
Rubinsztain Y, Ioselis P, Ikan R, Aizenshtat Z (1984) Investigations on the Structural Units of Melanoidins. Org Geochem (1983 Adv Org Geochem) 6:791–804
Serban A, Nissenbaum A (1981) Melanoidin Polymers as Possible Oxygen Sinks in the Pre-biotic Oceans. In: Origin of Life. Wolman Y (ed) D Reidel, Dordrecht, pp 151–156
Serban A, Nissenbaum A (1985) Biomimetic Catalysis Mediated by Humic Substances and Melanoidins: Geo-Enzyme Activity? In: Planetary Ecology, (Caldwell DE, Brierli JA, Brierli CL, eds), Reinhold, New York, pp 17–26
Serban A, Nissenbaum A (1986) Humic Acid Association with Peroxidase and Catalase. Soil Biol Biochem 18:41–44
Skujins J (1976) Extracellular Enzymes in Soils. CRC Crit Rev Microbiol 6:383–421
Slawinska D, Slawinski J, Sarna T (1975) The Effect of Light on the ESR Spectra of Humic Substances. J Soil Sci 26:93–99
Stadman FH, Chichester CO, McKinney G (1952) Carbon Dioxide Production in the Browning Reaction. JACS 74:3194–3196
Waksman SA (1938) Humus: Origin, Chemical Composition and Importance in Nature. Williams and Wilkins, Baltimore MD, 526 pp
Zafiriou O (1977) Marine Organic Photochemistry Previewed. Mar Chem 5:497–522
Zepp RG (1988) Environmental Photoprocesses Involving Natural Organic Matter. In: Frimmel FH, Christman RF (eds) Humic Substances and their Role in the Environment, Dahlem Workshop, Wiley, New York, pp 193–214
Ziechmann W (1988) Evolution of Structural Models from Consideration of Physical and Chemical Properties. In: Frimmel FH, Christman RF (eds) Humic Substances and their Role in the Environment, Dahlem Workshop, Wiley, New York, pp 113–132
Zika RG, Cooper WJ (1987) Photochemistry and Environmental Aquatic Systems. ACS Symposium Series 327. Am Chem Soc, Washington DC

# Global Change: Real Time Geochemistry

W. S. Fyfe

## 1 Introduction

It is a great pleasure to contribute to this volume in honour of Egon T. Degens, one of the great geochemists of this century. I recently listened to a lecture by John Polanyi of the University of Toronto, a recent Nobel Laurate. Polanyi suggested that great scientists choose the focus of their research by reflecting on the impact progress in a given area is likely to have. In this respect it is clear that Egon always chose well. His early work on sediments and particularly on the biogeochemistry of sediments, later work in the Red Sea and Black Sea, and more recently on the global carbon cycle show this wisdom of choice for all these topics are at centre stage of modern geochemistry; all are difficult but the impact of progress large.

A growing concern with change on this planet and the human impact on the surface environment which increases with the phenomenal growth of the human species (plus 86 million, 1987), and the technologies man requires to support this growing population, has led to an urgent need to understand geochemical processes on short time scales. The International Council of Scientific Unions new initiative, The International Geosphere Biosphere Programme, reflects such needs and concern (ICSU 1986). Scientists are being asked to obtain the knowledge necessary to develop strategies for sustainable development of a planet for a population of at least 10 billion. The situation is well summarized by Malone (1986) and the World Commission on Environmental Development (The Bruntland report 1987).

Historically, geologists and geochemists have been greatly concerned with longer-term features of geologic dynamics. We are now being asked the more demanding questions related to short term variation. We are not simply concerned with the major integrated fluxes but with the short term fluctuations in the system. There is no doubt that this will lead to greatly improved geologic models for such predictions demand that we understand the mechanistic details on large scales. Egon Degens was one concerned with exactly such approaches to geochemistry, what I will call "real time" geochemistry.

There is nothing new in this approach which goes back to the great pioneer Charles Lyell. In 1872 Lyell wrote:

It appeared to them far more philosophical to speculate on the possibilities of the past, than to patiently explore the realities of the present; and having invented theories under the influence of such maxims, they were consistently unwilling to test their validity by the criterion of their accordance with the ordinary operations of Nature .... It produced a state of mind unfavourable in the highest degree to the candid reception of the evidence of those minute but incessant alterations which every part of the earth's surface is undergoing, and by which the condition of its living inhabitants is continually made to vary .... The student, instead of being encouraged with the hope of interpreting enigmas presented to him in earth's structure – instead of being prompted to undertake laborious enquiries into the natural history of the organic world, and the complicated effects of the igneous and aqueous causes now in operation – was taught to despond from the first.

Today, we are being asked again to take stock of our knowledge of the geochemical cycle but with major emphasis on short term fluctuations in the major fluxes between the classic geospheres.

## 2 Observation

It would have been virtually impossible to approach the Geosphere-Biosphere project, with any chance of success, a few decades ago. What this project requires is a deeper understanding of processes at the microscale to guide the necessary integration of data on the macroscale and planetary scales. The system to be studied in the IGBP involves all significant mass-energy fluxes in the system:

Sun→ Living Cell← Earth.

Except for fluctuations in the Sun, the geochemist is involved in all the chemical processes in this, the ultimate system.

With satellites we can study an increasing spectrum of phenomena on the large scale. But for the geochemist the advances have been equally spectacular. The new array of laser and microbeam techniques coupled to the new generation of mass spectrometers enables us to produce chemical data with increased sensitivity, reliability, on samples of very small size. New techniques in electron microscopy allow us to essentially see single atoms in a structure.

Some of the most spectacular advances with great potential for geochemistry involve techniques for the analysis of thin films, surfaces, interfaces. Via techniques such as ESCA, Auger and SIMS, we can analyze the chemistry of a surface at the single atomic layer level. It is these techniques that allow us to quantify the interactions between any surface and its environment. Surface processes at the micro level, processes which ultimately control the environment, are no longer the terra incognita of our science. In the future we will see the same techniques increasingly applied to bio-surface.

At other levels, the techniques of observing the deep earth are improving. The new seismic tomography is revealing the convective patterns of earth down to the core boundary. At higher levels, seismic and electrical methods, coupled to deep drilling, reveal the temperature structures and the presence of active fluids. Magma chambers can be mapped along with active metamorphic fluid systems. We are now prepared to do "real time" geo-geochemical dynamics on most appropriate scales. I am sure that Lyell would be envious of our opportunities to study actual processes in action.

## 3 The Great Fluxes

Our surface phenomena, chemical and biological, are related to heat-mass transfer processes forced by the sun and heat production in the Earth (Figs. 1, 2). The thermal structure of the Earth is being increasingly resolved by seismic tomography (Dziewonski and Woodhouse 1987). Plate tectonic processes reflect the nature of such processes in the outer few hundred kilometers of the planet.

Rising convection cells drive the ocean floor spreading process and basalt production. We now know that about half the Earth's energy production is focussed into ridges by such convective flow. We have also discovered that the energy and mass involved in the ridge processes is partly transferred to the hydrosphere in the outer few kilometers of the new crust. This processs has a very significant influence on the chemistry of the oceans and even the nutrient supply to the oceans. In large part, the temperature structure in the outer layers of the crust is related to water cooling processes involving the hydrosphere with thermally driven flow in cracked and porous media. While we appreciate the role of such processes in terms of the formation of mineral deposits, such fluxes have not been adequately considered in terms of their influence on the biosphere. But magmatic processes are never steady state. Study of any volcanic terrain on earth reveals periodicity in mass and time on widely ranging scales. This in turn implies that cooling fluxes will fluctuate in a similar fashion. Recently the first observations of large hydrothermal plumes off the Juan de Fuca ridge have been observed (Baker et al. 1987). But surely such erratic behaviour must be quite normal. Will the new surface techniques allow us to detect the record of such events, particularly those of large magnitude? They will lead to local variations of temperature, and should be apparent in the geochemistry of species like $^{87}Sr/^{86}Sr$, $^{3}He/^{4}He$ and the like. Owen and Rae (1985) even suggested that such processes could substantially perturb atmospheric carbon dioxide by injection of large quantities of calcium chloride brines.

The greatest part of our planet's surface involves an ophiolitic crust with a thin sediment blanket. At the present time our knowledge of the dynamics of these vast areas is small.

Lister (1977) predicted that water cooling of oceanic crust could continue at considerable distances away from ridges. He predicted closed convection cells beneath the impermeable sediment cover. There is increasing evidence that this process occurs. Recent studies of selected sites for nuclear waste disposal have revealed considerable complexity. Energy to drive such flow processes can be provided in part by residual heat from ridges but perhaps more importantly by exothermic serpentinization of deep-level peridotites.

Is it possible that warm springs are common over the entire areas of such submarine systems? Further such fluids could be very saline resulting from hydration reactions concentrating the soluble salts. Particularly interesting situations might occur where a closed subsediment convective system is perturbed by faulting or off-ridge magma intrusion leading to overpressure in the deep system. Much exploration remains to be done in the off-ridge systems but could satellite observations detect major fluctuations by the influences on surface biological production?

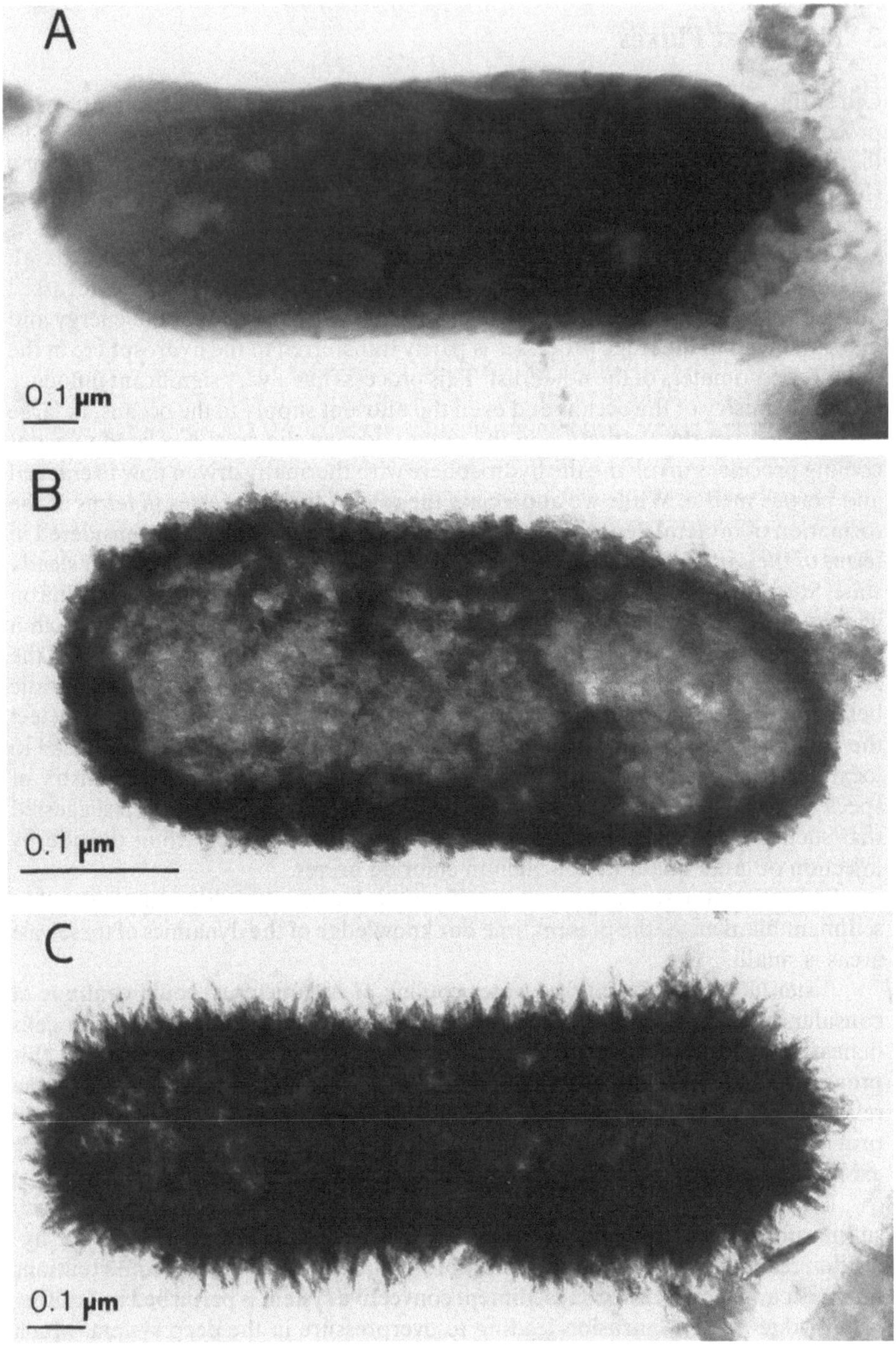


**Fig. 1A-C.** Bacteria from a core in a mine-polluted lake in Ontario. **A** near the top a living bacterium is metal opaque; **B** amorphous Fe-oxides forming and **(C)** at a few cm depth fully mineralized by crystalline iron oxy-hydroxides. We find that in many of our lake sediments the heavy metal mineralization is controlled by microorganisms. (Photo Dr. Henrietta Mann)

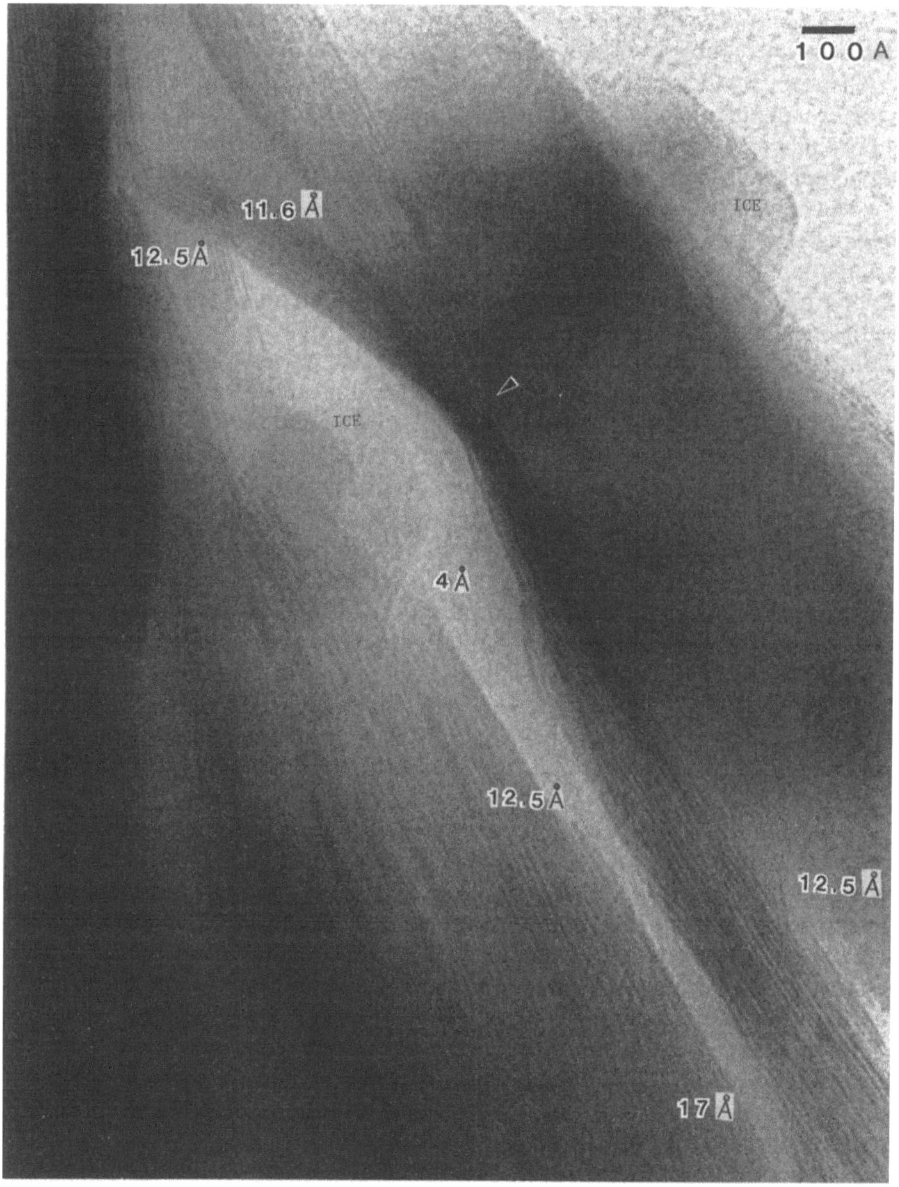


**Fig. 2.** Ice forming in a frozen smectite clay is shown by modern transmission microscopy. Ice formation deforms the clay structure and can even fragment the structure. It seems that this process may be partly responsible for accelerated wind erosion in arctic regions. (Photo Dr. K. Tazaki)

Work over the past decade has now begun to reveal details of the subduction processes. Direct observation in trenches (Lallemant et al. 1986) with appropriate geophysical techniques has clearly shown that recycling in subduction zones involves more than the removal of old, $H_2O$-$CO_2$-S-K-U . . . , loaded ocean crust. There is now clear evidence that massive quantities of sediments are subducted and can be tracked into the subduction-related volcanics by tracers such as $^{10}Be$.

The old views of one way evolution of volatiles and continents must now be challenged (cf. the views on the geochemical cycle of the 1950s, Mason 1958). Via solutions and sediments, continental debris is subducted, volatiles are subducted. Do continents and oceans grow continuously on a mixing and cooling planet? Could a cooling Earth evolve to a surface similar to Mars, almost devoid of $H_2O$? The return flow of fluids up subduction thrusts is now well documented. These fluids encourage exotic fauna similar to those found at ridge hot springs (Lallemant et al. 1986). Fluids released at greater depths, lower mantle viscosity and generate the mantle heat-mass flows which produce the basalt-andesite systems above the subduction zones. This process, frequently occurring inside continental margins, triggers the great array of events, basaltic-andesitic volcanism, granite plutonism and crustal fusion, deep crustal metamorphism, associated with subduction. Again it must lead to periodic water cooling processes at high levels and metamorphic fluid pumping at deep levels. Models of mineral deposit formations in such situations are well advanced but the total surface fluid flux regimes are poorly quantified. The environmental impact of giant acid-volcanic eruptions is recognized (Toba's, Tamboras and the like) but our knowledge is inadequate in terms of hydrothermal fluids. Deep electrical sounding has revealed the enormous extent of conductive anomalies associated with subduction along the west coast of North America, anomalies which extend over almost 30% of the continent (Gough 1981).

Recent work on magma compressibilities has clearly shown that at Moho pressures basaltic magmas can be denser than continental crust. Extensive underplating may be required to explain the huge areas of conductive anomalies and the extensive areas of high-grade deep metamorphism. In this connection it must be stressed that the Earth's crust contains in hydrated minerals about the same mass of water as the oceans. A major thermal perturbation near the Moho can thus liberate very large amounts of fluids. The "fine scale" distribution of water in hydrated minerals, a catalyst of the metamorphic reactions, will lead to efficient separation of this fluid. In comparison, massive carbonate units which may react only by $H_2O$ and $SiO_2$ corrosion processes may respond more slowly and explain the late enrichment of fluid inclusions in carbon dioxide.

The subduction processes may procede and be associated with continental collisions and thrusting events and strike slip motions. Both may lead to extensive thickening of the crust (e.g., the Himalayas and New Zealand Alps) and degassing of the thickened roots. Fyfe (1986) has shown that in an event of Himalayan scale, the mass of liberated metamorphic water is similar to that of the ice caps. Similar calculations for salt, hydrocarbons and carbon dioxide show potential for major environmental perturbations. And again, the fluxes cannot be steady state.

Where extensive crustal thickening by underplating of basic magmas occurs, foundering of dense crustal components, and their degassing, may produce very high-T gas phases rich in $CO_2$-CO-$H_2O$-$H_2$-HF . . . and may explain the frequent graphite introduction into granulitic basement and the slower carbonate reactions, the late $CO_2$ flushing.

Finally, mantle volcanism of any type and collision processes tends to produce the high elevations of the continents. Given high elevations, deep groundwater flow (e.g. the Artesian situation in the European Alps) must influence the chemistry of the discharge. Such gravity driven flow of deep water may change the chemistry of local surface reservoir and their biota. It will also be influenced by climatic fluctuations and the stratigraphic nature of uplifted regions (e.g. the presence of a major salt or hydrocarbon basin). We have little knowledge of such changes except for a few pointers like changing $^{87}Sr/^{86}Sr$ in ocean waters on long time scales. But there could be much finer scale oscillations for the fluid volumes involved are large.

All processes that produce elevations and metamorphism ultimately root in deep mantle convection which focusses energy into restricted surface regions. The process may be simple (ridges) or complex (subduction zones, collisions, strike slip faults, etc.) where mantle convection is induced by injection of small quantities of volatiles. This process leads to an immense amplification of energy-mass transfer.

## 4 The Surface

Finally I would like to return to the realms of the surface-weathering process and the biosphere. As stressed by Mason (1958) the integrated mass of living cells over time approaches the mass of the Earth. All life requires a wide spectrum of elements for survival (Mertz 1981). Organisms must be able to extract these elements against the limits of the controls of inorganic solubilities. As the average content of inorganics is around 1 to 2%, the total mass of materials cycled through the biosphere is large and similar to the mass of the crust.

Weathering and biological processes produce the high surface area materials of the surface sediments and soils. These are the ultimate buffers to change in the inorganic aquatic and atmospheric systems. In detail we are only beginning to appreciate how such systems function.

And here I must return to Egon Degens. I have heard him lecture on sediments, on life, on mineral templates for life. This science is beginning. As Vernadsky (1929) appreciated (cf. Lovelock 1986) our surface is a geosphere-biosphere system. Egon contributed to the development of the new geochemistry – real time geochemistry. His contributions will influence our science for a long time into the future.

## References

Baker ET, Massoth GJ, Feely RA (1987) Cataclysmic hydrothermal venting on the Juan de Fuca Ridge. Nature 329:149–151

Dziewonski AM, Woodhouse JH (1987) Global images of the Earth's interior. Science 236:37–48

Fyfe WS (1986) Fluids in deep continental crust. Am Geophys Union Geodynamics Ser 14:33–39

Gough DI (1981) Magnetometer arrays and geodynamics. Am Geophys Union Geodynamics Ser 5:87–95

ICSU (1986) The international geosphere biosphere project: a study of global change. ICSU Paris 1–21
Lallemant S, Lallemant S, Jolivet L, Huchon P (1986) Kaiko: L'exploration des fossess du Japan. Recherche (Paris) 182:1344–1357
Lister CRB (1977) Qualitative models of spreading centre processes, including hydrothermal penetration. Tectonophysics 37:203–218
Lovelock J (1986) Gaia: The world as a living organism. New Sci (Dec 18):25–28
Lyell C (1872) Principles of Geology, 11th edn. Murray, London, p 671
Malone T (1986) The international geosphere biosphere programme. Environment (Wash DC) 28:6–42
Mason B (1958) Principles of geochemistry. Wiley, New York, p 310
Mertz W (1981) The essential trace elements. Science 214:1332–1338
Owen RM, Ray DK (1985) Sea-floor hydrothermal activity links climate to tectonics: the Eocene carbon dioxide greenhouse. Science 227:166–170
Vernadsky VI (1929) The biosphere, new edition (1986). Synergetic, Arizona, p 82
World Commission on Environment and Development (1987) Our common future. Oxford University Press, p 383

# Biogeochemical Problems of Living Matter of the Present-Day Biosphere

E. A. ROMANKEVICH

## 1 Introduction

It is being increasingly recognized that the Earth's crust has undergone transformation not only by the action of geological agents (wind, water, ice, volcanic activity) but also by living matter. The composition of the atmosphere, hydrosphere and to a lesser extent the lithosphere has been transformed by the living organisms in bioinert bodies – ordered structures representing an inseparable unity of living and inert matter – and/or by biogenic, substances. The essential geological and geochemical role of living matter reduced to weight, composition, energy (and nowadays we also add information) was stressed by the founder of biogeochemistry, and the proponent of the theory of the biosphere, Vernadsky (1965, 1983).

Geological and paleontological data provide evidence for the conjugated nature of the transformation of the biosphere's living and non-living matter; the biosphere in this sense is the sphere of life, a complex veneer including all the hydrosphere and upper lithosphere down to a depth of some kilometers. The most vivid indicators of the biosphere's evolution should apparently include the acceleration of geological evolution (growing diversification of minerals, types of rock, their parageneses, facies, sedimentation rates), exponential character of ecosystem development, wider diversity of animal and plant taxa, increasing complication of the nervous system (cephalisation), increasing balancing of biogeochemical cycles of elements on Earth. To my mind, these problems relating to the evolution of the biosphere have not been adequately dealt with so far. The situation becomes still less clear when it comes to the evolution through the Earth's history of specific parameters of the biosphere governing, for example, the free oxygen released into the atmosphere, such as biomass of plants and animals, primary and secondary production organic matter, its fossilization coefficients and burial rates. Even for present-day environments many basic parameters of elemental cycling have been established only roughly, and in some instances only an order of magnitude is known. This work endeavours to discuss the available data on the amount and the production of Earth's biomass, group and elementary composition of living matter of our times. These data are necessary for tackling the problem of the evolution of the Biosphere.

## 2 Earth's Biomass and Production

The amounts of biomass and its production on land and in the oceans are critical characteristics of the biosphere. They represent the basic parameters required for the calculation of elemental fluxes in global biogeochemical cycles. Of utmost importance among them is the carbon cycle due to their controlling influence on the Earth's climate (greenhouse effect), on the processes of lithogenesis and on the formation of the global hydrocarbon reserves.

The progress in the estimation of the Earth's productivity and in the development of bioproductivity theory is related to the elaboration, during the 1930s and 1940s of an adequate methodology based on an "ecosystem production-energy approach", representation of totality of organisms from the viewpoint of their weight, energy, chemical composition, volume and character of relevant space (Vernadsky 1983).

The work accomplished in recent years has resulted in a quantitative expression of energy transformation and flows of substances at all major steps of the production process, the evaluation of statistical quantities of present biomass and of dynamic characteristics of feeding, growth (production), reproduction, expenditure on energy and constructive exchange common to all living matter. An important new development was the linking of a dynamical approach to a community with Vernadsky's views on the biogeochemical activity of living matter, and using this combination as a foundation for evolving a trophodynamical concept in ecology (for example Lieth and Whittaker 1975; Odum 1983; Bolin et al. 1979; Vinberg 1983; Degens et al. 1984; Vinogradov and Shushkina 1987).

Before the mid-seventies the researchers had arrived at average values of terrestrial ($450–1170 \times 10^{15}$g $C_{org}$) and oceanic biomass ($1–3.1 \times 10^{15}$g $C_{org}$), and primary production of land ($10–106 \times 10^{15}$g $C_{org}$ $y^{-1}$), ocean ($15–126 \times 10^{15}$g $C_{org}$ $y^{-1}$), and the Earth as whole ($28–135 \times 10^{15}$g $C_{org}$ $y^{-1}$. The values vary by an order of magnitude. The situation has somewhat improved for the land production estimates, but is still unsatisfactory for the ocean (Table 1 and 2). In spite of the progress currently achieved in sampling techniques and quantitative determinations, there is yet no commonly recognized combination of procedures for the determination of production and biomass with an accuracy of 30–40%. Our poor knowledge of the Earth's production and biomass is responsible for our inability to assess, with a high degree of certitude, the huge quantity of chemical elements drawn into the biogeochemical cycles by living matter. Table 3 gives global production characteristics using data on land and ocean biomass and their production (e.g. Moiseyev 1969; Bazilievich et al. 1970; Bogorov 1971; Lieth and Whittaker 1975; Dobrodeyev and Suetova 1976; Koblenz-Mishke 1977; Bolin et al. 1979; Degens et al. 1984; Ivanenkov 1985; Vinogradov and Shushkina 1987). These should be viewed as the author's own version obtained from a critical analysis and summing up of the various data.

The Earth's biomass appears to make up approx. $750 \times 10^{15}$ g $C_{org}$ (approx. $4{,}000 \times 10^9$ tons of wet mass), where annual production is about $120 \times 10^{15}$ g $C_{org}$ (approx. $800 \times 10^9$ tons of wet mass). (Hereafter all biomass and production figures are recalculated for organic carbon-$C_{org}$. This form of presentation is most

**Table 1.** Biomass and production of land

| Living matter | Biomass $10^{15}$g $C_{org}$ | Primary production $10^{15}C_{org}$ per year | $C_{org}$ per dry mass % | Area $10^{12}\,m^2$ | Source[a] |
|---|---|---|---|---|---|
| Living phytomass in toto including: | 560[b] | 60 | 45 | 149.3 | Ajtay et al. 1979 |
| Forests | 427.73 | 21.9 | 47 | 31.3 | Ajtay et al. 1979 |
| Temperate zone forests | 16.2 | 1.35 | – | 2 | Ajtay et al. 1979 |
| Bush | 7.88 | 0.9 | – | 2.5 | Ajtay et al. 1979 |
| Savannah | 65.56 | 17.71 | – | 22.5 | Ajtay et al. 1979 |
| Temperate zone meadows | 9.11 | 4.39 | – | 12.5 | Ajtay et al. 1979 |
| Arctic tundra | 5.87 | 0.95 | 42 | 9.5 | Ajtay et al. 1979 |
| Deserts and semi-deserts | 7.42 | 1.35 | – | 21 | Ajtay et al. 1979 |
| Extremely dry deserts | 0.35 | 0.06 | – | 9 | Ajtay et al. 1979 |
| Perennial ice | 0 | 0 | – | 15.5 | Ajtay et al. 1979 |
| Lakes and rivers | 0.02 | 0.36 | – | 2 | Ajtay et al. 1979 |
| Swamps | 11.81 | 3.26 | – | 2 | Ajtay et al. 1979 |
| Swampy meadows and bogs | 3.37 | 0.68 | – | 1.5 | Ajtay et al. 1979 |
| Cultivated lands | 2.99 | 6.77 | – | 16 | Ajtay et al. 1979 |
| Human habitation areas | 1.44 | 0.18 | – | 2 | Ajtay et al. 1979 |
| Phytomass | 1170[c] | 81[c] | – | 149.3 | Dobrodeyev and Suetova 1976 |
| Zoomass | 1[c] | – | – | 149.3 | Dobrodeyev and Suetova 1976 |
| Land living phytomass in toto | 592 | 62.7 | – | 148.8 | Bolin et al. 1979 |
| same | 972.7 | 69.0 | 40 | 149.3 | Bazilevich 1979 |
| same | 826.5 | 52.8 | – | 149.0 | Whittaker and Likens 1975; Woodwell et al. 1978 |
| Zoomass and microbiomass | 11.7 | – | 51 | 149.3 | Bazilevich 1979 |
| Zoomass and microbiomass | 4.33 | – | – | – | Ajtay et al. 1979; Bowen 1966 |
| including: | | | | | |
| Bacteria and fungi | 3.07 | – | 49 | – | Ajtay et al. 1979; Bowen 1966 |
| Zoomass | 1.26 | – | 45 | – | Ajtay et al. 1979; Bowen 1966 |
| Zoomass and microbiomass | 7.06 | – | – | – | Ajtay et al. 1979; Rosswall, 1976; Whittaker and Likens 1975 |
| including: | | | | | |
| Bacteria and fungi | 6.6 | – | 50 | – | Ajtay et al. 1979; Rosswall, 1976; Whittaker and Likens 1975 |
| Zoomas | 0.46 | – | 45 | – | Ajtay et al. 1979; Rosswall, 1976; Whittaker and Likens 1975 |

[a] References comprise generalized works.
[b] Dry trees and freshly fallen leaves also contain $30 \times 10^9$ t $C_{org}$ and $60 \times 10^9$ t $C_{org}$ respectively; fallen foliage contains $(45\text{–}50) \times 10^9$ t $C_{org}$.
[c] Calculation per $C_{org}$ effected basing on $C_{org}$ dry mass content of 45% (plants) and 50% (animals).

**Table 2.** Biomass and production of the ocean

| Living matter | Biomass $10^{15}$g $C_{org}$ | Production, $10^{15}$ $C_{org}$ per year | $C_{org}$ per wet mass | Source[a] |
|---|---|---|---|---|
| Phytoplankton | – | 25–30 | 6 | Bagorov 1971; Koblenz-Mishke 1977 |
| same | – | 31 | – | Platt and Subba Rao 1975 |
| same | – | 63 | – | Ivanenkov 1985 |
| same | – | 43.5 | – | De Vooys 1979 |
| same | – | 60 | 6 | Vinogradov and Shushkina 1987 |
| Marine plants in toto including: | 1.74 | 24.7 | – | Woodwell et al. 1978 |
| Bottom algae, including coral reef algae | 0.54 | 0.7 | – | Woodwell et al. 1978 |
| Estuary areas | 0.63 | 1.0 | – | Woodwell et al. 1978 |
| Upwelling areas | 0.004 | 0.1 | – | Woodwell et al. 1978 |
| Continental shelf | 0.12 | 4.3 | – | Woodwell et al. 1978 |
| Open ocean | 0.45 | 18.7 | – | Woodwell et al. 1978 |
| Living matter in toto including: | 2.82[b] | – | 8.5[b] | Bogorov 1971 |
| Phytoplankton | 0.08 | 30.2 | 5.5 | Bogorov 1971 |
| Phytobenthos | 0.015 | 0.015 | 7.5 | Bogorov 1971 |
| Zooplankton | 2.04 | 4.77 | 9.0 | Bogorov 1971 |
| Zoobenthos | 0.57 | 0.21 | 7.0 | Bogorov 1971 |
| Nekton | 0.12 | 0.024 | 12.0 | Bagorov 1971 |
| Phytobenthos (macro-phytes, seagrass) | 0.13 | 0.6 | 7 | Romankevich 1984 |
| Ocean living matter in toto | 2.2–3.1 | – | – | De Vooys 1979 |
| Phytoplankton | 0.40 | 60 | 6 | Vinogradov and Shushkina 1987 |
| Phytobenthos in toto (macrophytes, seagrass, bottom microalgae, coral reefs phototrophes) – 5–7% of phytoplankton primary production | 0.5–0.7 | 3–4 | – | Vinogradov and Shushkina 1987 |
| Bacteria[c] | 0.28 | 57 | 10 | Vinogradov and Shushkina 1987 |
| Bacteria, fungi | 0.36[b] | – | – | Bowen 1966 |
| Protozoa[c] | 0.07 | 12 | 8 | Vinogradov and Shushkina 1987 |
| Mesozooplankton[c] | 0.39 | 2.3 | 6 | Vinogradov and Shushkina 1987 |
| Plankton – feeding nekton | 0.05 | 0.27 | 10 | Vinogradov and Shushkina 1987 |
| including: fish and, squids | – | 0.16 | 10 | Vinogradov and Shushkina 1987 |
| Zooplankton (without protozoa) | 1.19 | – | 6 | Vinogradov and Shushkina 1987 |

[a] The sources reffered to are mostly generalized works.
[b] Calculated for $C_{org}$ by author.
[c] For layer 0–200 m.

**Table 3.** Biomass, production and composition of Earth's organic matter[a]

| Characteristic / Biotopes | Area, $10^6 km^2$ | Biomass, $10^{15}g\ C_{org}$ | Production, $10^{15}g\ C_{org}$ per year | Biomass composition, % of total: Wet: Water | Wet: Organic matter[b] | Wet: Ash | Dry: Organic matter[b] | Dry: Ash | Organic matter[e]: Proteins | Organic matter[e]: Carbohydrates[c] | Organic matter[e]: Lipids | Organic matter[e]: Lignin | Element composition of organic matter, % of total[e]: C | H | O | N | S |
|---|---|---|---|---|---|---|---|---|---|---|---|---|---|---|---|---|---|
| Land | 149 | 560–1170(738) / 4–11(8) | 53–69 (60) | 60 | 38 (18) | 2 | 95 (45) | 5 | 5/60 (5) | 62/20 (62) | 6/20 (6) | 27/– (27) | 48.2 | 7 | 41.3 | 2 | 0.5 |
| Ocean | 316 | 0.9–1.9(1.7) / 2.1–3.2(2.3) | 30–126 (60) | 77 | 13 ( 7) | 10 | 56 (30) | 44 | 51/71 (62) | 39/13 (25) | 10/16 (13) | n.d. | 50.1 | 7.4 | 29.1 | 10.4 | 2.0 |
| Earth total[d] | 510 | 740/10 (750) | 83–196 (120) | 60; 70 | 38; 25 | 2; 5 | 95; 83 | 5; 17 | 5; 28 | 62; 50 | 6; 8 | 27; 14 | 48.2; 48.2 | 7.0; 7.2 | 41.3; 37.2 | 2; 5.4 | 0.5; 0.9; |

[a] Above line – plants, below line – animals and bacteria, in brackets – average values.

[b] Bracketed figures are $C_{org}$ at 48–49% C content in organic matter.

[c] Including chitin.

[d] Semicolon stands to divide biomass composition (first figure) and annual primary production (second figure).

[e] The adduced data on water, ashes and organic matter enable one to calculate all elements (components per wet mass, dry mass and organic matter).

meaningful and internationally accepted. Calculation to "living" mass is easily done using ash content and organic matter (OM) content values in living matter in Table 3). The bulk of the Earth's living matter is contained in land plants. The mass of the ocean plants (phytoplankton, macrophytes, seagrass, bottom microalgae, symbiotic peridineans of coral reefs) is relatively small (approx. $1.7 \times 10^{15}$ g $C_{org}$, i.e. 0.2% of land plants mass). But latest estimates suggest that it is much greater than previously thought (approx. $0.1 \times 10^{15}$ g $C_{org}$, Bogorov 1971). The total zoomass of the Earth, including fungi and bacteria, is about $10 \times 10^{15}$ g $C_{org}$, of which approx. $8 \times 10^{15}$ g $C_{org}$ are on land and $2.3 \times 10^{15}$ g $C_{org}$ in the sea. Tropical forests and savanna phytocenoses make a major contribution to living biomass on land. The relationship phytomass:zoomass on land and in the ocean varies widely: 92:1 and 1.4:1 respectively. The variations of these ratios during different periods in the geological history and in different regions of ancient biospheres may be estimated using biogeochemical indices and organic-geochemical indicators, and may, given similar conditions, serve as a criterion for establishing the distribution of land and oceans on Earth.

Land production makes up on the average some $60 \times 10^{15}$g$C_{org}$ $y^{-1}$; from the available information the oceanic production can also be assumed to be about $60 \times 10^{15}$g$C_{org}$ $y^{-1}$. An earlier figure for the latter ($25 \times 10^{9}$t$C_{org}$, Romankevich 1984) has been revised because of new data and of the accomplished analyses of admissible errors of measurements. Recent trend from new data is a decrease in total land production from $80 \times 10^{15}$g$C_{org}$ $y^{-1}$ to $69 \times 10^{15}$g$C_{org}$ $y^{-1}$ (Table 1); in contrast the trend for ocean primary production values is an increase from 20–25 $\times 10^{15}$g$C_{org}$ $y^{-1}$ to 56–70 $\times 10^{15}$g$C_{org}$ $y^{-1}$ and even to $187 \times 10^{15}$g$C_{org}$ $y^{-1}$.

However attempts to correlate primary production with general heterotrophic destruction in the ocean have not been very successful, though of late there have been new developments in this field (Vinogradov and Shushkina 1987). A promising avenue lies in more accurate accounting for primary production of organic matter by picoplankton in vast oligotrophic oceanic regions. It must be noted that organic matter produced by picoplankton is subjected to rapid biological cycling, and practically does not leave this cycle: the pelagic bottom deposits do not bear traces of this finely dissipated, albeit living organic matter. It is nevertheless certain that the involvemennt of large masses of this organic matter and metals in the carbon cycle should be duly assessed by geochemists, however difficult the task may be. It is clear at present that various methods employed to evaluate production and destruction of organic matter contain numerous assumptions unavoidable in studying complex systems. Therefore reliable estimates of these parameters can be obtained only from two or three correlated determinations made by different methods (such as algological, oxygen and photosynthetic techniques).

An important point is a quantitative estimate of oceanic primary production contributed by chemolithotrophic organisms of deep-ocean hydrothermal springs discovered in 1977 in the Galapagos rift and, after that, in many other rift areas of the ocean. The notion of an insignificantly small (less than 1%) contribution requires additional confirmation. Different geochemical activity of living matter from land and the ocean is connected with a summary (total) active surface area of living matter rather than with the volumes of production. No reliable data exist on this problem. However it deserves further study that would consider fractional dimensions of living organisms which must have been changing in the course of evolution of living matter.

The quantitative estimate of productivity of the present-day ocean closely related with major problems of biogeochemistry, i.e. the reconstruction of paleoproductivity, the dynamics of carbon and the conjugate oxygen reservoirs of the biosphere, the determination of a fraction of the solar energy assimilated by living matter, and the variation of this fraction throughout the Earth's history. The elaboration of this problem, which has begun relatively recently, has proved to be beneficial for marine stratigraphy (Müller and Suess 1979; Stein 1986).

## 3 Composition of Living Matter

The biogeochemical approach considers the chemical composition of organisms as a whole, and not only their protoplasm and soft tissues, which are commonly the only objects of study in biology. In this case some organisms turn out to be still more powerful accumulators of chemical elements which are often concentrated in skeletal formations and support tissues.

The study of numerous data has shown that the content of $C_{org}$ in wet mass has on the average the following percentages: in land plants-18%, in phytoplankton, zooplankton-6%, in phytobenthos-7.5%, in bacteria, nekton-10%, protozoa-8%, zoobenthos-7%. Water content may assumed to be: in land plants-60%, phytoplankton, phytobenthos, zooplankton-63%. Ash content, calculated in the majority of cases by the direct procedure, and in other instances (bacteria, fungi-7%, protozoa-11%) from difference 100 – (OM + water), makes up 2% in land plants, 5% in phytobenthos, 7% in nekton, 8% in zooplankton, 23% in zoobenthos. These data together with organic matter contents (the content of proteins, carbohydrates, lipids, lignin) and the contribution of various groups of organisms to the biomass (for the ocean, in $10^{15}gC_{org}$: phytoplankton-1.4; phytobenthos-0.5; bacteria plus fungi-0.28; protozoa-0.07; nekton-0.05; zooplankton-1.19; zoobenthos-0.57; for land, in $10^{15}gC_{org}$: forests-608; savannah vegetation-89; meadows-12) were used to calculate the average figures given in Table 3.

From the totality of these data, land biomass has the following composition: 95% organic matter and 5% ash; 4.7% proteins; 58.9% carbohydrates; 5.7% lipids; 25.7% lignin. As distinct from land, the biomass of the ocean has 1.7 times smaller organic matter (56%), 9 times greater ash content (44%), 7 times higher proteins (34.7%), a markedly lower carbohydrates content (14%); the percentage of lipids is about equal (7.3%), and lignin is nearly absent. The content of proteins, carbohydrates, lipids in organic matter (Table 3) reflects these specifics. In contrast to plants, land zoomass is not so different from the zoomass of the ocean by the group composition of organic matter.

Based on the major elemental composition (C,H,O,N,S) land and the oceans are not identical: Only the content of C and H (48.2% and 7% on land and 50.1% and 7.4% in the ocean) are similar, differences exists in O,N and S contents as well as in microelements (see below). The organic matter of the ocean is poorer than land in oxygen (approx. 29% and 41% respectively), and richer in N (Ca. 10% and 2%) and

S (2% and 0.5%). This is due to the predominance of organic matter of animal origin in the living matter of the ocean.

The size and composition of the Earth's biomass are practically governed by terrestrial vegetation (if oceanic biomass is also considered, the result begins to tell only after the first decimal). In the case of annual production the situation is different. If equal contribution of land and ocean plants to annual production is assumed, the content of the Earth's annual production will be represented by different figures (Table 3). The transformation during the lithogenesis starts with this initial composition expressed through dry weight.

For the purpose of body-buiding and energy exchange the living matter makes use of the most widespread cosmic elements (Table 4). A great similarity exist with interstellar substances and the volatile fraction of comets. The differences manifest themselves in the selective absorption of carbon, the key element in all living things on Earth, and in smaller contents of oxygen and, apparently also sulphur. With respect to microelements, the similarities and differences are not yet known.

Besides the 5 elements (H,C,O,N,S) making up the bulk of living matter, it also includes over 80 other elements (Table 5), of which no less than 40 are indespensable for the normal functioning of organisms. The physiological, structural and energy functions of elements such as H, C, O, N, P, S, Ca, Cl, Na, Li, Mg, Si, Fe, Sr, Mn, B, J, F, Zn, Cu, V, Ni, Co, Mo, Se, Ti, Sn, Pb, Cd, As have been proved empirically.

The percentages of elements in living matter have been measured with a low accuracy. Differences may sometime attain an order of magnitude, which makes the results quite unsatisfactory. The inferior accuracy with which the percentages are determined is due to sharp variations in the content of elements in organisms of different systematic levels, spatial and temporal oscillations of concentrations in the environment, intrapopulational variability of concentrations and their relation to development stage, age of organisms and rhythmical processes of all duration scales. In most cases, the available data do not allow one to consider these factors.

It seems that in future the principle fact of marine vegetation exceeding terrestrial plants in contents of most microelements (sometimes by two orders of magnitude) will continue to be born out by experimental information, although the actual values may be subjected to refinement. The elements that are present in sea plants in greater concentration include: I, Na, Br, Li, Sr, N, S, K, Si, Cl, Fe, Ag, La,

**Table 4.** Relative occurrence of elements in cosmic and living matter (atomic percentage)

| Occurrence | Present work | Voitkievich and Bessonov (1986) | | |
|---|---|---|---|---|
| Elements | Earth's living matter, of dry biomass | Interstellar | Volatile fraction of comets | Cosmic |
| H | 50.9 | 55 | 56 | 76.5 |
| C | 29.2 | 13 | 10 | 0.34 |
| O | 18.8 | 30 | 31 | 0.82 |
| N | 1.0 | 1.0 | 2.7 | 0.12 |
| S | 0.1 | 0.8 | 0.3 | 0.0015 |

**Table 5.** Elementary composition of Earth living matter, $10^{-4}$ weight % of dry matter[a]

| Chemical elements | Plants | | Animals | | Earth crust | | |
|---|---|---|---|---|---|---|---|
| | Land | Ocean | Land | Ocean | Continental crust (without sedimentary cover) | Sedimentary (sandstone) | Carbonate |
| 1 | 2 | 3 | 4 | 5 | 6 | 7 | 8 |
| O | 39.6[b] | 31.1[b] | 26.8[b] | 23.0[b] | 46[b] | 51.5[b] | 49.2[b] |
| C | 46.3[b] | 45.3[b] | 51.0[b] | 46.2[b] | 0.017[b] | 1.3[b] | 11.0[b] |
| H | 6.7[b] | 6.7[b] | 7.4[b] | 6.7[b] | 0.10[b] | 0.25[b] | 0.09[b] |
| N | 1.9[b] | 8.3[b] | 9.8[b] | 10.4[b] | 0.0020[b] | 0.0135[b] | 0.007[b] |
| Ca | 1.5[b] | 1.0[b] | 0.02-8.5[b] | 1500-20000 | 4.3[b] | 2.68[b] | 32.5[b] |
| K | 1.1[b] | 5.2[b] | 0.74[b] | 0.5-3.0[b] | 1.8[b] | 1.32[b] | 0.28[b] |
| Si | 3000 | 50000 | 120-6000 | 70-1000 | 277000 | 347000 | 34000 |
| P | 2000 | 35000 | 17000-44000 | 4000-18000 | 1000 | 400 | 500 |
| S | 4800 | 12000 | 5000 | 8000 | 300 | 200 | 1200 |
| Mg | 3200 | 5200 | 1000 | 5000 | 24000 | 7300 | 46000 |
| Na | 1200 | 33000 | - | 4000-48000 | 23000 | 9200 | 2500 |
| Cl | 2000 | 4700 | 2800 | 5000-90000 | 100 | 10 | 150 |
| Fe | 200 | 630 | 160 | 400 | 57000 | 28000 | 8600 |
| Al | 200 | 170 | 4-100 | 10-50 | 81000 | 29000 | 9600 |
| Mn | 240 | 40 | < 1 | 3-12 | 900 | 400 | 400 |
| Zn | 50 | 1443 | 160 | 6-1500 | 87 | 16 | 20 |
| Sr | 40 | 545 | 14 | 20-500 | 380 | 20 | 610 |
| Ti | 32 | 119 | < 0.2 | 0.2-20 | 6000 | 3000 | 1200 |
| B | 25 | 113[c] | 0.5 | 20-50 | 70 | 35 | 20 |
| Ba | 22 | 30 | 0.75 | 0.2-3 | 450 | 10n | 10 |
| Cu | 10 | 152 | 2.4 | 4-50 | 65 | 1 | 4 |
| Zr | 7.5 | 1-15 | < 0.3 | 0.1-1 | 130 | 220 | 20 |
| Rb | 5.0 | 7.4 | 17 | 20 | 90 | 60 | 3 |
| Br | 4.0 | 730[c] | 6 | 60-1000 | 2.0 | 1.0 | 6.2 |
| F | 3.5 | 4.5[c] | 150-500 | 2 | 600 | 270 | 330 |
| Pb | 2.5 | 25 | 2 | 0.5 | 9 | 7 | 9 |
| Ni | 2.0 | 17 | 0.8 | 0. 4-25 | 95 | 2 | 2 |
| Cr | 1.8 | 11 | 0.07 | 0. 2-1 | 120 | 3500 | 1100 |
| V | 1.5 | 4.2 | 0.15 | 0.14-2 | 190 | 20 | 20 |
| Li | 1.5 | 5 | < 0.02 | 1 | 20 | 15 | 5 |
| Co | 1.0 | 6.4 | 0.03 | 0. 5-5 | 34 | 0.3 | 0.1 |
| La | 0.8 | 10 | 0.0001 | 0.1 | 25 | 30 | 1n |
| Y | 0.8 | - | - | - | 26 | 40 | 30 |
| Mo | 0.6 | 0.9 | 0.2 | 0.6-2.5 | 1.3 | 0.2 | 0.4 |
| J | 0.3 | 600 | 0.43 | 1-150 | 0.5 | 1.7 | 1.2 |
| Sn | 0.25 | 32 | 0.15 | 0.2-20 | 1.9 | 0.n | 0.n |
| As | 0.12 | 13[c] | 0.2 | 0.005-0.3 | 1.9 | 1.0 | 1.7 |
| Cs | 0.12 | 0.06[c] | 0.064 | - | 2.0 | 0.n | 0.n |
| Be | 0.10 | 0.001-7.1 | 0.0003-0.002 | - | 1.5 | 0.n | 0.n |
| Se | 0.05 | 0.7 | 1.7 | - | 0.10 | 0.05 | 0.08 |
| Ga | 0.05 | 0.07-0.05 | 0.006 | 0.5 | 17 | 12 | 4 |
| Ag | 0.03 | 0.028 | 0.006 | 3-11 | 0.09 | 0.0n | 0.0n |
| U | 0.02 | - | 0.013 | - | 1.5 | 0.45 | 2.2 |
| Hg | 0.012 | 0.45[c] | 0.046 | - | 0.046 | 0.03 | 0.04 |
| Sb | 0.005 | 0.04[c] | 0.006 | 0.2 | 0.2 | 0.0n | 0.2 |
| Cd | 0.005 | 3.0 | 0.5 | 0.15-3 | 0.19 | 0.0n | 0.1 |

**Table 5.** *(Continued)*

| Chemical elements | Plants | | Animals | | Earth crust | | |
|---|---|---|---|---|---|---|---|
| | Land | Ocean | Land | Ocean | Continental crust (without sedimentary cover) | Sedimentary (sandstone) | Carbonate |
| 1 | 2 | 3 | 4 | 5 | 6 | 7 | 8 |
| Sc | 0.008 | – | 0.00006 | – | 24 | 1 | 1 |
| Au | < 0.002 | 0.022 | 0.00023 | 0.0003–0.008 | 0.0017 | 0.00n | 0.00n |
| Nb | < 0.02 | – | 0.00002 | – | 19 | 0.1n | 0.3 |
| W | 0.07 | 0.035 | – | 0.0005–0.05 | 1.1 | 1.6 | 0.6 |
| Bi | 0.06 | – | < 0.004 | 0.04–0.3 | 0.008 | – | – |
| Ru | 0.005 | – | 0.002 | – | – | – | – |
| Jr | < 0.02 | – | 0.00002 | – | 0.0002 | – | – |
| Th | – | – | 0.003–0.2 | 0.003–0.03 | 7.3 | 1.7 | 1.7 |
| Hf | < 0.01 | < 0.4 | 0.04 | – | 2.6 | 3.9 | 0.3 |
| Yb | 0.0015 | – | 0.00012 | – | 2.6 | 4.0 | 0.5 |
| Tm | 0.0015 | – | 0.00004 | – | – | 0.3 | 0.04 |
| Re | – | 0.014 | – | 0.0005–0.006 | 0.0007 | – | – |

[a] Main sources: land and ocean plants (Vinogradov 1954; Dobrovolsky 1983; Bowen 1966; Morozov 1981; Morozov, Petukhov 1986; Eisler 1981; Vinogradov 1953; present work and other sources); land and ocean animals (Vinogradov 1935; Vinogradov 1953; Pokarzhevsky 1985; Bowen 1986; present work and other sources). The phytoplankton/phytobenthos biomass ratio is assumed 3:1.
[b] Percentage.
[c] Phytobenthos.

Sn, B, Zn, Pb, Co, V, Se, Ga, Cd and possibly, Au. Alone the concentrations of Al, Mn and possibly W appear to be greater on the average in land plants than in ocean plants. Enhanced contents of elements in the latter seem largely due to the concentrating power of phytoplankton. When compared to this, the content of metals in phytobenthos is usually lower and resembles land plants. The total amount of ash elements yearly involved in cycles by ocean plants (chiefly, by phytoplankton) approx. $66 \times 10^{15}$g $y^{-1}$, is by an order of magnitude higher than in land plants (approx. $6.7 \times 10^{15}$g $y^{-1}$). However, biomass on land contains 33 times as much ash elements as ocean biomass (approx. $28 \times 10^{15}$g and approx. $2.5 \times 10^{15}$g, respectively).

Among the chemical elements of living matter one could single out those that play a vital part (energy and structural), and those diffused elements, i.e. admixtures with strongly variable concentrations and as yet uncertain physiological role. In contrast to the above category, the occurrence of elements from the second group does not seem to obey the general law of concentrations increasing in the series: whole organism – separate organs – cell structures – molecular forms (Sayenko 1986).

Percentages of microelements of living matter are drastically different from those of the lithosphere, of granite and sedimentary veneers. A general trend of decreasing accumulation of chemical elements in the biomass exists with their growing numerical order in the periodic systems and diminished environmental concentration. The selective concentration of microelements by living matter (biogeochemical selection) is conditioned by external and internal factors that remain largely unknown. That some of them play an especially important role can be specifically noted. This concerns, firstly, the property of chemical elements to occur in natural conditions in various forms as to dispersion, solubility, etc. having different biological accessibility (this is the reason for large descrepancies in the accumulation rates of microelements in natural and laboratory conditions caused by non-observance of hydrodynamical, chemical and ecological similitude criteria); secondly, the physiological regulation of optimum content of microelements depends on the composition of the environmental medium at varied requirements of organisms for some or other elements (including their demand for the formation of skeleton and other supporting tissues); thirdly, the ability of chemical elements to form lasting complex compounds that would be stable both inside the organism and on contact with the environment, as well as physicochemical (sorption) ties in the organism-environment system, which govern the subsequent state (free, strongly or weakly bound) of elements in an organism. All of this in toto determines the ability of organisms to accumulate microelements to levels by far exceeding their needs, including reserve storage.

The established phenomena of selective and group concentration of elements by various organisms have been numerically assessed in recent years, which is of great significance for protecting the environment from chemical pollution, for reclaiming useful components from natural and waste waters, and for developing biogeochemical techniques of mineral prospecting. Group concentration effect consists in the ability of certain plant species to accumulate enhanced concentrations not of one or two but a whole group of polyvalent metals. The mechanism of this phenomenon is not clear. It would seem that the critical factors here are the permeability of cell membranes and the formation of stable complex compounds of metals with bioligands – on the one hand, and related passive accumulation of some elements caused by a physiological requirement for enhanced concentration of one or two elements on the other.

Lately the ability of living cells (some bacteria strains) to interact in multiple ways with colloidal forms of microelements by means of their metabolic products has been established (Ovcharenko et al. 1985). The discovered phenomenon of heterocoagulation enhances our knowledge of the geochemical activity of living matter and has important consequences for the technology of reclaiming elements from natural and waste waters, for understanding the processes taking place at the interfaces of natural media and their phases in the presence of living matter, as well as mechanisms resulting in the formation of stratiform deposits of some metals.

Lastly, it should be pointed out that the solution of urgent problems of the biosphere is more closely tied with tackling the problem: "Living matter and its geological activity". To achieve real progress in this field one should realize the full importance of the problem and overcome the usual inertia.

## References

Ajtay GL, Ketner P, Duvigneaud P (1979) Terrestrial primary production and biomass. In: Bolin B, Degens ET, Kempe S, Ketner P (eds) The Global Carbon Cycle. SCOPE Rept 13. Wiley, New York, pp 129–181

Bazilievich NI (1979) Productivity and biogeochemistry of the present day biosphere and functional models of ecosystems. Pchvovedeniye 2:5–21 (in Russian)

Bazilievich NI, Rodin LE, Rozov NN (1970) Geographic aspects in studying the biological productivity. Proc 5th Congr USSR Geogr Soc Leningrad, p 28 (in Russian)

Bogorov VG (1971) On the quality of materials in living organisms of the world ocean. In: Vassoyevich NB (ed) Organic matter in recent and fossil sediments. Nauka, Moscow, pp 12–15

Bolin B, Cook RB (eds) (1979) The major biogeochemical cycles and their interactions. SCOPE Rept 21, Wiley, New York, 532 pp

Bolin B, Degens ET, Duvigneaud P, Kempe S (1979) The global biogeochemistry of carbon cycle. In: Bolin B, Degens ET, Kempe S, Ketner P (eds) The Global Carbon Cycle. SCOPE Rept 13. Wiley, New York, pp 1–56

Bowen HJM (1966) Trace elements in biochemistry. Academic Press, London, 246 pp

Burdin KS (1985) The basic concepts of biological monitoring. Moscow State Univ, Moscow, 185 pp (in Russian)

Christoforova NK, Maslova LM (1983) A comparative estimate pollution of coastal areas of the Atlantic and west Pacific with heavy metals as determined from the mineral content of fucus algae. Biol Morya 1:3–11 (in Russian)

Degens ET, Kempe S, Spitzy A (1984) Carbon dioxide: a biogeochemical portrait. In: Hutzinger O (ed) Handbook of environmental chemistry 1:127–215

De Vooys CGN (1979) Primary production in Aquatic Environments. In: Bolin B, Degens ET, Kempe S, Ketner P (eds) The Global Carbon Cycle. SCOPE Rept 13. Wiley, New York, pp 259–292

Dobrodeyev OP, Suetova UA (1976) The living matter of the earth (mass, production, geography, geochemical importance and possible impact on climates and glaciation). In: Problems in geography and paleogeography. Moscow State Univ, Moscow, pp 26–58 (in Russian)

Dobrovolsky VV (1983) Geography of microelements; global dispersion. Mysl, Moscow, 272 pp (in Russian)

Eisler R (1981) Trace metal concentrations in marine organisms. Pergamon Press, New York, 687 pp

Ivanenkov VN (1985) The cycle and balance of biogenic elements in the ocean. In: Biological resources of the ocean. Agropromizdat, Moscow, pp 40–48 (in Russian)

Koblenz-Mishke OI (1977) Primary production. In: Vinogradov ME (ed) Biology of the ocean, 1. Biological structure of the ocean. Nauka, Moscow, pp 62–64 (in Russian)

Kovalsky VV (1974) Geochemical ecology. Nauka, Moscow, 299 pp (in Russian)

Lieth H, Whittaker RH (1975) Primary productivity of the biosphere. Springer, Berlin Heidelberg New York (Ecological studies, vol 14)

Moiseyev PA (1969) Biological resources of the ocean. Izd Pishchevaya Prom, Moscow, 340 pp (in Russian)

Morozov NP, Petukhov SA (1986) Microelements in fishery ichtyofauna of the ocean. Agropromizdat, Moscow, 160 pp (in Russian)

Müller PJ, Suess E (1979) Productivity, sedimentation rate and sedimentary organic carbon content in the ocean. Deep-Sea Res 26:1347–1362

Odum EP (1983) Basic Ecology. Saunders, Philadelphia

Ovcharenko FD, Ulberg ZP, Garbara SV, Kogan BS, Pertsov NV (1985) The mechanism of biogenic formation of autigenous inclusions of gold in fine grained sediments. Dokl AN SSSR 284:711–712

Patin SA, Morozov NP (1981) Microelements in marine organisms and ecosystems. Legkaya I Pishchevaya Prom, 153 pp (in Russian)

Perelman AI (1979) Geochemistry. Vysshaya Shkola, Moscow, 424 pp (in Russian)

Platt T, Subba Rao (1975) Primary production of marine microphytes. In: Cooper JP (ed) Photosynthesis and productivity in different environments. Cambridge University Press, Cambridge, pp 249–280

Pokarzhevsky AD (1985) Geochemical ecology of land animals. Nauka, Moscow, 300 pp (in Russian)

Romankevich EA (1984) Geochemistry of organic matter in the ocean. Springer, Berlin Heidelberg New York Tokyo, 334 pp
Rosswall T (1976) The internal nitrogen cycle between microorganisms, vegetation and soil. In: Svensson BH, Soderlund R (eds) Nitrogen, Phosphorus and Sulphur. SCOPE Rept 7. Ecol Bull (Stockholm) 22:157-167
Sayenko GN (1986) Concentrating function of present day marine organisms and problems of paleobiogeochemistry. In: Sultanov KT (ed) The problems in paleobiogeochemistry. Gos Univ, Baku Azerbaidjan, pp 30-56 (in Russian)
Stein R (1986) Surface water paleo-productivity as inferred from sediments deposited in oxic and anoxic deep-water environment of the Mesozoic Atlantic Ocean. In: Degens ET, Meyers PA, Brassel SC (eds) Biogeochemistry of black shales. Mitt Geol Palaeontol Inst Univ Hamburg, SCOPE/UNEP Sonderbd 60:55-70
Vernadsky VI (1965) Chemical structure of the earth's biosphere and its surrounding. Nauka, Moscow, 374 pp (in Russian)
Vernadsky VI (1983) Geochemical essays, 7th edn. Nauka, Moscow, 422 pp (in Russian)
Vinberg GG (1983) The biological productivity of water basins. Ecology 3:1-12 (in Russian)
Vinogradov AP (1935) Elementary chemical composition of organisms and Mendeleyev's periodic system. Trudy Biochim Lab, vol 3. Izd AN SSSR, Leningrad Moscow, pp 5-30 (in Russian)
Vinogradov AP (1953) The elementary chemical composition of marine organisms. Mem 2. Sears Found Mar Res, New Haven
Vinogradov AP (1954) Editor's notes to the 1st volume of VI Vernadsky's selected works. Izd AN SSSR, Moscow, pp 361-367 (in Russian)
Vinogradov ME, Shushkina EA (1987) The functioning of planktonic communities in the oceanic epipelagial. Nauka, Moscow, 242 pp (in Russian)
Voitkievich GV, Bessonov OA (1986) The chemical evolution of the earth. Nedra, Moscow, 215 pp (in Russian)
Whittaker RH, Likens GE (1975) The biosphere and man. In: Lieth H, Whittaker RH (eds) Primary productivity of the biosphere. Springer, Berlin, Heidelberg, New York, pp 305-328 (Ecological studies, vol 14)
Woodwell GM, Whittaker RH, Reiners WA, Liken GE, Delwiche CC, Botkin DB (1978) The biota and the world carbon budget. Science 199:141-146

# The Geologic Enigma of the Red Sea Rift

E. UCHUPI and D.A. ROSS

## 1 Introduction

The Red Sea is a narrow 1920 km long trough extending from the Sinai Peninsula at 27°40′N to the Straits of Bab al Mondab at 12°40′N. It is part of an extensive rift system that extends from southern Africa to the Afar depression from which the Red Sea and the Gulf of Aden bifurcate (Fig. 1; Hötzl 1984). The Afar depression is partially blocked from the Red Sea by a northwest trending basement high, the Danakil horst which Le Pichon and Francheteau (1978) believe is a microplate. It probably formed as a result of reorganization of the plate boundaries at the triple junction of the African, Arabian, and Somalian plates (see Engeln et al. 1988; Schouten et al., in press, for microplate formation resulting from spreading center reorganization.) In the Afar depression, tectonic and volcanic activity occurred in taphrogenetic phases of short duration at the boundaries of the Oligocene and Miocene, Miocene and Pliocene, during middle and upper Pliocene, and early Pleistocene (Pilger and Rösler 1976). Within the depression are a series of northwest trending en echelon active tectono-volcanic axes that may be considered emerged sea-floor spreading centers (Stieltjes 1973).

At its northern end the Red Sea splits into the Gulf of Suez, a tensional rift structure, and the Gulf of Aqaba, part of the Dead Sea left lateral shear. The edges of the Red Sea are marked by a pronounced 1000 to 3000-m-high basement hinge facing seaward along which Pan-African granites, metamorphics, and volcanic rocks and accreted exotic terrains including ophiolites are exposed (Stern et al. 1984; Stoeser and Camp 1985). From the mouth of the Gulf of Suez at 27°50′N to 24°N for a distance of about 460 km the coastlines of the Red Sea are straight and parallel and are about 190 km apart. Farther south the coastlines are more irregular, widening to 350 km between 16°N and 17°N narrowing from there to 40 km at the Straits of Bab al Mandab. The floor of the Red Sea consists of narrow shelves, a main trough at a depth of 600 to 1200 m which at its southern end from 20°N to 15°N is entrained by a less than 20 km wide and up to 2000 m deep axial trough. South of about 20°N on the east and 17°N on the west side, reefs and carbonate banks have prograded seaward almost completely filling in the main trough (Martinez and Cochran, in press). Toward the northwest the main trough terminates on a series of northeast trending asymmetric rhombal half grabens in an en echelon pattern formed by motion along the "leaky" Dead Sea transform (Ben-Avraham 1985).

The axial trough is underlain by oceanic crust and well-developed magnetic anomalies as old as anomaly 3 (about 5 Ma old) from 16°N to 23°N (Roeser 1975). Marginal to the axial trough are sediments accumulations 4 to 5 km thick including 3 to 4 km of middle and late Miocene evaporites. The nature of basement beneath

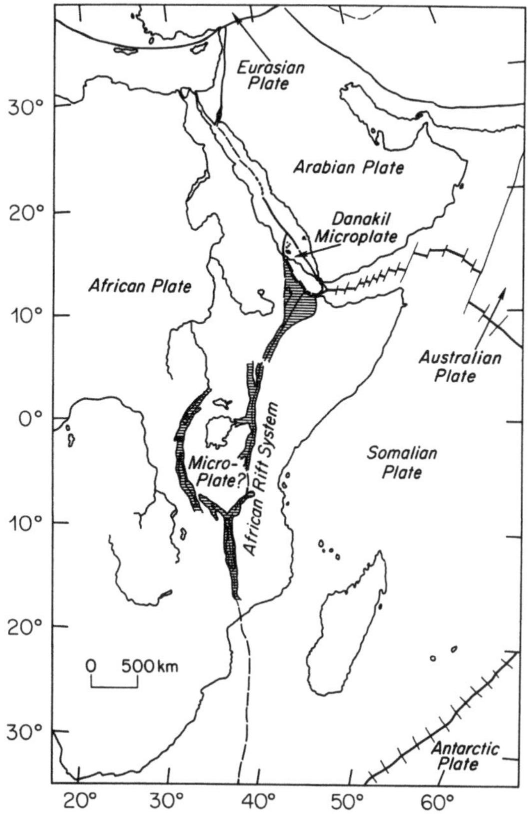


**Fig. 1.** Plate boundaries and associated structures in the general area of the Red Sea. Modified from Pilger and Rösler (1976) using data from Fig. 2, Emery and Uchupi (1984, p 113) and Rosendahl (1987)

the sediments is the subject of considerable controversy, a controversy illustrated by the three models that have been proposed for the origin of the Red Sea. One hypothesis, which we call the oceanic crustal model, suggests that the whole Red Sea is underlain by oceanic crust. In the second hypothesis, the intermediate model, the main trough is underlain by oceanic crust and the area landward of the trough by intruded attenuated continental crust. Proponents of the third hypothesis, which we call the attenuated continental crustal model, believe that most of the Red Sea is underlain by continental crust and that oceanic crust is restricted to the axial trough, and a few isolated depressions immediately north of the trough.

## 2 The Oceanic Crustal and Intermediate Models

Advocates of the oceanic crustal model for the Red Sea believe that the rift is underlain by oceanic crust from shore to shore (Girdler and Styles 1974, 1976; Styles and Hall 1980); others have even proposed that oceanic crust extends inshore in southern Arabia (Gettings 1977; Girdler and Underwood 1987). Such an interpretation appears to be supported by seismic refraction velocities of about 6.7

km/s in the inner half of the main trough between 22°N and 23°N (Tramontini and Davies 1969; Davies and Tramontini 1970) and the presence of linear magnetic anomalies in parts of the main trough which are believed to be of seafloor spreading origin (Girdler and Styles 1974, 1976; Styles and Hall 1980). Girdler (1983) has described the plate setting of the Gulf of Aden, Red Sea and the Gulfs of Suez (Clysmic rift), and Aqaba to the propagation of rifting westwards through the Gulf of Aden and then northwards through the Red Sea terminating north of the present Gulf of Suez, seafloor spreading and separation of the Gulf of Aden and Red Sea beginning 25 Ma ago and 62 km left lateral motion along the Dead Sea shear, and a second cycle of seafloor spreading in the Red Sea that began about 5 Ma ago with a further 45 km movement along the Dead Sea shear. In a combined study of gravity, magnetic and seismic refraction data from the southeast margin of the Red Sea extending from 16.3° to 17.3°N and 41.6°E to 43.1°E, Girdler and Underwood (1985) state that a model where the oceanic-continental boundary might be 35 km inland from the coast and 8 km from the Precambrian shield satisfies all the geological and geophysical measurements. The implication from their model is that little continental crustal extension took place and the decoupling of the continental lithosphere was fairly clean. For example, the presence of positive Bouguer anomalies on the coastal plain in a region of large unconsolidated sediment thickness where one would expect negative values suggests that the sediments might be underlain by heavy oceanic lithosphere rather than light sialic continental crust. The magnetic field over the region displays high frequency small anomalies over the Precambrian shield and long wavelength, large amplitude linear anomalies in the coastal plain; a similar pattern also is found on the opposite side of the Red Sea. Gravity data would indicate that these are not due to basement relief. Thus they postulate that the magnetic lineations were formed by seafloor spreading. Seismic refraction measurements also suggest that the crustal transition may take place some 25 km inland from the coast. Bohannon (1986) states that the geometric configuration of the Arabian passive margin constrained by seismic refraction measurements predicted by an oceanic model indicates a 35 to 40 km shoreline overlap. A continental crustal model predicts a minimum divergence of 320 km resulting in a narrow gap between the shorelines. According to Bohannon this minimum is not realistic because oceanic rocks are known from the margin. Thus, the shoreline overlap described by Girdler and Underwood (1985) may be real.

In another publication Girdler (1985) suggested that the Red Sea may have evolved in three phases. In phase one in the Oligocene during which continental rifting took place, the Gulf of Suez opened up. In the second phase from ?latest Oligocene to early Miocene the first 62-km movement along the Dead Sea left-lateral shear occurred and sea-floor spreading began in the Red Sea. The third phase took place during the second 45-km motion of the Dead Sea shear in the Plio-Pleistocene. Lack of clear magnetic lineations of seafloor spreading in origin in the northern Red Sea is believed to be due to a combination of large thickness of salt, high temperatures, and slow spreading rates as the region is near the pole of rotation at 36.5°N, 18.0°E.

In a recent publication based on bathymetric, gravity and magnetic profiles recorded between the Sinai Peninsula and 19°N Girdler and Southern (1987) also proposed a similar three stage development for the Red Sea. They are (1) a "Gulf of Suez" stage in the Oligocene during which the continental crust was attenuated by faulting to form an early "Red Sea-Gulf of Suez" graben 30 Ma ago with an average crustal extension of 60 km; (2) an "early Dead Sea" stage in the Miocene 25 to 15 Ma ago accompanied by slow seafloor spreading and 62 km left lateral motion along the Dead Sea shear; and (3) a recent "Dead Sea" stage again with slow seafloor spreading and 45 km left lateral motion along the Dead Sea shear from 4.5 Ma ago to the present. They further suggested that gabbro on the borehole 55 km off the Egyptian coast at about 25.8°N would indicate that the oceanic/continental crystal boundary lies between 25 and 55 km offshore. During the 10 Ma long seafloor spreading hiatus from 15 to 4.5 Ma ago circulation in the Red Sea was restricted leading to the accumulation of the middle and late Miocene evaporites. When seafloor spreading began again the earlier oceanic crust was cut in two to form the deep axial trough and evaporite deposition ended as a connection with the Gulf of Aden was established.

Labrecque and Zitellini (1985) also believe that sea floor spreading extends the length of the Red Sea. In their reconstruction, (intermediate model) however, oceanic crust is restricted to main trough. They state that the oldest magnetic anomaly that they can correlate with the reversal scale is anomaly 5C along the landward flank of the main trough. The 50 km wide zone between anomaly 5C and the continental basement hinge basement they state is composed of stretched continental crust intruded by basaltic dikes. Basement seaward of anomaly 5C is a quasi-oceanic crust composed of a higher percentage of lava flows than normal oceanic crust, a stratoid crust comparable to the crust in the Afar depression (Barberi et al. 1976).

## 3 Continental Crust Attenuation Model

The first seismic refraction measurements (Drake et al. 1959; Drake and Girdler 1964; Girdler 1969) appear to verify the concept that the Red Sea is underlain by continental crust (Drake and Girdler 1964; Girdler 1969; Coleman 1974). More recent geophysical measurements, however, have been interpreted as indicating that the regions is underlain by oceanic crust. This new interpretation has not met with universal approval. As pointed out by Cochran (1983): (1) the magnetic anomalies in the marginal areas of the Red Sea are smooth, low-amplitude without any distinguishing characteristics and are unlike the seafloor spreading anomalies within the axial trough; Cochran was also able to demonstrate that the magnetic anomaly sequence found in the marginal areas could be generated as a result of normal faulting of continental basement using a susceptibility value of 0.002 emu; well within the range of values reported for granites; (2) the presence of pre-rift crystalline rocks on St. John and possibly on The Brothers suggest that basement beneath the main trough is continental; the fresh peridotite on St. John and the gabbro in the borehole 55 km off Egypt may represent intrusives into attenuated continental crust. Based on a structural and kinematic analyses of the peridotites in Zabargad Island, Nicolas et al. (1987) believe that they represent asthenospheric diapirism emplaced during the early phases of rifting of the Red Sea; (3) a seismic reflection line along the western edge of the axial trough outside the region of high-amplitude magnetic anomalies shows that the crust with a velocity of 5.91 km/s is block-faulted in a manner characteristic of attenuated continental crust; (4) crustal seismic velocities considered diagnostic of continental rocks are found in a number of locations under the main trough. The "oceanic" crustal velocities encountered in one region in the main trough may be Neogene intrusions or Precambrian or early Paleozoic mafic and ultramafics such as those described from Arabia and Nubia. Similarly, the ophiolite complex at base of the basement hinge near 17°N (Tihama Asir igneous complex; Coleman et al. 1975, 1979) in southern Saudi Arabia and several locations between the Yemen border and Ad Bard at 17° 45′N may represent massive intrusions into continental crust.

The model described by Girdler and Underwood (1985) for the southern Red Sea also implies that little continental crustal attenuation took place prior to seafloor spreading, a concept that appears to be verified by a crustal geometric configuration constrained by seismic refraction measurements (Bohannon 1986). Yet all passive margins of the Atlantic pull-apart type (not translation margins such as the southwest Grand Banks margin, the Gulf of Guinea margin and the Benue trough, and the margin of southern Africa where crustal extension is minimal) appear to have undergone long periods of crustal extensions (Uchupi, in press; references therein), an extension that lasted 30 to 47 Ma in the northern North Atlantic, 75 Ma in the southern North Atlantic, 68 to 53 Ma years in the Equatorial Atlantic and 77 to 68 Ma in the South Atlantic. This extension resulted in a 100 to 250 km wide rift system. The fundamental structure of the Atlantic rift is a half graben with a geometry ranging from overlapping, partially overlapping, to non-overlapping half-grabens. In the non-overlapping pattern the half graben is fronted by a platform (i.e., Baltimore trough, Reguibat Massif). Where the boundary faults overlap a set of cross trending normal faults they form accommodation faults. When continental decoupling occurred the plane of separation apparently was closer to the platform. Thus most of the extended continental crust is only found next to the boundary fault. Only when two half-grabens face each other or where the fundamental structure is graben did continental de-coupling take place in the center of the rift.

It is these reservations that have led some geologists to propose the attenuated continental crustal model for the origin of the Red Sea (Lowell and Genik 1972; Ross and Schlee 1973). In this model the basement hinge bordering the Red Sea is a series of master normal faults that define a chain of half-grabens and opposing platforms. The main boundary faults are not restricted to one hinge, but probably occur along both yielding a tectonic style comparable to the East African rift of nonoverlapping, partially overlapping, and facing half-grabens. Along the master normal faults are subsidiary faults dipping toward or away from the master fault breaking up basement into a series of blocks. The bathymetric terraces in the marginal areas of the northern Red Sea represent such a series of rotated fault blocks (Cochran and Martinez, in press). A set of cross-trending normal faults connecting partially overlapping master faults form accommodation zones in the manner described by Rosendahl (1987) for the East African rift system. Such a geometry is present in the Gulf of Suez with the accommodation zones being oriented oblique and orthogonal to the boundary faults (M. Steckler in Martinez and Cochran, in press).

Rifting in the Red Sea may have been preceeded by arching in the Oligocene (Lowell and Genik 1972). Using fission track ages from Precambrian apatites, Kohn and Eyel (1981) determined that domal uplift in the Sinai peninsula began 26.6 ± 3 Ma ago and continued for most of the Miocene. The uplift amounted to at least 5 km, 3 km of which has taken place since 9 Ma ago. Uplift was greater parallel to the Gulf of Suez, an area of extension, and less parallel to the Gulf of Aqaba along the Dead Sea shear along which there has been 105 km of left lateral motion during the last 22 Ma. Steckler (1985), however, believes that this doming was not a pre-rift event, but developed during the main phase of rifting. Absence of Oligocene sediments, generally cited as evidence of uplift, Steckler (1985) states is probably the consequence of a major northwards regression. Jarride et al. (1986) postulate that the tectonic style in the Gulf of Suez and the northwest edge of the Red Sea is controlled by two events; a strike-slip displacement that produced antithetic tilted blocks 24 to 20 Ma ago, and a second phase that began 20 Ma ago and still developing now when synthetic normal movements formed a horst and graben pattern.

Rifting in the Red Sea region was underway by the start of the Miocene and initially the Gulf of Suez formed part of the Red Sea. Subsequent 105 km left lateral motion along the Dead Sea shear and opening of the Gulf of Aqaba has led to further extension in the Red Sea proper. Extension in the central Gulf of Suez has been estimated to be 27 to 25 km (Steckler 1985) and in the northern Red Sea 135 km (Cochran 1983). On the rift were deposited a thick sequence of Miocene volcanics and clastics and middle and late Miocene evaporites. About 5 to 4 Ma ago a spreading axis was established near Lat. 17°N and that axis propagated to the north and south until it extended from Lats. 20°N to 15°30′N, the drift zone of Fig. 2. From 20°N to 23°30′N is a series of large deeps spaced at intervals of about 50 km floored by oceanic crust. This area is presently changing from continental extension to a sea-floor spreading mode. Lineated magnetic anomalies indicate that seafloor spreading in this transitional zone (Fig. 2) was possibly initiated in the southern deeps and is propagating northward (Martinez and Cochran, in press). One, possibly two, of the deeps in the rifted continental terrain have high amplitude dipolar magnetic anomalies which are believed to be the result of intrusions into attenuated continental crust (Pautot et al. 1986; Cochran 1983; Cochran et al. 1986). Intrusion may be a consequence of extension and crustal thinning being concentrated in the main trough during the late phase of rifting with points of injection being controlled by periodically spaced gravitational instability of the asthenosphere (Bonatti 1985; Whitehead et al. 1984). With continuing intrusion and extension these deeps may develop into isolated cells of seafloor spreading which in time will coalesce to form a continuous seafloor spreading axis (Martinez and Cochran, in press).

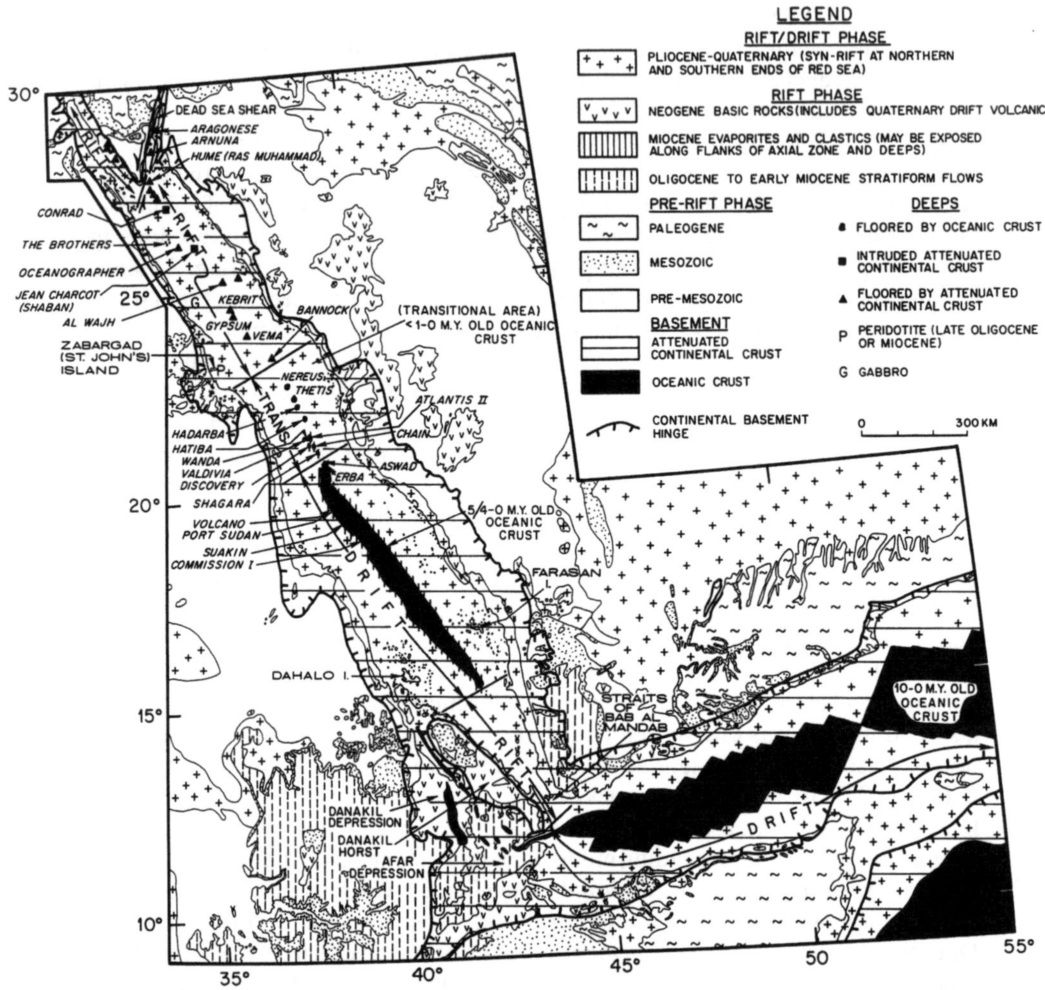


**Fig. 2.** Tectonic map of the Red Sea region based on the attenuated continental crustal model. In the oceanic crustal model most of the area between the basement hinges will be floored with oceanic crust, and in the intermediate model the main trough is underlain by oceanic crust landward of which is a intruded rifted continental crust. Compiled from data obtained during the present investigation, Nesteroff (1955); Association of African Geological Societies and UNESCO (1963, 1968); Laughton (1970); Gass et al. (1973); Ross and Schlee (1973); Barberi et al. (1974); Searle and Ross (1975); Civetta et al. (1978); Heezen et al. (1978); Ben-Avraham et al. (1979); Cochran (1981); Beydoun (1982); El Shazley (1982); Bartholomew and Son (1983); Bonatti et al. (1983); Cochran (1983); Pautot and Guennac (1984); Girdler and Southern (1987); Nicolas et al. (1987); Martinez and Cochran (in press)

# 4 Conclusion

The data available to date does not make it possible to determine the validity of the three models that have been proposed for the origin of the Red Sea rift. No borehole data are available as to the nature of the crust beneath the main trough. As a result of the acoustic impedence of the Miocene evaporites multi-channel seismic profiles have provided little information as to the tectonic style of basement in the main trough from which inferences could be made regarding its origin. Gravity and magnetic measurements are too ambiguous and can be interpreted as due to either continental or oceanic crustal sources. The oceanic crustal model implies that rifting in the Red Sea was minimal to non-existent, a conclusion contrary to observations in the Atlantic Ocean where seafloor spreading was preceded by a long period of continental crustal extension. If continental rifting did take place in the Red Sea, it was of much shorter duration lasting less than 20 Ma ago, whereas in the Atlantic rifting may have lasted in places as much as 77 Ma. Rift basins in the Atlantic also rarely exceed 200 km (from basement hinge to basement hinge) in width, yet in the Red Sea the rift basin in places exceeds 400 km in width. The two regions also may vary in the tectonic style. In the Atlantic, continental separation appears to have been asymmetrical with the plane of separation between accomodating zones being located closer to the platforms opposing the half-grabens. Only in regions of facing half-grabens and grabens is the plane of separation located along the axis of the rift. In the Red Sea, however, the plane of separation delineated by the axial deep is along the axis of the rift. This would suggest that the Red Sea lost its assymetry during its evolution as described by Martinez and Cochran (1988).

*Acknowledgments.* Funding for this study was provided by NOAA Office of Sea Grant award (NA-86-AA-D-SG 090 to Woods Hole Oceanographic Institution Project M/O-1). Pam Foster types the manuscript and Ruth Davis drafted the illustrations. Contribution No. 6767 of the Woods Hole Oceanographic Institution.

# References

Association of African Geological Societies and UNESCO (1963) Geologic map of Africa, scale 1:5,000,000 9 sheets. UNESCO, Paris

Association of African Geological Societies and UNESCO (1968) International tectonic map of Africa, scale 1:5,000,000, 9 sheets. UNESCO, Paris

Barberi F et al. (1974) Traverse tectonics during the split of a continent: data from the Afar rift. Tectonophysics 23:17–29

Barberi F et al. (1976) Structural evolution of the Afar triple junction. In: Pilger A, Rosler A (eds) Between continental and oceanic rifting. Schweizerbartische Verlagsbuchhandlung, Stuttgart, p 38

Bartholomew J and Son (1983) The Times atlas of the world, comprehensive edition. Times, New York

Ben-Avraham Z (1985) Structural framework of the Gulf of Elat (Aqaba). J Geophys Res 90:703–726

Ben-Avraham Z et al. (1979) Continental breakup by a leaky transform: Gulf of Elat (Aqaba). Science 2096:214–216

Beydoun ZR (1982) The Gulf of Aden and northwest Arabian Sea. In: Nairn AEM, FG (eds) The ocean basins and margins, vol 6. The Indian Ocean. Plenum, New York, p 253
Bohannon RG (1986) How much divergence has occurred between Africa and Arabia as a result of the opening of the Red Sea? Geology 14:570–513
Bonatti E (1985) Punctiform initiation of seafloor spreading in the Red Sea during the transition from continental to oceanic rift. Nature 316:33–37
Bonatti E et al. (1983) Zabargad (St. John's Island): an uplifted fragment of sub-Red Sea lithosphere. Geol Soc Lond J 140:677–690
Civetta L, LaVolpe L, Lirer L (1978) K-Ar ages of the Yemen Plateau. J Volcanol Geothern Res 4:307–314
Cochran JR (1981) The Gulf of Aden: structure and evolution of a young ocean basin and continental margin. J Geophys Res 86:263–288
Cochran JR (1983) A model for development of Red Sea. Am Assoc Pet Geol Bull 67:41–69
Cochran JR, Martinez F (In press) Evidence from the normal northern Red Sea on the transition from continental to oceanic rifting. Tectonophysics
Cochran JR et al. (1986) Conrad Deep: a new northern Red Sea deep. Origin and implications for continental rifting. Earth Planet Sci Lett 78:18–32
Coleman RG (1974) Geological background of the Red Sea. Initial Rep Deep Sea Drilling Project 23:813–820
Coleman RG et al. (1975) The volcanic rocks of southwest Arabia and the opening of the Red Sea. Red Sea research 1970–1975. Saudi Arabia Dir Gen Mineral Res Bull 22:D1-D30
Coleman RG et al. (1979) The Miocene Tihama Asir Ophiolite and its bearing on the opening of the Red Sea. In: Al-Shanti AMS (ed) Evolution and mineralization of the Arabian-Nubian shield. King Abdulaziz Univ Inst Appl Geol Bull 3:173–186
Davies D, Tramontini C (1970) The deep structure of the Red Sea. R Soc Lond Philos Trans A267:181–189
Drake CL, Girdler RW (1964) A geophysical study of the Red Sea. R Astron Soc Geophys J 8:473–495
Drake CL, Girdler RW, Landisman M (1959) Geophysical measurements in the Red Sea. In: Sears M (ed) Preprints of abstracts of papers to be presented at afternoon sessions. International Oceanographic Congress. Am Assoc Adv Sci p 20
El Shazley EM (1982) The Red Sea. In: Nairn AEM, Stehli FG (eds) The ocean basins and margins, vol 6. The Indian Ocean. Plenum, New York, p 205
Emery KO, Uchupi E (1984) The geology of the Atlantic Ocean. Springer, Berlin Heidelberg New York
Engeln JF, Stein S, Werner J, Gordon RG (1988) Microplate and shear zone models for oceanic spreading center reorganizations. J Geophys Res 93:2839–2856
Gass IG, Mallick DIJ, Cox KG (1973) Volcanic islands of the Red Sea. Geol Soc Lond J 129:275–310
Gettings ME (1977) Delineation of the continental margin of the southern Red Sea from new gravity evidence. Red Sea research, 1970–1975. Saudi Arabian Dir Mineral Res Bull 2:K1-K11
Girdler RW (1969) The Red Sea. In: Degens ET, Ross DA (eds) Hot brines and recent heavy metal deposits in the Red Sea. Springer, Berlin Heidelberg New York, p 38
Girdler RW (1983) The evolution of the Gulf of Aden and Red Sea in space and time. In: Angel MV (ed) Marine science of the northwest Indian Ocean and adjacent waters. Pergamon, New York, p 747
Girdler RW (1985) Problems concerning the evolution of oceanic lithosphere in the northern Red Sea. Tectonophysics 116:109–122
Girdler RW, Southern TC (1987) Structure and evolution of the northern Red Sea. Nature 330:716–721
Girdler RW, Styles P (1974) Two stage Red Sea spreading. Nature 247:1–11
Girdler RW, Styles P (1976) Opening of the Red Sea with two poles of rotation – some comments. Earth Planet Sci Lett 33:169–172
Girdler RW, Underwood M (1985) The evolution of early oceanic lithosphere in the southern Red Sea. Tectonophysics 16:95–108
Heezen BC, Lynde RP Jr, Fornari DJ (1978) Geological map of the Indian Ocean. In: Heirtzler JR (ed) Indian Ocean geology and stratigraphy. Am Geophys Union
Hötzl H (1984) General geology of the western Saudi Arabia. 1.2 The Red Sea. In: Jado AR, Zötl JG (eds) Quaternary period in Saudi Arabia, vol 2. Springer, Berlin Heidelberg New York, p 13
Jarride JJ et al. (1986) Inherited discontinuities and Neogene structure: the Gulf of Suez and the northwest edge of the Red Sea. Philos Trans R Soc Lond Math Phys Sci A 317:129–139

Kohn BP, Eyel M (1981) History of uplift of the crystalline basement of Sinai and its relation to the opening of the Red Sea as revealed by fission track dating of apatites. Earth Planet Sci Lett 52:129–141
Labrecque JL, Zitellini N (1985) Continuous seafloor spreading in Red Sea: an alternative interpretation of magnetic anomaly pattern. Am Assoc Pet Geol Bull 69:513–524
Laughton AS (1970) A new bathymetric chart of the Red Sea. R Soc Lond Philos Trans A226:243–248
Le Pichon X, Francheteau J (1978) A plate tectonic analysis of the Red Sea-Gulf of Aden area. Tectonophysics 46:369–406
Lowell JD, Genik GJ (1972) Sea-floor spreading and structural evolution of the southern Red Sea. Am Assoc Pet Geol Bull 56:247–259
Martinez F, Cochran JR (in press) Structure and tectonics of the northern Red Sea: catching a continental margin between rifting and drifting. Tectonophysics 150:1–32
Nesteroff W (1955) Les recifs coraliens du banc Farsan nord. Inst Ocean Ann 30:7–54
Nicolas A, Francoise B, Montigny R (1987) Structure of Zabargad Island early rifting of the Red Sea. J Geophys Res 92:461–474
Pautot G, Guennoc P (1984) Les fosses à saumures et sediments métalliferes de la Mer Rouge: apports de l'analyse morphostructurale effectuée par les équipes Francaises. Germinal 2.84.29:543–556
Pautot G et al. (1984) Discovery of a large brine deep in the northern Red Sea. Nature 310:133–136
Pautot G et al. (1986) La dépression axiale du segment nord mer Rouge (de 25° a 28°): nouvelles données géologiques et géophysiques obtenues du cours de la campagne Transmerou 83. Bull Soc Geol France 8:381–399
Pilger A, Rösler A (1976) General aspects with special reference to Afar. Temporal relationship in the tectonic evolution of the Afar depression (Ethiopia) and the adjacent Afro-Arabian rift system. In: Pilger A, Rösler A (eds) Afar between continental and oceanic rifting. Schweizerbartische Verlagsbuchhandlung, Stuttgart, p 1
Roeser HA (1975) A detailed magnetic survey of the southern Red Sea. Geol Jahrb 13:131–153
Rosendahl BR (1987) Architecture of continental rifts with special reference to East Africa. Ann Rev Earth Planet Sci 15:445–503
Ross DA, Schlee J (1973) Shallow structure and geologic development of the southern Red Sea. Geol Soc Am Bull 184:3827–3848
Schouten H, Gallo DG, Klitgord KD (in press) Microplate kinematics of the second order. Spring meeting, Am Geophys Union
Searle RC, Ross DA (1975) A geophysical study of the Red Sea axial trough between 20.5° and 22°N. R Astron Soc Geophys J 43:555–572
Steckler MS (1985) Uplift and extension at the Gulf of Suez: indications of induced mantle convection. Nature 317:135–139
Stern RJ, Gottfried D, Hedge CE (1984) Late Precambrian rifting and crustal evolution in the north eastern desert of Egypt. Geology 12:168–172
Stieltjes L (1973) Evolution tectonique récente du rift d'Asal. T.F.A.I. Rev Géogr Phys Géol Dynamique 15:425–436
Stoeser DB, Camp VE (1985) Pan-African microplate accretion of the Arabian Shield. Geol Soc Am Bull 96:817–826
Styles P, Hall SA (1980) A comparison of the seafloor spreading histories of the western Gulf of Aden and the central Red Sea. Geodynamic evolution of the Afro-Arabian rift system. Accad Naz Lincei, Rome pp 587–606
Tramontini C, Davies D (1969) A seismic refraction survey in the Red Sea. R Astron Soc Geophys J 17:225–241
Uchupi E (in press) The tectonics style of the Atlantic Mesozoic rift system. J Afr Earth Sci
Whitehead JA Jr, Dick HJB, Schouten H (1984) A mechanism for magmatic accretion under spreading centres. Nature 384:518–520

# Ocean Particles and Fluxes of Material to the Interior of the Deep Ocean; The Azoic Theory 120 Years Later

S. HONJO

## 1 Introduction

About 120 years ago the influential Azoic theory, that no life existed on the deep ocean floor, was finally defeated by Charles Wyville Thomson. He described many types of animals collected from the North Atlantic abyssal floor using dredges during the HMS Lightning and HMS Porcupine cruises in 1868 and 1870. Without doubt, the establishment of a lively deep sea realm had a serious impact on the philosophy of science at that time, and this was a powerful encouragement to wider and deeper recognition of Charles Darwin's "Origin of Species," published about a decade before Thomson's cruises (Mills 1983). For the past century since Thomson's discovery and the legendary HMS Challenger voyage, scientists have collected innumerable specimens from every corner of the oceanic abyss. Deep-sea biology is now well established, resulting from an increase in the critical mass of benthic biologists, better quality and quantity shiptime, and improved equipment (Rowe 1983).

Ironically, during this seemingly glorious century of deep sea research, the important question of what supported life in the abyssal environment was answered by sheer speculation, particularly with reference to the quantities and mechanisms of energy and nutrient supply. Until very recently this "knowledge" remained essentially unchanged from the late 19th century views of Alexander Agassiz who postulated that slowly descending plankton detritus nourished the benthos when, or if, it arrived at the abyssal sea floor. The great insight of Agassiz was that he also mentioned that the benthic population was positively related to the abundance of plankton in the ocean's surface (cited from Uda 1955).

The demand for a more dependable understanding of the earth's carbon cycle and its role in the ocean became very strong in the early 1970s with the threat of rising levels of atmospheric $CO_2$ due to man's consumption of fossil fuels; this was clearly seen in the Mauna Loa experiment (Keeling et al. 1976). The raining rate of particulate organic and carbonate carbon into the interior of the ocean was recognized as one important key to understand the complex carbon cycle throughout the biosphere and lithosphere, and, in turn, to build a useful model of the processes and fate of fossil fuel $CO_2$ on this planet (Global Ocean Flux Study Committee 1984).

Scientists have come to realize that understanding the absolute mass accumulation rates of sediments such as carbon is an essential key to reconstructing past ocean environments as realistically as possible. Geologists have struggled to assess the mass accumulation rates of skeletal remains, refractory particles, and other environmental tracers on the ocean floor. By the early 1970s, a number of marine geologists and paleoceanographers began to feel strongly that we needed a bridge to relate past and present environments (Lisitzin 1972), and that this could be done by examining settling particles as they leave their living habitat and investigating them before burial in a permanent sink. Quantification of particle fluxes in the modern ocean has thus become an urgent need for all geologists. In turn, we have re-evaluated the old axiom of geological sciences: "the Present is the key to the Past," stated by Charles Lyell in the mid-19th century.

## 2 Ocean Particles and Their Settling Mechanisms

Our friend Egon Degens has been a great advocate of ocean particle flux studies since the earliest days of this revival. Together with him, we have been intrigued by the fascinating biogeochemical role of particles in oceans and lakes since the early 1970s when we were both at the Woods Hole Oceanographic Institution. Egon's superbly successful Atlantis II expedition to the Black Sea in 1969 with David Ross (Degens and Ross 1974) and his series of African lake studies in the early 1970s (Degens et al 1972) led us to feel strongly that the knowledge of particle flux was the missing link needed to put together the evolution of those environments. In reality, Egon, Erol Izdar, and I had to wait until the early 80s to actually begin experiments together in the Black Sea. This was stimulated by the concept in my laboratory of automated sediment trap technology which can be applied in virtually any oceanographic condition and for any period of time (Honjo and Doherty 1988).

Today, led by Egon, the University of Hamburg maintains one of the largest ocean flux study programs in the world; this program covers the North Sea, Black Sea, Indian Ocean, and South China Sea, and is accomplished with excellently organized international cooperation among the Woods Hole Oceanographic Institution, USA, the University of Izmir, Turkey, the National Institute of Oceanography in Goa, India, and various institutions in the Republic of China. During the course of our joint study, he, his colleagues, and I have made a number of important findings and discoveries, such as an assessment of the immediate impact of the Chernobyl radionuclide fallout on the deep sea environment in 1986 (Kempe et al. 1987).

## 3 Ocean Particles; Realities

Oceanic particle studies are relatively new, and many new concepts and interpretations have emerged only in the last decades. Some of these are still controversial and others are unconventional compared to more established oceanographic common sense. For example, I present the fact that ocean particles settle fast, orders of magnitude faster than the speed of independent fine ocean particles, and without an efficient mechanism of particle transportation the deep ocean bottom might really have been azoic for lack of nutrients.

There are two major sources of particles to the ocean layers (Honjo 1984). In this essay, "particle" means passively sinking, small, ubiquitous detritus and does not include large objects such as a marine mammal corpse or a large sunken log. Biogenic (ocean) particles are formed in the upper ocean as the result of planktonic metabolism and biogeochemical reactions. Lithogenic fine particles (mostly clay and fine rock detritus) are transported from continents by rivers and coastal erosion. In the pelagic ocean, lithogenic particles are also supplied as fallout from wind-driven aerosols. Also, a large volume of fine lithogenic particles are transported within the ocean interior as a result of resuspension of sediments from all sides of the oceans.

Particle settling occurs when a resisting (or dragging force) and a net gravitational force have reached a terminal velocity that is steady. The dragging force depends upon the viscosity of the liquid, settling velocity, the volume of the particle, and its shape factor. Naturally occurring ocean particles may have very complex shapes, differing diameters of so-called "hydrated halos" and other factors which increase the drag force; thus, the settling speed is far smaller than that of a quartz particle of equivalent size. The majority of independent ocean particles (in number) do not settle in the practical sense because the gravitational descending speed is smaller than the local eddy velocity and many particles belong to the Brownian motion resume. For example, if one applies Stokesian settling rates to an individual coccolith of a few to several micrometers, it will take thousands of years to settle through a long, calcium carbonate ($Ca\ CO_3$) undersaturated water column, and no cocolith should actually reach the abyssal ocean floor. Organic tissue remains are often lighter than seawater, in particular oil and wax droplets from plankton tissue even have positive buoyancy in seawater. In reality, coccolith oozes cover the deep ocean floor and oil droplets are common on the sea floor serving the benthos as a high-quality energy source. Why is this true in violation of gravitational law? The accelerated sinking of fine particles by aggregation was proposed as a hypothesis (Smayda 1971) and this was proved by using coccoliths as a tracer (Honjo 1976).

## 4 Fecal Pellets and Marine Snow; Accelerated Settling

A number of mechanisms exist in the ocean to accelerate the sinking speed of fine particles. One is fecal pellets of metazoan zooplankton, typically of copepods in the open ocean (Pilskaln and Honjo 1987). These animals graze, by filter feeding, mostly on phytoplankton, other detritus, and fine lithogenic particles and eject membrane covered pellets into the water column. Organic matter is usually only partially digested by the host; thus, a pellet itself can provide relatively high food value (Honjo and Roman 1978). Pellets, which contain a variety of fine and light particles, settle at relatively high speeds. A typical oceanic adult *Calanus* pellet descends at the rate of about 200 m per day, taking only a few to several weeks to reach the deep ocean bottom (Small et al. 1979).

Many metazoan plankton fecal pellets are covered with a thin membrane, called a peritrophic membrane, that protects the contents of the pellet from dispersion. Laboratory experiments suggest that the peritrophic membrane is colonized by microbes, and microbe growth is a function of seawater temperature. For example, in the Sargasso Sea the membrane is almost completely degraded before a pellet can sink through the upper layers. On the other hand, a pellet produced in Polar Oceans may not be degraded while traveling down through the water column. Degrading of the peritrophic membrane is thus important since it allows dissolved nutrients to return to the upper ocean which would otherwise become depleted by a one-directional flow of nutrients. Meanwhile, fine, individual particles are exposed, some sloghed off from the surface and, in turn, regenerated as suspended particles in the depths.

The other form of biogeochemical aggregate which also accelerates the fine and light individual ocean particles is "marine snow." Marine snow is a rather ambiguous term used to define aggregates of millimeter size made up of numerous fine particles and they are not usually seen by normal reflective light, but will show up as a light-scattered image (white on a dark background) under certain illumination conditions. For centuries, marine snow aggregates have been recognized by many maritime nations where fishermen dive and are able to see light scattering in seawater. It is almost impossible to see marine snow through the air from a ship. For example, marine snow and related phenomena are called "nuta" among diving fishermen in southern Japan. It was Tsujita (1949) who first described the oceanography of nuta from the neritic environment off Kyushu. The term "marine snow" came from Suzuki and Kato (1953) who reported on it after using an early submersible for observations. Full recognition of the oceanographic importance of large aggregates in the ocean was made by Riley (1963). This was followed by Milliman et al. (1967) who made the first detailed semi-quantitative observations throughout the deep sea column from DSRV Alvin on the continental margins off the southeast U.S. Amorphous aggregates received more attention on its importance in upper ocean oceanography through studies in the late 1970s and early 1980s by Alldredge (1976), Silver and Alldredge (1981), and Silver et al. (1978), for example.

One well known fact from these previous studies is that a marine snow aggregate is extremely fragile if physically disturbed. A scientist who dives to collect marine snow aggregates will find that they disperse like a "puff of smoke" in eddies caused by approaching objects, such as the diver, and it is difficult to bring them to the surface intact for laboratory examination. Aggregate populations and sinking speeds have been measured successfully in a number of instances, but this work is limited to SCUBA depths.

The application of a "marine snow camera," which is based on a simple light scattering principle, caused a quantum leap in depiction of spatial distribution, size, volume and sinking speeds of marine snow (Honjo et al. 1974). My group at the Woods Hole Oceanographic Institution and Vernon Asper at the University of Southern Mississippi have deployed this optical equipment in many areas of the deep open ocean, including waters off California, near Hawaii, Gulf Stream regions, the Nordic Seas, and the Black Sea. We have obtained over 100,000 photographs taken throughout the full length of the water column and, to our surprise, images of aggregates larger than a half-millimeter were seen in almost 100% of those photographs. We believe that marine snow is almost ubiquitous in the world's oceans, except, probably, in areas where organic productivity is severely limited such as in sea-ice-covered polar seas during the winter period, although no camera deployment has been made in high Polar Oceans to confirm this.

The role of marine snow in dealing with ocean particles is similar to that of fecal pellets. Amorphous detritus of organic matter, often produced from the decomposition processes of macroplankton such as jelly fish, remains of larvacean houses, and entangled diatoms, etc., serves as a matrix of the aggregate. Although it is speculative, we believe that independent suspended particles are caught by this hydrated, adhesive object with a relatively large diameter. Tsujita's early work on "nuta," when he examined them under an electron microscope, showed a higher diversity of particles in an aggregate than among suspended particles in water (Tsujita 1953), which indicated that an aggregate "collects" independent particles. The collection of fine but dense particles, such as clay minerals and calcite (coccoliths), makes the total specific gravity of an aggregate larger until it gains settling velocity. While an aggregate descends through the upper water column, it agglutinates more independent particles. When it leaves the upper ocean, an aggregate thus gains a specific steady speed.

Analogous to fecal pellets which have lost their surface cover, an aggregate will also loose particles while settling through the deep water column and supply fine particles to the suspended particle reservoir. Suspended particles, either sloughed off from fecal pellets or marine snow aggregates, can be scavenged by another settling particle and gain settling speed again. This switching between suspended and settling status may happen many times while an ocean particle settles through a long water column. Recent studies on the efficiency of particle trapping using a radiochemical tracer showed nearly 100% of the transuranium elements generated in the water column above were found in a sediment trap (Bacon et al. 1985). Therefore, we suspect all suspended particles eventually settle down with large particles after changing back and forth from suspended to settling with all types of aggregates. In this model, another important point is that a suspended particle can be injected into any depth by being sloughed off from a host; therefore, no time-stratigraphic relationship will be found among suspended particles distributed throughout a water column. In other words, the age distribution of suspended particles is completely random in the ocean layers and this hypothesis was proved by measuring the extent of dissolution in suspended coccoliths in an undersaturated Pacific Ocean water column (Honjo 1975).

Another proof of the hypothesis that ocean sedimentation occurs through relatively large diameter settling particles came from Vernon Asper's "flux camera" experiment in the Panama Basin. He set a flat, horizontal, transparent plate just above the nepheloid layer and took timelapse macrophotographs of the particles arriving on the surface of the plate which was lit by a back light. As expected, he found only large aggregates reaching the plate at a rate close to the particle flux sinking speeds measured by a sediment trap deployed near by.

## 5 Settling Speeds of Aggregates in the Deep Ocean

A time-series sediment trap collects samples for a defined period of time, from one to several weeks. If a pair of sediment traps are separated vertically at a given distance and the collection periods are synchronized, then the residence time of a descending particle can be estimated from the offset of arrival times at the shallow and deeper trap. We have done this synchronized time-series trap experiment a number of times, some of which we did together with Egon's group in the Black Sea and Gulf of Arabia. From the results of those experiments, we can now estimate the bulk residence time of ocean particles and the speed of settling in the ocean interior.

An example is illustrated in Fig. 1. There occurred an unusually large peak of fluxes which coincided with the early stage of the 1982–83 El Nino at this sub-boreal station, Ocean Station P. The highest flux peak recorded at the 1,000 m deep trap appeared one collection period later, 2 weeks, in the 3,800 m deep trap moored along the same tautline. The estimated bulk settling velocity was thus 200 m $day^{-1}$. The settling velocity estimated for a number of species of diatom was also the same as the bulk rate (Fig. 2 from Takahashi, 1986). Unfortunately, the unit length of

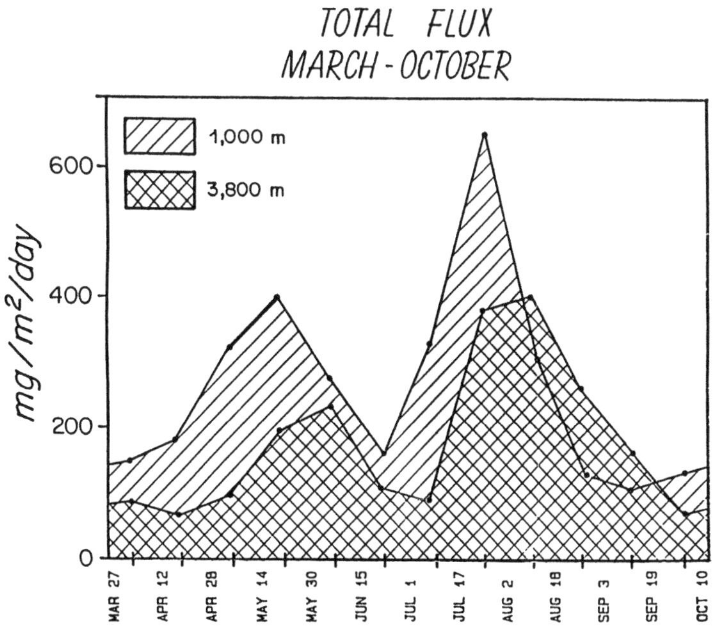


**Fig. 1.** Off-set of the arrival time of ocean particles at a pair of time-series sediment traps, 2,800 m apart, along the same mooring at Ocean Station P, Gulf of Alaska. The maximum and minimum flux period observed in the trap deployed at 3,800 m are uniformly retarded for one collecting period (two weeks) from the one recorded in the 1,000 m trap. This indicates that the residence time of particles is about a few weeks and the bulk settling speed of a particle during this period was about 200 m $day^{-1}$. This figure was enlarged from the right half of the 1982–1983 flux chart (solid line) in Fig. 3. This figure was cited from Honjo, 1984

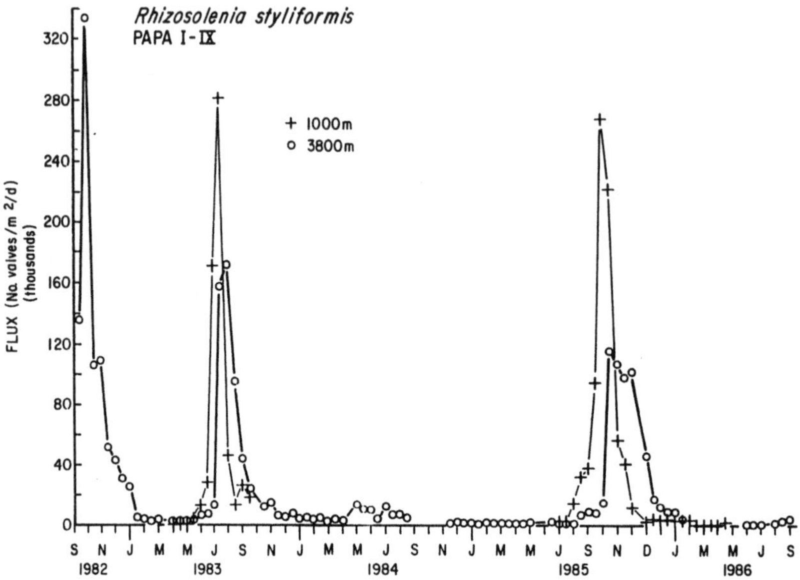


**Fig. 2.** Flux of individual frustules of *Rhizosolenia styliformis* at Ocean Station P from the summer of 1982 to 1986. Three peaks of the frustule flux were observed. At each peak, the arrival time of *R. styliformis* frustules was delayed for 2 to 4 weeks (the collecting intervals; the time resolution was 2 weeks, the same as in Fig. 1), indicating that a fragile particle with high surface area/mass ratio also settles at a higher speed than estimated for bulk particles, as indicated in Fig. 1. This figure was cited from Takahashi, 1986

time of collection periods has been set for 2 weeks in most experiments in the oceans, so the resolution of this estimate is limited and accepts a large error range. Nevertheless, the bulk settling rate of ocean particles is a few hundred meters per day and this shows an acceleration of several orders of magnitude over independent particle settling speeds. An important finding was that the bulk settling speed of ocean particles is less irregular and is within a range of 100 to 400 m $day^{-1}$, as least as far as our observations have told us in the Gulf of Arabia and the Black Sea in addition to Ocean Station P.

## 6 Rhythms in Oceanic Sedimentation

Oceans contain rhythms and cycles in many wavelengths and amplitudes. The smaller end of visible rhythms include diurnal or tidal cycles, then there are geological cycles, such as glacial/inter-glacials, and on up to million- or billion-year celestial rhythms involving planetary evolution. Among all of these, the annual seasonality and inter-annual cycles of several years are most relevant to our daily life and can accommodate the research strategies of today.

If the settling speeds of particles are fast and the particles arrive at the ocean bottom quickly after being produced in the surface water, then biological activities in the upper oceans (e.g. seasonal events such as spring plankton blooms or barren periods of productivity in late summer) should greatly influence the fluxes of particles and the rate of sedimentation. Therefore, one cannot estimate yearly flux rates from an experiment done for a short time. Particularly for a geologist, flux information is not very useful unless, at least, it provides a reliable yearly sedimentation rate.

To set up an experiment which will accurately depict the rhythm of ocean sedimentation in the far interior of the deep ocean and compare it with that in the upper ocean, a number of requirements must be met. First, the experiment should continue as long as possible to pick up longer, which are often larger, cycles; thus it is desirable to keep an experiment going for several years so that one can pick up at least the shortest visible interannual cycle, such as El Nino. Second, in contradiction to the first requirement, in order to record the shorter cycles, the settling material should be monitored in time series with the resolution of sequential collection as high as possible (that is, the period of unit time of sample collecting should be as short as possible). Third, in order not to miss short but very important pulses of ocean sedimentation (Honjo 1982), monitoring of flux must be carried on without a break. Such an ocean experiment must also be carried on automatically because, simply, otherwise the shiptime costs alone would be astronomical.

Considering these requirements for an ideal experiment in light of the reality of today's technology, funding availability, and the number of researchers with this particular interest, we designed a flux measurement experiment around PARFLUX time-series sediment traps (Honjo and Doherty 1988) and started a joint experiment with C.S. Wong at Ocean Station P in the northern Pacific in the summer of 1982. A few months later, due to the strong initiative of Egon, a joint research program on particle flux in the anaerobic Black Sea began with an international team from Hamburg University, Izmir University, and Woods Hole Oceanographic Institution. The initial results were published in the Izmir Symposium report: "Particle Flux in the Oceans," 1987, edited by Degens, Izdar, and Honjo.

Long-term time-series sediment trap experiments have kept on growing since 1982. Today, a sediment trap experiment usually means, among the ocean research community, a time-series deployment of over one year, at least. Of course, there are short-term, non-time-series experiment still generating useful data for specific objectives. Up to the spring of 1988, more than 50 long-term, year-round time-series sediment trap research experiments have been completed which have lasted for more than a year or are ongoing at this writing.

The most significant result of long-term time-series experiments up to date, probably, is that we now have started to know the annual flux of essential elements which support the biosphere of this planet, such as organic carbon, nitrogen, biogenic calcium carbonate, biogenic silica (opal), and others. For example, we know that the annual flux of carbon to the pelagic deep ocean interior, that is an approximation of "new carbon" (Eppley and Peterson 1979), ranges from 3 mmolm$^{-2}$ (North Central Weddell Sea) to over 500 mmolm$^{-2}$ (Southern Black Sea), and the norm is around 50 to 150 $mmolm^{-2}$. It was a surprise to find that the annual flux in the Greenland Sea (about 100 $mmolCm^{-2}$) turned out to be twice as much as in the Sargasso Sea (50 $mmolCm^{-2}$). However, the annual flux data in the world ocean is still very far from the level at which one could draw a regional model of the biogeochemical cycles of critical elements. An attempt today to create a comprehensive picture of global flux in relation to surface productivity would be like trying to draw a geological map of North America by examining one location in each of the fifty states.

Long-term sediment trap experiments have revealed that the amplitude of flux between maximum and minimum periods can be as large as up to 10 times in a temperate pelagic ocean, and 100 to 1,000 times in the Polar Oceans. Sometimes particle flux almost completely ceases during a certain time of the year. This happened, for example, in the Weddell Sea while the ice covered the ocean during the austral winter, and also during a period of a few months after a monsoon in the Arabian Sea for different reasons. Gerald Wefer and his colleagues observed very large annual particle flux at a station in the Bransfield Strait, near the Antarctic Peninsula, which was 107 $gm^{-2}$ (note that is "grams"!). Virtually all of the annual flux at this station occurred during December and January with a total flux as high as 2 $gm^{-2}$ $day^{-1}$ and half of the particles which settled were biogenic.

In cooperation with C.S. Wong, a long-term (several years) particle flux monitoring experiment at Ocean Station P has provided useful sets of data to understand the intra- and interannual variability of fluxes in a station in the North Pacific. At this station the particle flux fluctuates from winter, 25 $mgm^{-2}$ $d^{-1}$, to summer, 150 $mgm^{-2}$ $d^{-1}$, (right after the spring bloom in May/June) with very small interannual differences. However, during 1982–1983 there were 3 peaks of high flux and the maximum flux in August reached 400 $mgm^{-2}$ $d^{-1}$ (Fig. 3). The annual carbon flux during this year was about 3 times as high as in ordinary years: 200 $mmolm^{-2}$ vs 60 $mmolm^{-2}$ (Fig. 4). This coincided with the record El Nino of 1983 which devastated coastal areas of South and Central America and brought unusual droughts to Australia. However, the relationship between the El Nino and the occurrence of the large particle flux observed in the Gulf of Alaska is not known.

An important discovery from time-series trap experiments is that the rate of arrival of food matter at the deep ocean floor changes in parallel with the variability of the upper ocean through "filters" with a fixed retardation of a few months. This retardation has been observed, as explained earlier in this essay. Our efforts must now be concentrated on decoding this filter. If we can solve the function details of the filter, we can depict primary production of the ocean by simply measuring the amount of new product which is approximately the biogenic flux at the ocean interior. The results can be extended back into geological records.

Lastly, life in a benthic community is not boring and monotonous but full of regular or irregular events which are probably directly coupled to the busy events in the upper ocean, although actual measurements of these effects, such as influence on the bottom and respiration, are still rare (Smith and Hinga 1983). Such strong variability in time and space in the abyssal environment may have stimulated the diversity and evolution of the community which has adapted to the deep sea environment. Sediment traps do not measure the descent of a huge object, such as

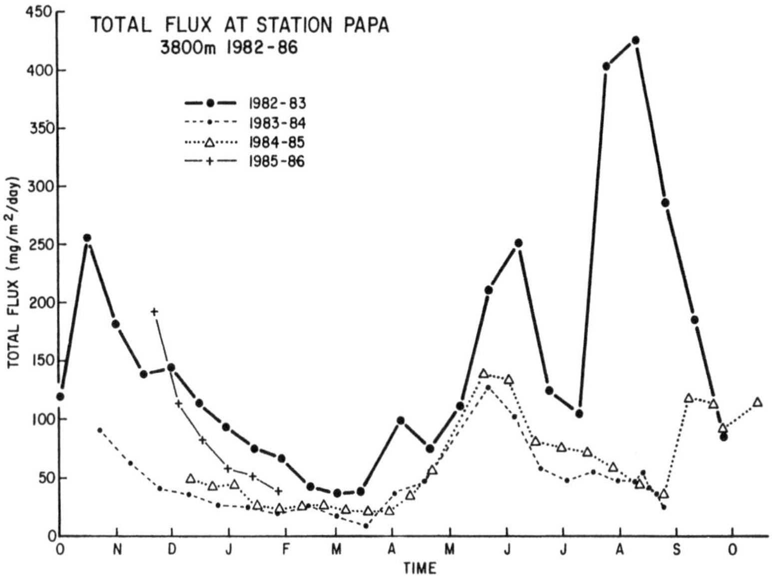


**Fig. 3.** Seasonal and interannual variabilities of total flux recorded in a deep time-series trap deployed at Ocean Station P, for 3-½ years from the summer of 1982 to 1986. A major peak was observed annually in late May to early July, with a secondary peak, although not showing as clearly as the main peak, taking place in mid-summer of each year. The interannual difference during this period from 1983 to 1986 (August to August) was relatively small. However, the mass flux during 1982–1983 was 2 to 10 times higher at any season of the year than during other years in this figure. This data is from a joint program between Woods Hole Oceanographic Institution, USA, and the Institute of Ocean Sciences, Sidney, B.C. Canada

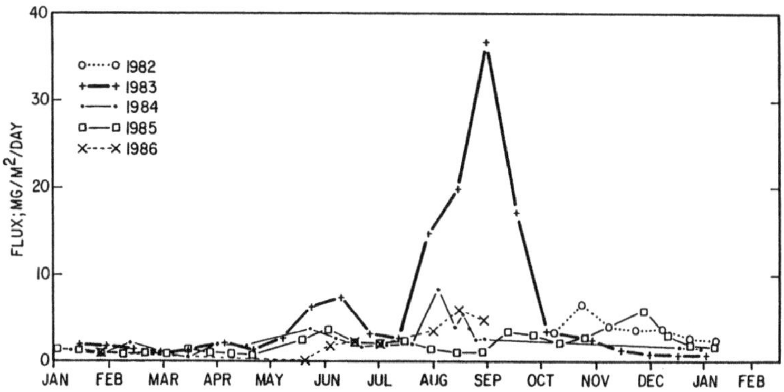


**Fig. 4.** The organic carbon flux at Ocean Station P recorded at 3,800 m deep by a time-series sediment trap. The organic carbon flux during 1982–1983, particularly that observed in August, 1983, was large compared to more "normal" years as indicated in Fig. 3. This period of large organic carbon flux coincides with an El Nino event in the tropical Pacific Ocean. This data is from a joint program between Woods Hole Oceanographic Institution, USA, and the Institute of Ocean Sciences, Sidney, B.C. Canada

a marine mammal corpse or a tree log, nor the active transportation of animals to the deep sea bottom (Vinogradov and Tseitlin 1983). Research should be continued on the "benthic events" related to the food supply.

One hundred years after the denial of the Azoic theory of the deep ocean, Egon and I with numerous colleagues have made modest progress toward understanding the quality and quantity of material which is arriving at the interior of deep oceans and the deep ocean floor. In particular, we have spent at exciting 10 years deploying and recovering large mooring arrays with sediment traps all over the world in a truly cooperative spirit. There were also a lot of difficulties which we shared together and we have learned a lot about running large ocean experiments successfully with international support. We are still far, far from our common goal: understanding the biogeochemically cycles of the essential elements in the global ocean, present and past. At this time in Egon's superbly productive research career, age 60 seems to me to be a perfect milestone for Egon to accelerate the speed of research which will go far to narrow the serious gap between our goal and our realities.

## References

Alldredge AL (1976) Discarded appendicularian houses as sources of food, surface habitats, and particulate organic matter in planktonic environments. Limnology and Oceanography, 21:14–23

Bacon MP, Huh C-A, Fleer AP, Deuser WG (1985) Seasonality in the flux of natural radionuclides and plutonium in the deep Sargasso Sea. Deep-Sea Res 32(3):273–286

Degens ET, Okada H, Honjo S, Hathaway JC (1972) Microcrystalline sphalerite in resin globules suspended in Lake Kivu, East Africa. Mineral. Deposita (Berl) 7:1–12

Degens ET, Ross DA (ed) (1974) The Black Sea – geology, chemistry, and biology. Amer Assoc Petrol Geol Memoir 20, Tulsa, Oklahoma, 633 pp

Degens ET, Izdar E, Honjo S (ed) (1987) Particle flux in the ocean, Hamburg University Press, 308 pp

Eppley RW, Peterson BJ (1979) Particulate organic matter flux and planktonic new production in the deep ocean. Nature 282:677–680

Honjo S (1975) Dissolution of suspended coccoliths in the deep-sea water column and sedimentation of coccolith ooze. In: Sliter WV, Be AWH, Berger WH (eds) Dissolution of deep-sea carbonates. Cushman Foundation for Foraminiferal Research Spec Publ No 13:114–128

Honjo S (1976) Coccoliths: production transportation and sedimentation. Mar Micropaleont 1(1976):65–79

Honjo S (1982) Seasonality and interaction of biogenic and lithogenic particulate flux at the Panama Basin. Science, 218:883–884

Honjo S (1984) Study of ocean fluxes in time and space by bottom-tethered sediment trap arrays: a recommendation. In: Maynard N (ed) Global ocean flux study, Proc Workshop, National Academy Press, Washington, D.C.

Honjo S, Roman MR (1978) Marine copepod fecal pellets; production, preservation and sedimentation. J Mar Res, 36:45–57

Honjo S, Manganini SJ, Cole JJ (1982a) Sedimentation of biogenic matter in the deep ocean. Deep-Sea Res 29:609–625

Honjo S, Spencer DW, Farrington JW (1982b) Deep advective transport of lithogenic particles in Panama Basin. Science 216:516–518

Honjo S, Manganini SJ, Poppe LJ (1982c) Sedimentation of lithogenic particles in the ocean. Mar Geol, 50:199–220

Honjo S, Doherty KW, Agrawal YC, Asper VL (1984) Direct optical assessment of large amorphous aggregates (marine snow) in the deep ocean. Deep-Sea Res, 31:67–76

Honjo S, Doherty KW (1988) Large aperture time-series sediment trap: design objectives, construction and application. Deep-Sea Res, 35:133–149

Keeling CD, Bacastow RB, Bainbridge AE, Ekdahl, Jr. CA, Guenther PG, Waterman LS, Chin JFS (1976) Atmospheric carbon dioxide variations at Mauna Loa observatory, Hawaii. Tellus 28:538–551

Kempe S, Nies H, Ittekkot V, Degens ET, Buesseler KO, Livington HD, Honjo S, Hay BJ, Manganini SJ, Izdar E, Konuk T (1987) Comparison of Chernobyl nuclide deposition in the Black Sea and in the North Sea. In: Degens ET, Izdar E, Honjo S (ed) Particle flux in the ocean. Hamburg University Press (308 pp):165–178

Lisitzin AP (1972) Sedimentation in the world oceans. SEPM, Spec Publ 17

Milliman JD, Manheim FT, Pratt RM, Zarudzki EFK (1967) Alvin dive on the continental margin of the southeast United States, July 2–23, 1967. Woods Hole Oceanographic Institution Technical Report 67–80, 64 pp

Mills EL (1983) Problems of deep-sea biology: an historical perspective. In: Rowe GT (ed) Deep-sea biology; 8. The sea: 1–79, Wiley-Interscience Publication, John Wiley & Sons, New York, 560 pp

Pilskaln CH, Honjo S (1987) The fecal pellet fraction of biogeochimcal particle fluxes to the deep-sea. Global Biogeochemical Cycles 1:31–48

Riley GA (1963) Organic aggregates in seawater and the dynamics of their formation and utilization: Limnology and Oceanography 8:372–381

Rowe GT (1983) Biomass and production of the deep-sea macrobenthos. In: Rowe GT (ed) Deep-sea biology; 8. The sea: 97–121, Wiley-Interscience Publication, John Wiley & Sons, New York, 560 pp

Silver MW, Alldredge AL (1981) Bathypelagic marine snow: deep sea algal and detrital community. Jour Mar Res 39:501–530

Silver MW, Shanks AL, Trent JD (1978) Marine snow: microplankton habitat and source of small scale patchiness in pelagic populations. Science, 201:371–373

Small LF, Fowler SW, Unlu ML (1979) Sinking rates of natural copepod fecal pellets. Mar Biol 51:239–241

Smayda TJ (1971) Normal and accelerated sinking of phytoplankton in the sea. Mar Biol, 11:105–122

Smith KL, Hinga KR (1983) Sediment community respiration in the deep sea. In: Rowe GT (ed) Deep-sea biology; 8. The sea: 331–370, Wiley-Interscience Publication, John Wiley & Sons, New York, 560 pp

Suzuki N, Kato K (1953) Studies on suspended materials. Marine snow in the sea. Part 1. Source of marine snow. Bull Faculty Fisheries, Hokkaido University 4:132–135

Takahashi K (1986) Seasonal fluxes of pelagic diatoms in the subarctic Pacific, 1982–1983. Deep-Sea Research 33:1225–1251

Tsujita T (1949) Nippon kinkai ni hassei-suru yuki kendaku bustu ni tuite. Saikai Suisan Kenkyushyo Gyoseki 4:1–5

Tsujita T (1953) Studies on naturally occurring suspended organic matter in waters adjacent to Japan (II). On an application of the suspended organic matter for the analysis of water masses. Records of Oceanographic Works in Japan, 2:94–100

Uda M (1955) Sekai kaiyo tanken-shi (History of exploration of the world's ocean). Kawade-Shobo, Tokyo, 436 pp

Vinogradov ME, Tseitlin VB (1983) Deep-sea pelagic domain (aspects of bioenergetics). In: Rowe GT (ed) Deep-sea biology; 8. The sea: 123–165, Wiley-Interscience Publication, John Wiley & Sons, New York, 560 pp

Wiebe PH, Boyd SH, Winget C (1976) Particulate matter sinking to the deep-sea floor at 2,000 m in the Tongue of the Ocean, Bahamas, with a description of a new sedimentation trap. J Mar Res 34:341–354

# Use of Multivariate Statistical Analysis in Geology – Two Examples

H. Kin Wong

## 1 Introduction

Geology is the study of the planet Earth: its origin, constituents, morphology, history and the processes that act to shape its form, both past and present. Thus, any geological system or any subsystem thereof is necessarily multivariate (i.e., it can only be described adequately by the use of more than two variables). The application of uni- or bivariate statistical analysis to a geological system is hence often an over-simplification. Multivariate analytical techniques must be applied. In the present article, two examples from the literature will be used to illustrate this point. Since the purpose is not to present any up-to-date data compilations, the original data sets as published are retained, so that a direct comparison with the original conclusions may be made. In addition, this article should not be understood as a critique of the original works, both of which represent significant contributions to their respective fields. Their choice is made merely to render this article more instructive and, not entirely by coincidence, because they deal with Lake Van in eastern Turkey and the East African rift valley lakes, two geographic areas dear to Prof. Egon Degens, with whom I had the good fortune to become associated two decades ago and to whom this article is dedicated on the occasion of his sixtieth birthday.

## 2 Example 1 – Discrimination Between Limnic and Oceanic Sediments Using Their Sugar Contents

Since ribose is a phosphate carrier and is therefore released rapidly to the sediment after cell lysis (Degens and Mopper 1975), it is often considered a useful parameter in the characterization of the origin (terrestrial or marine) of a sediment. Kempe (1977), for example, established a scheme using the glucose/ribose and glucose/galactose ratios whereby he concluded that all oceanic (marine) samples and plankton have glucose/ribose ratios of less than 20 and glucose/galactose ratios less than one. Values much higher than 20 and one respectively imply a limnic or terrestrial origin. He then classified sediment samples from Lake Van using this scheme and found somewhat surprisingly that these lacustrine samples fall between the oceanic and limnic fields.

The weakness of his method lies in the fact that while eight sugars were quantified in all samples (Tables 9 and 16 in Kempe 1977), only two sugar ratios were used in the discrimination scheme. Much of the information contained in the original data was therefore forfeited. To overcome this difficulty, a multivariate, linear discrimination analysis could be employed. The samples are divided into groups according to their origin (terrestrial or marine; Table 1) and then plotted as points in eight-dimensional space with the eight sugars as independent variables. Because the sugar contents of sediments having a common origin are more similar than those having different origins, the points fall naturally into two clusters in this multivariate space. The procedure is to search for a multi-dimensional, planar surface onto which the clusters project with the greatest separation but with the smallest spread. Analytically, this is equivalent to finding a transform which gives the minimum ratio of the difference between the multivariate means to the multivariate variance within the two groups. The resulting surface or transform is the discrimination function R which in our case is given by:

$$R = -1.30(\text{rhamnose}) + 8.33(\text{fucose}) + 2.67(\text{ribose}) + 8.67(\text{arabinose}) + 6.09(\text{xylose}) + 8.82(\text{mannose}) + 6.91(\text{galactose}) + 5.57(\text{glucose})$$

Here all sugar contents are in promille of the total content.

**Table 1.** Discriminant scores of the three groups of sediment samples

| Sample | | Discriminant score |
|---|---|---|
| *Terrestrial (limnic)* | | |
| Peat | | 6048 |
| Wood | | 6088 |
| Black Sea, oxidizing, 125 cm below surface | | 6010 |
| Lake Kivu, 125 cm below surface | | 6053 |
| Lake Kivu, 240 cm below surface | | 5957 |
| Lake Kivu, 530 cm below surface | | 5989 |
| Lake Kivu, 930 cm below surface | | 6210 |
| Oyster Pond | | 6058 |
| | Mean | 6052 |
| *Marine (oceanic)* | | |
| Black Sea, reducing, 15 cm below surface | | 827 |
| Black Sea, intermediate conditions, 65 cm below surface | | 858 |
| Walrus Bay | | 914 |
| Argentina Basin | | 725 |
| Bermuda | | 786 |
| Oceanic plankton, average | | 856 |
| | Mean | 828 |
| *Lake Van* | | |
| Sample 1 | | 5854 |
| Sample 9 | | 5922 |
| Sample 16 (turbidite) | | 5613 |
| Sample 23 | | 6365 |
| Sample 32 | | 4816 |
| Sample 45 | | 5467 |
| Sample 49 (soil) | | 6254 |
| | Mean | 5674 |

The F-test using a value for F obtained by a transformation of the Mahalanobis distance suggests that the linear discrimination function given above is statistically highly significant. This conclusion is confirmed by an examination of the discriminant scores (Table 1), which yield a mean of 6052 for limnic (terrestrial) sediments (range: 5957 to 6088) and of 828 for oceanic (marine) sediments (range: 725 to 914), two values that differ distinctly from each other. The discriminant index is 3440. A sediment sample is classified as limnic if substitution of its sugar contents into the discriminant function yields a score above 3440; it is classified as oceanic if the corresponding discriminant score is less than 3440. The greater the difference between the score and the discriminant index, the less equivocal is the classification.

For the Lake Van samples, the discriminant scores all lie considerably above 3440. The lowest value is 4816 (sample 32), and the highest is 6365 (sample 23). (Table 1). The mean is 5674, a value similar to the limnic group mean. Thus, all sediment samples from this lake carry a terrestrial sugar signature as expected. This is true also for samples 16 (a turbidite) and 49 (soil), although their absolute total sugar concentrations are lower by factors of 2 and 5 respectively (Table 16 in Kempe 1977).

That the scheme of Kempe (1977) misclassifies many of the Lake Van samples is a result of at least two factors. The first factor, as already mentioned, is the selective use of only two sugar ratios instead of all 8 sugar concentrations. The second, as implicated in his Fig. 25, is the tacit assumption of a more-or-less linear relationship between these sugar ratios. Neither a linear nor a quadratic least squares analysis could confirm a statistically significant correlation between these two ratios.

It would be interesting to perform a corresponding discrimination analysis, perhaps even on multiple groups, using a more complete and up-to-date data compilation. The results shown here suggest that such a linear discrimination scheme appears promising.

## 3 Example 2 – Hydrochemical Classification of Rift Lakes in East Africa

Lakes are often classified according to their origin (Hutchinson 1957) or hydrochemical composition (see, e.g., Tse and Wong 1987). For lakes in tropical and sub-tropical Africa, Talling and Talling (1965) proposed a convenient although admittedly arbitrary scheme based on conductivity at 20°C ($K_{20}$). Their Class I lakes have a low total ionic concentration ($K_{20} < 600$ μmho), a pH value between 7.0 and 8.7, and includes many lakes fed by surface run-off or by rivers of low salinity. Their Class II lakes are characterized by 600 μmho $< K_{20} <$ 6000 μmho and $8.8 < \text{pH} < 9.5$, and are located in closed basins or are drained by salt-rich rivers. Their Class III lakes are very saline and shallow, and are for the most part situated in closed basins. The rate of evaporation is high and deposition of trona is frequent. The characteristic features are $jk_{20} > 6000$ μmho and $\text{pH} > 9.5$.

To obtain an objective classification scheme, the data of Talling and Talling (1965) were re-analysed using multivariate statistical techniques. However, because values for many parameters were unavailable for a number of lakes, the data matrix had to be reduced so that all entries are significant. Moreover, values considered unreliable by the original authors were discarded and averages were used whenever multiple measurements were listed. The final data set consisted of 30 lakes and 9 parameters (Table 2).

Q-mode factor analysis suggests a statistically-based hydrochemical classification of the African lakes into five types. The deciding factors (from factor scores) in addition to conductivity at 20°C are the contents of calcium and magnesium. In retrospect, this is to be expected because in lakes of high alkalinity (and practically all of the 30 lakes included are alkaline or highly alkaline), the intrinsic cationic composition of lake water is to a large extent controlled by the precipitation of carbonates of calcium and magnesium. Total anionic and cationic concentrations as well as chloride content are directly relatable to salinity and hence to $K_{20}$. Therefore they do not play a significant, independent role. Bicarbonate is almost always the principal anion (Talling and Talling 1965), with the consequence that $HCO_3 + CO_3$ is likewise correlatable to $K_{20}$. Silica, abundant because of its high content in tropical soil, and sulphate, generally low due to paucity in the water inflow, do not appear to be important parameters in the classification.

Type 1A (Table 3) lakes have low to intermediate salinities and normal hydrochemical compositions. They include graben lakes with outflow (Malawi, Zwei, Baringo, Margherita), volcanic lakes created by damming (Mutanda, Bunyoni), as well as lakes in a shallow tectonic basin (Victoria) or in a volcanic field (Mulehe, Mutanda). Type 1B lakes also have low to intermediate salinities, but their Ca and Mg contents are low. Examples are Awassa and Rudolf (rift lakes), Naivasha (a crater lake) and Bangweulu (a shallow tectonic lake).

Lake Mohasi is the only example of a Type 2A (Table 3) lake on the list. Its waters have an intermediate salinity, but they are enriched in Ca and Mg, as well as in Cl. The deep, graben lakes of the Western Rift, the water bodies of which are strongly influenced by the Virunga volcanic field (Talling and Talling 1965), are representatives of Type 2B lakes. They are moderately saline and have a water chemistry characterized by a high Mg/Ca ratio, as well as high K/Na and $Cl/SO_4$ ratios. These marked features in ionic composition are a result of leaching of the Mg and K-rich volcanic rocks, and of precipitation of carbonates within the water column. Type 2B lakes include Tanganyika (Degens et al. 1971; Hecky and Degens 1973; Craig et al. 1974), Kivu (Degens et al. 1973; Stoffers and Hecky 1978), Edward and Albert (Wong and Degens 1984). The Kivu and Virunga area plays a central role in the water chemistry of these lakes. From here, a considerable amount of salt is contributed to the Congo and the Nile, as well as to Lake Tanganyika via the Ruzizi River and to Edward and Albert via the rivers Ruchuru and Semliki respectively. Thus, prior to the Virunga activity and hence the establishment of the current water source and the volcanic hydrochemical imprint, the lake chemistry must have been more saline (closed basin) and the lake waters cooler. Perhaps the exceedingly low temperatures found below the hypolimnion and the vertical distribution of ionic concentrations in Lake Tanganyika today are remnant expressions of its pre-volcanic (pre-Virunga) past.

Type 3 of this multivariate statistical classification corresponds to Class III of Talling and Talling (1965). The lakes Abiata, Elmenteita, Magadi, Manyara, Nakuru and Shala belong to this type. They are supersaline because they have no

**Table 2.** Chemical composition of African lake waters. (After Talling and Talling 1965)

| Lake | $K_{20}$ μmho | Sum of cations (meq/l) | Sum of anions (meq/l) | Ca (mg/l) | Mg (mg/l) | $HCO_3$ + $CO_3$ (meq/l) | Cl (mg/l) | $SO_4$ (mg/l) | $SiO_2$ (mg/l) |
|---|---|---|---|---|---|---|---|---|---|
| Bangweulu | 35 | 0.31 | 0.29 | 1.9 | 0.85 | 0.24 | 0.55 | 1.0 | 16.9 |
| Mweru | 76 | 1.03 | 1.05 | 10.6 | 5.55 | 0.85 | 12.4 | 3.7 | 5.9 |
| Nkugute | 86 | 1.01 | 1.01 | 7.8 | 4.8 | 0.92 | 1.5 | 2.2 | 10.0 |
| Victoria | 96 | 1.09 | 1.10 | 8.17 | 3.43 | 0.83 | 4.76 | 2.1 | 6.5 |
| Kariba | 99 | 1.05 | 1.14 | 12.57 | 2.63 | 1.00 | 1.65 | 3.5 | 10.0 |
| George | 186 | 2.26 | 2.28 | 19.63 | 5.44 | 1.95 | 8.27 | 13.9 | 21.4 |
| Malawi | 224 | 2.45 | 2.54 | 18.84 | 7.24 | 2.51 | 3.94 | 5.5 | 5.1 |
| Mutanda | 223 | 2.53 | 2.60 | 14.43 | 11.3 | 1,783 | 10.14 | 9.8 | 11.5 |
| Bunyoni | 228 | 2.78 | 2.52 | 17.82 | 10.5 | 1.9 | 24.5 | 4.1 | 4.4 |
| Mulehe | 256 | 2.94 | 3.02 | 21.4 | 13.47 | 2.14 | 12.73 | 27.0 | 26.8 |
| Naivasha | 346 | 3.85 | 3.68 | 1.85 | 6.08 | 3.48 | 13.47 | 10.3 | 36.4 |
| Zwei | 385 | 4.46 | 4.82 | 10.2 | 9.8 | 3.92 | 18.0 | 29.0 | 47.0 |
| Baringo | 416 | 6.22 | 6.49 | 16.8 | 2.6 | 6.3 | 30.5 | 19.7 | 19.2 |
| Tanganyika | 612 | 7.56 | 7.60 | 13.47 | 42.4 | 6.64 | 27.89 | 11.3 | 6.0 |
| Mohasi | 640 | 7.47 | 7.19 | 29.65 | 27.1 | 3.09 | 144.0 | 1.1 | 8.8 |
| Albert | 713 | 8.84 | 8.87 | 10.52 | 31.7 | 7.86 | 28.29 | 36.5 | 1.5 |
| Kitangiri | 785 | 8.6 | 9.15 | 12.2 | 6.7 | 6.65 | 64.0 | 34 | 34.5 |
| Margherita | 806 | 10.5 | 10.6 | 24.1 | 6.0 | 8.5 | 52.5 | 28 | 40.3 |
| Edward | 930 | 11.3 | 12.1 | 14.0 | 39.4 | 9.47 | 26.43 | 43 | 6.3 |
| Awassa | 938 | 12.0 | 12.0 | 16.62 | 4.7 | 10.5 | 34.0 | 27 | 72 |
| Langano | 2060 | 22.7 | 22.5 | 2.5 | 2.7 | 15.0 | 216 | 68 | 54 |
| Kivu | 2210 | 17.7 | 17.7 | 19.08 | 97.0 | 22,86 | 35.85 | 20.9 | 29.2 |
| Rudolf | 3020 | 35.3 | 35.4 | 5.5 | 3.5 | 22.04 | 408 | 67 | 12.8 |
| Rukwa N | 5120 | 51.7 | 73 | 0.4 | 0.4 | 53.5 | 383 | 165 | 115 |
| Shala | 24950 | 273.0 | 306 | 6.2 | 7.9 | 288 | 3218 | 650 | 130 |
| Abiata | 20350 | 189.7 | 316 | 6.0 | 5.7 | 123.8 | 1914 | 720 | 55.7 |
| Elmenteita | 33125 | 420 | 371 | 5.0 | 1.4 | 220.8 | 4523 | 2200 | 232 |
| Manyara | 94000 | 937 | 1097 | 6.0 | 20 | 806 | 5795 | 2280 | 19 |
| Magadi | 160000 | 1664 | 1869 | 6.0 | 20 | 1180 | 22600 | 2400 | 250 |
| Nakuru | 162500 | 961 | 1072 | 8.0 | 20 | 552 | 5442 | 4270 | 376 |

**Table 3.** Hydrochemical classification of rift lakes in East Africa

| Type | Lakes | | Characteristics | | |
|---|---|---|---|---|---|
| | | | Conductivity | Ca and Mg | Other |
| 1 A | Mutanda | Baringo | Low to intermediate | Normal | |
| | Margherita | Kitangiri | | | |
| | Kariba | Malawi | | | |
| | Bunyoni | Zwei | | | |
| | Mweru | George | | | |
| | Nkugute | Victoria | | | |
| | Mulehe | | | | |
| 1 B | Rukwa-N | Naivasha | Low to intermediate | Low in Ca and Mg | |
| | Langano | Bangweulu | | | |
| | Awassa | Rudolf | | | |
| 2 A | Mohasi | | Intermediate | High in Ca and Mg | |
| 2 B | Kivu | Tanganyika | Intermediate | High in Mg, low in Ca | High in K and $Cl/SO_4$ |
| | Albert | Edward | | | |
| 3 | Nakuru | Elmenteita | High | Proportionally low | High in Na, Cl and $HCO_3$ + $CO_3$ |
| | Magadi | Manyara | | | |
| | Shala | Abiata | | | |

drainage outlet, are subjected to high rates of evaporation and are fed by hot, saline ground water or receive surface inflow salt-enriched by leaching. Their water chemistry is characterized by a dominance of monovalent (Na and K) over bivalent cations, and by an abundance of bicarbonates and carbonates.

The analysis presented here demonstrates that an objective hydrochemical classification of African lakes is possible. Moreover, it provides statistical support to the insight of Talling and Talling (1965), and permits a more detailed partition scheme.

## References


Craig H, Dixon F, Craig VK, Edmond J, Coulter GW (1974) Lake Tanganyika geochemical and hydrographic survey, 1973 expedition. Scripps Inst Oceanography, Univ California San Diego, Rep 75:65

Degens ET, Mopper K (1975) Early diagenesis of organic matter in marine soils. Soil Sci 119(1):65–72

Degens ET, Herzen RP von, Wong HK (1971) Lake Tanganyika: Water chemistry, sediments, geological structure. Naturwissenschaften 58:229–240

Degens ET, Herzen RP von, Wong HK, Deuser WG, Jannasch HW (1973) Lake Kivu: Structure, chemistry and biology of an East African rift lake. Geol Rundsch 62:245–277

Hecky RE, Degens ET (1973) Late Pleistocene-Holocene chemical stratigraphy and paleolimnology of the rift valley lakes of Central Africa. Woods Hole Oceanogr Inst Tech Rep 73–28:114

Hutchinson GE (1957) A treatise on limnology I. Geography, physics, and chemistry. Wiley, New York, p 1015

Kempe S (1977) Hydrographie, Warven-Chronologie und organische Geochemie des Van Sees, Ost-Türkei. Mitt Geol-Paläont Inst Univ Hamb 47:125–228

Stoffers P, Hecky RE (1978) Late Pleistocene-Holocene evolution of the Kivu-Tanganyika basin. In: Matter A, Tucker M (eds) Modern and Ancient Lake Sediments. Spec Publ Intern Assoc Sediment 2:43–55

Talling JF, Talling IB (1965) The chemical composition of African lake waters. Int Rev Hydrobiol 50:421–463

Tse P-H, Wong HK (1987) Lake Warder: Sediment properties and phosphorus geochemistry. In: Degens ET, Kempe S, Gan W-B (eds) Transport of carbon and minerals in major world rivers, Pt 4. Mitt Geol-Paläont Inst Univ Hamb 64:387–406

Wong HK, Degens ET (1984) The crust beneath the Red Sea-Gulf of Aden-East African Rift System: a review. Mitt Geol-Paläontol Inst Univ Hamb 56:53–94

Chapter 3

# Geochemistry of River Systems and Coastal Areas

# River Discharge of Water and Sediment to the Oceans: Variations in Space and Time

J. D. MILLIMAN

## 1 Introduction to Introduction

As a graduate student in marine geology and carbonate sedimentology in the early and mid 1960's, the name Egon Degens had particular significance to me, perhaps because some of my work (with Cesare Emiliani) concerned carbon and oxygen isotopes. Egon's papers at Penn State and his first book, *Geochemistry of Sediments*, had served as basic sources of both data and ideas during my graduate school years.

When I moved to Woods Hole in 1966 to begin what has been a 22-year stay at the Woods Hole Oceanographic Institution, I was particularly intrigued with the possibility of interacting with the "great man" himself. Having never met him before, I expected to see a rather distinguished, middle-aged Germanic professor, certainly not the wild-eyed, long-haired Egon that I actually met and grew to know.

One thing that can always be said about Egon during his days at WHOI, as well, I am sure, as his days before and since, is that things were never dull when he was around. I remember one noontime seminar at which Egon changed a working hypothesis literally 180° in mid-sentence, although his delivery did not miss a beat. It was as if the thought processes in answering a question from the audience had caused an instantaneous shift in his interpretation of the data at hand.

With Egon, ideas and data jumped out at alarming rates. I often thought that Egon should have a team of specialists following him around, furiously scribbling down his ideas and insights, later to follow-up the most promising – a sort of narrow-band filter, as it were. What has happened at Hamburg was been even better – Egon's co-workers provided the perfect symbiosis for him and his ideas by providing data and their own insights that have aided greatly in our understanding of the geology and geochemistry of the aqueous environment (I say "aqueous" because a considerable amount of their work has involved lakes and rivers).

In the last ten years, I didn't see Egon nearly as much as I would have liked, the last time being a delightful few hours on a train in Norway. But Egon's influence on me and my work is perhaps greater than he would have thought. The present paper relates to some of the things that Egon has been interested in the past few years, and as such, I hope it might help initiate or redirect further research on rivers and fluvial-oceanic processes.

## 2 Introduction

Geologists and geochemists have had a long interest in defining the amount (and quality) of water and sediment reaching the oceans from rivers. Such data have been used locally to estimate water resources and globally to calculate such parameters as denudation rates. Marine scientists, on the other hand, are more interested in determining the amount of solid and dissolved loads reaching the oceans.

In this short paper it is not possible nor necessary to list all previous compilations of fluvial fluxes to the oceans. Perhaps the best source for water discharge is Livingstone (1963), although this compilation is more than 25 years old. The sediment flux estimates by Strakov (e.g., Lisitzin 1972) and Holeman (1968) were perhaps the most quoted until a more recent compilation by Milliman and Meade (1983).

The purpose of this paper is to discuss the flux of both water and sediment to the oceans, not only in terms of actual fluxes, but also in terms of sources and sinks. I also discuss (albeit briefly) variations in these trends with time, as there is a great temptation, even amongst workers who should know better (I include myself), to extrapolate modern sediment fluxes (in particular) to pre-modern times. Many of these data have been presented elsewhere, but with the exception of Fig. 1, the illustrations and much of the discussion are new.

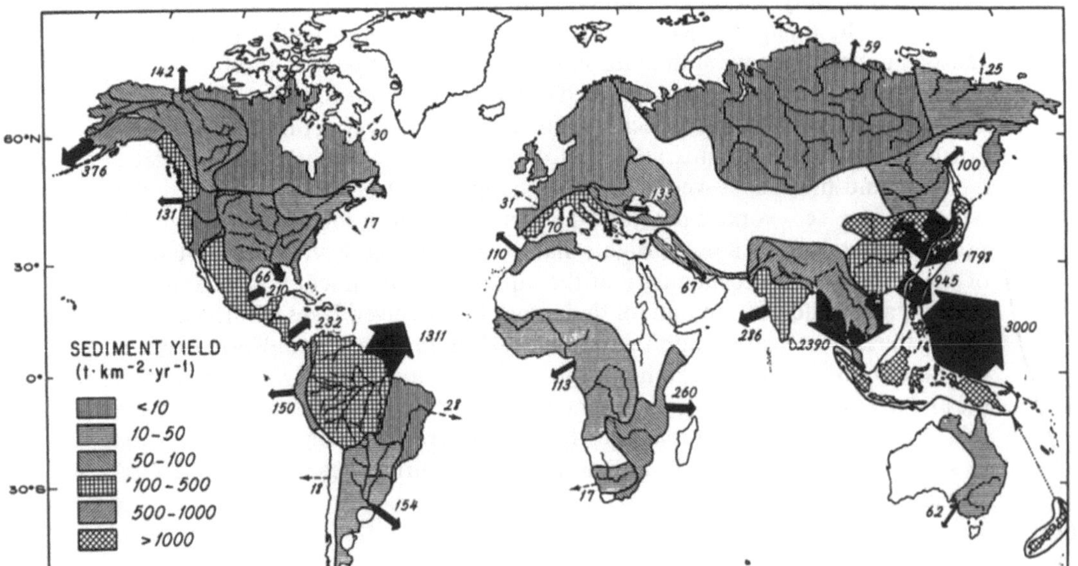


**Fig. 1.** Annual fluvial sediment flux from large drainage basin areas to the oceans. Numbers in millions of tons (per year); *arrows* proportional to the numbers. (Milliman and Meade 1983)

## 3 Sediment Flux from Rivers

The present fluvial suspended sediment flux to the oceans is approximately $13.5 \times 10^9$ t/a, with an additional $1\text{-}2 \times 10^9$ t/a coming from flood discharge and bed load sediment (Milliman and Meade 1983). Southern Asia and Oceania, areas with high rain fall, high relief and (seasonally) heavy rainfall, contribute about 70 percent of the fluvial sediment although they account for only about 15 percent of the land area draining into the oceans (Fig. 2); northeastern South America (i.e., Amazon, Orinoco and Magdalena rivers) contributes another 11 percent. In contrast, Arctic rivers, draining basins of similar area to southern Asia and Oceania (combined), contribute about two orders of magnitude less sediment, the result of relatively low-lying topography and considerable erosion during Pleistocene glaciations. The mean sediment yield (that is, the load per unit area of drainage basin per year) of world rivers is 150 t/km$^2$/a, but the range of values for individual river basins varies by three to four orders of magnitude, from 5 to 6 (e.g. Ob, Lena and Yenisei) to 28,000 t/km$^2$/a (Tsengwen River in Taiwan) (Milliman and Meade 1983; see also Fig. 3).

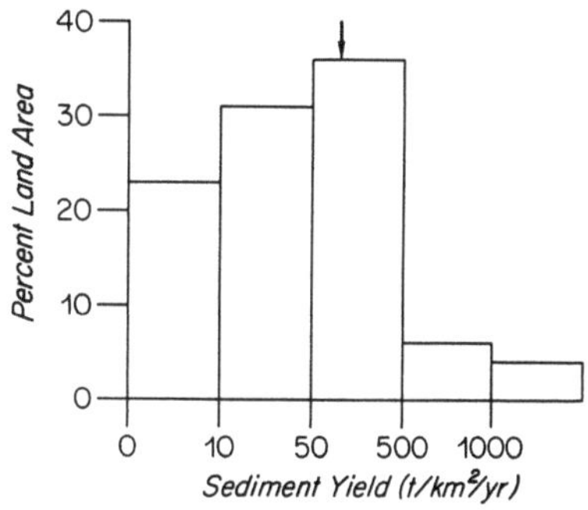


**Fig. 2.** Distribution of sediment yields for all land draining into the ocean (Data from Milliman and Meade 1983)

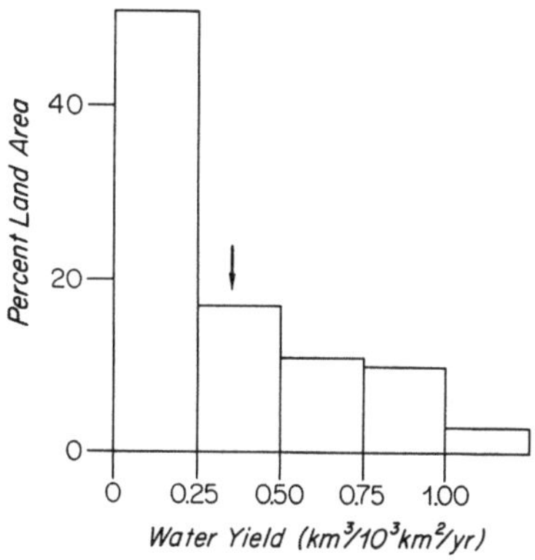


**Fig. 3.** Distribution of water yields for all land draining into the ocean. (Most data from Milliman and Meade 1983)

With the exception of the Chinese rivers, most large rivers (mostly in Asia and South America) are poorly documented, whereas many of the small rivers in North America and Europe are well documented. Many large rivers (the Ganges-Brahmaputra, Irrawaddy and Magdalena are perhaps the best examples) have scant data bases that may not be seasonally averaged. Moreover, loads for rivers are changing continually, increasing in response to poor soil conservation (e.g., the deforestation of Nepal has increased dramatically the sediment loads in the upper reaches of the Ganges River) or decreasing by river damming. Several examples of damming are well known – for example, the complete cessation of sediment flux from the Colorado and Nile rivers. Recent analysis of unpublished data indicates that the present load of the Indus River is less than $50 \times 10^6$t/a, an 80 percent decrease from river loads before construction of barrages in the late 1940s (Milliman et al. 1984). This value is half that estimated by Milliman and Meade, suggesting that values for many other Asian and South American rivers may be similarly erroneous, either because of sparse or erroneous data or because new dams have changed the loads dramatically since the quoted data were obtained. As new numbers become available our estimate of world-wide flux probably will change, perhaps by 10 percent or more.

## 4 Water Flux from Rivers

The world rivers discharge an estimated $33 \times 10^3$ km$^3$ of water annually to the oceans. As with sediment flux, however, this number changes annually in response to both fluctuating weather patterns and the continued construction of dams across many large and small rivers. The average water yield is about 35 km$^3$/10$^3$ km$^2$/a (Fig. 3), but a large portion of the land mass (most of North America, Europe, the Eurasian Arctic) exhibits water yields, between 15 and 20 km$^3$ (Fig. 4). Africa and Australia have lower yields, and, in fact, with the exception of the Zaire and Niger Rivers, Africa contributes practically no fluvial water to the oceans. As with sediment flux, southern Asia, Oceania and northeastern South America contribute most of the riverine water – about 65 percent. The Amazon River itself contributes between 15 and 20 percent of the world total, but the water yield is less than for the rivers draining the highstanding oceanic islands (Fig. 4).

## 5 Sources and Sinks

Another way of viewing these data is in terms of the sources and sinks of fluvial water and sediment (Figs. 5 and 6). Asia and Oceania are clearly the major sources of sediment, while the Eurasian Arctic, Europe, and Africa contribute relatively little. South America and Asia are the largest sources of water, although in terms of water yield, Oceania also must be considered a prime source.

Not surprisingly, the West Pacific and East Indian Ocean (adjacent to southern Asia and Oceania) receive more than 60 percent of the sediment reaching the oceans

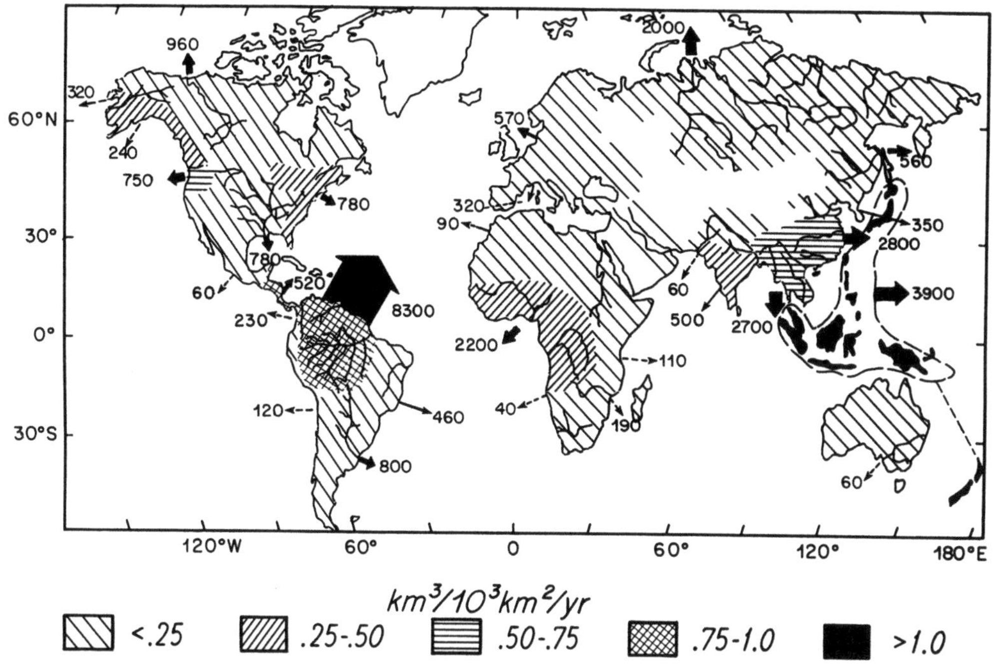


**Fig. 4.** Annual discharge of fluvial water to the oceans. Numbers in $km^3/a$; *arrows* proportional to the numbers. (Most data from Milliman and Meade 1983)

and about 35 percent of the water (Fig. 5). The equatorial and South Atlantic receive somewhat more water (mostly from northeastern South America and tropical Africa), but less than 15 percent of the world budget of fluvial sediment. The North Atlantic and Arctic each receive more than 5 percent of the world fluvial water but a much smaller percentage of sediment.

To many conversant oceanographers and marine geologists, these trends are not surprising, although the actual numbers and percentages may be somewhat different than expected. Clearly if one is looking for fresh water, one does not look in the Southeast Pacific, whereas if one is searching for an area dominated by fluvially derived sediment, the western Pacific and East Indian oceans would be good places to look.

## 6 Changes with Time

These numbers represent the world condition as of about 1980, and in some instances (where data are poor or old) they do not even do that. They cannot necessarily be used as indicators of river influx either before or since. One can assume that within the next 50 years many large rivers will have gone the way of the Colorado, Nile and Indus. Continued deforestation in tropical and mountainous areas, on the other hand, may accentuate water and sediment transport locally.

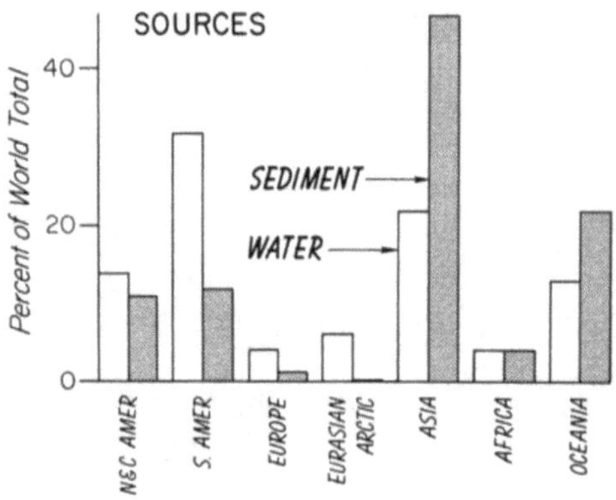


**Fig. 5.** Land sources for fluvial sediment and water discharged into the ocean

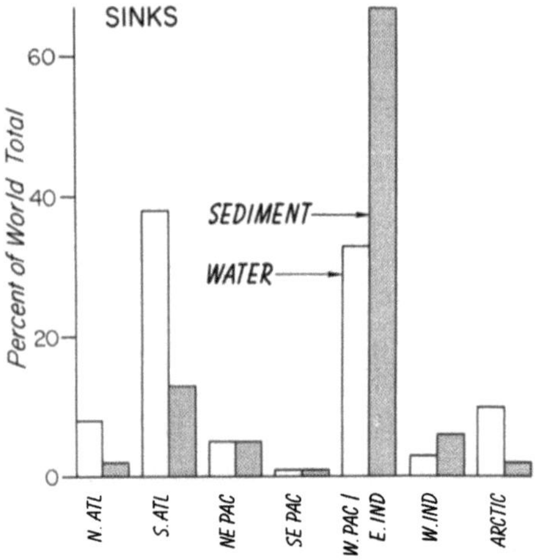


**Fig. 6.** Marine sinks for fluvial sediment and water discharged into the ocean

Moreover, these mid to late 20th century numbers cannot predict water and sediment transport in former centuries and millenia. Reasonable estimates of former sediment loads of large rivers are not easy to make, particularly because of the lack of sufficient geological and geophysical data combined with poor documentation of river flow over long periods. I know of only one successful attempt (perhaps because it is my own), and that is for the Yellow River, where the sediment loads are very large ($1.1 \times 10^9$ t/a), the Yellow Sea acts as a quasi-sediment trap, and the data base is sufficient (if not good). We conclude that the sediment load of the Yellow River over the past 2300 years has been approximately 10 times greater than it was before intensive farming of the easily eroded loess plateau in northern China (Milliman et al. 1987). Although our data do not permit further estimates, preliminary interpretation of geophysical data suggests that sediment flux from the Yellow River during the early Holocene was higher than it was in the mid-Holocene.

While it is not entirely fair to extrapolate this human factor to all of southern Asia (primarily because the Chinese loess represents one of the most easily erodable materials on Earth), one can infer that the very high sediment loads in modern Asian rivers represent a considerable influence from human activities, particularly poor conservation practices. Assuming that present-day Asian and Oceania river loads are five times greater than before man began deforestation and farming, then worldwide fluvial sediment reaching the ocean 2500 years ago might have been less than $7 \times 10^9$ t/a, and the percentage from Asia and Oceania would have accounted for less than 30 percent of the world total (assuming that sediment flux from the other world rivers had not changed during that period – a most unlikely assumption, however).

The present-day sediment load of the Hudson River (draining New York State) is less than $1 \times 10^6$ t/a, or more than three orders of magnitude smaller than the present-day Yellow River. Yet seismic data of sediment wedges on the New Jersey shelf (Milliman et al., in press) indicate that during the late Pleistocene and earliest Holocene, the Hudson periodically may have had sediment loads between 45 and $100 \times 10^6$ t/a. As the Yellow River during this period had loads only somewhat higher than $100 \times 10^6$ t/a, the sediment fluxes from these markedly different rivers may have been quite similar, albeit for a very short time (Fig. 7).

Going back somewhat farther, one can imagine that severe climatological changes may have changed both sediment and water flux to the ocean. For instance, a possible arid climate in the Amazon River basin during glacial times alone could have drastically altered the sedimentological and oceanographic regime of the adjacent equatorial Atlantic.

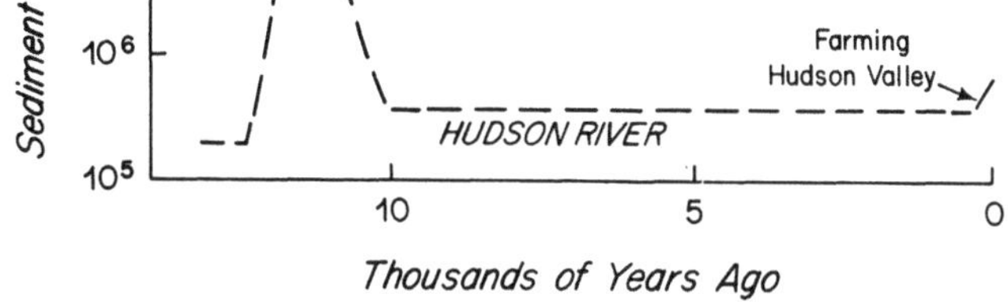


**Fig. 7.** Schematic variation of sediment discharge from the Yellow River and Hudson River during the late Quaternary. Note the dramatic shifts in sediment discharge in response to farming (particularly of the Yellow River basin) and glacial melt (in the Hudson valley). (After data presented in Milliman et al. 1987; in press)

# 7 Concluding Remarks

In recent years, considerable scientific interest and effort have been placed in measuring vertical and horizontal fluxes in the sea. One of the professed aims of these studies is to extrapolate modern processes and rates to those present during earlier periods of the Earth's history. In making such extrapolations, however, rivers can not be assumed to have had similar water and sediment discharge; see Meade (1969) for examples of historical (i.e., the last several hundred years) changes in sediment loads from North American rivers. For pre-Quaternary times, the loads of rivers are particularly difficult to estimate.

More detailed study of the shallow marine stratigraphy near large rivers may help considerably in extrapolating the character of rivers during earlier periods. Palynological, mineralogical and geochemical studies of these sediments can help determine subtle changes in the terrestrial climate and environment.

*Acknowledgments.* In a paper that re-evaluates relatively straightforward data, acknowledgments probably are neither needed nor wanted. However, many of these ideas have emerged from many years of working with colleagues at WHOI and other institutes, particularly in the People's Republic of China. Support for much of the data collection and contemplation of their significance, has been provided by the Office of Naval Research and the National Science Foundation.

# References

Lisitzin AP (1972) Sedimentation in the world ocean. Soc Econ Paleontol Miner Spec Publ 17

Livingstone DA (1963) Chemical composition of rivers and lakes. US Geol Surv Prof Pap 440-G

Meade RH (1969) Errors in using modern stream-load data to estimate natural rates of denudation. Geol Soc Am Bull 80:1265–1274

Milliman JD, Meade RH (1963) World-wide delivery of river sediment to the oceans. J Geol 91:1–21

Milliman JD, Quraishee GS, Beg MAA (1984) Sediment discharge from the Indus River to the ocean: Past, present and future. In: Haq BU, Milliman JD (eds) Marine Geology and Oceanography of the Arabian Sea. van Nostrand Reinhold, New York

Milliman JD, Qin YS, Ren ME, Saito Y (1987) Man's influence on the erosion and transport of sediment by Asian rivers: The Yellow River (Huanghe) example. J Geol 95:751–762

Milliman JD, Zhaung JZ, Li AC, Ewing J (in press) Late Quaternary sedimentation on the outer and middle New Jersey continental shelf: Result of two local deglaciations? J Geol

Holeman JN (1968) The sediment yield of major rivers of the world. Water Resources Res 4:737–747

# Minerals in Soils and in Suspended Matter of Rivers and Their Climatic Zonation

J. Konta

## 1 Introduction

In June, 1982, Professor Egon T. Degens invited me to join the SCOPE/UNEP (Scientific Committee on Problems of the Environment/United Nations Environmental Programme) project "Transport of Carbon and Minerals in Major World Rivers". He offered me the possibility to study the mineralogy of suspended matter collected from rivers by colleagues in several continents. I accepted this generous offer most happily, since I had been studying argillaceous residual and sedimentary rocks for nearly 30 years. Many data were already available on the inorganic constitution of suspended matter (e.g., Post and Sloane 1971; Owens et al. 1974; Potter et al. 1975; Weir et al. 1975; Tomadin 1979) or appeared later (Meade et al. 1985; Tomadin et al. 1985; Kronberg et al. 1986) and many data have been collected additionally during the project (Martins 1982; Emeis and Stoffers 1982; Koch et al. 1982; Irion 1983; Emeis 1985). However, one of the main questions which I had been repeatedly asking myself had not been answered: would it be possible to formulate an equation which describes the intensity of weathering for any type of rock in any natural environment?

The influence of climate on weathering is a widely accepted concept already developed by Strakhov (1962). Reiche (1943) formulated the potential index of weathering and Konta (1984a) suggested an index of corrosion describing the relative stability of either a mineral or rock in an aqueous environment by using chemical composition as the only variable. These indices, however, did not include other important qualities of rocks, e.g., their permeability or their capacity to soak up water.

Furthermore, the relation between the authigenic minerals of soils and the minerals transported in river suspensions was not as yet clear. Ideally, this material should be homogenized due to the high turbulence of the river and therefore should yield samples representing the most recent soils and weathering processes in the respective river basin.

The material collected during the SCOPE/UNEP Project could answer some of these questions.

This paper is dedicated to Egon T. Degens on his 60th birthday. It serves as a witness to Egon's extraordinary talent to stimulate international research and to create favorable working environments for his coworkers.

## 2 Intensity of Weathering and Minerals in Soils

### 2.1 Macromilieu

Soils contain remnants of parent rock constituents of various stability plus authigenic minerals. Soil air, moisture and organic matter are also integrate parts of the soil system. Sheet silicates and $Fe^{3+}$-oxyhydroxides dominate among the authigenic minerals. The structure of the newly formed crystalline and amorphous minerals is controlled by the chemistry of the soil (Fig. 1) which may change vertically in a soil profile. Normally, chemical maturity increases with the vertical distance from the parent rock. The average chemical maturity, however, of soil profiles changes with climate and is highest in the soils of the tropical forests and lowest in tundras and deserts.

The chemical maturity ($Ch_M$) of suspended matter (SM) in rivers, which is derived from soil by erosion, was described by Konta (1985):

$$Ch_M = Al\%: (Na + Mg + Ca)\%. \quad (1)$$

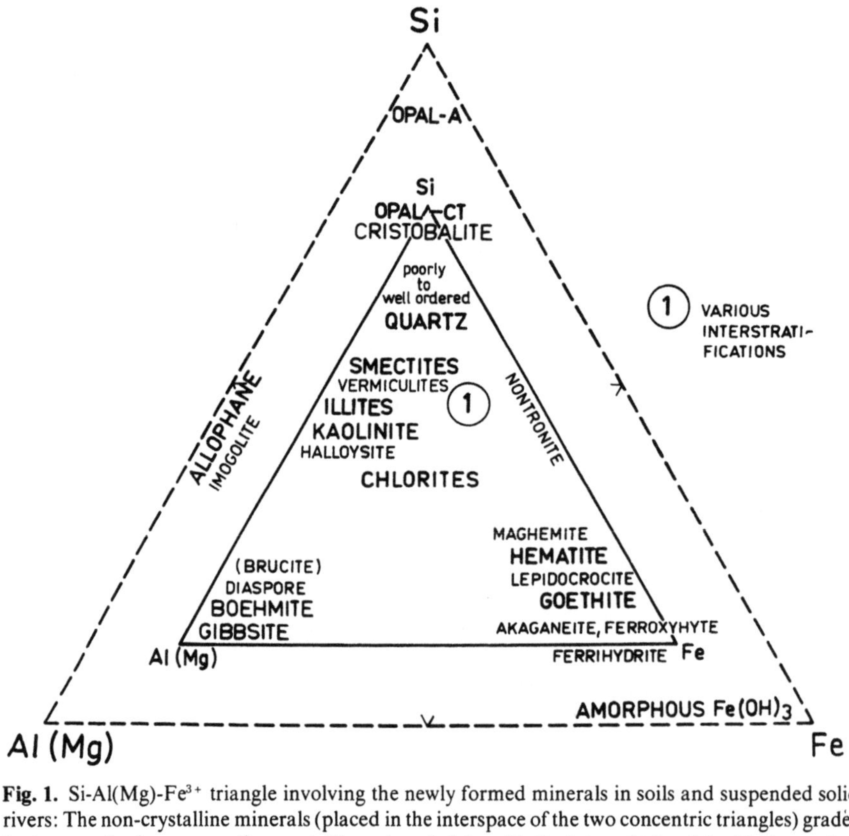


**Fig. 1.** Si-Al(Mg)-$Fe^{3+}$ triangle involving the newly formed minerals in soils and suspended solids in rivers: The non-crystalline minerals (placed in the interspace of the two concentric triangles) grade into better organized corresponding crystalline minerals (placed in the internal triangle) with time and under the influence of the environment

In this formulation chemical maturity is given by the ratio between the least soluble element (Al in solutions of pH 5–9) and the sum of the three most mobile elements (Na, Mg and Ca). This ratio proved to be quite sensitive for SM samples from various climatic regions (Konta 1985).

The weathering of rocks or minerals and the genesis of soils is controlled by various factors. These factors can be described by a series of primary variables: (1) the physico-chemical nature of the parent rock (R), (2) the amount of acting solar energy ($E_s$) and thus the resulting climate, (3) the action of hydrosphere and atmosphere (H + A), (4) the energy of tectonic and other movements ($E_m$) and (5) the duration of the various actions, i.e., time (t). The intensity of weathering ($I_w$) is thus a function of all these variables:

$$I_w = f(R, E_s, H + A, E_m, t) \tag{2}$$

(Konta 1984b, or formulated differently by Bridges 1978). The biosphere enters as a secondary variable only, whose chemical and physical interaction is included in the variables H + A and $E_m$.

Sometimes, the micromilieu has also a strong influence on the resulting products of weathering because it modifies the regional parameters $E_s$, H + A or $E_m$ locally, causing different environments for the various minerals involved

macromilieu (M) for all minerals

$$I_w = f(R, E_s, H + A, E_m, t) \tag{3}$$

micromilieu (i) for each mineral individually.

The question is whether numbers can be assigned to these variables meaningfully. Investigations of various argillaceous residual rocks (Ginzburg and Rukavishnikova 1951; Grim 1953, 1968; Keller 1957; Petrov 1967; Loughnan 1969; Konta 1981, 1984b), of weathered stones in historical monuments (de Quervain 1967; Winkler 1973; Peschel 1977; Niesel 1979; Niesel and Schimmelwitz 1982; Konta 1984c, 1987) and of SM in major world rivers (Irion 1983; Konta 1985) make the solution of this question possible.

The rock properties (R) can be described by:

1. the axial compressive strength of dry rock (which is mainly influenced by the porosity and the sizes of the textural components), expressed in MPa (1 megapascal = 10 kg $cm^{-2}$);
2. the porosity, expressed in %;
3. the imbibition capacity for water, measured in mg/s perpendicularly to the stratification in sedimentary rocks and by the calculated $m_z$ (= graphic mean of the imbibition curve, i.e., the mg $H_2O$ taken up during the first 15 min on 1 $cm^2$ of smoothed rock surface at 20°C) (nearly all fresh igneous and metamorphic rocks have porosities and imbibition capacities equal to zero);
4. the index of corrosion, $I_{KO}$, which is calculated from the chemical composition of the rock by:

$$I_{KO} = \frac{200[(\Sigma A_i(Z_i/r_i) - C(Z_{H^+}/r_{H^+})/10]}{\Sigma A_i(Z_i/r_i) + \Sigma B_i(Z_i/r_i)}, \tag{4}$$

where $A_i$ is the mol concentration of oxides of those cations which are easily removed, i.e., alkaline and alkaline earth and other divalent cations plus some of the smaller cations ($C^{4+}$, $P^{5+}$, $S^{6+}$) which easily form complex anions, while $B_i$ are the mol concentrations of oxides of the more insoluble tri-, tetra- and pentavalent cations such as $SiO_2$, $ThO_2$, $ZrO_2$, $TiO_2$, $Al_2O_3$, $Fe_2O_3$, $Cr_2O_3$, $B_2O_3$, $Ti_2O_3$, $V_2O_3$, and C is the mol concentration of the chemically bound water (OH plus zeolitic $H_2O$ multiplied by 2.703); $Z_i$ is the cation charge and $r_i$ the ionic radius of the respective ion i, Z/r for $H^+/10 = 2.703$ (for more detail see Konta 1984a).

The climatic parameters for the macromilieu (M) are given by:

5. the amount of annual rainfall, in mm/a;
6. the pH value of the reacting water (normally between pH 3 and 10);
7. the degree of continentality (K), given by (Konta 1985):

$$K = [(1.7 \times T_m) / \sin(\text{phi} + 10°)] - 4 \quad (5)$$

where $T_m$ is the annual mean temperature deviation and phi the geographic latitude (for Verkhoyansk, Siberia, K is equal to 100, while for the equatorial low lying sites K becomes zero, hot arid areas have K equal to 45).

The intensity of weathering ($I_w$) increases with decreasing factor 1, increasing factors 2, 3, 4, 5, decreasing or increasing factor 6 (depending on whether the pH value is above or below pH = 7) and a decreasing or increasing factor 7 (from K = 45 for most arid hot regions to K = 0 for equatorial ones, or to K = 100 for those with the highest degree of continentality).

Authigenic clay minerals in soils include: illite, vermiculite, montmorillonite (smectite) including nontronite, chlorite, kaolinite, and sometimes their respective mixed-layer structures and allophanes (amorphous clay minerals). The minerals chlorites→ Mg,Fe, Al-montmorillonites, Mg-vermiculites→ Al-montmorillonites, Al-vermiculites→ Al-illites→ kaolinites represent a series indicative of increasing $Ch_M$ and $I_w$ (Konta 1984a) for a variety of different rocks and under the condition of effective drainage. Figure 2 illustrates the global distribution of major soil types showing their general coincidence with the climatic zones.

$Fe^{3+}$-oxyhydroxides are also common authigenic constituents of soils, but compared to clay minerals, of less importance. They mostly occur finely dispersed but can also fill cracks or cover the surfaces of primary minerals or precipitate in the form of nodules and schlieren. $Fe^{3+}$-oxyhydroxides form different phases, amorphous $Fe(OH)_3$, poorly or well ordered crystalline species, or hydroxyl-free minerals such as hematite or maghemite (Fig. 1). Free Al-hydroxides, mostly gibbsite, occur in the overmature uppermost horizon of some tropical soils.

Chemical elements of lower abundance are mostly included in the structure of the clay minerals or $Fe^{3+}$-oxyhydroxides or are adsorbed to their surfaces. Titanium, however, may occur as authigenic anatase or as a mixture of anatase plus rutile and manganese forms a series of $Mn^{4+}$-oxyhydroxides or oxides (Chukhrov and Gorshkov 1981). Relatively soluble minerals like carbonates and sulfates may form in soils authigenically only under special conditions, mostly under a hot and arid climate.

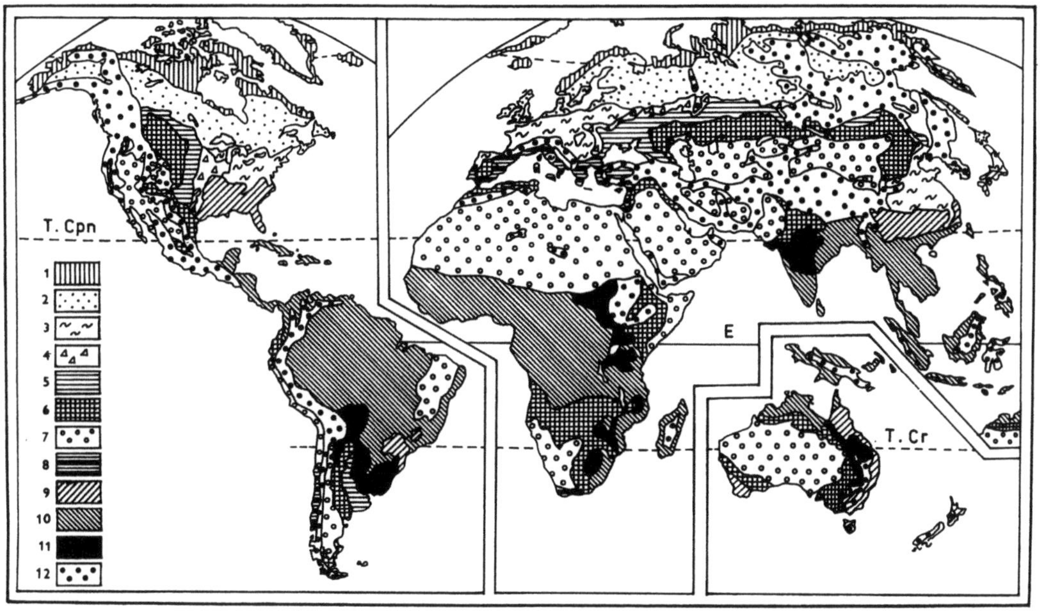


**Fig. 2.** Simplified world soil map reflecting the relation of soil type to climatic zones (Bridges 1978, slightly changed). *1* Soils of the tundra; *2* podzols and related soils of the boreal forest; *3* brown earth and leached soils of the deciduous forest; *4* gray forest soils of the forest-steppe transition; *5* chernozems of the temperate grasslands; *6* chestnut soils and brown soils of the semiarid grasslands; *7* red and grey soils of the deserts; *8* red and brown soils, cinnamon soils of the "Mediterranean" woodlands; *9* red-yellow podzolic soils of the sub-tropical woodlands; *10* red and yellow tropical rain forest and savanna soils (ferrallitic, ferruginous, ferrisols); *11* dark gray and black soils of the tropics and sub-tropics (vertisols); *12* soils of mountainous areas

## 2.2 Micromilieu

The micromilieu plays a very important role in weathering (Konta 1981). Different micromilieus arise in the same soil profile because of the different crystallochemical properties of the individual parent minerals. Figure 3 shows, for example, that different, but adjacent parent feldspars yield different mixtures of secondary clay minerals. In a granite the primary K-feldspars tend to produce illite or IL > MO interstratifications under medium or poor drainage conditions, while Na-rich plagioclase forms Al-montmorillonite at the same time. The better drainage in the plagioclase results in the formation of pure kaolinite, further increasing the pore space. Thus, well-ordered kaolinite forms relatively fast in place of the Na-rich plagioclase while the kaolinite which replaces the K-feldspar is less ordered, and less mature, usually forming an argillaceous aggregate which still contains remnants of illite (Konta 1981).

Investigations with the scanning electron microscope (SEM) or by EDAX can help to discern chemical micromilieus and their long-lasting influence on the evolution of a soil profile. Sometimes, however, the original space of a primary

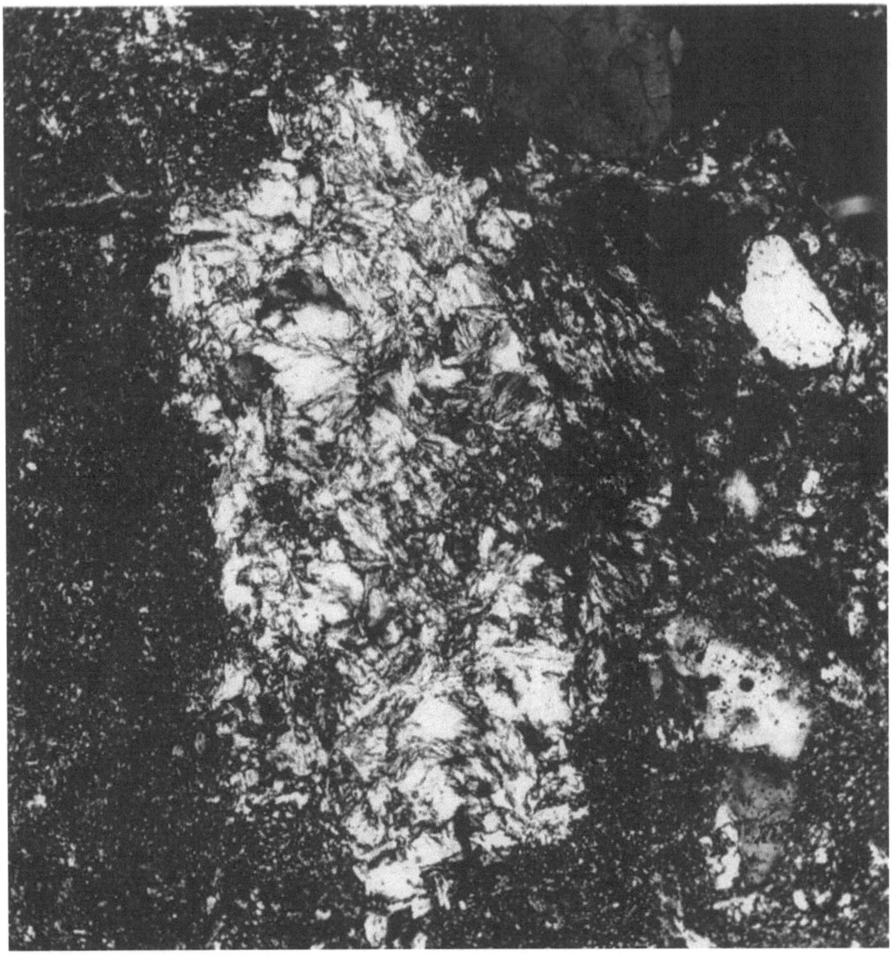

**Fig. 3.** Coarse-crystallized kaolinite pseudomorph rich in illite in place of a potassium feldspar, surrounded by pseudomorphs of pure fine-crystallized kaolinite after sodium-rich plagioclase; upper horizon of the kaolinized granite, Otovice near Carlsbad, Bohemia. X nicols, 60 ×

mineral is so large that it is overlooked under the high magnification of a SEM, causing a wrong assessment of the role of the micromilieu in the sample.

The limitation of a specific micromilieu during weathering or hydrothermal alteration to a certain grain of a parent mineral has the consequence that the secondary clay mineral forms a pseudomorph after the parent mineral. These pseudomorphs support the idea that the movement of water and dissolved solids through the weathering rock is rather slow and that the secondary minerals form in thermodynamic equilibrium with their parent mineral within the micromilieu (e.g., Garrels 1960; Lippmann 1981; Aagaard and Helgeson 1983). There is no doubt that the joints of a rock represent the main input and output passages for solutions but the countless cleavage cracks in feldspars, micas, amphiboles, or other primary minerals represent a large volume to be filled with weathering solutions. These cleavage cracks and their capillary force counterbalance the gravitational drainage and promote weathering maturity compared to materials free of cleavages, such as volcanic glass. Although amorphous silicates are usually less stable thermodynamically than the corresponding crystalline phases and dissolve more rapidly on a geologic time scale, they very often form only smectites or even zeolites. In glasses of basic and intermediate magmas argillization often ends, due to the limited drainage capacity, with secondary hydrosilicates rich in alkaline and alkaline-earth elements.

Due to the existence of micromilieus and due to the ongoing, albeit slow, erosion, chemical equilibrium can never be reached completely even at the surfaces of well-drained soils (Konta 1981). Nevertheless, the topmost layer of the soil, which is subject to erosion throughout most of the continental surfaces, is most intensively influenced by the macromilieu and its factors $E_s$ and H + A which determine the resulting association of secondary clay minerals and their chemical maturity.

Other factors influencing the micromilieu, e.g., tectonic movements and the resulting mylonitization, the action of plant roots and of animals burrowing in the soil, the anomalous influence of blowing wind or of running water, and the increased insolation on southern slopes compared to their northern counterparts, are locally important but do not, in general, override the deterministic role of the macromilieu.

## 3 Minerals in Suspended Matter of Rivers

It follows from the previous chapter that soils are loose accumulations on the surfaces of continents produced by weathering. In general, they are unconsolidated and cohere much less than their respective parent rocks and are therefore more susceptible to erosion than the latter. Eroded topsoil is collected, transported, and sorted by streams which also collect the cations and anions mobilized during weathering. Suspended matter outweighs the dissolved load of rivers by roughly 5.8 times globally (Meybeck 1984; Milliman and Meade 1983).

The investigations of the mineral composition of SM in rivers in a variety of different latitudes and climates clearly showed that the suspended matter inherits the mineral composition of the average composition of the finest size fraction of topsoil in the respective drainage basin (Irion 1983; Konta 1983, 1985). Under normal transport conditions, the SM in rivers contains clay- and silt-sized particles consisting of illite/muscovite, kaolinite, and montmorillonite, often along with chlorite, vermiculite, and sometimes interstratifications of these minerals. Amorphous $Fe^{3+}$-oxyhydroxide was found in 80 of the 86 rivers studied and clastic quartz is an omnipresent mineral. Sometimes very small amounts of K- or Na-rich feldspars and rarely amphibole are found. Calcite and/or dolomite are only subordinate constituents of SM. Opal-A with an admixture of cristobalite and opal-CT are associated with diatom tests in the SM.

The chemical maturity, as expressed by Eq. (1), of SM was found to increase with increasing annual precipitation, with the decrease of K below 45 or its increase above 45, with the decreasing relief gradient and with the increasing density of the vegetation cover in the respective river basin (Konta 1985, 1988). Additionally, an increase in permeability of the surface rocks also causes an increase in $Ch_M$. According to the scheme of climatic and soil zones as published by Strakhov (1962), it was expected to find mostly kaolinite in the lowland rivers of the equatorial forests. This was not always the case. Montmorillonite has also been found as the dominant mineral in the below 2-$\mu$m size fraction in, for example, the Amazon and in three of its 15 southern low-land tributaries (Irion 1983). Similarly, montmorillonite dominates in nine of the 29 rivers studied in New Guinea (Irion 1983) and in six of the 13 rivers studied in India (Naidu et al. 1985). It is important to note that all these rivers flow through large areas of vertisols (see map Fig. 2), mostly in the region of their headwaters. Vertisols are dark-colored, seasonally poorly drained soils rich in expandable clay minerals, chiefly montmorillonite. They develop on less permeable, basic parent rocks under a climate of alternating wet and dry seasons in tropical and subtropical regions.

These results allow to amend Strakhov's scheme of the climatic zonation of soils (Fig. 4). The revised scheme includes the position of recent global soil zones (1 to 11 on Fig. 4) and includes shallow soils on poorly drained parent rocks rich in bases which tend to develop montmorillonites, as in vertisols. This scheme may help in the future to better interpret paleosoils as to their former climatic environment.

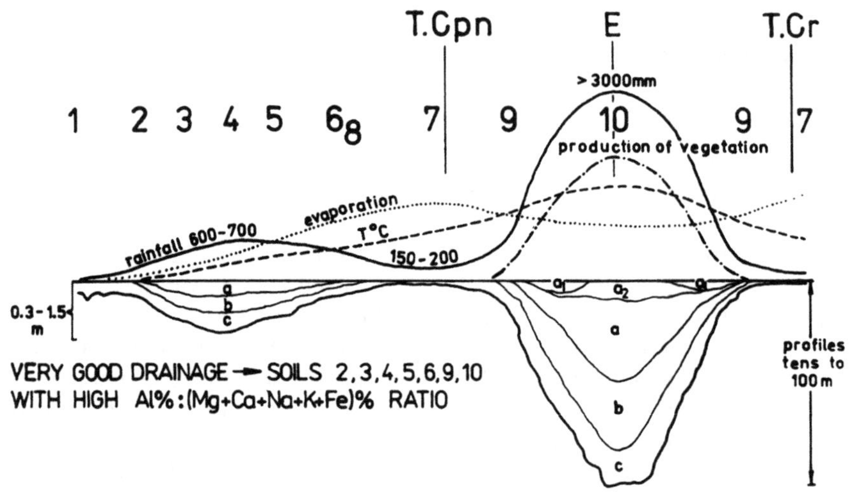


**Fig. 4.** Generalized scheme of the zonal distribution of the continental weathering crust depending on climatic zones in tectonically inactive regions. *Above* Relatively permeable rocks or those rich in minerals with dense cleavage cracks according to Strakhov (1962). *Below* Rocks of limited permeability and rich in bases, weathered to vertisols. The numbers 1 to 11 denote soil types (see Fig. 2). Vertical sequence in horizons: *a* rich in kaolinite; $a_1$ laterites; $a_2$ ochres; *b* rich in three-sheet clay minerals; *b'* vertisols, rich in swelling three-sheet clay minerals; *c* rich in primary relatively unstable constituents

The admixtures of primary minerals of low stability, such as feldspars, calcite, dolomite, and amphibole, to the SM reduce its overall $Ch_M$ value. This is indicative of rivers draining hot or cold arid basins, or regions of high relief or regions with diminishing vegetation cover. Pyrite, detected in very low amounts in the fraction below 2 μm separated from river sediments as well as in SM, occurred in the Elbe, Weser, and Ems (Koch et al. 1982) and in the Changjiang River (Emeis 1985). It probably formed diagenetically in the muds of those rivers. Similarly, gypsum and other sulphates form authigenically in recent river muds (Konta unpubl.).

## 4 Conclusion

The suspended matter of the world's major rivers is representative of the fine-grained (below 4 μm) topsoil in their respective basins. Its mineral components are determined by the climatic variables, i.e., by the interaction between atmosphere, hydrosphere, and lithosphere. Clay minerals are the most important product of weathering, they therefore are also the most common sediment being deposited. It is consequently of basic importance to understand the close relation between authigenic minerals of soils, suspended matter in rivers, and lutites which prevail among sediments.

According to the above discussion one should expect that the SM of relatively small rivers, which belong to only one climatic zone, is even more suitable to define the interaction of climate and soils than the inorganic constituents of the SM of larger rivers which intersect several climatic zones (Ernst and Konta 1987). If this assumption could be confirmed, then we would have another proof that the macromilieu is the most important factor in determining the composition of the fine material of topsoils.

Only those rivers, which drain rocks including older sediments without a developed soil mantle, may contain in their SM considerable portions of primary minerals. The eroded rock material accompanying soils in many regions of the world represents the main source of gravels, sands, and coarse silts in rivers. The finest part of this material, containing mostly fragments of disintegrated phyllosilicates, also contributes to the suspended matter of rivers. It is remarkable that this fraction, however, does not significantly affect chemical or mineral maturity of the SM as it is also subject to the influence of all the variables given in Eq. (3).

## References

Aagaard P, Helgeson HC (1983) Activity/composition relations among silicates and aqueous solutions: II. Chemical and thermodynamical consequences of ideal mixing of atoms on homological sites in montmorillonites, illites and mixed-layer clays. Clays Clay Min 31:207–217

Bridges EM (1978) World soils. University Press, Cambridge

Chukhrov FV, Gorshkov AI (1981) Iron and manganese oxide minerals in soils. Trans R Soc Edinburgh, Earth Sci 72:195–200

Emeis K (1985) Particulate suspended matter in major world rivers II: Results on the rivers Indus, Waikato, Nile, St. Lawrence, Yangtse, Parana, Orinoco, Caroni and Mackenzie. In: Transport of carbon and minerals in major world rivers, Pt. 3 (eds ET Degens et al.) Mitt Geol-Paläont Inst Univ Hamburg, SCOPE/UNEP Sonderbd 58:593–617

Emeis K, Stoffers P (1982) Particulate suspended matter in the major world rivers: EDAX analysis, scanning electronmicroscopy and X-ray diffraction study of filters. In: Transport of carbon and minerals in major world rivers. Pt. 1 (ed ET Degens) Mittl Geol-Paläont Inst Univ Hamburg, SCOPE/UNEP Sonderbd 52:529–554

Ernst A, Konta J (1987) Mineralassoziationen in den Schlicken einiger böhmischer Flüsse. Acta Univ Carol, Geologica, No 2, 117–144

Garrels RM (1960) Mineral equilibria at low temperature and pressure. Harper, New York

Ginsburg II, Rukavishnikova IA (1951) Mineraly drevnej kory vyvetrivaniya Urala. Izdat AN SSSR, Moskwa

Grim RE (1953) Clay mineralogy. McGraw-Hill, New York

Grim RE (1968) Clay mineralogy. 2nd edn. McGraw-Hill, New York

Irion G (1983) Clay mineralogy of the suspended load of the Amazon and of rivers in the Papua-New Guinea Mainland. In: Transport of carbon and minerals in major world rivers, Pt. 2 (ed ET Degens et al.) Mitt Geol-Paläont Inst Univ Hamburg, SCOPE/UNEP Sonderbd 55:483–504

Keller WD (1957) The principles of weathering. Revised edn. Lucas, Columbia

Koch R, Schoer J, Sioulas A (1982) Semiquantitative mineral analysis of grain size fractions $< 63\,\mu$, $< 2\,\mu$, and suspended matter from the rivers Elbe, Weser, Ems and the North Sea. In: Transport of carbon and minerals in major world rivers, Pt. 1 (ed ET Degens) Mitt Geol-Paläont Inst Univ Hamburg, SCOPE/UNEP Sonderbd 52:703–718

Konta J (1981) The products arisen from acid plagioclase and potassium feldspar during the kaolinization of Karlovy Vary granite. In: Eighth Conf Clay Mineralogy and Petrology in Teplice 1979 (ed J Konta) pp 173–180, Univ Karlova, Praha

Konta J (1983) Crystalline suspended particles in the Niger, Parana, Mackenzie and Waikato Rivers. In: Transport of carbon and minerals in major world rivers, Pt. 2 (eds ET Degens et al.) Mitt Geol-Paläont Inst Univ Hamburg, SCOPE/UNEP Sonderbd 55:505–523

Konta J (1984a) A new way to express the relative stability of silicates during weathering in aqueous environment. In: 9th Conf on Clay Mineralogy and Petrology in Zvolen 1982 (ed J Konta) pp 11–22, Univ Karlova, Praha

Konta J (1984b) Clay substance in the geological history of the Earth. Acta Univ Carol, Geologica, No 1: 19–54 (1st Geol-Petrological Colloq, Charles Univ and Univ of Hamburg, 1982, Climate in geological history (ed J Konta)

Konta J (1984c) Die Verwitterung der Bau- und Bildhauersteine an historischen Baudenkmälern. Acta Univ Carol, Geologica, No 2: 137–165 (2nd Colloq of Earth Sciences, Univ of Hamburg and Charles Univ, 1983, Air – water – life – lithosphere system (ed J Konta)

Konta J (1985) Mineralogy and chemical maturity of suspended matter in major rivers sampled under the SCOPE/UNEP project. In: Transport of carbon and minerals in major world rivers, Pt. 3 (eds ET Degens et al.) Mitt Geol-Paläont Inst Univ Hamburg, SCOPE/UNEP Sonderbd 58:569–592

Konta J (1987) Decay of building stones in historical monuments: first quantitative approach. Introductory lecture in Round Table on: The decay of building stone (eds E Galán, MA Vázquez) The Sixth Meeting of the European Clay Groups, Cádiz, Soc Españ de Arcillas, pp 7–9

Konta J (1988) Minerals in rivers. In: Transport of carbon and minerals in major world rivers, Pt. 5 (eds ET Degens et al.) Mitt Geol-Paläont Inst Univ Hamburg, SCOPE/UNEP Sonderbd Heft 66:341-365

Kronberg BI, Nesbitt HW, Lam WN (1986) Upper Pleistocene Amazon deep-sea fan muds reflect intense chemical weathering of their mountainous source lands. Chem Geol 54:283–294

Lippmann F (1981) Stability diagrams involving clay minerals. In: Eighth Conf on Clay Mineralogy and Petrology in Teplice 1979 (ed J Konta) pp 153–171, Univ Karlova, Praha

Loughnan FC (1969) Chemical weathering of the silicate minerals. Elsevier, New York

Martins O (1982) Geochemistry of the Niger River. In: Transport of carbon and minerals in major world rivers, Pt. 1 (ed ET Degens) Mitt Geol-Paläont Inst Univ Hamburg, SCOPE/UNEP Sonderbd 52:397–418

Meade RH, Dunne T, Richey JE, Santos Ude M, Salati E (1985) Storage and remobilization of suspended sediment in the lower Amazon River of Brazil. Science 228:488–490

Meybeck M (1984) Les fleuves et le cycle géochimique des éléments. Thèse Doc d'Etat 84-85, Univ P et M Curie, Paris 6

Milliman JD, Meade RH (1983) World-wide delivery of river sediments to the oceans. J Geol 91:1–21

Naidu AS, Mowatt TC, Somayajulu BLK, Sreeramachandra Rao K (1985) Characteristics of clay minerals in the bed loads of major rivers of India. In Transport of carbon and minerals in major world rivers, Pt. 3 (eds ET Degens et al.) Mitt Geol-Paläont Inst Univ Hamburg, SCOPE/UNEP Sonderbd 58:559–568

Niesel K (1979) Zur Verwitterung von Baustoffen in schwefel-oxid-haltiger Atmosphäre. Literaturdiskussion. Fortschr Min 57:68–124

Niesel K, Schimmelwitz P (1982) Zur quantitativen Kennzeichnung des Verwitterungsverhaltens von Naturwerksteinen anhand ihrer Gefügemerkmale. Forschungsbericht 86, Bundesanstalt Materialprüfung (BAM), Berlin

Owens JP, Stefansson K, Sirkin LA (1974) Chemical mineralogic and palynologic character of the Upper Wisconsinan-Lower Holocene fill in Chesapeake estuaries. J Sed Petrol 44:390–408

Peschel A (1977) Natursteine. VEB Deutscher Verlag Grundstoffindustrie, Leipzig

Petrov VP (1967) Osnovy ucheniya o drevnikh korakh vyvetrivaniya. Izdat Nedra, Moskwa

Post JL, Sloane RL (1971) The nature of clay soils from Mekong delta, An Giang province, South Vietnam. Clays Clay Min 19:21–29

Potter PE, Heling D, Shimp NF, van Wie W (1975) Clay mineralogy of modern alluvial muds of the Mississippi river Basin. Bull Ctr Rech Pau SNPA 9:353–389

de Quervain F (1967) Technische Gesteinskunde. 2. Auflage: von Moos A und de Quervain F. Birkhäuser, Basel

Reiche P (1943) Graphic representation of chemical weathering. J Sed Petrol 13:58–68

Strakhov NM (1962) Osnovy teorii litogeneza. Tom 1. Izdat AN, Moskwa

Tomadin L (1979) Clay mineralogy of recent sediments around the PO River Delta. Giornale Geol, Bologna (2) XLIII fasc I: 249–275

Tomadin L, Gallignani P, Landuzzi V, Oliveri F (1985) Fluvial pelitic supplies from the Apennines to the Adriatic Sea. I – The rivers of the Abruzzo Region. Proc Clays Clay Min, 1st Ital-Spanish Congr, Seiano di Vico Equense and Amalfi 1984 (ed A Pozzuoli), Min Petrogr Acta, Bologna 29-A: 277–286

Weir AH, Ormerod EC, Mansey IMI (1975) Clay mineralogy of sediments of the western Nile Delta. Clay Min 10:369–385

Winkler EM (1973) Stone: properties, durability in man's environment. Springer, Berlin Heidelberg New York

# Assessment of P, K, Ca Dynamics During Land-Use Changes

G. Esser, H. Lieth, and M. Clüsener Godt

## 1 Introduction

Carbon dioxide is one of the most important trace gases which has increased substantially since preindustrial times. The contribution of $CO_2$ to the greenhouse effect has been thoroughly investigated in recent years [3]. Models were developed [6,7,2] for the estimation of the carbon content in the various pools such as atmosphere, ocean, biosphere, and soils by balancing the fluxes of net primary productivity, depletion of litter production, litter and soil organic carbon, leaching from soils [10], absorption by the ocean, sedimentation, as well as the main anthropogenic activities such as clearing of forests, and fossil fuel burning.

Absorption of $CO_2$ by the ocean is partly influenced by the productivity of the marine primary producers, which in turn are limited by the available amounts of nutrients such as phosphorus, and others. The import of nutrients into marine systems may have increased due to anthropogenic processes such as industrial production and mineral leaching from cleared areas. The same is true for terrestrial systems where the regrowth of woody vegetation on cleared areas may be limited by the loss of nutrients during the agricultural and early fallow periods, especially on ferralic or sandy soils. Thus mineral dynamics during the clearing of forests may be expected to reduce carbon binding in terrestrial systems and enhance absorption of carbon by the sea. The availability of minerals is, therefore, an important driving force of the global carbon cycle.

Esser and Kohlmaier [10] tried to make an inventory of the terrestrial sources of some minerals, including an estimate of mineral leaching from cleared areas in the tropics. Their study was based on a limited number of data and, as far as minerals were concerned, was not yet regionalized. In this paper we present a new regional model approach for leaching of phosphorus, potassium and calcium which we consider to be most important for either terrestrial regrowth or ocean fertilization. We use the clearing and land use submodel of the Osnabrück Biosphere Model which is regionalized on a 2.5 degree grid together with data for biomass mineral content from approx. 1500 papers collected for our new data base "MINDAT".

## 2 Methods and Data

The Osnabrück Biosphere Model has already been described in several other papers [6,7]. Recently, the procedure for modelling land use changes and natural vegetation clearing was improved [8]. This land use submodel and scenario is used in this paper together with the productivity and biomass functions of the Osnabrück Biosphere Model to calculate the amount of biomass cut during land use changes. The cut biomass is decomposed by the decomposition submodel and the minerals of this biomass move via litter into the soil. From there they are leached by rain into ground water and finally into brooks and rivers.

The leaching rate depends on the amount of precipitation and the type of soil. Chemical reactions between elements and element species are not yet considered explicitly, neither are absorption and dispersion processes in the soil.

The crucial point for the depicted modeling procedure is the need for a representative data pool of mineral concentration in the various plant organs such as leaves, branches, stem-wood, roots, and others.

The model does not consider the uptake of minerals by vegetation regrowth, because land use changes in the model are net changes.

## 2.1 Carbon Balance Model

The aims of the carbon balance model are firstly to provide data for sites and change of land use areas and secondly to calculate the changes in vegetation biomass which occur along with such land use changes. The Osnabrück Biosphere Model was used by the authors for this purpose. Since model structure and equations have already been published several times [6,7] in easily available journals, we have restricted ourselves to describing only the main features relevant for our present study. The model uses an empirical approach on a global 2.5 degree grid of the terrestrial biosphere which results in 2433 grid elements for all the continents except Antarctica.

### *2.1.1 Biomass*

The calculation of biomass is done on a one-square meter basis as a mean for each integer grid element. The function which yields biomass as dependent variable considers *net primary productivity* and *mean stand age* as independent variables.

The *mean stand age* for each grid element is calculated as a weighted mean of all vegetation units present in the grid element. The area covered by natural vegetation was digitized from the Atlas for Biogeography of Schmithüsen [24], which distinguishes 172 vegetation units. The information for field crops was obtained from the World Atlas of Agriculture [15] and the FAO production yearbooks [13].

The *net primary productivity* is calculated by using the empirical functions of the Osnabrück Biosphere Model The independent variables are *mean annual temperature* and *average annual precipitation, soil fertility, yield* of field crops, and *carbon dioxide concentration* of the atmosphere. The values for each grid element are again weighted means of the input data which originate from Walter and Lieth [26], Müller [17], NCAR [18], a digitization of the Soil Map of the World [12], and the FAO Production Yearbooks [13]. The *carbon dioxide concentration* of the atmosphere was calculated by the model itself on a basis of a global balance of C-pools and fluxes including an ocean submodel (box diffusion ocean after Oeschger et al. 1975 [21]).

### *2.1.2 Litter Depletion*

It is assumed in the model that the amount of decomposed material depends on the pool of material and the depletion coefficient $k$. Woody and herbaceous material were calculated separately. Coefficient $k_h$ and $k_w$ are dependent on temperature and precipitation. The shared factors for woody and herbaceous were weighted means for Schmithüsen's 172 vegetation units for each grid element.

### *2.1.3 Land Use Changes*

The land use areas are calculated differently for the periods 1860–1980 and 1981–2400 [8].

For 1860–1980 the areas were calculated from the World Atlas of Agriculture [15] assuming its values to be valid for 1970. The other years' values were calculated using factors for each grid element and year from a global data base collected by Richards et al. [22] and the evaluation of 934 landsat images by the authors [11].

For 1981-2400 the probable development of agriculture was calculated by a scenario. On grid element level clearings occur due to logistic functions. The year of the turning point of each logistic function is calculated by the global clearing function. The sequence of the grid elements cleared is determined by an individual clearing probability for each grid element. Each value is the product of four separate probabilities considering the intensity of land use in neighbouring grid elements, the natural productivity, the soil fertility, and the historical land use changes.

## 2.2 Submodel for Mineral Leaching

The mineral leaching submodel MINER was developed for this paper. The submodel calculates the amount of leached material of the elements phosphorus, potassium, and calcium per square meter on a grid element basis. It was assumed that the change of the pool of each element in the soil is the balance of the related fluxes, with the output fluxes having a negative prefix:

$$\frac{dE}{dt} = F_z - F_l. \tag{1}$$

The variables correspond to:

| | | |
|---|---|---|
| $E$ | element pool in soil | $g \cdot m^{-2} \cdot a^{-1}$, |
| $F_z$ | input flux from litter | $g \cdot m^{-2} \cdot a^{-1}$, |
| $F_l$ | leached flux from soil | $g \cdot m^{-2} \cdot a^{-1}$. |

The flux $F_z$ is assumed to be proportional to the decomposed amount of litter either woody or herbaceous. The flux leaching is assumed to be proportional to the pool of the element $E$ in soil:

$$\frac{dE}{dt} = k_z \cdot LD(t) - k_l \cdot E(t) \tag{2}$$

$$= c_w \cdot LD_w(t) + c_h \cdot LD_h(t) - k_l \cdot E(t). \tag{3}$$

The variables, coefficients, and indices correspond to:

| | | |
|---|---|---|
| $w$ | index woody material, | |
| $h$ | index herbaceous material, | |
| $c$ | concentration of the element $[g \cdot 100g^{-1}]$ in the respective plant material, | |
| $LD$ | depleted amount of litter | $[g \cdot m^{-2} \cdot a^{-1}]$, |
| $k_l$ | leaching coefficient | $[a^{-1}]$. |

It was further assumed that the leaching coefficient $k_l$ is a function of either soil quality and precipitation. The lower the soil quality (i.e. sandy soils or well drained ferralic soils), and the higher the precipitation, the higher is the leached portion per time unit of a given element pool in the soil:

$$k_l = f(Soil, Pp). \tag{4}$$

To quantify Eq. (4) we used the matrix of soil fertility factors $f(o)$ of the Osnabrück Biosphere Model together with the precipitation matrix $Pp(m)$. The subscript $m$ refers to a grid element of the Osnabrück Biosphere Model ($m = 1, \ldots, 2433$). The subscript $o$ refers to one of the soil units of the Soil Map of the World [12] ($o = 1, \ldots, O(m)$). The matrix $f(o)$ is given in Table 1.

**Table 1.** Soil factors $f(o)$ of the Osnabrück Biosphere Model which characterize the fertility of the main soil units of the world. Names of soil units according to FAO-Unesco [12]

| Soil unit | $f(o)$ | Soil unit | $f(o)$ | Soil unit | $f(o)$ |
|---|---|---|---|---|---|
| Gleyic Acrisol | 0.87 | Lithosol-Yermosol | 1.14 | other Podzols | 0.55 |
| Humic Acrisol | 0.22 | Fluvisol | 0.49 | Calcaric Regosol | 1.61 |
| Orthic Acrisol | 0.70 | Eutric Fluvisol | 0.61 | Eutric Regosol | 1.14 |
| other Acrisol | 0.60 | other Fluvisol | 0.55 | Gelic Regosol | 0.91 |
| Dystric Cambisol | 0.94 | Haplic Kastanozem | 1.96 | other Regosol | 1.20 |
| Eutric Cambisol | 1.69 | Luvic Kastanozem | 1.61 | Orthic Solonetz | 0.59 |
| Humic Cambisol | 1.58 | other Kastanozem | 1.80 | Vitric Andosol | 1.65 |
| Gelic Cambisol | 0.76 | Albic Luvisol | 0.34 | Haplic Xerosol | 0.42 |
| Luvic Chernozem | 0.99 | Chromic Luvisol | 1.04 | Yermosol | 0.30 |
| Dystric Podzoluvisol | 0.83 | Ferric Luvisol | 1.65 | Haplic Yermosol | 0.66 |
| Xanthic Ferralsol | 0.55 | Gleyic Luvisol | 2.78 | Luvic Yermosol | 0.23 |
| Humic Gleysol | 0.47 | Orthic Luvisol | 0.85 | Takyric Yermosol | 0.09 |
| Gelic Gleysol | 0.57 | Dystric Histosol | 1.39 | Orthic Solonchak | 0.44 |
| other Gleysol | 0.50 | Humic Podzol | 0.56 | Takyric Solonchak | 0.03 |
| Lithosol | 0.52 | Orthic Podzol | 0.61 | other Solonchak | 0.20 |

With the soil fertility factors $f(o)$ and the digitized Soil Map of the World global information for using Eq. (4) was made possible.

For the parameterization of Eq. (4) we used the well documented fact [4,19] that during shifting cultivation at 2000 mm annual precipitation, approximately 90% of the initial mineral pool will be leached within 2 to 9 years after clearing, depending on soil quality ($0.5 \leq f(o) \leq 1.5$). Considering an earlier observation that leaching of organic compounds may linearly depend on precipitation [9] (see Fig. 1), we

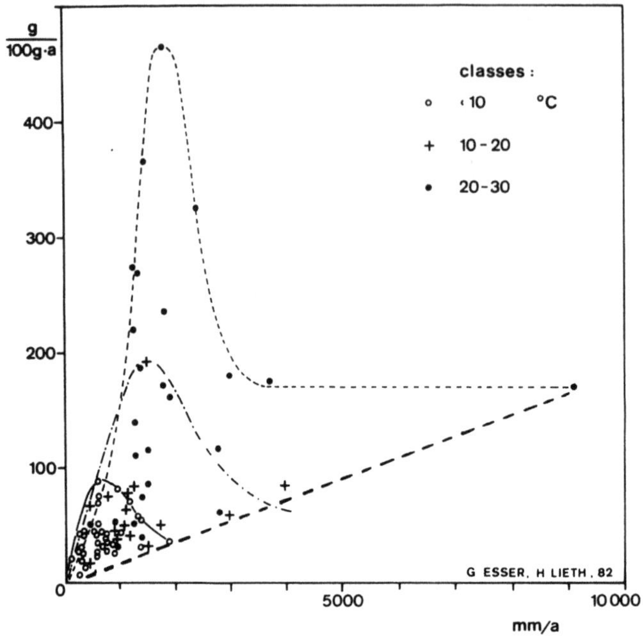


**Fig. 1.** The correlation of the amount of average annual precipitation and the depletion rates of fresh herbaceous litter for three classes of mean annual temperature. The lower dotted line indicates the lowest observed values regardless of the temperature class. This lower limit of values linearly depends on precipitation and may be considered as a leaching rate (figure from Esser et al. [9])

assume for a first approximation the same to be true for leaching of minerals. The parameterized form of Eq. (4) with these assumptions then is:

$$k_l(m) = 1.2 \cdot \exp[-1.4(F(m) - 0.5)] \cdot 0.0005\ Pp(m). \quad (5)$$

$F(m)$ is the weighted mean of all $f(o)$ present in a grid element:

$$F(m) = \frac{\sum_{o=1}^{o(m)} f(o) \cdot AS(m,o)}{\sum_{o=1}^{o(m)} AS(m,o)}. \quad (6)$$

$AS(m,o)$ is the share of area of soil unit $o$ in grid element $m$.

As one can see, Eqs. (5) and (6) can easily be modified as more information becomes available about absorption, dispersion and percolation of individual elements across the top soil. The Eqs. (3), (5), and (6) were implemented into the subroutine MINER of the Osnabrück Biosphere Model. The parameters $c_w$ and $c_h$ of Eq. (3) were derived from our data base "MINDAT".

In the model run, the equations were integrated by using the numerical routine DO2BAF of the NAG library which uses a Runge-Kutta-Merson algorithm.

### 2.3 The Mineral Data

The determination of the parameters $c_w$ and $c_h$ of Eq. (3) is based on the evaluation of approx. 8500 data for mineral content of plant material of the natural vegetation from all parts of the world. These values were extracted from approximately 1500 papers of our "MINDAT" file concerned with element concentrations in natural vegetation.

In a first global approach the element concentrations of certain plant parts (leaves, trunks, branches, bark, root) of natural vegetation were found to be significantly different ($p = 0.05$) independend of the vegetation formation, climate, and soil unit. In contrast the element concentrations of a given plant part in various vegetation formations were not significantly different. This result may suggest a renewed global mapping of mineral circulation first published 1967 by Rodin and Bazilevich [23].

In Fig. 2 the number of reported values for the concentration of the considered elements P, K, and Ca in plant parts are plotted against concentration intervals in the form of histograms. The maxima of the histograms rather than the means were used as globally valid element concentrations in plant material (see Table 2), since the concentration values obviously are not normally distributed.

The parameters $c_w$ and $c_h$ in the model Eq. (3) were calculated from the concentrations listed in Table 2. The parameter $c_w$ is the weighted mean of the values for wood, roots, bark, and branches in Table 2.

## 3 Results

Two model runs were carried out using two different scenario assumptions for the period after the year 1980, while the period 1860–1980 was calculated once. The scenarios 4 and 6 (Esser 1988 [8]) were used. The assumptions made in the scenarios are shown in Table 3.

As one can see from the assumption given in Table 3 is scenario 4 the more "economic" approach while scenario 6 is the "imaginative case" with 1/2 of each grid element remaining uncleared.

Each model run took 19 hours of CPU time of the mainframe TR440 of the University of Osnabrück[1].

### 3.1 Period 1860–1980

The globally integrated annual fluxes of estimated element export from cleared areas together with the carbon flux due to clearings are given in Table 4. The global pattern of element leaching rates for 1980 is given in Fig. 3.

[1] We are very much obliged to our colleagues in the Computer Center of the University of Osnabrück for their cooperation.

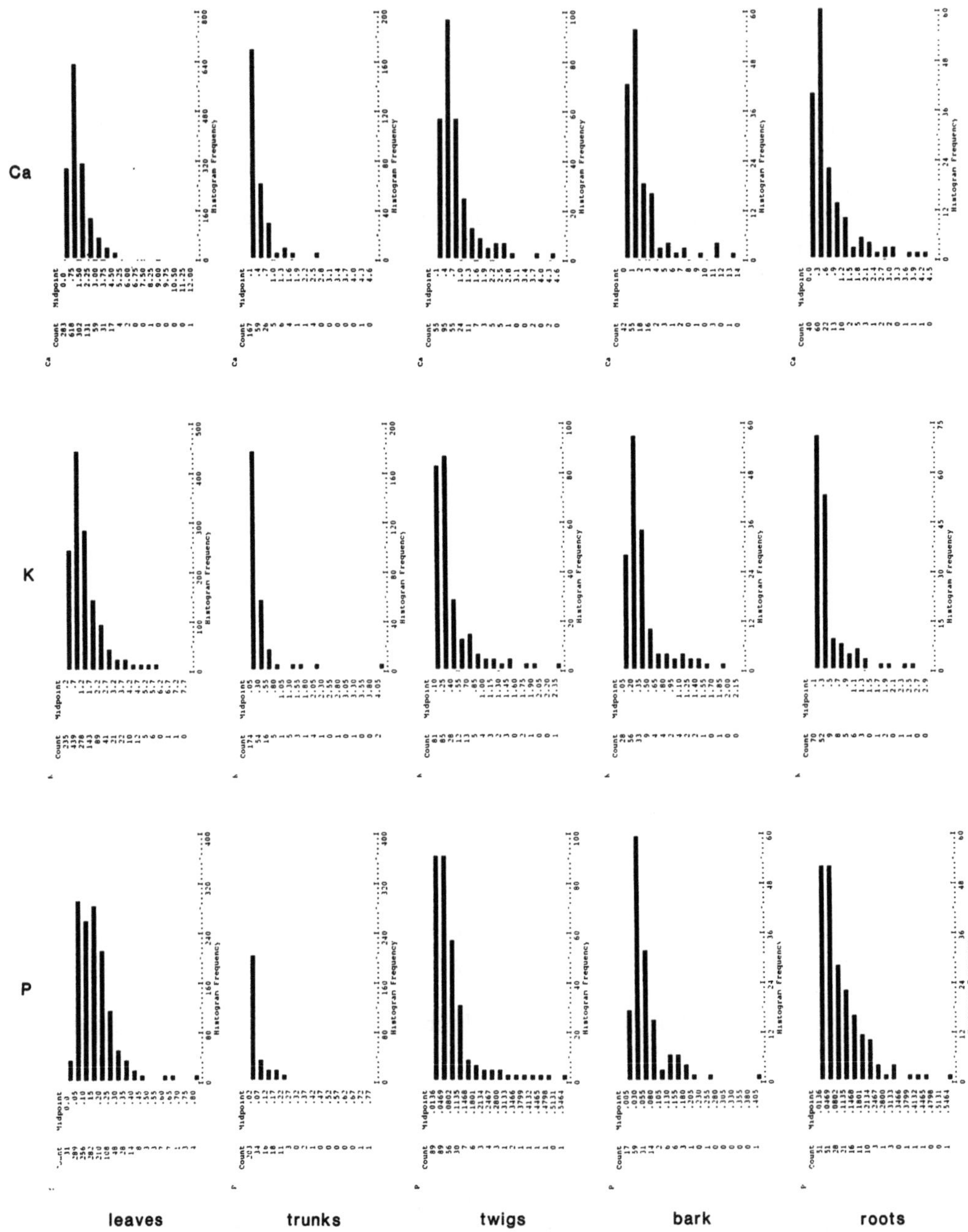


**Fig. 2.** Frequencies of concentration classes of calcium (above), potassium (center), and phosphorus (below) in various plant parts. Global data from our database "MINDAT". Values for each element are given in grams per 100 grams dry matter of the respective plant part, from Clüsener Godt 1989 [5]

**Table 2.** Concentrations of the elements P, K, and Ca in various plant materials derived from the histograms in Fig. 2. The data were assumed to be globally valid regardless of vegetation type. All values are in grams per 100 grams dry matter. The concentration ranges given below the values delimit the 80% intervals of the histogram data, from Clüsener Godt 1989 [5]

| Element | Leaves | Trunks | Branches | Bark | Root |
|---|---|---|---|---|---|
| Ca | 0.75<br>0.25–2.44 | 0.10<br>0.05–0.75 | 0.40<br>0.19–1.29 | 1.00<br>0.15–3.28 | 0.30<br>0.08–1.53 |
| K | 0.70<br>0.33–2.35 | 0.05<br>0.04–0.68 | 0.20<br>0.07–0.70 | 0.20<br>0.09–0.75 | 0.20<br>0.08–0.90 |
| P | 0.13<br>0.05–0.27 | 0.02<br><0.01–0.18 | 0.03<br>0.02–0.14 | 0.03<br>0.01–0.14 | 0.03<br>0.02–0.20 |

**Table 3.** Future land use conditions assumed for the two scenarios. $NPP_{lim}$ is the minimum productivity for economical use. $PART_{max}$ is the maximum share of agriculturally used areas in a grid element

| Scenario | $NPP_{lim}$ [$g \cdot m^{-2} \cdot a^{-1}$] | $PART_{max}$ |
|---|---|---|
| 4 | 500 | 1.0 |
| 6 | 0 | 0.5 |

**Table 4.** The globally integrated annual fluxes to groundwaters and rivers of the elements Ca, K, P due to land use changes in the period 1860 to 1980 as calculated by the subroutine MINER of the Osnabrück Biosphere Model. Values are given in ten-years steps and are in Tg per year of the respective element. The column clearings means annual global reduction of natural phytomass by human influence in Gt of carbon

| Year | Clearings $Gt \cdot a^{-1}$ | Calcium | Potassium $Tg \cdot a^{-1}$ | Phosphorus |
|---|---|---|---|---|
| 1860 | 0.65 | 1.05 | 0.90 | 0.17 |
| 1870 | 0.68 | 5.83 | 4.99 | 0.97 |
| 1880 | 0.72 | 6.60 | 5.66 | 1.09 |
| 1890 | 0.75 | 6.49 | 5.58 | 1.08 |
| 1900 | 0.70 | 7.25 | 6.22 | 1.20 |
| 1910 | 0.72 | 7.13 | 6.13 | 1.18 |
| 1920 | 0.90 | 7.04 | 6.05 | 1.17 |
| 1930 | 0.93 | 7.42 | 6.39 | 1.23 |
| 1940 | 0.87 | 8.95 | 7.69 | 1.48 |
| 1950 | 0.84 | 9.36 | 8.04 | 1.55 |
| 1960 | 0.80 | 8.65 | 7.47 | 1.45 |
| 1970 | 0.77 | 8.37 | 7.23 | 1.39 |
| 1980 | 0.71 | 7.61 | 6.58 | 1.27 |

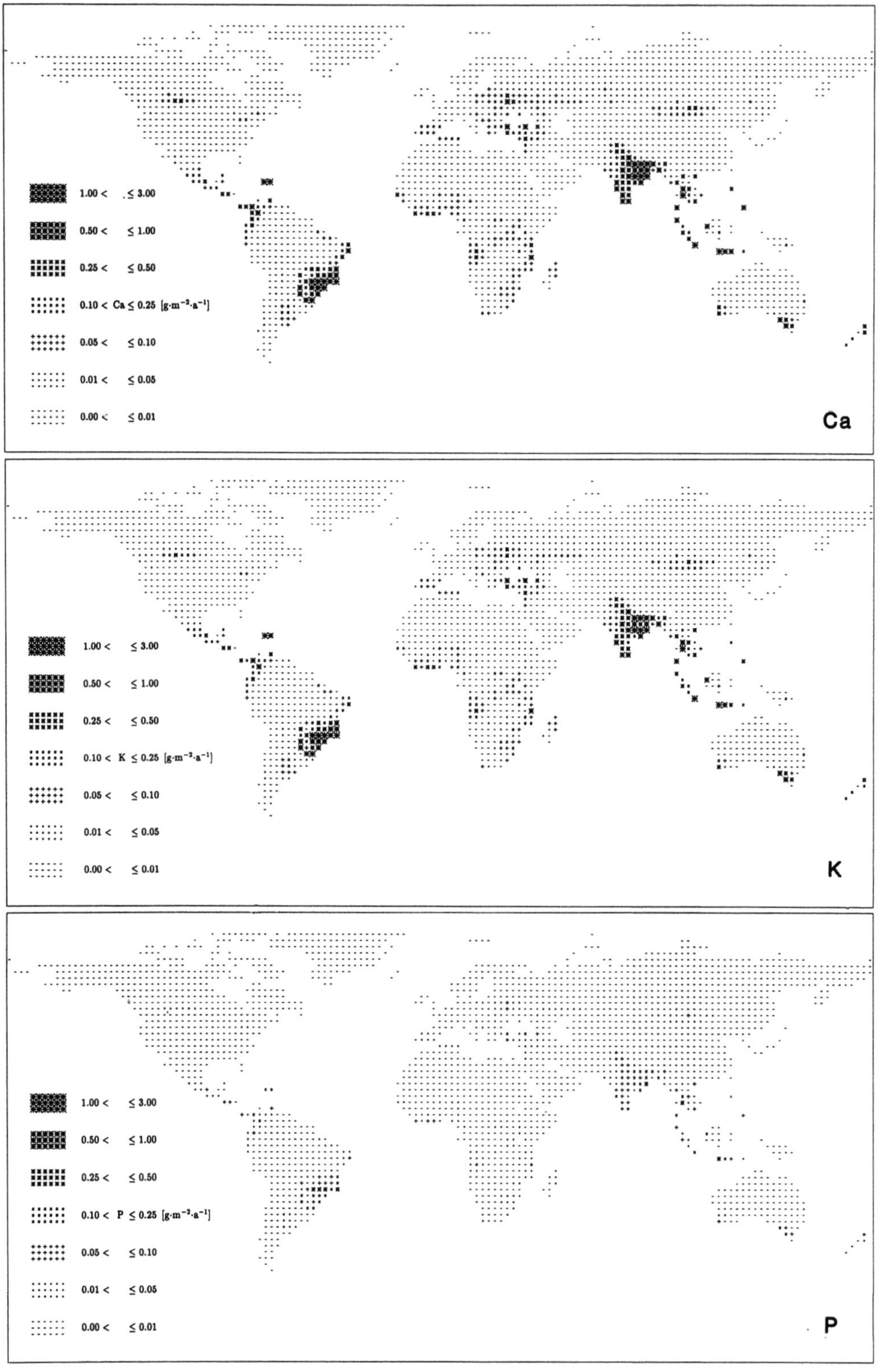
1.00 < ≤ 3.00
0.50 < ≤ 1.00
0.25 < ≤ 0.50
0.10 < Ca ≤ 0.25 [g·m⁻²·a⁻¹]
0.05 < ≤ 0.10
0.01 < ≤ 0.05
0.00 < ≤ 0.01
Ca
1.00 < ≤ 3.00
0.50 < ≤ 1.00
0.25 < ≤ 0.50
0.10 < K ≤ 0.25 [g·m⁻²·a⁻¹]
0.05 < ≤ 0.10
0.01 < ≤ 0.05
0.00 < ≤ 0.01
K
1.00 < ≤ 3.00
0.50 < ≤ 1.00
0.25 < ≤ 0.50
0.10 < P ≤ 0.25 [g·m⁻²·a⁻¹]
0.05 < ≤ 0.10
0.01 < ≤ 0.05
0.00 < ≤ 0.01
P

**Fig. 3.**

The clearings run through a maximum around 1930 A.D. when 0.93 Gt of carbon per year are turned over globally. After this time the value declines again to 0.71 Gt per year in 1980. The leached amounts of elements reflect this trend but the maximum appears as late as 1950. The mean time lag between the input into and the output from the mineral pool of soil is thus 20 years in the model. The model yields, for the year 1980, a total global export from soils due to clearings of 7.61 $Tg{\cdot}a^{-1}$ for calcium, 6.58 $Tg{\cdot}a^{-1}$ for potassium, and 1.27 $Tg{\cdot}a^{-1}$ for phosphorus.

The highest 1980 leaching rates for each of the three elements are calculated for regions, where a considerable amount of clearing takes place, and the annual precipitation is high. Leaching rates up to 3 $g{\cdot}m^{-2}{\cdot}a^{-1}$ for calcium are found in India (Bengal), in parts of South-East Asia, in South-East Brazil (São Paulo through Bahia), and parts of the Caribbean. Slightly lower rates appear in Central and NW South America, in Eastern Europa and W Asia, S of the Black Sea, in the Western Mediterranean, and S Australia and New Zealand. The leaching rates for the elements potassium and phosphorus are high in these zones as well, but the maximum values are lower. While potassium reaches 3 $g{\cdot}m^{-2}{\cdot}a^{-1}$ in restricted areas, phosphorus does not exceed 0.5 $g{\cdot}m^{-2}{\cdot}a^{-1}$ in these regions.

## 3.2 Period 1981-2400

The results are included in Table 5. The two scenarios differ widely in the progress of clearings and leaching rates of elements. In scenario 4 the clearings approach

**Table 5.** Prediction for the period 2000 to 2400 of the globally integrated annual fluxes to groundwaters and rivers for the elements Ca, K, P due to land use changes as calculated by the subroutine MINER of the Osnabrück Biosphere Model. In scenario 4 no clearings take place if the natural productivity of a region is less than 500 $g{\cdot}m^{-2}{\cdot}a^{-1}$ dry matter. In scenario 6 clearings will not exceed 50% of a given grid element area. Values are in 50-years steps in Tg per year of the respective element. Clearings are in Gt of carbon per year

| | Scenario 4 | | | | Scenario 6 | | | |
|---|---|---|---|---|---|---|---|---|
| Year | Clearings | Ca | K | P | Clearings | Ca | K | P |
| | $Gt{\cdot}a^{-1}$ | $Tg{\cdot}a^{-1}$ | | | $Gt{\cdot}a^{-1}$ | $Tg{\cdot}a^{-1}$ | | |
| 2000 | 1.53 | 12.53 | 10.84 | 2.09 | 1.97 | 18.32 | 15.88 | 3.05 |
| 2050 | 2.57 | 24.32 | 21.01 | 4.05 | 2.50 | 25.21 | 21.78 | 4.19 |
| 2100 | 4.30 | 43.33 | 37.43 | 7.21 | 1.48 | 14.42 | 12.31 | 2.38 |
| 2150 | 3.91 | 36.59 | 31.22 | 6.05 | 0.27 | 3.26 | 2.83 | 0.54 |
| 2200 | 1.79 | 18.25 | 15.64 | 3.02 | 0.05 | 0.81 | 0.71 | 0.14 |
| 2250 | 0.43 | 4.60 | 3.97 | 0.77 | 0.01 | 0.21 | 0.19 | 0.04 |
| 2300 | 0.18 | 2.02 | 1.74 | 0.34 | 0.00 | 0.04 | 0.03 | 0.01 |
| 2350 | 0.06 | 0.78 | 0.66 | 0.13 | 0.00 | 0.01 | 0.01 | 0.00 |
| 2400 | 0.00 | 0.45 | 0.38 | 0.07 | 0.00 | 0.01 | 0.01 | 0.00 |

**Fig. 3.** Regional distribution on a 2.5° grid of the annual leaching rates due to land use changes for the elements calcium (above), potassium (center), and phosphorus (below) as calculated by the subroutine MINER of the Osnabrück Biosphere Model. The values are weighted means of each entire grid element and are given in [$g{\cdot}m^{-2}{\cdot}a^{-1}$]. The reference model year is 1980

4.3 Gt•a$^{-1}$ around 2100 while in scenario 6 the rate will not exceed the maximum of 2.5 Gt•a$^{-1}$ in 2050. That means that the "economic case" (cleared are sites having productivities above 500 g•m$^{-2}$•a$^{-1}$) is much more incisive than the "imaginative case" which conserves 50% of the natural vegetation in each grid element.

The leaching rate for calcium comes up to 43 Tg•a$^{-1}$ in scenario 4 around 2100. In contrast the maximum in scenario 6 is 25 Tg•a$^{-1}$ in 2050. The respective rates for potassium are 37 Tg•a$^{-1}$ and 22 Tg•a$^{-1}$ and for phosphorus 7 Tg•a$^{-1}$ and 4 Tg•a$^{-1}$. The leaching rates of 1980 are reached again before 2250 in scenario 4 and before 2150 in scenario 6.

## 4 Discussion

Three items are crucial for the reliability of the results:

- The quality of the carbon balance model which estimates the cleared amount of biomass.
- The mineral concentration values in the biomass.
- The representation of the leaching processes and their chemical interaction with soil components and climate.

These three points are presented in different states of validity as we explain below. Especially the lack of erosion and water percolation models compiled to our biosphere model restrict the validity of our model exercise. Nevertheless the results appear in the correct order of magnitude globally in many respects. We expect to refine our approach considerably in the near future.

### 4.1 The Carbon Balance Model

The validation of large models is always a critical point since all available data are necessarily used for the model construction. Validation may be achieved by comparison of model results with the current assumptions of the scientific community and, more effectively, by comparison of derived model results with environmental data which were not used for calibration. Both methods were employed with the Osnabrück Biosphere Model [6,7]. The comparison of the model predictions for atmospheric $CO_2$ with the Mauna Loa data [1] and the ice-core data supplied by Oeschger [20] showed excellent correspondence. The model predictions for global carbon pools and fluxes do not contradict the currently accepted estimates for the respective variable [7].

The flux "clearings" of Tables 4 and 5 may be compared with the recent estimate of Richards et al. [22] since only the areal data of them were used in the Osnabrück Biosphere Model. They calculate a global net carbon release from land use conversion of 21.5 Gt for the period 1860–1920 and 17.9 Gt for 1920–1978. If a linear development is assumed the annual clearing rates are 0.36 Gt•a$^{-1}$ for the first and 0.31 Gt•a$^{-1}$ for the second period. The authors calculate the additional releases from soils to 0.13 respective 0.26 $Gt{\cdot}a^{-1}$ so that the total annual releases are 0.49 and 0.57 $Gt{\cdot}a^{-1}$. In comparison the mean values for the respective periods calculated by the Osnabrück Biosphere Model are 0.73 and 0.82 $Gt{\cdot}a^{-1}$. The difference is caused partly from rather low C-content estimates used by Richards et al. while in the Osnabrück Biosphere Model the biomass of vegetation is calculated dynamically, and from the fact that the Osnabrück Biosphere Model considers the $CO_2$-fertilization effect.

Houghton et al. [14] calculated a net flux due to use of land of 1.8 $\pm$ 0.9 $Gt{\cdot}a^{-1}$ of carbon in 1980 but found this high flux inconsistent with most geochemical models of the carbon cycle. Siegenthaler and Oeschger [25] estimated the biospheric $CO_2$ emissions by means of deconvolution of concentration records from Siple Station ice-core and from Mauna Loa Observatory and found a net release of 0–0.9 $Gt{\cdot}a^{-1}$ as mean for the period 1959–1983 which is in good agreement with our model results.

In studies based on areal conversion estimates the possibility should be considered that the effected areas of natural vegetation which is converted to agricultural use annually may have been overestimated. This presumption is supported by the results of our evaluation of 934 Landsat images of South America which gave for the time investigated a lower result for areal conversion than commonly assumed [11].

### 4.2 Mineral Content in Biomass

The "MINDAT" data base considers most of the relevant publications dealing with mineral data in plants and soils. In the present state of evaluation it is not yet possible to define the parameters which are responsible for the mineral concentration in given plant parts. There are indications that the mineral content of plant parts is influenced, among others, by the mineral composition of the soil and by systematic relationships of the species. In this paper we only consider the share of woody and herbaceous biomass of the 172 vegetation units of the Osnabrück Biosphere Model to regionalize mineral concentrations since both fractions have significantly different contents. It still can not be satisfactory to have no explanation for the large variance in mineral content as represented in Fig. 2. One may overcome the problem by using models for a better understanding of the system in the future.

### 4.3 Leaching Process and Soil Chemistry

No soil chemistry is included in the model. Chemical reactions with soil components, adsorption on clay and humus fractions as well as precipitations occurring in situ need to be considered in the future since their effects differ on various soil types. The presently used Eq. (5) is an empirical approach to the problem. Violent reactions which may occur after biomass burning during shifting cultivation are not considered either. Therefore the model may overestimate the leaching rate rather than underestimate.

In a conservative estimate based on areal transfers and standing biomass data Esser and Kohlmaier [10] estimated the phosphorus export from tropical deforestation to be 1.3 Tg•a$^{-1}$. If one agrees that tropical deforestation is the main source of phosphorus from biomass and that extratropical deforestation may be compensated by reforestation in developed countries, the result in this paper of 1.27 Tg•a$^{-1}$ for global phosphorus export (Table 4) is in good agreement.

Kempe [16] collected available river transport data for a global estimate of long-term transport rates of nutrients and carbon in major world rivers. Due to his study the element ratios calcium : potassium : phosphorus in the two important streams influenced by clearings of forests in South America were 6.25 : 1 : 0.017 in the Amazon and 3.33 : 1 : 0.009 in the Paraná. In comparison the ratio for leached minerals in this paper is 1.11 : 1 : 0.167. The calcium content of river water is relatively higher and the phosphorus content relatively lower than suggested by the mineral fluxes from cleared biomass in our model. A possible reason may be the excess calcium import from calcareous soils in the Andes and phosphorus precipitation in soils and river sediment.

Rivers of the temperate and boreal zones which are less influenced by forest clearings in their watersheds have even higher ratios Ca : K : P probably due to higher Ca import from weathering material. The element ratios, also calculated from Kempe [16], in the Mississippi (New Orleans) were 12.99 : 1 : 0.069, in the St. Lawrence were 25.64 : 1 : 0.015, and in the Yukon were 21.74 : 1 : 0.002.

## References


1. Bacastow RB, Keeling CD (1981) Atmospheric carbon dioxide concentration and the observed airborne fraction. In: Bolin (ed) Carbon Cycle Modelling, SCOPE 16:103–112
2. Bolin B (1986) How much $CO_2$ will remain in the atmosphere? In: Bolin, Warrick, Döös, Jäger (eds) The greenhouse effect climatic change and ecosystems. SCOPE 29:93–155
3. Bolin B, Jäger J, Döös BR (1986) The greenhouse effect, climatic change, and ecosystems – a synthesis of present knowledge. In: Bolin, Warrick, Döös, Jäger (eds) The greenhouse effect climatic change and ecosystems. SCOPE 29:1–32
4. Bormann FH, Likens GE, Siccama TG, Pierce RS, Eaton JS (1974) The export of nutrients and recovery of stable conditions following deforestation at Hubbard Brook. Ecol Monogr 44:255–277
5. Clüsener Godt M (1989) Statistical analyses to the relation of nutrients in plant and soil in natural stands at a global scale. Dissertationes Botanicae 135. Gbr. Borntraeger, Stuttgart (in German with English Summary)
6. Esser G (1986) The carbon budget of the biosphere – structure and preliminary results of the Osnabrück Biosphere Model (in German with extended English summary). Veröff Naturf Ges Emden von 1814 vol 7 160 pages and 27 figures
7. Esser G (1987) Sensitivity of global carbon pools and fluxes to human and potential climatic impacts. Tellus 39B:245–260
8. Esser G (1989) Global land use changes from 1860 to 1980 and future projections to 2500. Ecological Modelling 44:307–316
9. Esser G, Kohlmaier GH (1988) Modelling terrestrial sources of nitrogen, phosphorus, sulfur, and organic carbon to rivers. In: Degens, Kempe, Richey (eds) Biogeochemistry of major world rivers. SCOPE/UNEP
10. Esser G, Lieth H (1986) Evaluation of climate relevant land surface characteristics from remote sensing. Proc. ISLCP Conference Rome, ESA Publication SP-248:205–211
11. Esser G, Aselmann I, Lieth H (1982) Modelling the Carbon Reservoir in the System Compartment "Litter". Mitt Geolog-Paläontolog Inst Univ Hamburg. SCOPE/UNEP Sonderb Heft 52:39–58

12. FAO-Unesco (1974 ff.) Soil Map of the World 1:5,000,000. Vol I-X, Unesco Paris
13. FAO (1980 ff.) Production Yearbooks. FAO Statistics Series No. 28 ff. Food and Agricultural Organization Rome
14. Houghton RA, Boone RD, Fruci JR, Hobbie JE, Melillo JM, Palm CA, Peterson BJ, Shaver GR, Woodwell GM, Moore B, Skole DL (1987) The flux of carbon form terrestrial ecosystems to the atmosphere in 1980 due to changes in land use: geographic distribution of the global flux. Tellus 39B:122–139
15. Instituto Geographico de Agostini (1969, 1971, 1973) World Atlas of Agriculture. Novara (Italy)
16. Kempe S (1982) Long-term records of $CO_2$ pressure fluctuations in fresh waters. In: Degens ET (ed) Transport of carbon and minerals in major world rivers, Mitt Geol-Paläont Inst Univ Hamburg SCOPE/UNEP Sonderb Heft 52:91–332
17. Müller MJ (1982) Selected climatic data for a global set of standard stations for vegetation science. In: Lieth H (ed) Tasks for vegetation science 5, Junk, The Hague
18. NCAR (1983) TD-9645 tape documentation of National Center for Atmospheric Research. Boulder, Colorado
19. Nye PH, Greenland DY (1960) The soil under shifting cultivation. Commonw Agric Bur Tech Commun 51, Harpender U.K.
20. Oeschger H (1986) Investigation of climatic and environmental systems by analysis of ice cores (in German). Ann Meteorolog 23:1–3
21. Oeschger H, Siegenthaler U, Schotterer U, Gugelmann A (1975) A box diffusion model to study the carbon dioxide exchange in nature. Tellus 27:168–192
22. Richards JF, Olson JS, Rotty RM (1983) Development of a data base for carbon dioxide releases resulting from conversion of land to agricultural uses. Inst Energy Anal Oak Ridge Assoc Univ ORAU/IEA-82-10(M) ORNL/TM-8801
23. Rodin LE, Bazilevich NI (1967) Production and mineral cycling in terrestrial vegetation. Oliver and Boyd, London
24. Schmithüsen J (1976) Atlas for Biogeography. Meyers großer Physischer Weltatlas 3, Bibl Inst, Mannheim
25. Siegenthaler U, Oeschger H (1987) Biospheric $CO_2$ emissions during the past 200 years reconstructed by deconvolution of ice core data. Tellus 39B:140–154
26. Walter H, Lieth H (1960 ff.) World Atlas of Climate Diagrams. Fischer, Jena

# Nutrients in the Turbidity Zone of the Elbe River

U.H. Brockmann and B. Onken

## 1 Introduction

Three rivers drain directly into the German Bight: the Elbe, Weser, and the Ems of which the dominating one is the Elbe. It is 1143 km long and drains an area of 146500 $km^2$, and has a mean fresh water discharge of 1150 $m^3 s^{-1}$. The tidal estuary of the Elbe is about 150 kilometers long with tidal currents of up to 4 knots. The river exhibits seasonal variations in its water discharge with high flow rates during spring ($> 1000$ $m^3 s^{-1}$) and low flow rates ($< 500$ $m^3 s^{-1}$) during summer.

The Elbe river represents a significant source of nutrients, in the German Bight (Anonymus 1980; Gerlach 1987); it carries a nitrogen load of about 250,000 t per year, dominated by nitrate. The nutrient concentrations are dependent on the variable freshwater flux (Brockmann and Eberlein 1986). Roughly, nitrate shows significant negative linear correlation with salinity, giving the impression that nitrate is a conservative parameter in the river. But this appears not to be the case in the upper estuary, as seen from monitoring measurements of the ARGE Elbe (1978–1987) with monthly transects: During winter – until April – high ammonium concentrations reach the river mouth. With increasing temperatures during May, nitrification is enhanced in the estuary, and ammonium is oxidized via nitrite to nitrate. As a consequence of intensified nitrification, the location of ammonium decomposition moves in an upstream direction resulting in low ammonium concentrations at the river mouth. This area of maximum nitrification is marked by an ammonium decrease, a nitrite maximum, and only sometimes by a small, local nitrate increase within the generally high nitrate level. With decreasing temperatures during autumn the turnover of ammonium is reduced and consequently direct discharge of ammonium becomes a significant contributor to the nitrogen export of the Elbe river.

The significance of changes of nutrient concentrations in the vertically mainly well mixed estuary is difficult to estimate by single measurements along transects because gradients in the frequently changing tidal currents will dominate over many biological-chemical processes. Variable gradients will be controlled by discontinuous mixing.

In 1986 within an interdisciplinary research cooperation (Sonderforschungsbereich 327 "Tide-Elbe") nutrient gradients were investigated in detail in the turbidity zone of the Elbe river (Brockmann 1986; 1987). Besides sampling of longitudinal profiles especially time series at anchor stations in the turbidity zone were measured. Generally the turbidity zone in the salinity gradient is assumed to be an important site of chemical and biological processes (Schubel and Kennedy 1984).

## 2 Methods

Transects between the city of Hamburg and the German Bight (54°00′N, 08°00′E) (Fig. 1) were sampled from the R.V. VALDIVIA during spring, summer, fall and winter of 1986 and 1987 (Table 1). Along these sections gradients and qualities of the interacting endmembers in the river and in the sea were estimated. Mainly time series were measured at anchor stations in the turbidity zone of the Elbe River at km 695 (Fig. 1) three to four days with frequencies up to 30 min in order to discriminate the effects of tidal-controlled current speed and salinity gradients.

Temperature and salinity were measured with a probe-system (ME). Water samples were taken at the surface near the bottom and in between (mostly 6 m depth) with Niskin-bottles, and were filtered through glassfiber filters (GF/C, Whatman) at a constant low pressure (0.2 atm). Turbidity was measured with a Nephelometer (Turner) and suspended matter calculated after calibration. Nutrients were analysed by an Auto Analyzer using adapted Technicon-Methods (Eberlein et al. 1983) after dilution (1:5 to 1:25) within the Analyzer-System. Chlorinity was estimated by $AgCl_3$ – titration.

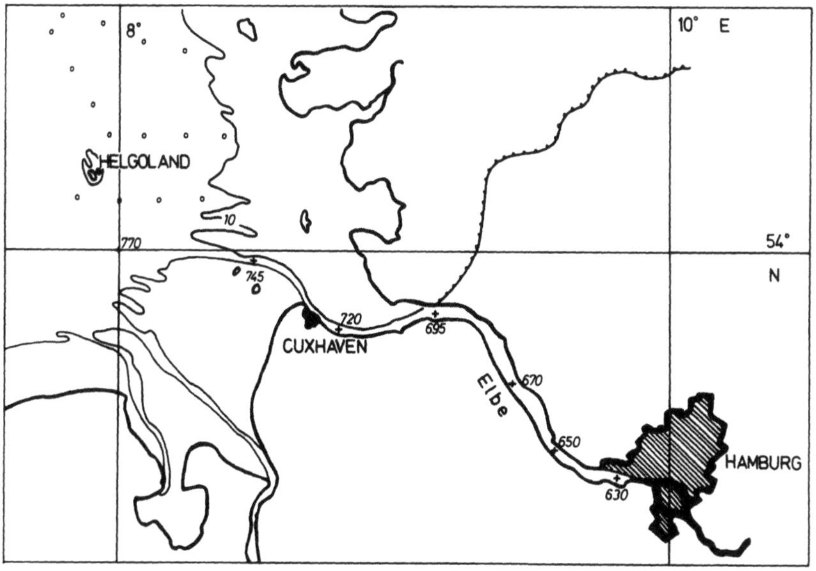


**Fig. 1.** Map of the Elbe estuary from the city of hamburg to the German Bight. Sampling stations in the river are indicated at positions of stream – *km* by *crosses*, in the German Bight by circles

**Table 1.** Cruises of R.V. VALDIVIA in the Elbe river

| Cruise | Date |
|---|---|
| VA 37 | 20/01 – 26/01/1986 |
| VA 47 | 04/07 – 09/07/1986 |
| VA 56 | 26/04 – 02/05/1987 |
| VA 64 | 26/09 – 02/10/1987 |

## 3 Results and Discussion

During all measurements, the turbidity zone was normally located near the river mouth, at the beginning of the salinity gradient (Fig. 2A-D). In the turbidity zone at a depth of about 6 m maxima of 70–240 mg/l of suspended material were found. The position of the turbidity zone is mainly controlled in addition to tidal influences by the rates of freshwater discharge, and by wind forces and directions in the German Bight. For instance at low discharge periods or strong westerly winds the turbidity zone will move upstream. Since turbidity is mainly caused by tidal action in the salinity gradient (weak salt wedge), extreme variations in the concentration of suspended matter occur near the bottom with current speed.

The nitrate concentrations reaching the river mouth were always high between 250 and 400 μmol/l (Fig. 2E-H) and mainly decreased below km 695 (position of anchor stations) due to dilution with sea water. Ammonium reached in the estuary during winter maximum concentrations above 200 μmol/l and decreased during summer/autumn to values below 10 μmol/l at the river mouth.

### *Winter*

During January (Table 1) at water temperatures between 1.5–2.5 °C nitrification in the Elbe was rather low. As a consequence, besides nitrate (270 μmol/l) also high ammonium concentrations (150 μmol/l) were found at the river mouth (Fig. 2E) (Brockmann 1986). During this time of the year high ammonia concentrations were observed along the whole continental coast of the North Sea (Brockman et al. 1989). That the nitrification of this ammonia load would occur in the marine environment, was indicated already in the river mouth by a nitrite maximum of 0.9 μmol/l in the haline water masses. Since the abundance and activity of different species of nitrifying bacteria changes in the salinity gradient, the nitrification is performed by species in the coastal water, other than those in the freshwater environment (Bock, pers. comm. 1988).

### *Spring*

In April ammonium concentrations were still high at the river mouth (Fig. 2F). Generally, there was a decrease along the transect within the salinity gradient, due to dilution with marine water. A distinct nitrite maximum of 11 μmol/l and a relatively small nitrate maximum also occurred at 0.3 g/l $Cl^-$ indicating nitrification.

Frequent nutrient measurements at the anchor station within the salinity gradient and the turbidity zone showed that ammonium was converted to nitrate causing a local increase in the incoming tidal water (Fig. 3). Nitrate, plotted against chlorinity, increased exponentially by about 20 μmol/l from 390 to 410 μmol/l (Fig. 4). The ammonium concentrations, on the other hand, decreased exponentially by the same amount (60 to 40 μmol/l): The effect of dilution with seawater can be neglected because it will effect both nutrients by a parallel shift of haline

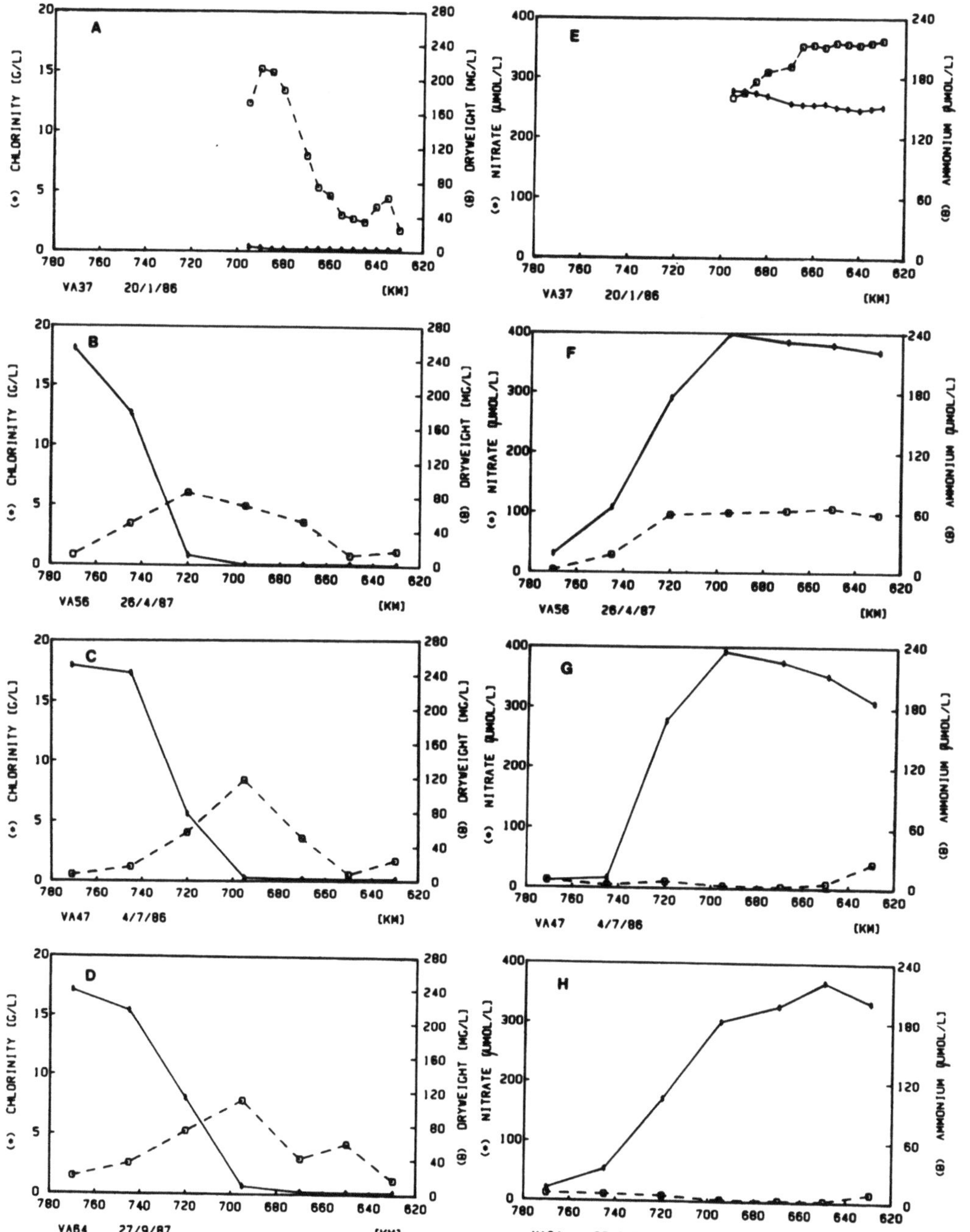


**Fig. 2A-E.** Transections of chlorinity, suspended material (**A** – **D**), nitrate and ammonium concentrations (**E** – **H**) in the Elbe estuary (0 and 6 m depths) from the city of Hamburg (km 620) to the German Bight (see Fig. 1)

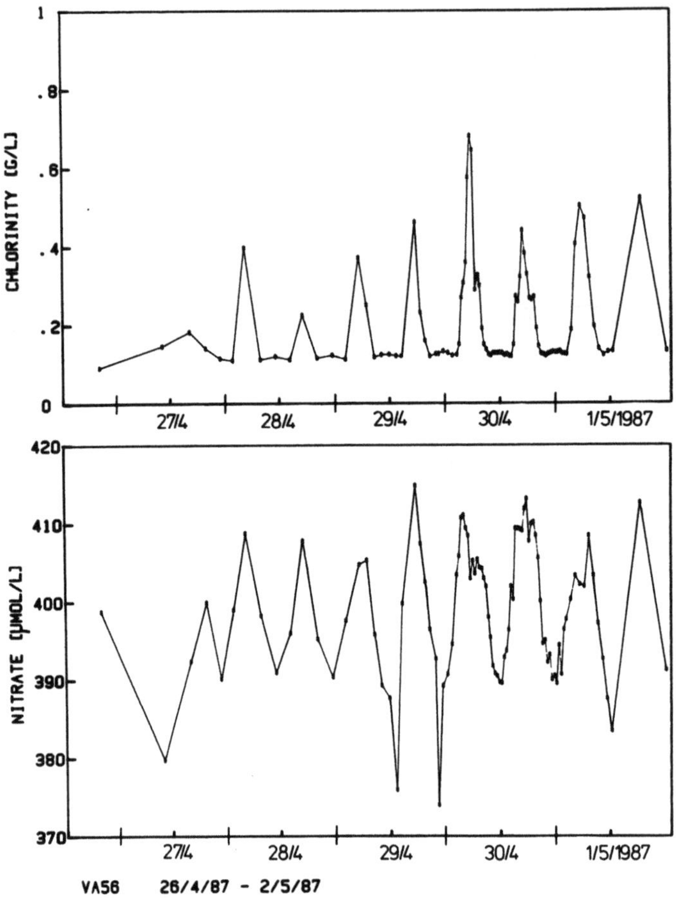


**Fig. 3.** Chlorinity and nitrate concentrations at an anchor station in the turbidity zone of the Elbe river (km 695; see Fig. 1) during April/May 1987 (VA 56)

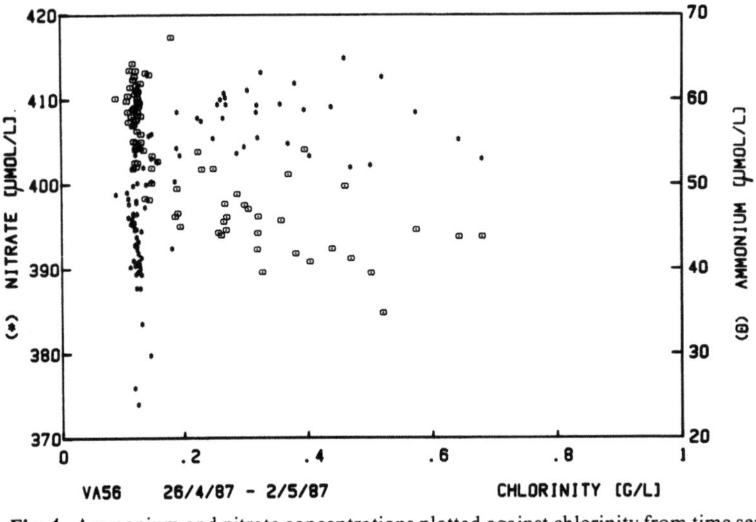


**Fig. 4.** Ammonium and nitrate concentrations plotted against chlorinity from time series at an anchor station (km 695), during April/May 1987 (compare Fig. 3)

endmember concentrations: The real nitrate increase will be higher and the ammonium decrease lower with the same differences.

At temperatures between 12 and 14 °C, the nitrification process will occur already at faster rates in the Elbe (ARGE 1978–1987). The time series measurements at the anchor station indicate that the annual process of estuarine nitrification starts in the turbidity zone. With its high load of particles this zone provides a maximum of reactive sites which move differently in the nutrient solution enhancing microbial activity (Gocke 1977).

Relative maxima of silicate and phosphate at the same position show that besides nitrification, decomposition occurred in the turbidity zone. A possible parallel release of ammonium was not detected, but could have been compensated by losses via nitrification. All nutrients, except ammonium, were higher in the marine influenced water masses than in the limnic water. But at the fixed station lack of significant correlations of phosphate and silicate e.g. with chlorinity were not significant, indicated the independence of remineralization from the transformation of nutrients by nitrification.

*Summer*

During July the ammonium concentrations were low ($< 3$ μmol/l) downstream ($> 670$ km) due to the advanced nitrification upstream. But a small secondary ammonium maximum was found within both transects at the outer front of the turbidity zone (Fig. 2G and 5). At the same position a silicate maximum was detected. Any parallel phosphate release was probably compensated by uptake processes.

This increase in nutrients downstream was confirmed by the time-series measurements at the anchor station during 4 days, shown here for ammonium (Fig. 6). The shift of tidal minima or maxima of ammonium peak concentrations at the anchor station further revealed either changes of endmembers, or shift of reactive position or reaction progress themselves. Since there was a parallel increase in chlorinity and ammonium, the positive trend of ammonium can have only been caused by changing endmembers. Within the haline water of the incoming tide, in addition to ammonium, also silicate, nitrite and phosphate increased, all of them exhibited high-significant correlations with chlorinity. These findings are interpreted to result from common release process following (sediment) remineralization.

During summer with water temperatures above 20 °C, decomposition processes proceed in the river and coastal zone. Therefore the release of remineralization products from the sediment by leaching of interstitial water must be taken into consideration. This is further suggested by the increase of silicate parallel to the ammonium increase, silicate remineralizations proceed slowly, and at shallow sites, like rivers, mainly in sediment.

Ammonium and nitrate concentrations, plotted against chlorinity, showed within the gross linear correlations either a negative (ammonium) or a positive (nitrate) deviation at a chlorinity of 2 g/l (Fig. 7). This indicates that : (1) nitrification of released ammonia occurred immediately and (2) the site of maximum nitrification in frequently oscillating water masses near the anchor station

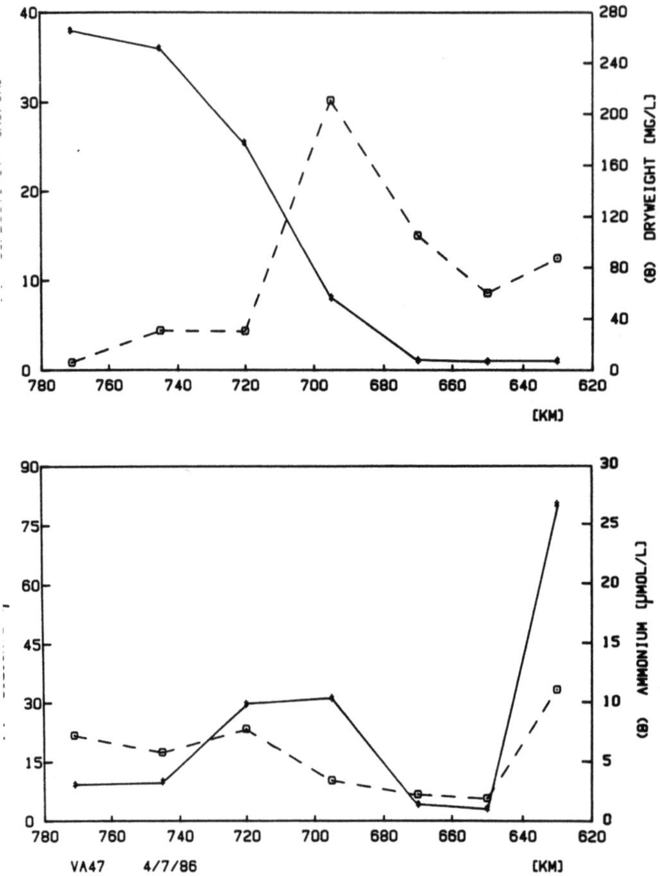


**Fig. 5.** Conductivity, turbidity, silicate and ammonium concentrations along an Elbe estuary-transect, July 4–5th, 1986 (surface) (see Fig. 1)

occurred in the turbidity maximum (Fig. 2C). The deviations from the linear trend were for nitrate about 8 µmol/l, but for ammonia only 0.5 µmol/l. This discrepancy can only be explained by a local source of ammonium which could not be detected. However, positive deviations from linear correlations with chlorinity, which were for silicate 10 µmol/l and phosphate 0.1 µmol/l at a chlorinity of 2 g/l indicate the existence of such a local source of remineralization products including ammonium.

Parallel to the chlorinity increase at flood (Fig. 6), from 2 to 4 g/l within 3 days, the ammonium increase from 3.5 to 5.5 µmol/l indicates that the main source of ammonium was more downstream, in the coastal water.

This interpretation was confirmed by simultaneous measurements in the German Bight. Here increasing ammonium concentrations were measured towards the northwest, with maximum values above 10 µmol/l (Fig. 8). This ammonium patch was not simultaneously coupled with the Elbe river discharge, because the positions of ammonium and nitrate nutriclines deviated significantly. High nitrate concentrations are characteristic for the Elbe river plume throughout the year.

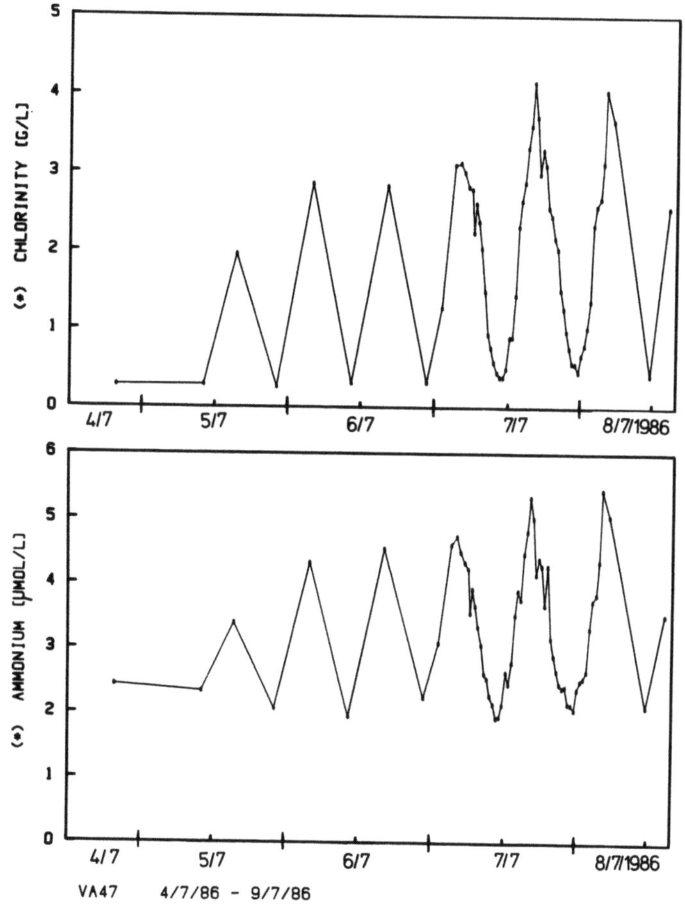


**Fig. 6.** Time series of chlorinity and ammonium concentrations at an anchor station (km 695) in the Elbe estuary (July 1986)

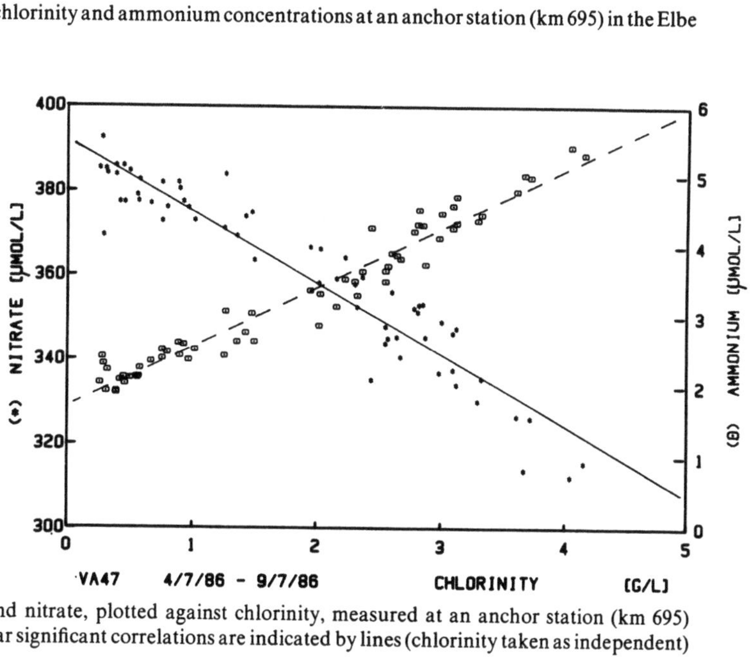


**Fig. 7.** Ammonium and nitrate, plotted against chlorinity, measured at an anchor station (km 695) during July 1986. Linear significant correlations are indicated by lines (chlorinity taken as independent)

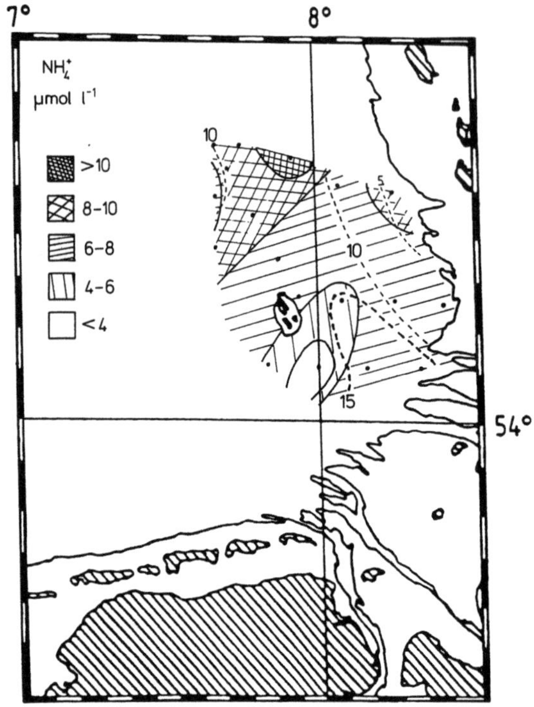


**Fig. 8.** Ammonium and nitrate concentrations in the German Bight (7–9 July, 1986). Nitrate isolines are shown as *dashed lines, numbers* = µmol/l. Sampling positions are indicated by *dots*.

*Autumn*

During autumn (September/October) the water temperature had dropped below 15.5 °C in the Elbe river mouth, and continued to decrease during the measurements. Along the transect, ammonium concentrations were still below 2 µmol/l downstream in the freshwater (Fig. 2H) and increased in the seawater gradient to values above 6 µmol/l at the beginning of the cruise (see Table 1). This gradient had disappeared at the end of the cruise, but for both transects in the salinity gradient still increasing nitrite concentrations (up to 2 µmol/l) were detected. This as well as the coupled concentration maxima of ammonium (mean 2 µmol/l) and nitrite (0.4 µmol/l) at the anchor station in the more haline water indicated again a local, but for a longer period lasting remineralization process downstream with following nitrification.

During measurements at the anchor station in addition to a decrease in salinity during flood, also an inversion of nutrient dominance in the different water regimes could be followed: Silicate (55 µmol/l) and phosphate (4.3 µmol/l) first occurred with maximum values in the haline water. Subsequently, for a few tidal cycles intermediate secondary maxima in the freshwater (ebb) appeared which then increased and finally dominated (Fig. 9). Comparison of transects at the beginning and at the end of the cruise (Brockmann 1987) show that these shifts were not caused by strong concentration changes of the endmembers far upstream or in the German Bight. The inversion of gradients were only a result of a small shift of concentrations in the middle of the estuary transect.

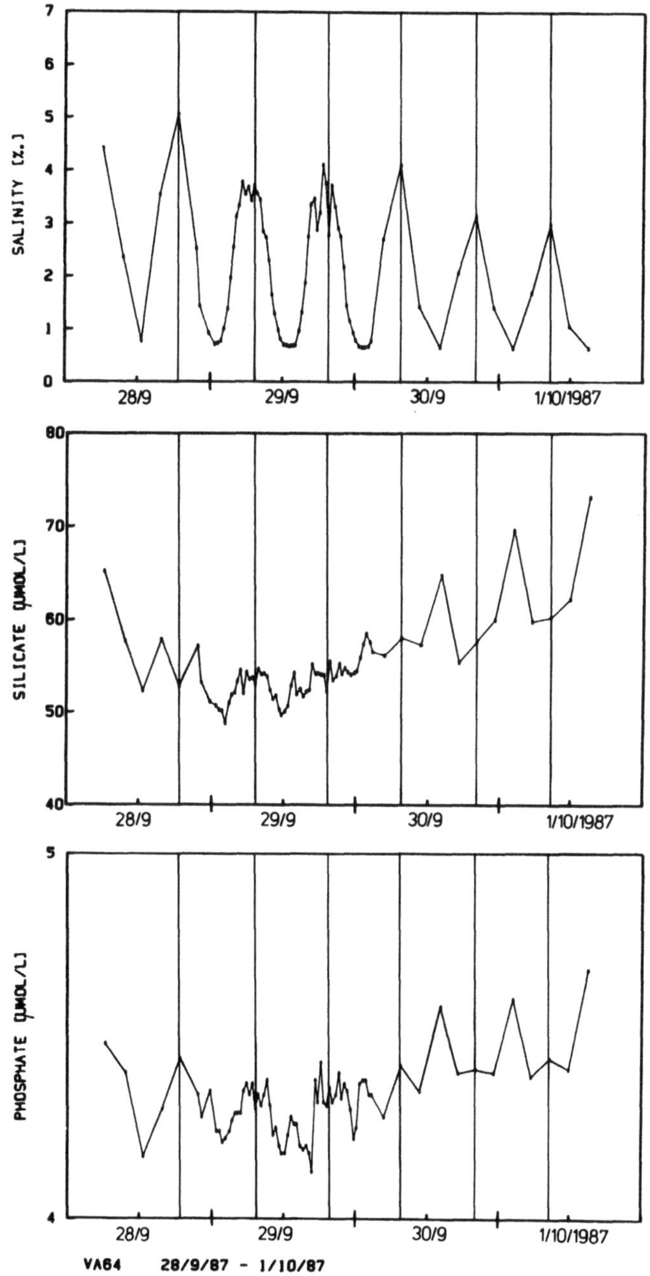


**Fig. 9.** Chlorinity, silicate and phosphate concentrations at an anchor station (km 695) in the Elbe estuary (September/October 1987)

Whereas nitrite maxima remained in the haline water mass, the ammonium maxima shifted to the freshwater domain. This indicates that the nitrification site did not change together with the site of dominating remineralization, probably due to the salinity adaption of nitrifying bacteria. An increase of nitrate concentrations could not be detected within the high level of discharge (310 μmol/l) at the river mouth.

# 4 Conclusions

Our findings show that the area of the turbidity zone is an important site for nutrient turnover which can influence the seasonal pattern of nutrient discharge to the sea significantly. These processes can only be detected by chance from transects with single stations over longer distances. Time series at anchor stations for longer periods are more suitable and indicated that estuarine nutrient turnover can be very persistent.

*Acknowledgements.* We thank for assistance and cooperation the SFB 327 group, especially Ilse Büns, Renate Lucht and Monika Schütt and the VALDIVIA crew. Fundings were provided by the Deutsche Forschungsgemeinschaft supporting the research project within the SFB 327 "Tide-Elbe Hamburg", and by the Hansestadt Hamburg.

# Literature

Anonymous (1980) Umweltprobleme der Nordsee. Sondergutachten Juni 1980: 503 pp

ARGE Elbe (1978-1987) Wassergütedaten der Elbe. Arbeitsgemeinschaft für die Reinhaltung der Elbe, Hamburg

Brockmann UH (1986) Cruise reports VA 37, 47, Hamburg, University

Brockmann UH (1987) Cruise reports VA 56, 64, Hamburg, University

Brockmann UH, Eberlein K (1986) River input of nutrients into the German Bight. In: Skreslet S (ed) The role of freshwater outflow in coastal marine ecosystems. NATO ASI Series G 7. Springer, Berlin Heidelberg New York Tokyo, pp 232-240

Brockmann UH, Laane R, Postma H (1989) Cycling of nutrient elements in the North Sea. Neth J Sea Res (in press)

Eberlein K, Brockmann UH, Hammer KD, Kattner K, Laake M (1983) Total dissolved carbohydrates in an enclosure experiment with unialgal Skeletonema costatum culture. Mar Ecol Prog Ser 14:45-58

Gerlach SA (1987) Pflanzennährstoffe und die Nordsee – ein Überblick. Seevögel, Z V Jordsand 8:49-62

Gocke K (1977) Comparison of methods for determining the turnover times of dissolved organic compounds. Mar Biol 42:131-141

Schubel JR, Kennedy VS (1984) The estuary as a filter: an introduction. Academic Press, London, pp 1-11

# Dispersal of Mahakam River Suspended Sediment in Makasar Strait, Indonesia

D. Eisma

## 1 Introduction

The Mahakam River is one of the largest rivers in Kalimantan, Indonesia. It flows from the Kapuas Mountains to the southeast and drains into Makasar Strait where it has built up a large fan-shaped delta of the same type as the deltas of the Niger and the Nile. In contrast to the large Chinese and Indian rivers, the outflow of the larger rivers in S.E. Asia has been little studied. Best known is the estuary of the Chao Phya, that flows into the Gulf of Thailand near Bangkok (NEDECO 1965). Since the east-Asian rivers, and particularly those draining the east-Asian islands, supply an important amount of suspended matter to the world ocean (Milliman and Meade 1983), the outflow of the Mahakam was studied when the opportunity to do so presented itself within the Indonesian-Dutch Snellius-II program that was carried out in 1984/85 in east-Indonesian waters. Shallow seismic profiling on the shelf off the river mouth was combined with sampling of suspended matter in the same area and in Makasar Strait. The location of the area of study and the sampling stations are shown in Fig. 1, the seismic profiling tracks and the sampling stations at the river mouth in Fig. 2A and C. This paper is concerned with the dispersal of the river-supplied sediment, with the fate of the organic matter coming from the river, and with the process of flocculation of the suspended particles, three subjects which were part of the SCOPE/UNEP program during recent years. This small paper is therefore dedicated to the memory of Egon Degens, who took part in the program off the Mahakam, and whose untimely death, not long after reaching the age of 60, is so keenly felt.

## 2 Methods

Water samples were collected with 30-l Niskin bottles mounted on a Rosette sampler together with a Guildline CTD. Filtration was done over pre-weighed 0.4 $\mu$m pore size Nuclepore filters, by suction filtration at concentrations above approx. 1 mg/l and directly from the Niskin bottles by pressure filtration at lower concentrations, using $N_2$ at 2.5 atm. The filters were stored in plastic petri dishes and weighed in the laboratory on a Cahn microbalance after drying at 70°C. Organic content was determined by ashing at 500°C. Suspension particles were observed through a microscope directly on board, and in the laboratory with a SEM, for which a small amount of water sample (5 to 100 ml depending on the concentration)

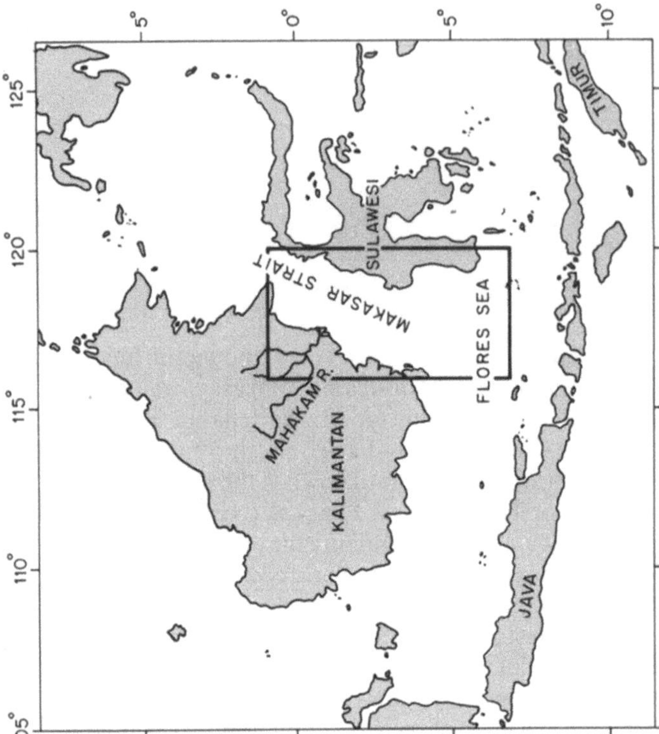


Fig. 1. Location of study area and sampling stations

Fig. 1. Location of study area and sampling stations

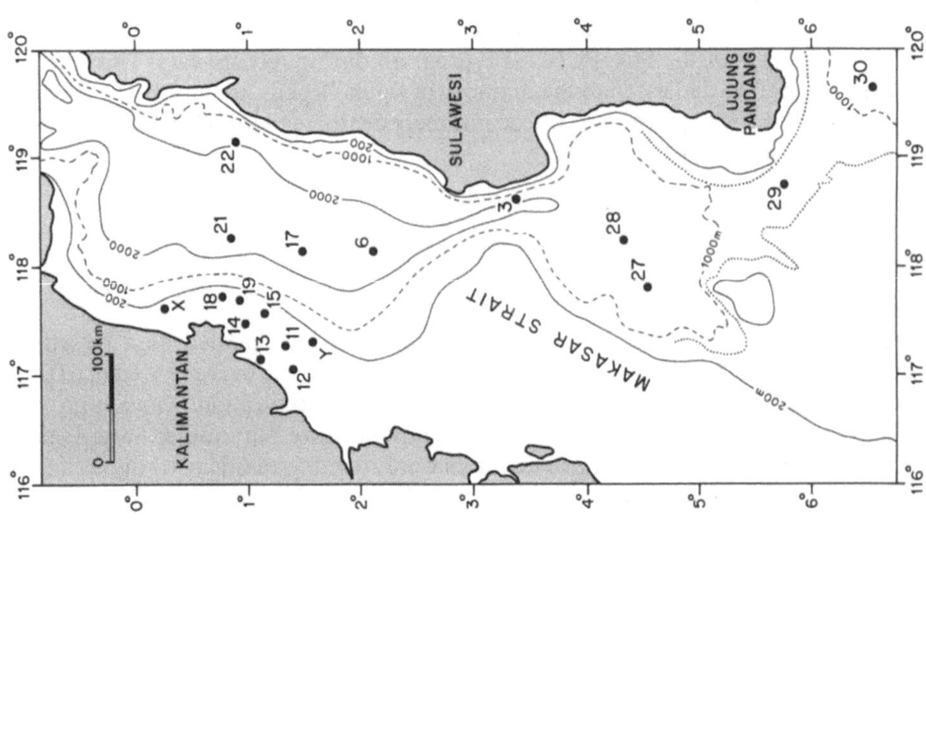

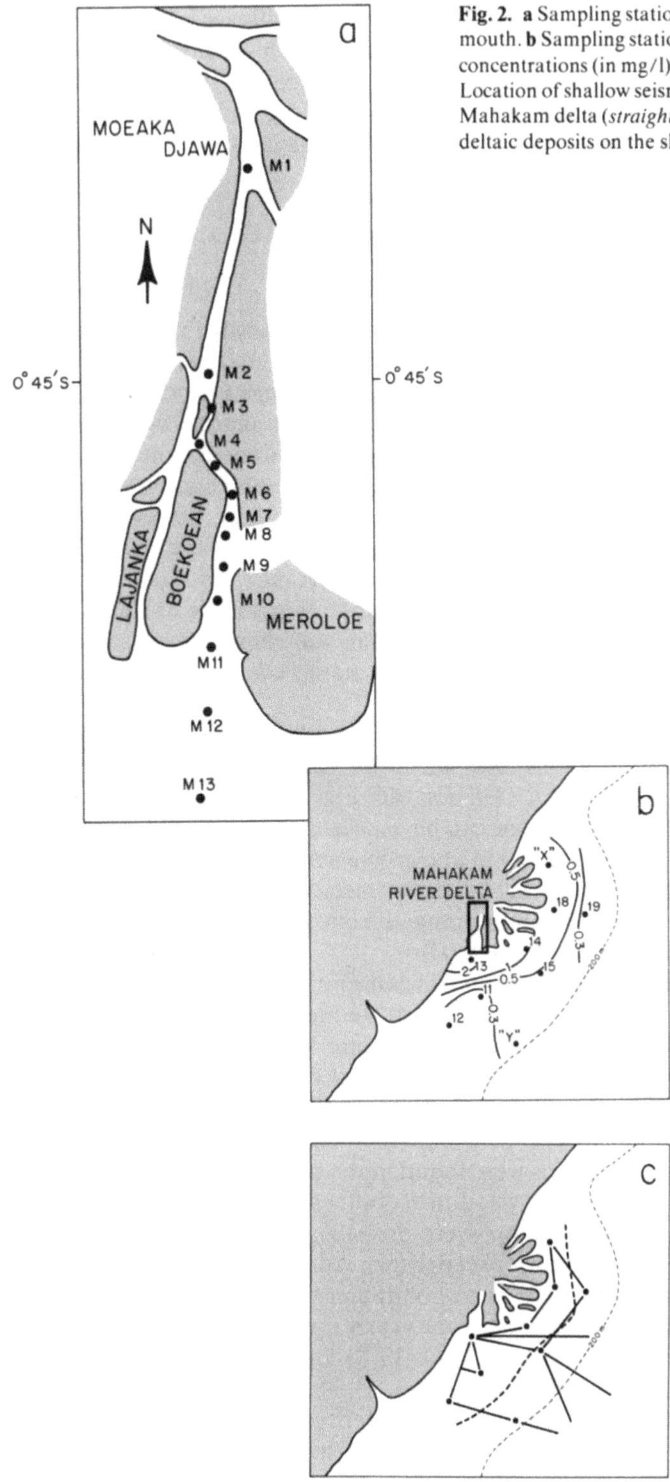


**Fig. 2.** **a** Sampling stations in the Mahakam River mouth. **b** Sampling stations and suspended matter concentrations (in mg/l) off the Mahakam delta. **c** Location of shallow seismic profiling tracks off the Mahakam delta (*straight lines*) and the limit of the deltaic deposits on the shelf (*broken lines*)

was filtered over a 0.4 μm pore size Nuclepore filter. A small piece was cut from the centre of the filter, mounted on a stubb and coated with gold. Particle size was determined directly on board with a TAII Coulter Counter. The stable carbon isotope data used in this paper were determined at the Isotope Laboratory of Groningen University by mass spectrometry after removal of the carbonate.

## 3 The Mahakam River Outflow

The discharge volume of the Mahakam River is not well known but varies between 1000 and 3000 $m^3/s$ (Allen et al. 1979). Of the suspended load even less is known. During the Holocene the sediment output was in the order of $10 \times 10^6$ t/a on the average, which was estimated from the volume of sediment that has been deposited (Allen et al. 1979). This gives an average suspended sediment concentration of approx. 200 mg/l. In November 1984 the suspended matter concentration in the river (at 0.04‰ Cl′) was 24.8 to 24.9 mg/l. These values, however, are not representative for the whole year, since they were collected at the end of the dry season. It can be expected that during the beginning of the rainy season, when the discharge rapidly increases, much larger amounts of sediment are transported seaward. The total supply of suspended matter will therefore be somewhere between $1.5 \times 10^6$ t/a (based on the suspended matter concentration of the river water in November 1984) and $10 \times 10^6$ t/a.

In the estuary there was in November 1984 a small turbidity maximum where suspended matter concentrations reached, in the surface water at approx. 1 m waterdepth, 77.8 mg/l at 0.39‰ Cl′ (station M6; Fig. 2A). Considerable variation in the turbidity maximum, however, can be expected since the tidal range varies from less than 1 m at neap tides to 3 m at extreme spring tides (Allen et al. 1979). At the river mouth (stations M11–M13) the concentration was around 20 mg/l (18.7 to 21.4 mg/l); the turbidity maximum ranged from station M3 (0.06‰ Cl′, 32.9 mg/l) to station M8 (1.07‰ Cl′, 31.4 mg/l).

Off the river mouth suspended matter concentrations rapidly dropped to 2 mg/l and to less than 1 mg/l about halfway down the shelf (Fig. 2B). From there on the river plume stretched in a SE direction into Makasar Strait (Fig. 3). The salinity gradually increased from around 20‰ S off the river mouth to more than 34‰ S south of station 3, while the suspended matter concentrations decreased to less than 0.3 mg/l outside the plume. On the shelf off the river mouth the highest suspended matter concentrations were found in the surface water and near to the bottom (Fig. 4). The surface water was directly influenced by the river outflow; the high concentrations near the bottom were probably caused by the tides, which either resuspend bottom sediment, or keep suspended matter in suspension that has settled out from the surface. Stirring up of bottom material by surface waves at 40 to 60 m water depth is less likely because the yearly median significant wave height in this area is less than 60 cm (Allen et al. 1979) and also during the period of sampling waves were very small.

Not only the distribution of suspended matter concentrations but also the 3.5 KC profiles indicated that only a very small part of the suspended matter supply of

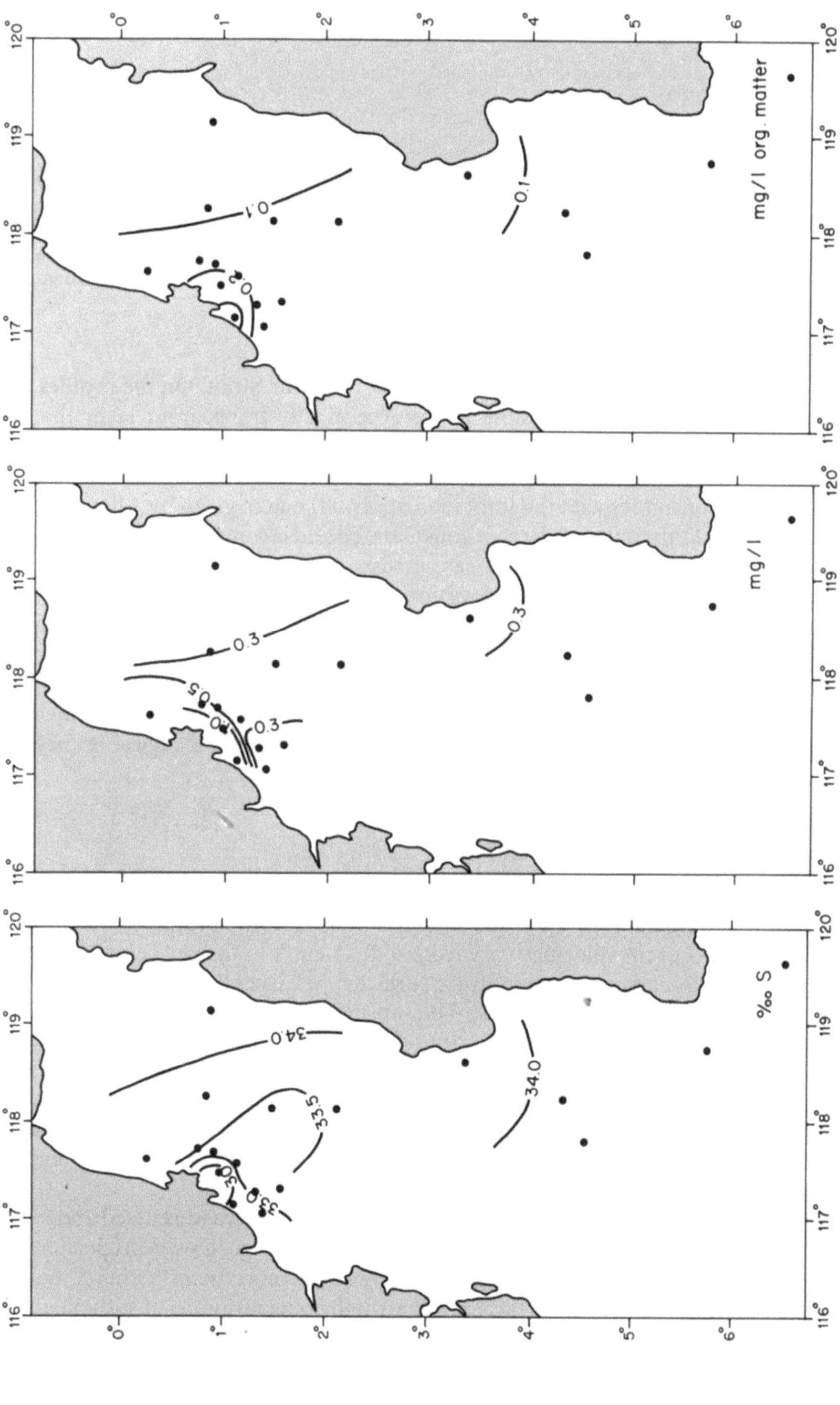


**Fig. 3.** Distribution of salinity (‰ S), suspended matter concentration (mg/l) and particulate organic matter concentration in suspension (mg/l) in the surface water off the Mahakam delta and in Makasar Strait

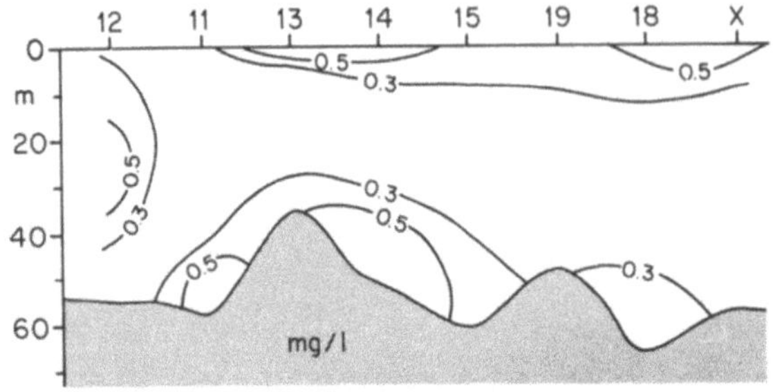


**Fig. 4.** Profile of suspended matter concentration off the Mahakam delta. The isolines indicate concentrations in mg/l. The numbers above indicate the sampling stations

the Mahakam River reaches the deeper waters of Makasar Strait. On the profiles, the river-supplied sediment is visible as an acoustically transparent layer that decreases in thickness from the river mouth in a seaward direction (Fig. 5A). The outer limit of these deposits lies approximately halfway the shelf, as indicated on Fig. 2C. Its position agrees with the limit indicated on the map given by Allen et al. (1979). Seaward of this limit older, more consolidated and layered deposits form the seafloor and elevations are present of an as yet unknown composition that show up in the profiles as peaks (Fig. 5B). The elevations are provisionally interpreted as old reefs. At some places between the elevations an acoustically transparent infill is present that may have been recently supplied by the river. The sharp decrease in suspended matter concentrations over the shelf and the presence of relict deposits on the outer shelf indicate that only a very small part of the river supplied suspended sediment – in the order of a few percent on less – reaches the shelf edge (Fig. 5C).

## 4 Dispersal in Makasar Strait

The flow through Makasar Strait in the surface waters is from north to south (Fig. 6; Wyrtki 1961). Current velocities vary from 6 to 25 cm/s in the wet season to 12 to 38 cm/s in the dry season. The southward surface flow causes the deflection of the Mahakam River plume to the southeast. The subsurface flow in Makasar Strait has not been studied before. The salinity distribution (Fig. 7A) indicates inflow from the south over the sill that separates Makasar Strait from the Flores Basin as well as inflow from the north. As mentioned above, the salinity data are open to doubt and the distribution of temperature (Fig. 7B) hardly indicates such an inflow, but the distribution of dissolved oxygen and of total $CO_2$ concentration (Fig. 7C,D) support the conclusion based on the salinity distribution.

The distribution of suspended matter in the surface water was discussed above. The suspended matter concentrations in the deeper water (Fig. 7E) were high in the water below the plume, where concentrations reached more than 0.5 mg/l, but remained below 0.1 mg/l outside this area. The high concentrations of suspended matter below the plume are most likely caused by particles settling from the plume, which would mean that the horizontal flow in the deeper water of Makasar Strait is very small, either northward or southward.

The particle size of the suspended matter was generally less than 80 to 100 μm; only rarely larger particles were found. Plotting particle volume in $mm^3/l$ against log D, two extreme size distribution types were present (Fig. 8): one with a sharp peak in the 20 to 50 μm range (sample 3–400) and one with a peak at sizes smaller than 6 μm (sample 28–100). These extremes were not common; usually the size distributions had two or more peaks (sample 6–100; 28–400; Fig. 9) and on the shelf off the Mahakam delta often a number of peaks (sample 12–1). The size distributions reflected the variability of the particles in suspension: supplied by the river, resuspended particles, planktonic origin, flocculated, unflocculated. The particle size distributions are used below to estimate the average distance between the suspended particles.

## 5 Distribution and $\delta^{13}C$ of Organic Matter

The content of organic matter in the suspended matter of the Mahakam River was 20 to 24%, decreasing a little to 18% in the turbidity maximum and increasing to 36% at the river mouth (stations M11-M13). Off the mouth the organic content became more variable ranging from 6.3 to 100%, with an average of 43.3%. Taking only the surface water samples gives an average of 49.5% and a range of 21.7 to 92.7%. The lowest values were found near the bottom where probably bottom material poor in organic matter had been resuspended. In the surface waters the average content was 48.5% (15.6 to 91.1%) in the plume and 38.6% (17.3 to 62.9%) outside the plume.

The $\delta^{13}C$ in the suspended organic matter in the river and estuary varied from -28.52‰ to -29.50‰ (average -29.22‰; samples M1-M11). In the river mouth (samples M12, M13) it increased to -17.00‰ and -26.52‰ and off the mouth on the shelf to -23.62‰ to -25.40‰. In the deeper water in Makasar Strait, in the river plume and outside the river plume a similar range of values was found (-23.72‰ to -26.26‰). The change in $\delta^{13}C$ at the river mouth reflects a change in composition of the organic matter and/or a replacement of the particulate terrestrial (river-) organic matter by particulate organic matter of marine origin. Primary production on the shelf off the river mouth was evident from the presence of plankton as observed by microscope, and from the nutrient data. In the river $H_4SiO_4$ concentrations were approx. 210 μM at the river mouth. Off the mouth they became almost 0, although the salinity (23‰ S) indicated that on the basis of conservative mixing it would be approx. 42 μM. Similarly the nitrate concentration should be approx. 1.95 μM and the phosphate concentration 0.33 μM, where the actual concentrations were almost 0 μM and 0.14–0.21 μM, respectively. The average organic content of the suspended matter on the shelf (43.3%) was higher than in the river (20 to 24%). Also the $\delta^{13}C$ indicated the dominance of marine organic matter, directly off the river mouth as well as further offshore, the $\delta^{13}C$ being the same as found everywhere in Makasar Strait and in the Flores Basin. This does not exclude the possibility that organic matter of fluvial origin has been transformed in such a way, e.g., by dissociation and subsequent loss of lighter compounds with relatively much $^{12}C$, that part of the original material has remained, has become stable under marine

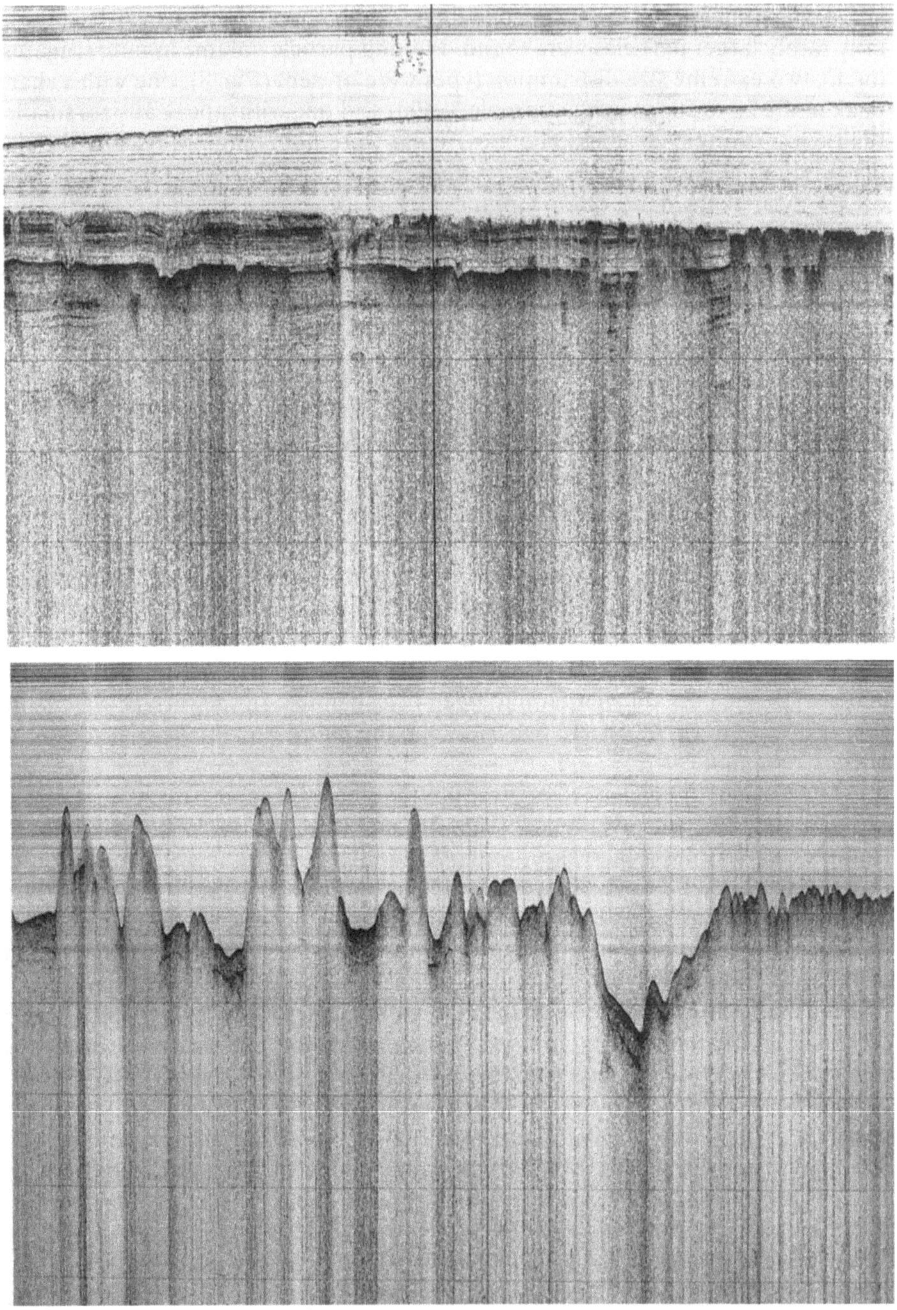

**Fig. 5a-c.** 3.5 KC profiles off the Mahakam delta. **a** Acoustically transparent mud deposits on older sediments. **b** Elevations (peaks) on the outer shelf interpreted as old reefs. **c** Pre-Holocene deposits at the shelf edge

c

conditions, and still contributes to the organic matter in suspension off the mouth. If this possibility is discarded, however, it must be assumed that directly off the mouth almost all of the particulate organic matter of fluvial origin has been lost, through consumption and/or oxidation.

## 6 Flocculation

Observation of the suspended matter with a SEM indicated a large variation in the number and the size of flocs. The observed flocs (microflocs) are most probably the fragments of much larger flocs (macroflocs) which are the flocs commonly found in situ in the water in those rivers and estuaries (mostly in temperate climates) that were examined for them: macroflocs are fragile and usually break apart during sampling and handling the samples. They are ca. one order of magnitude larger than the microflocs (Eisma et al. 1983; Eisma 1986). In all the shelf samples large microflocs ($>$ 10 $\mu$m; maximum diameter approx. 100 $\mu$m) were found as well as in the river and the river mouth, and in the deeper water in Makasar Strait in the river plume and below the plume (Fig. 10A). Fibers, most probably from land plants, were present in low numbers in most of these samples, indicating the presence of land-derived material. They constitute only a small part of the total organic matter in suspension, which is also indicated by the $\delta^{13}$ $C_{org}$ data. Small microflocs ($<$ 10 $\mu$m) occurred in the same samples as the large ones, as well as in the samples collected at station 21 at the edge of plume, and in the surface waters (0 to 200 m water depth) at station 30. In the other samples they constituted only a

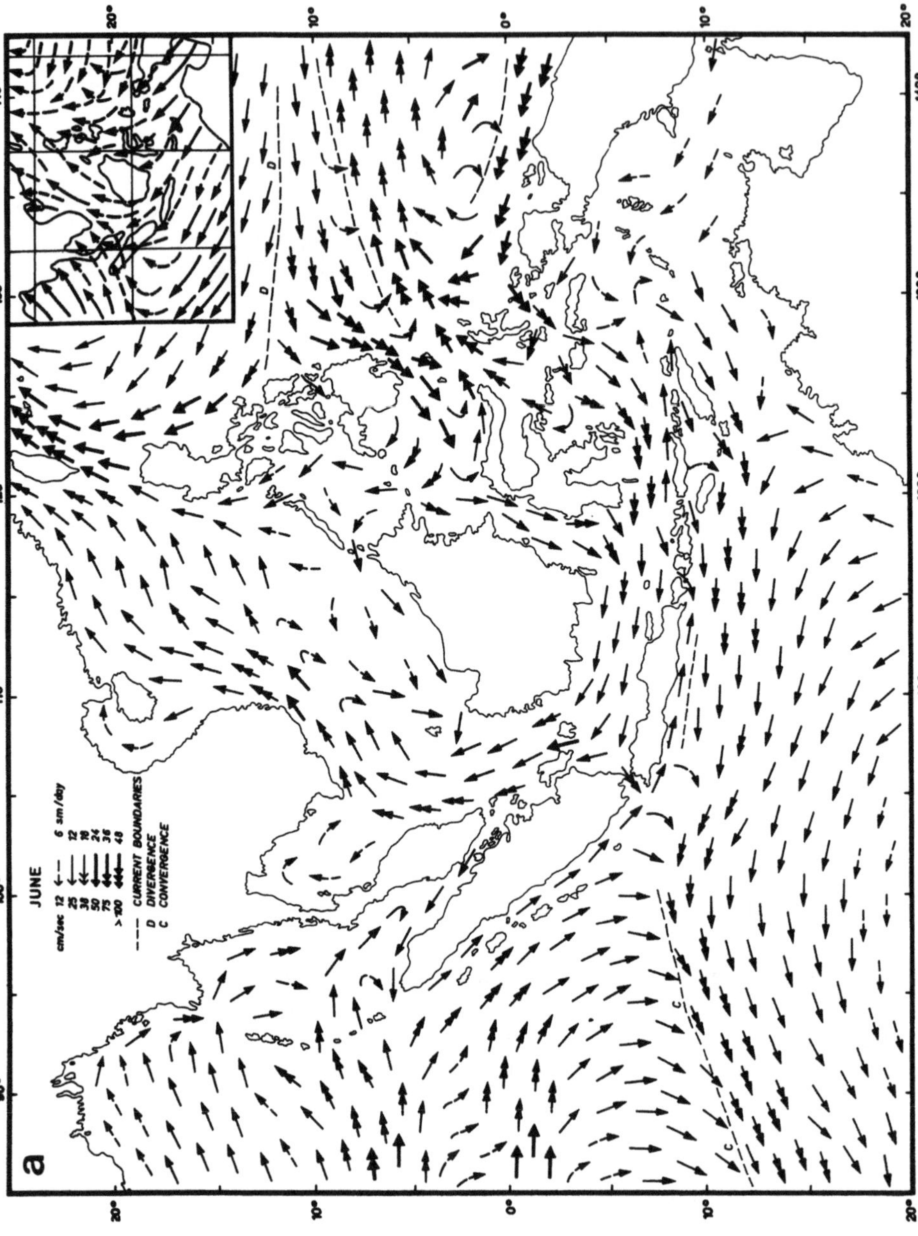
a
JUNE
cm/sec
sm/day
CURRENT BOUNDARIES
D DIVERGENCE
C CONVERGENCE

**Fig. 6a,b.** Flow in the surface water in the Indonesian waters and adjacent sea areas. (Wyrtki 1961). **a** in June; **b** in February

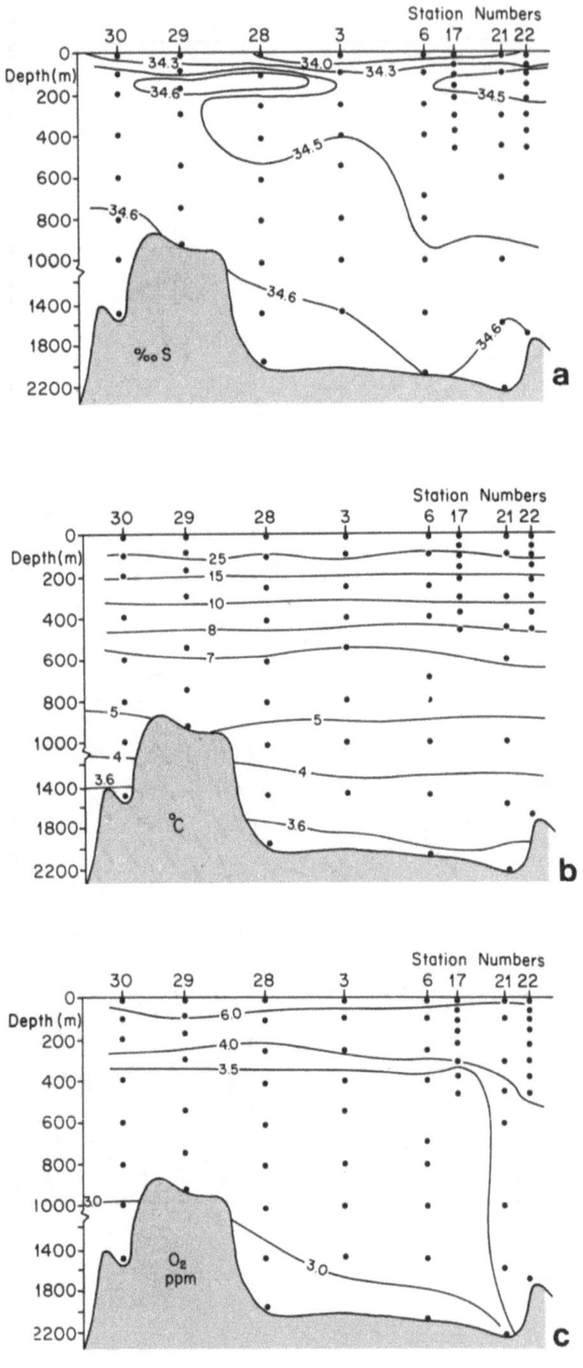


**Fig. 7a-e.** Profiles through Makasar Strait. The numbers above indicate the sampling stations (see Fig. 1). **a** Salinity (‰ S). **b** Temperature (°C). **c** Dissolved oxygen concentration (ppm). **d** Total dissolved $CO_2$ concentration (μmol/l). **e** Suspended matter concentration (mg/l).

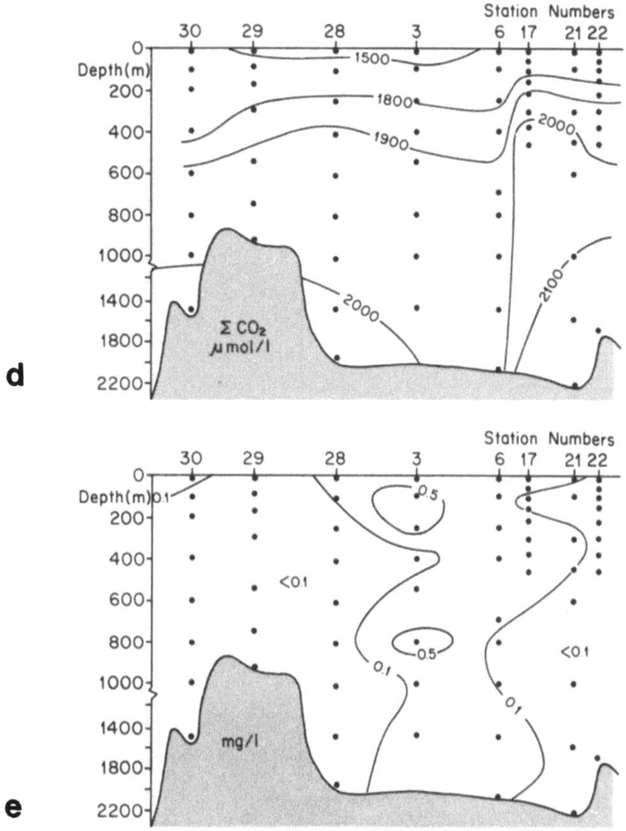


Fig. 7d,e

very minor part of the suspended particles or were absent, virtually all particles being single grains. The distribution of large microflocs coincided with the distribution of higher suspended matter concentrations in and below the river plume, the limit being in the surface water approx. 0.3 mg/l (Fig. 3) and in the deeper water ca. 0.1 mg/l (Fig. 7).

In earlier work (Eisma 1986; Eisma and Kalf 1987) it was found that when living plankton is present, the bulk organic content of the suspended matter does not influence very much the flocculation or the settling of the suspended particles. Many planktonic cells apparently do not flocculate or do not induce flocculation. In the absence of living plankton, however, as in the North Sea during the winter, small microflocs were present where the organic content of the suspended matter was low, and large microflocs where the organic content was high. Off the Mahakam delta in the plume and below the plume the concentration of organic matter (in mg/l) was higher than outside the plume and the waters below it (Fig. 10B). The organic content of the suspended matter (in %), however, did not show such a distribution (Fig. 10C). Where the larger microflocs (> 10 µm) occur, the organic content varies from less than 20% to more than 80%. Where no large microflocs occur, the organic content varies from less than 20% to almost 80%. There is no

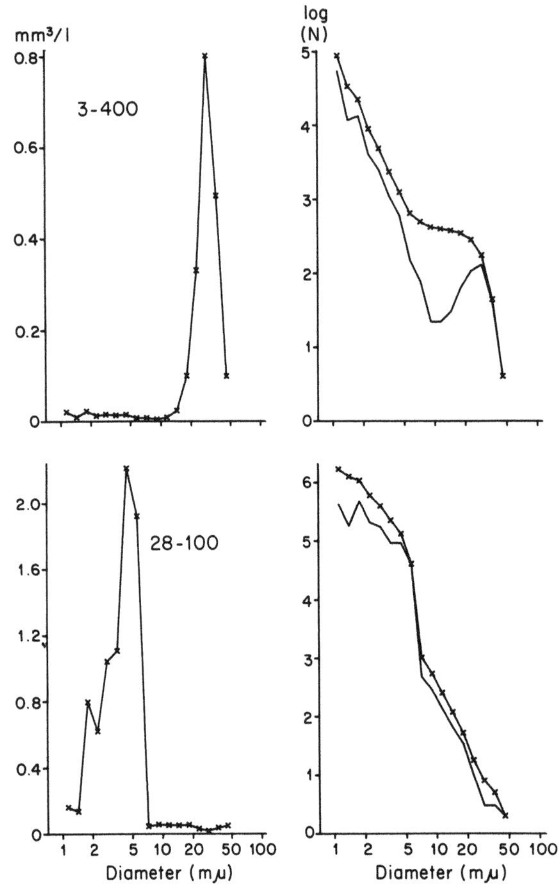


**Fig. 8.** Particle size distributions. *Left* particle volume ($mm^3/l$) against log diameter (μm). *Right* log N against log diameter. *N* number of particles counted in 2 $cm^3$. *3–400* station 3 at 400 m water depth; *28–100* station 28 at 100 m water depth

relation between the presence of large microflocs and the organic content of the suspended matter, but there is a clear relation with suspended matter concentration. The mechanisms for bringing particles together into flocs include Brownian motion, turbulence, differential particle settling, and organisms aggregating particles (McCave 1984). Sticky substances enhance the effect of particle collisions by glueing particles together. A higher particle concentration results in more frequent particle collisions.

Taking the microflocs and single mineral grains to be the units that are brought together in macroflocs, the influence of the particle concentration on the formation of flocs can be estimated from the particle size distribution as measured by Coulter Counter. Since at all stations the same methods were used for sampling and handling the samples, the results obtained at different stations and water depths can be regarded as comparable. To use the microfloc data in this way is less arbitrary than it may seem: it is difficult to break them further apart, for which severe ultrasonic treatment is necessary, so that they can be regarded as stable (Eisma and Kalf 1979; Eisma et al. 1980; see also Kranck 1973). Assuming the microflocs and

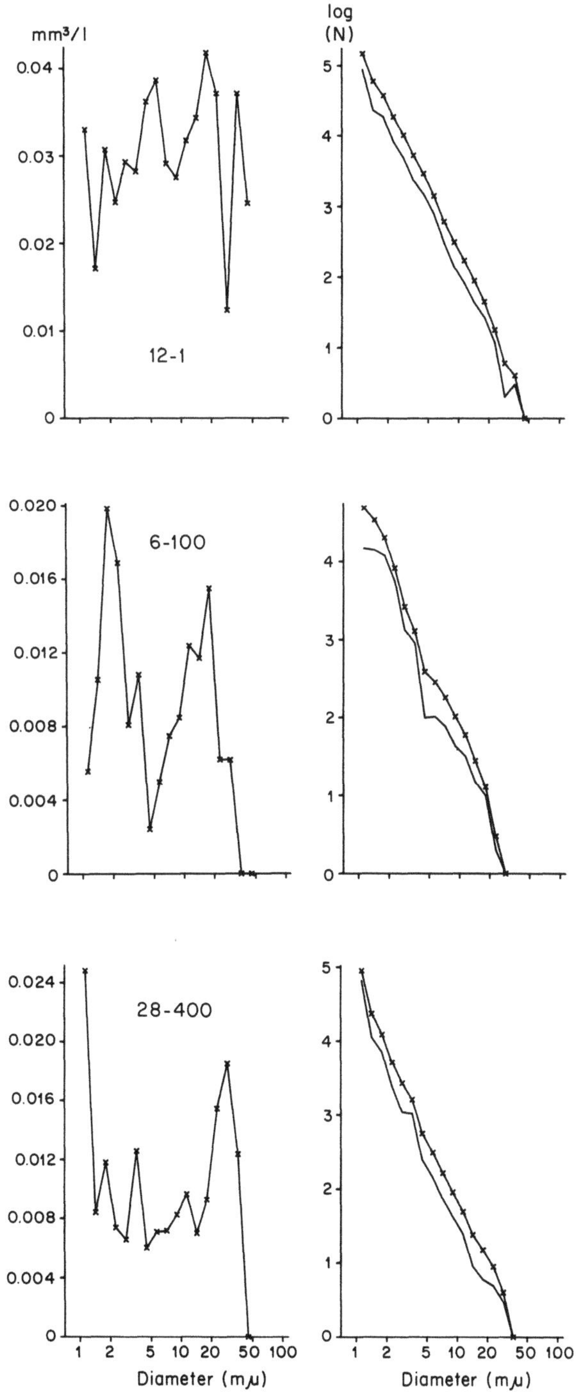


**Fig. 9.** Particle size distributions (see Fig. 8). *6–100* station 6 at 100 m water depth; *28–400* station 28 at 400 m water depth; 12-1: station 12 at 1 m water depth

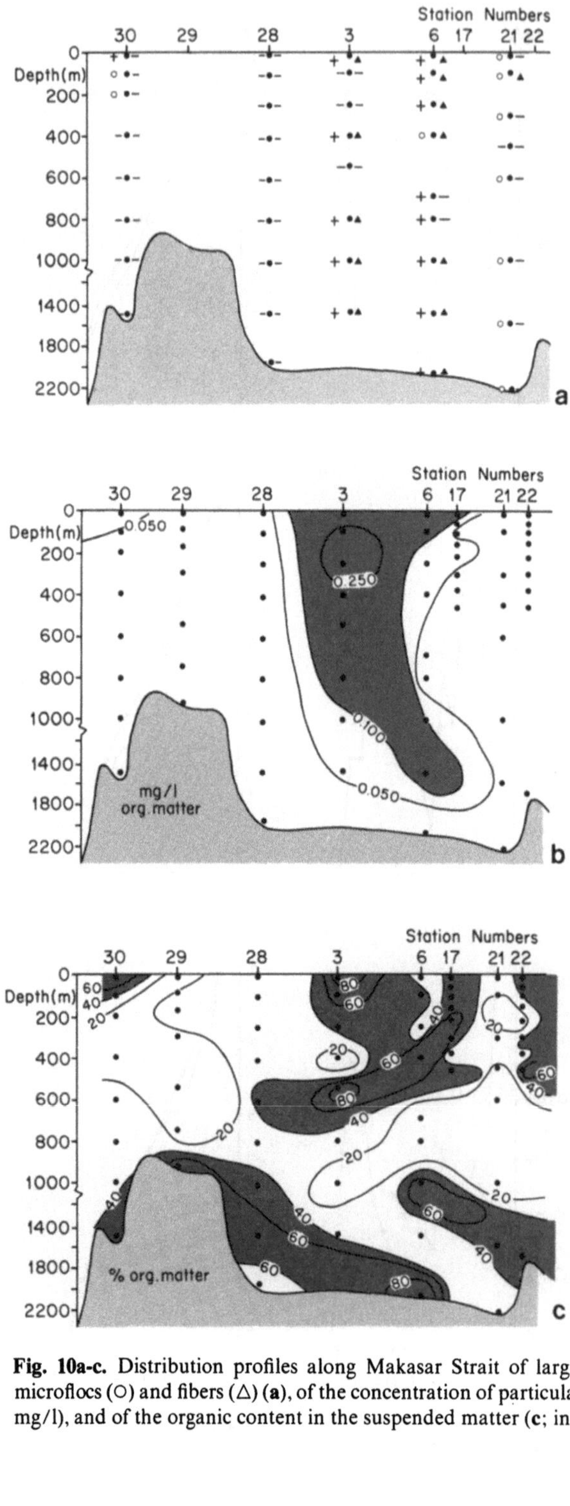


**Fig. 10a-c.** Distribution profiles along Makasar Strait of large microflocs (> 10 μm) (+), small microflocs (○) and fibers (△) (**a**), of the concentration of particulate organic matter in suspension (**b**; in mg/l), and of the organic content in the suspended matter (**c**; in %)

single mineral grains (both from here on referred to as 'particles') to be evenly distributed in the water, the average distance $\bar{d}$ between the centre of the particles is $\sqrt{\frac{V}{N}}$ where V is the total volume of the water that contains the particles (2 $cm^3$) and N is the total number of particles in this volume in the range of 1.6 $\mu$m to 50 $\mu$m as measured by Coulter Counter. The volume of the particles can be neglected as the total particle volume is in the order of $10^1$ $mm^3$ and the total water volume is $2 \times 10^3$ $mm^3$. The lowest size limit in the Coulter Counter measurements is determined by the reliability of the instrument, the upper limit by the number of particles that was found to be present: when only one or no particle is measured in a size fraction the error becomes very large. Within the plume, $\bar{d}$ was found to be 220 to 450 $\mu$m, outside the plume 460 to 630 $\mu$m. In deeper water (> 200 m water depth) under the plume $\bar{d}$ is, in six samples, between 270 and 400 $\mu$m, in twelve samples, between 510 and 790 $\mu$m, whereas in the deep water outside the area below the plume $\bar{d}$ in all the samples is between 460 and 740 $\mu$m. From these data it follows that the presence of large microflocs in the plume is related to values of $\bar{d}$ being approx. 450 $\mu$m or less. Below the plume, where the large micro flocs are settling out from the surface, large micro flocs occur also in samples where $\bar{d}$ is larger.

Exceptions were found in the surface waters. Small microflocs and some larger ones were found in the surface water of station 30. The organic content of the suspended matter was high (almost 80%), the suspended matter concentration was slightly above 0.1 mg/l, the samples were found to contain diatoms as well as flocs, but $\bar{d}$ was 500 to 580 $\mu$m. It is probable that flocculation was induced by the diatoms, which produce sticky substances, or by organisms grazing on the diatoms and producing feces. In the surface waters of stations 28 and 29 concentrations were below 0.1 mg/l, but $\bar{d}$ was 350 to 400 $\mu$m while no flocs were present. The low values for $\bar{d}$ were caused by the presence of large numbers of small particles of only a few $\mu$m diameter, that were not particularly numerous in the SEM samples, indicating that they were mostly not mineral grains but probably plankton that does not flocculate. Off the river delta at two stations (17 and 19) $\bar{d}$ was 500 to 600 $\mu$m although there were flocs present. These flocs, however, had the character of fecal pellets and the organic content was high in these samples (45 to 92%). These exceptions indicate that the relation between flocculation and particle concentration may be strongly influenced by organisms in the surface water: through the presence of large numbers of plankton "particles" that do not flocculate, through the flocculating effect of sticky substances that are produced, or by glueing particles together through the formation of pellets and faeces.

In this discussion the size of the particles was not considered. The largest particles measured were approx. 100 $\mu$m but they were very rare and the largest particles included in the calculation were 50 $\mu$m. When $\bar{d}$ is 450 $\mu$m, it will make a difference whether two particles of 50 $\mu$m are next to each other: the actual distance between the particles is then reduced to 400 $\mu$m, or by 12.5% (when $\bar{d}$ is 220 $\mu$m the reduction is 21%), whereas two particles of 2 $\mu$m next to each other has a negligible effect, reducing the distance between the particles by less than 1%. The probability that a particle of size x occurs at a distance $\bar{d}$ of a particle of size y is $\frac{N_y}{N} \times \frac{6n_x}{N} \times$

100% where $n_y$ and $n_x$ are the total number of particles of size y, resp. of size x and N the total number of particles of 1.6 to 50.0 μm in suspension.

The probability that a particle of size x is at a distance $\bar{d}$ of another particle of size x is $\frac{n_x}{N} \times \frac{6(n_x - 1)}{N} \times 100\%$. The probability that two large particles are next to each other is very small because in their case n is very small (in the order of 1 to 5). The probability that two very small particles are next to each other is very large because in their case n is very large (in the order of $10^3$ to $10^4$). The size distribution of the suspended particles can be described as a function of $\frac{n_s}{N}$, where s ranges from 1.6 to 50 μm: when the particles are evenly distributed, both probabilities (of a particle of size x being next to a particle of size y, or to another particle of size x) follow a distribution which is a function of the particle size distribution. Therefore, when only the distance between the particles determines whether particles flocculate together, the particles in the flocs will follow the size distribution of the unflocculated particles.

Kranck (1980), comparing the size distributions of populations of stable microflocs and mineral grains with the size distributions of the unflocculated grains, has shown that the size composition of the particles in each individual (micro)floc is approximately the same as that of all the particles in suspension finer than the floc itself. Turbulence, differential particle settling and Brownian movements, all three produce particle collisions but at different speeds and efficiencies, which also depends on the particle size. Turbulence acts at a velocity of $10^3$ to $10^4$ μm/s, settling of mineral grains of 1 to 100 μm diameter at $10^1$ to $10^4$ μm/s, and Brownian movements, affecting particles of only a few mm diameter, at $10^0$ μm/s. This makes turbulence and the settling of the larger mineral particles (here 30 to 50 μm, involving only relatively few particles) by far the most efficient processes to bring particles together that are at an average distance of $10^2$ to $10^3$ μm from each other.

In situ in the water, the microflocs and single mineral grains are most likely to be united in macroflocs, with the probability that some of the mineral grains remain unflocculated (Kranck 1980). The size distribution of the macroflocs in situ in the water is not known, but an estimate of their size in relation to the average distance D between the macroflocs in the water can be obtained from the Coulter Counter size distributions. For the samples for which $\bar{d}$ was found to be larger than 450 μm the average total volume of the particles is 0.22 $mm^3/l$ (range 0.11 to 0.46 $mm^3/l$) and for the samples for which $\bar{d}$ was found to be smaller than 450 μm the average total particle volume is 0.73 $mm^3/l$ (range 0.28 to 1.98 $mm^3/l$). A total particle volume of 0.22 $mm^3/l$ can be flocculated into one macrofloc of 1.6 mm diameter containing 90% water, into 10 macroflocs with a diameter of 760 μm and the same water content or into 100 macroflocs of the same water content with a diameter of 340 μm, assuming all particles and macroflocs to be spheres. In all three cases the diameter of the macroflocs is 1.6% of the average distance D between them. At the average total particle volume of 0.73 $mm^3/l$ this percentage becomes 2.4% and at the maximum total particle volume found (1.78 $mm^3/l$) it becomes 3.4%. It follows that also in nature, where probably almost all suspended material is flocculated into macroflocs, the size of the macroflocs, at the particle concentrations considered here (< 1 mg/l), is very small in relation to the average distance D between the macroflocs. This changes at higher particle concentrations: the distance between the particles becomes smaller and differential settling and Brownian motion become more efficient in producing particle collisions. It is concluded therefore that in deep water at low particle concentrations, and where the influence of the bottom is negligible, the average distance between the particles (controlled by the suspended particle concentration), the conditions of turbulence, the presence of sticky substances or surfaces, and the activities of organisms in the surface water largely determine the flocculation process. It was assumed, however, that the particles are evenly distributed throughout the water, which in reality is probably not so, at least not always: suspended matter can be seen to move in clouds or streaks and, because the waterflow is not uniform, particles of a certain size and density tend to become concentrated. Because of this the distances between the particles become smaller than $\bar{d}$ or D, so that more collisions may occur, produced by differential settling and Brownian movements.

Besides physical processes, organisms producing stable pellets and feces also produce mixtures of organic matter and mineral particles that can be regarded as flocs. As mentioned above, a relation between the organic content of the suspended matter and floc-formation is suggested by data from the North Sea, where relatively large microflocs ($> 20\ \mu m$) in relatively deep water are associated with a high organic content of the suspended matter, no or only little exchange of suspended particles with the bottom sediment, and relatively long residence times (in the order of 1 year). Small microflocs in the relatively shallow waters are associated with a low organic content of the organic matter, regular exchange of suspended particles with the bottom sediment, and relatively short residence times in the water (Eisma and Kalf 1987). The larger microflocs in the Mahakam plume are not associated with a higher organic content of the suspended matter, but are associated – outside the shelf at least – with no exchange of suspended particles with bottom sediment and with residence times in the order of 10 to 20 days (estimated from the length of the plume – approx. 400 km – and the average current velocity during the dry season – 12 to 38 cm/s – allowing for the fact that the sampling was done at the end of the dry season). This indicates that ageing of the flocs during a longer residence time in the water and no exchange with the bottom may result in flocs with a firmer structure, and that the bulk organic content is not an important factor. This agrees with the results obtained in West-European estuaries indicating that the bulk organic matter composition is not important in forming flocs (Eisma 1986) but that specific organic compounds (polysaccharides, fulvic acids) are. Since both physical processes and organisms may produce flocs, the observed microflocs and the macroflocs they are part of, may be of different origin with a distinctive morphology. Such a distinction was made by Pierce and Nicholson (1986) in the Rappahannock estuary (flocs and aggregates) and there are indications that a similar distinction can be made in the Ems estuary (Eisma et al. 1983) and off the Mahakam delta (stations 17 and 19).

*Acknowledgments.* I am much indebted to the participants of the cruise on which this program was carried out as part of a larger program covering also the Flores Basin and including bottom sampling. The results of this are being published elsewhere. Participants came from the Netherlands Institute for Sea Research and the Geologisch-Paläontologisches Institut of Hamburg University. To Captain de Jong of the R.V. TYRO and his crew I am much indebted for their pleasant cooperation. My thanks also go to Dr. S. Kempe (Hamburg) for the use of some of his data collected in Makasar Strait, to J. Kalf (NIOZ) for suspended matter analyses (total concentration, organic content, particle size), to W.G. Mook (Groningen) for the $\delta^{13}C$ data, and to R.T.P. de Vries (NIOZ) for some nutrient data. As stated above, the program was carried out as a part of the Snellius-II Expedition, organized by the Netherlands Council of Oceanic Research (NRZ) and the Indonesian Institute of Science (LIPI).

## References

Allen GP, Laurier D, Thouvenin J (1979) Etude sédimentologique du delta de la Mahakam. Notes Mém Total Fr Pétrol (Paris) 15:156

Eisma D (1986) Flocculation and deflocculation of suspended matter in estuaries. Neth J Sea Res 20:183–199

Eisma D, Kalf J (1979) Distribution and particle size of suspended matter in the southern bight of the North Sea and the eastern Channel. Neth J Sea Res 13:298–324

Eisma D, Kalf J (1987) Dispersal, concentration and deposition of suspended matter in the North Sea. J Geol Soc (Lond) 144:161–178

Eisma D, Kalf J, Veenhuis M (1980) The formation of small particles and aggregates in the Rhine estuary. Neth J Sea Res 14:172–191

Eisma D, Boon J, Groenewegen R, Ittekkot V, Kalf J, Mook W (1983) Observations on macro-aggregates, particle size and organic composition of suspended matter in the Ems estuary. In: Degens ET, Kempe S, Soliman H (eds) Transport of carbon and minerals in major world rivers Part 2. Mitt Geol-Paläont Inst Univ Hamb 55:295–314

Kranck K (1973) Flocculation of suspended sediment in the sea. Nature 246:348–350

Kranck K (1980) Experiments on the significance of flocculation in the settling of fine-grained sediment in still water. Can J Earth Sci 17(11):1517–1526

McCave IN (1984) Size spectra and aggregation of suspended particles in the deep ocean. Deep-Sea Res 31:329–352

Milliman JD, Meade RH (1983) World-wide delivery of river sediment to the oceans. J Geol 91:1–21

NEDECO (1965) A study on the siltation of the Bangkok Port Channel, II, the field investigation. Den Haag, 474 pp

Pierce JW, Nichols MM (1986) Change of particle composition from fluvial into an estuarine environment: Rappahannock River, Virginia. J Coastal Res 2:419–425

Wyrtki K (1961) Physical oceanography of the southeast Asian waters. NAGA Rep Scripps Inst Oceanogr 2:195

# Transport of Water and Sediment in the Strait of Dover

H. Postma

## Introduction

The Strait of Dover or, on the other side, Pas de Calais forms an important passage between the Channel and the North Sea for water, suspended materials, pollutants, fish, and shipping (Fig. 1). This paper was started with sediment transport in mind, as appropriate for a contribution in honor of Egon Degens, but as always transport of particles cannot be fully understood without a good insight into water transport. This aspect will, therefore, be discussed first, with special attention to residual water movements.

## 2 Water Movements

The existence of very strong flood and ebb currents through the strait has been documented since roman times but a quantitative insight was only obtained in this century. The first estimate of a residual transport from the Channel into the North Sea was made by Gehrke (1907). This estimate was based on the difference in salinity between a section across the Strait (average 34.9 ‰) and a section from Lands End to Britanny (average 35.2‰), and by computing net fresh water input into the Channel from French and English drainage basins.

Current measurements from the lightvessel Varne, which is very conveniently located in the middle of the Strait, somewhat to the south-west of the shortest traverse, gave a residual transport of 2000 km$^3$ per year into the North Sea and this value has been used since by a generation of oceanographers.

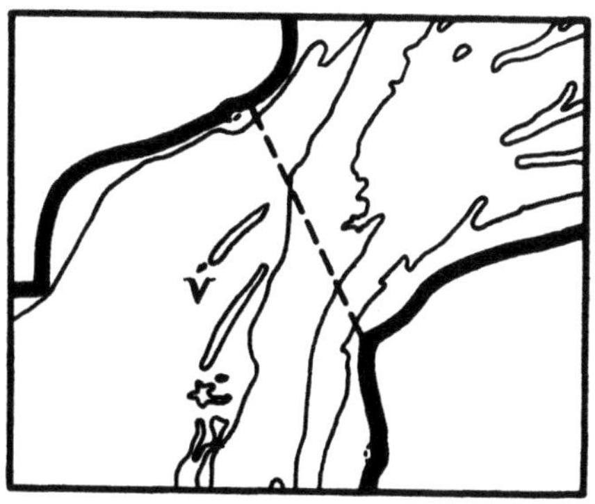


**Fig. 1.** The Strait of Dover showing van Veen's section and the Varne lightship

Regular measurements from the lightship were started much later, in 1926, by Carruthers and continued until 1938. For these measurements this investigator developed a sturdy instrument, called a "vertical log", including a device for current direction, which was suspended at a depth of 10 meters. The lightship lies on the southwestern tip of the Varne sandbank, which in the middle rises to about 5 m below sea level.

In a first publication Carruthers (1928) arrived at a residual current towards the North Sea of 5.5 cm/s, in a second (Carruthers 1935) at 6.8 cm/s, considering the latter value as the most accurate one. This value, if valid for the whole cross-section of the Strait (1.37 $km^2$) corresponds with a water transport of about 3000 $km^3$ per year.

The cross-section just mentioned was measured with great accuracy by van Veen (1936; Carruthers assumed 1.2 $km^2$) and refers to mean sea level; van Veen is known world-wide for the van Veen grab, a strong and very useful piece of equipment, but his chief scientific accomplishments are in the field of coastal hydraulics and engineering. At the end of his career he was secretary of the commission preparing the Delta works, for which he had, long before the storm surge of 1953, already laid the scientific foundation.

Van Veen was the first, and perhaps the last, to execute very extensive tidal measurements. He did this at 12 points across the Strait in the summer period of 1934. At present, because of heavy shipping traffic, such measurements are hardly possible any more. For these measurements and additional measurements in 1935 and 1936 he calculated an average flood transport of $19.15 \times 10^9$ $m^3$ and an ebb transport of $16.9 \times 10^9$ $m^3$, the difference being $2.24 \times 10^9$ $m^3$ per tide or 50 000 $m^3$ per second. To arrive at an annual value he made a correction for seasonal changes by means of Carruthers' current data and thus found a slightly higher value of $2.4 \times 10^9$ $m^3$ per tide, or 1750 $km^3$ per year.

This is obviously much lower than Carruthers' value, but even more surprising is that van Veen, who measured in very quiet summer periods with little or no wind, found no excess flood stream of importance and sometimes even a slight ebb surplus (van Veen 1938). The difference between flood and ebb volume transport is mainly caused by a difference in profile between flood and ebb of 8.5%. The so-called Stokes drift through the Strait generated by this difference is on an average 33 000 $m^3/s$ (Otto 1984). Subtracting this value, this leaves for the whole of van Veen's measurements a residual current towards the North Sea of only 1.2 cm/s or 16 000 $m^3/s$. Because Carruthers did not include the Stokes drift into his measurements, this effect must be added to his annual transport of 3000 $km^3$. This yields a value of about 4000 $km^3$ per year. The difference between Carruthers and van Veen of about 2000 $km^3$ must mainly be caused by the different wind stress.

A first analysis of the influence of wind on the residual transport through the Strait was made by Wyrtki (1952a). This writer derived a wind field for the southern North Sea and the Channel from air pressure data of a triangle of land stations around the Strait (Tynemouth, Paris and Emden), because insufficient sea data were available. The SW component of the wind over the southern North Sea has a good relation with the residual current (Fig. 2). The fact that this relation is linear excludes a direct influence of local wind stress which would imply a relation with the square of wind velocity. In a second paper Wyrtki (1952b) studied the influence

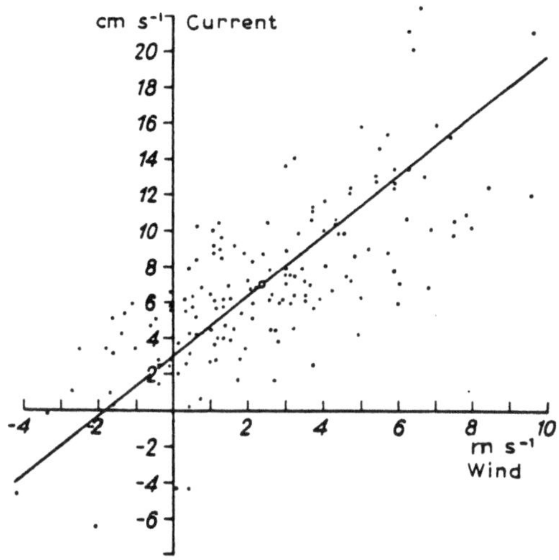


**Fig. 2.** Relation between current and wind at the Varne lightship. (Wyrtki 1952)

of wind on water levels over the whole North Sea showing that in most cases, but not always, changes in monthly flow through the Strait correspond with changes in flow between Scotland and Scandinavia, so that the monthly variation in total water volume in the North Sea remains relatively small. This effect, of course, has been discussed and modelled in much greater detail in recent years.

An interesting detail of Wyrtki's analysis is that with a zero SW wind component there is still a current through the Strait towards the North Sea of 3 cm/s or 41 000 $m^3/s$. This is the "basic current", subsequently also derived from the Varne data by others (Carruthers et al. 1950; Veley 1960), which must be chiefly caused by an average slope of water level from the Channel into the North Sea.

Figure 2 further gives an average wind driven current of 3.8 cm/s or 52 000 $m^3s$. Adding the Stokes drift, the slope current and the wind driven current we find a total flow of $(33 + 41 + 52) \times 10^3 = 126 \times 10^3$ $m^3/s$.

The next step in measuring currents together in the Straits of Dover was recording the electric potential in cross-channel submarine cables induced by the flow of water through the Strait. This method, of which the possibility had already been pointed out by Faraday, was carried out by a group of scientists from the Institute of Oceanographic Sciences. A difficulty in discussing the results is that this English group was not fully aware of the results of van Veen and the analysis of Wyrtki. In this respect, the Strait appeared to be still an important barrier between England and the Continent. A summary of results is given by Otto (1984). The analysis of residual flows measured by the Dover cable yielded a mean residual flow between 7400 $km^3/a$ (Cartwright 1961) and 4900 $km^3/a$ (Prandle 1978).

A strong point of the cable measurements is the length of the record, one series now being analyzed for over 40 years, but a weak point is that the results have to be calibrated with real current measurements. Unfortunately, in the post-war years traffic through the Strait has become so intense that a full row of current stations across the narrowest part of the Strait could no longer be occupied. Prandle and Harrison (1975), for example, use for their calibration four stations covering only half the crossing on the English side in a line displaced several miles to the North Sea. Because the residual current pattern in and around the Strait is complicated (see below) it seems questionable whether this is allowed.

Also on the basis of Prandle's analysis the total flow can be subdivided into three components: a Stokes drift, a slope current and a wind driven current, giving together $(33+90+32)\times 10^3 = 155\times 10^3$ $m^3/s$. For the wind factor measurements on Dutch lightvessels were used, a method completely different from that of Wyrtki. The results are, therefore, not comparable, but other publications (Oerlemans 1978; Pingree and Griffiths 1980) confirm the larger value. On the basis of Wyrtki's results, it is even more difficult to accept a basic current of 90 instead of 41 000 $m^3/s$. We prefer, therefore, to accept the value of $126 \times 10^3$ m/s for the total residual flow.

More important than the actual value of the total flow is for what follows in the next paragraph the pattern of residual transport in the cross-section. The data of van Veen (1936, 1938) show that a considerable flood excess exists in the middle, deepest part of the Strait and an ebb excess on both sides, especially on the continental side (Fig. 3). Maximum values are about 20 cm/s in both directions. Van Veen, as already stated, measured during very quiet weather but the excesses are significantly larger than the average residual current of about 6 cm/s, so that the pattern will persist most of the time.

Of course, the cable measurements themselves do not show differences in residual flow and flow direction in the cross-section; however, they suggest that this flow is strongest in the deepest part of the channel. It can be expected that both the tidal residual flow and the wind current are strongest in the deepest part of the Strait.

The return flow measured by van Veen on the continental side can be considered as a southward extension, over the Flemish Banks, of the current gyre along the Belgian coast. The existence of opposite residual currents in one tidal passage was certainly no surprise for van Veen who in the same period found similar, and even more complicated patterns in Dutch tidal inlets. The direction of the strongest of the two tidal phases, flood or ebb, can be reflected on a sandy sea floor by flood or ebb parabolas. An important ebb parabola in the Strait of Dover is located in front of the harbour of Calais (van Veen 1936, p 114), confirming the existence of an ebb surplus.

## 3 Sediment Movements

No appreciable sand transport through the Strait was measured by van Veen (1936). Stride (1973) considers the area as a region of bed load convergence which no net sand movement. Our discussion will, therefore, be restricted to fine-grained material with an effective grainsize below 50 to 60 micrometers.

Older data (Ploix 1876, van Veen 1936) indicated concentrations of the order of 10 mg/l near the harbour of Boulogne, but offshore during van Veen's measurements the water was extremely clear. This is not surprising because of the fine weather prevailing during these measurements. Unfortunately he refrained from an extensive collection of suspended matter data.

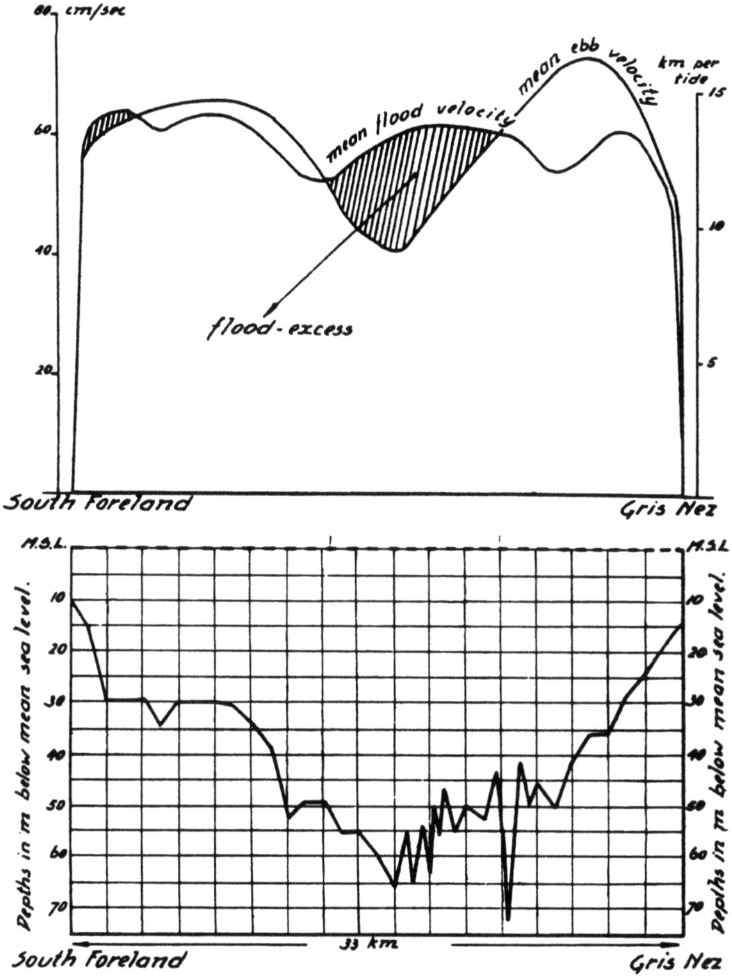


**Fig. 3.** *Upper graph* Ebb and Flood currents in the Strait, indicating ebb and flood excesses in different parts of the Dover-Calais section; *lower graph* depth profile of this section. (van Veen 1938)

Systematic investigations of suspended matter concentration, in fact covering the whole North Sea and the eastern part of the Channel, were carried out by Eisma and coworkers from 1977 (for a review, see Eisma and Kalf 1987). For our study only data from the region of the Strait are relevant. Using an average concentration of 1.9 mg/l Eisma (1981), taking Prandle's mean residual current value, calculates an input into the North Sea of about $10 \times 10^6$ tons, dry weight, per year, or 300 kg/s. In passing, it must be remarked that Eisma's samples were taken in a period of enhanced residual flow, which according to Prandle (1984), in December 1979, was $279 \times 10^3$ m³/s and in January 1980, $226 \times 10^3$ m³/s. Consequently, salinity in the Strait was also above normal (35.2‰ against 34.9‰), indicating the presence of undiluted Atlantic water.

Close to the shores on both sides of the Channel concentrations can rise to about 10 mg/l. Information (CNEXO 1981), brought to our attention by Dr. Eisma, gives comparable or somewhat higher values: 10.6 mg/l nearshore and 3.4 mg/l offshore. These are average data from a much longer series which in addition shows considerable differences, with a factor of about two, between summer and winter. A seasonal change is to be expected when in winter material is resuspended by waves in relatively shallow water.

The presence of ebb surplus water transport near the shores and flood surplus water transport in the centre of the Strait, combined with gradients in suspended matter, means that transport of finegrained sediment towards the North Sea must be less than assumed on the basis of over-all one-way flow. Table 1 gives a calculation based on the information of Fig. 3. The Strait is divided into four parts, according to the four ebb and flood sections. For every section surface area, flood volume and suspended matter transport is calculated.

**Table 1.** Water and sediment transport in the Strait of Dover. The Strait is divided into four parts according to four ebb and flood sections

| Sections | Water transport cm/s | m³/s | Sediment transport mg/l | kg/s |
|---|---|---|---|---|
| I flood excess | + 4 | + 3 × $10^3$ | 10 | + 30 |
| III flood excess | + 12 | + 67 × $10^3$ | 3 | + 202 |
| Flood total | | + 70 × $10^3$ | | + 232 |
| II ebb excess | - 6 | - 23 × $10^3$ | 3 | - 69 |
| IV ebb excess | - 12 | - 41 × $10^3$ | 10 | - 410 |
| Ebb total | | - 64 × $10^3$ | | - 479 |
| Net transport | | + 6 × $10^3$ | | - 247 |

In the example net water transport is very small, for reasons already given in the previous paragraph, and net sediment movement, –247 kg/s, is from the North Sea towards the Channel. However, to the net transport of water of 6 × $10^3$/s, m³/s, 120 × $10^3$ m³/s has to be added to obtain average residual transport of 126 × $10^3$ m³/s. Combined with an average suspended matter concentration of 3 mg/l this gives 320 kg/s. The mean sediment transport towards the North Sea then amounts to 320–247 = 73 m³/s or 2.5 × $10^6$ tons/a. The numbers used in this calculation could, within reasonable limits, be changed in such a manner that a larger or even smaller residual transport toward the North Sea would result. However, this transport would in any case be significantly smaller than 10 × $10^6$ tons per year.

## 4 Conclusions

No firm conclusions can at present be drawn about the amount of suspended matter that enters the North Sea from the Channel. A considerable amount must be carried back and forth through the Strait, however. What is indeed heeded is an extensive programme measuring suspended matter concentrations over an extended period across the Strait. It is also necessary to determine the pattern of residual flow under different weather conditions and in the four seasons.

The question of finegrained sediment input into the North Sea from the Channel is not academic because of the amounts needed in estuaries of the region to cope with the gradual rise of sea level and a possible acceleration of this rise in the future. Other sources of sediment supply around the North Sea are quite accurately known (Eisma 1981) and at present certainly do not amount to more than $7 \times 10^6$ tons/a. It is also known that there are great losses from the shallow part of the region into the deeper part, especially the Skagerak and Norwegian Deep. A sensible budget seems hardly possible without mobilizing large quantities, of the order of $10 \times 10^6$ tons per year, from the North Sea bottom or without net input through its northern boundary.

## References


Carruthers JN (1928) The flow of water through the Straits of Dover as gauged by continuous current-meter observations at the Varne Lightvessel (50°56′N 1°17′E), Part 1. Methods employed, with a preliminary survey of the results. Fishery Invest Lond Ser 2, 11:1–109

Carruthers JN (1935) The flow of water through the Straits of Dover as gauged by continuous current-meter observations at the Varne Lightvessel (50°56′N 1°17′E), Part 2. Second report on results obtained. Fishery Invest Lond Ser 2, 14:1–67

Carruthers JN, Lawford Al, Veley VFC (1950) Studies of water movements and winds at various lightvessels in 1938, 39 and 40. The Varne Lightship and her successors. Ann Biol Copenh 6:115–120

Cartwright DE (1961) A study of currents in the Straits of Dover. J Inst Navig London 19:130

Eisma D (1981) Supply and deposition of suspended matter in the North Sea. Spec Publ Int Assoc Sediment 5:415–428

Eisma D, Kalf J (1987) Dispersal, concentration and deposition of suspended matter in the North Sea. J Geol Soc (Lond) 144:161–178

Gehrke J (1907) Mean velocity of the Atlantic currents running north of Scotland and through the English Channel. Publ Circ 40:18

Oerlemans J (1978) Some results of a numerical experiment concerning the wind-driven flow through the Straits of Dover. Dtsch Hydrogr 5:182–189

Otto L (1984) Flushing times of the North Sea; 8, Straits of Dover. ICES Coop Res Rept 123:79–86

Pingree Rd, Griffiths DK (1980) Currents driven by a steady uniform wind stress on the shelf seas around the British Isles. Oceanol Acta 3:227–236

Ploix E (1876) Rapport sur la reconnaissance de Boulogne. Rech hydr sur le regime des cotes, 5e cahier, V, Paris (cited by van Veen)

Prandle D (1978) Monthly-mean residual flows through the Dover Strait, 1949–1972. J Mar Biol Assoc UK 58:965–973

Prandle D (1984) Monthly-mean residual flows through the Dover Strait, 1949–1980. J Mar Biol Ass UK 64:722–724

Prandle D, Harrison AJ (1975) Relating the potential difference measured on a submarine cable to the flow of water through the Strait of Dover Dtsch Hydrogr Z 28:207–226

Stride AH (1973) Sediment transport by the North Sea. In: Goldberg ED (ed) North Sea Science, MIT Press, Cambridge, pp 101–130

Veen Van J (1936) Onderzoekingen in de Hoofden in verband met de gesteldheid der Nederlandse kust. Algemene Landsdrukkerij 's-Gravenhage, p 252
Veen Van J (1938) Water movements in the Straits of Dover. J Conseil 13:7-36
Veley VFC (1960) The relationship between local wind and water movement in coastal waters of the British Isles. In: Pearson EA (ed) Proc 1st Conf on Waste Disposal in the Marine Environment, Univ Calif Berkeley July 22-25, 1959. Pergamon, London, pp 285-295
Wyrtki K (1952a) Der Einfluss des Windes auf die Wasserbewegungen durch die Strasse von Dover. Dtsch Hydrogr Z 5:21-27
Wyrtki K (1952b) Der Einfluss des Windes auf den mittleren Wasserstand der Nordsee und ihren Wasserhaushalt. Dtsch Hydrogr Z 5:245-252

# New Concepts in Patch Recognition of Suspended Matter in Coastal Areas

K.-H. Szekielda D. McGinnis, and R. Carey

## 1 Introduction

Over the last years it has been demonstrated that the visible data obtained with the NOAA Advanced Very High Resolution Radiometer (AVHRR) can be used for detection of water masses with high turbidity and/or plankton levels (Szekielda 1982; Stumpf 1987).

Although the AVHRR was not designed for this application, Stumpf (1987) showed that with algorithms that he developed sediment concentration may be estimated with an accuracy of $\pm 30\%$ under the assumption of constant and uniform sediment grain size, and chlorophyll with an error of $\pm 60\%$. In an effort to elaborate more on the presently available data sets of the AVHRR, the following study had the objectives to use additional information from the channels of the AVHRR and identify better display techniques of multispectral data. In addition, simple procedures to partially eliminate the atmospheric contribution and to correct for sun angle and/or sun glint have been evaluated.

In considering multispectral data for applications in oceanography, it is necessary to keep in mind that besides the Coastal Zone Color Scanner (CZCS) no other system has been spectrally designed to extract information obtainable especially in the visible and infrared part of the electromagnetic spectrum over the ocean. This is important since the existing remote sensing systems on satellites such as the MSS or TM on Landsat, monitor, according to their selection of spectral bands, images in different bands and are the basis for contrast RGB composite pictures. As in many multispectral recordings over water, *strongly correlated* data sets are obtained yet their display is either limited or tends to diminish certain features. Therefore, several techniques, to be shortly described in the following, have been applied for enhancing the display of AVHRR data.

As has been shown previously by Szekielda and McGinnis (1987) the use of ratioing two channels eliminates to a certain degree, the interference of sun glint and the atmosphere to the signal recorded at satellite altitudes. A pair of image channels can be displayed in the form of a normalized index

$$NI = \frac{CH_2 - CH_1}{CH_1 + CH_2} \times F,$$

where $CH_1$ represents the data recorded at 0.6–0.7 μm and $CH_2$ data recorded at the spectral interval of 0.7–1.0 μm. F is a factor introduced in order to bring the NI values into the best range for data display. The interpretation of NI values is a nonconservative approach as the same values may be caused by different processes. As Gillespie et al. (1987) have stated, some complexities arise in ratio images

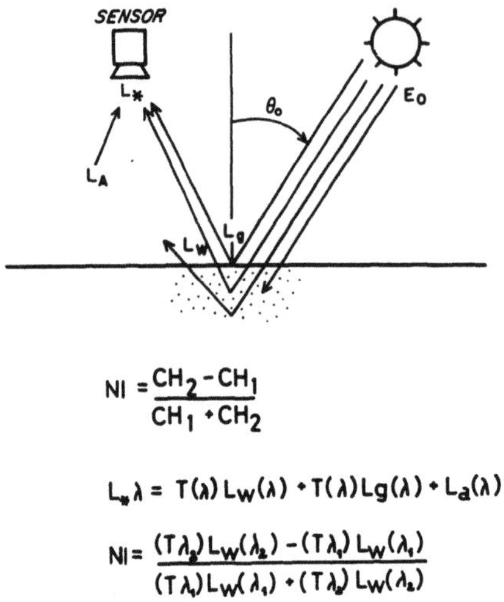


**Fig. 1.** Simplified presentation of the processes involved to interpret the normalized index from the wavelengths $Ch_1$ = 0.6–0.7 μm and $CH_2$ = 0.7–1.1 μm

because of the effects of atmospheric path radiance. Also, such ratioing decreases the effect of increasing satellite scan angle (Gallo and Eidenshink 1988).

The radiance obtained at two wavelengths $L_* \lambda_1$ and $L_* \lambda_2$ with a satellite can be written in a very simplistic way as

$$L_* \lambda_1 = T(\lambda_1)\,Lw(\lambda_1) + T(\lambda_1)\,Lg(\lambda_1) + L_A(\lambda_1)$$
$$L_* \lambda_2 = T(\lambda_2)\,Lw(\lambda_2) + T(\lambda_2)\,Lg(\lambda_2) + L_A(\lambda_2),$$

where

$L_A$ = path radiance;
T = atmospheric transmission coefficient;
Lw = radiance leaving water column;
Lg = radiance reflected from water surface.

A simplistic description of these processes is shown in Fig. 1. The assumptions for using the ratio NI is that the contributors to the radiance received by the radiometer from the atmosphere and sun glint are building constants throughout the scene and build values which are outside the dynamic range of the NI values for turbidity patterns. The NI can be rewritten and has the meaning of

$$NI = \frac{(T\,\lambda)\,Lw(\lambda_2) - (T\,\lambda_1)\,Lw(\lambda_1)}{(T\,\lambda_1^2)\,Lw(\lambda_1) + T(\lambda_2)\,Lw(\lambda_2)} \times F.$$

## 2 Display Methods

### 2.1 AVHRR Tape and Format Data

The digital data acquired by the AVHRR are available in two formats. The 10 bit data may be purchased in either 1600- or 6250-BPI (Bytes per inch) format containing information from all channels (Table 1). Each line of data contains 2048

**Table 1.** Advanced very high resolution radiometer sensor characteristics for polar orbiting satellites (After Kidwell 1986)

| Channel No. (1) | Band width (μm) TIROS-N (2) | NOAA -6,-8,-9 (3) | NOAA -7,-10 (4) |
|---|---|---|---|
| 1 | 0.55- 0.90 | 0.58- 0.68 | 0.58- 0.68 |
| 2 | 0.73- 1.00 | 0.73- 1.00 | 0.73- 1.00 |
| 3 | 3.55- 3.93 | 3.55- 3.93 | 3.55- 3.93 |
| 4 | 10.50-11.50 | 10.50-11.50 | 10.50-11.30 |
| 5 | Ch 4 repeated | Ch 4 repeated | 11.50-12.50 |

pixels or bits of information having a resolution of approximately 1.1 km at nadir. Data are scanned at the rate of six lines per second, with a satellite orbit nodal period close to 101.50 min. A complete description of the entire NOAA satellites series may be found in Kidwell (1986).

## 2.2 VICOM Processing Description

The scatter plots used to identify individual sediment clusters were generated on a VICOM digital image processor. The VICOM system was useful for manipulating the AVHRR data in a variety of ways. The raw data tape was first unpacked into two separate digital frames corresponding to coincident views of the study area as sensed by the AVHRR channels 1 and 4. Each frame was stored as a 512 × 512 image at full 10 bit resolution in the frame buffer. An analyst then constructed a mask to separate the land and clouds in the scene from the sedimentation in the water. This was done by using a trackball to trace the mask border in an overlay graphics plane and then filling in the area to be ignored by the scatter plotter routine. The routine then examined the unmasked area in each of the two AVHRR frames and constructed a scatter plot with the density color coding. This colorization was useful in determining the extent of individual clusters. The separate target plots were generated with similar mini-masks and made at the same resolution as the original full scene scatter plot. Both plots and image frames were then downloaded to a color film recorder for hardcopy.

## 2.3 Image Processing

The normalized index and thermal information were based on the 1.1 km, local area coverage (LAC) data sets and processed with a ground resolution or pixel size of about 2 km. Both data sets have been displayed as hue with intensity varying as functions of data recorded in the spectral range of 0.58–0.68 μm and stretching to full range of recorded counts for water features. Land contouring and identifying most of the clouds could be accomplished by using a 85 count threshold on Channel 2 (0.7–1.0 μm). This was found to be a critical threshold level, as water features disappeared in the images. Clouds, on the other hand, could be distinguished by patterns, nearness to masked clouds, and brightness values.

A special program written for this analysis has been applied. The outputs from the displays are being specified.

1. Raw data were displayed with color annotations; Channel 1 as green, Channel 2 as red, and Channel 4 as blue.
2. The normalized index was displayed as hue, using the visible brightness for land and clouds as black.
3. The thermal channel was displayed as hue, also using the visible brightness and with land and bright clouds black as with the NI. Keys have been added to the NI and Channel 4 images to relate hue to data values. All of the hues are the brightness color on the key, but brightness of the images is based on Channel 1 brightness which has to be kept in mind when identifying colors on the images for the Somali coast.

## 2.4 Data Interpretation

For interpretation of the data obtained with the AVHRR, the spectral properties of particulate matter and the spectral response of the channels in the radiometers have to be compared. A set of data from algae suspension with additions of clay has been recorded with a spectroradiometer over a wavelength range from 0.4–0.85 $\mu$m with a resolution of 0.002 $\mu$m. The spectroradiometer was installed above a container with 12 liters in capacity and 20 cm deep and painted with low reflectance paint. The observation angle was $\angle 20°$ from nadir in order to avoid surface glint and to view only the samples rather than the walls. For standardization and reference for the spectral, a standard white disk held horizontally at the surface of the sample container was used.

Standard suspended clay mixtures were used in this study as reflective sediment load. An algae mixture was composed of a concentration of $7.93 \times 10^2$ mg/m$^3$ Chl. *a* containing *Carteria, Phaeodactylum, Isochrysis*, and other marine diatoms. The total direct count of the algae standard was $138.5 \times 10^7$ cells/liter while the white clay standard had $140 \times 10^9$ counts/liter (c/l) with 95% of the size distribution in the 1–42 $\mu$m range. The red clay standard had a total concentration of $108 \times 10^9$ c/l with 95% in the 1–60 $\mu$m range (see Table 2).

**Table 2.** Algae and clay suspensions used in Fig. 2 and resulting effective radiance recorded in Channels 1 and 2 with resulting normalized index

| NR | Chlorophyll *a* (mg m$^{-3}$) | Phytoplankton (cells l$^{-1}$) | Phytoplankton + clay (counts l$^{-1}$) | 0.55–0.75 $\mu$m $CH_1$ | 0.7–1.0 $\mu$m $CH_2$ | $CH_2$-$CH_1$ | NI |
|---|---|---|---|---|---|---|---|
| 1 | 0 | 0 | 0 | 0.015 | 0.0096 | -0.054 | 0.22 |
| 2 | 64.18 | $11.21 \times 10^7$ | $11.21 \times 10^7$ | 0.054 | 0.140 | 0.086 | 0.44 |
| 3 | 118.76 | $20.75 \times 10^7$ | $20.75 \times 10^7$ | 0.065 | 0.181 | 0.116 | 0.47 |
| 4 | 118.31 | $20.67 \times 10^7$ | $72.97 \times 10^7$ | 0.071 | 0.255 | 0.184 | 0.56 |
| 5 | 117.44 | $20.52 \times 10^7$ | $164.24 \times 10^7$ | 0.094 | 0.362 | 0.268 | 0.58 |

The base water used in this investigation was designed to be an optically dense quasi-model of clear, deep seawater. Water-soluble food dyes were added until a transmittance curve of the water standard was obtained, which was a close reproduction of the seawater transmittance curve, for clear Mediterranean seawater.

The conclusions from the analysis of the spectra shown in Fig. 2 are:

1. the major albedo from suspended matter is in the red and reflected near-infrared part of the electromagnetic spectrum;
2. the spectra of the pure algae suspension show no specific absorption bands as compared to the well-pronounced spectral characteristics of pure chlorophyll solutions;
3. the spectral region between 0.60–0.65 μm shows minimal changes in albedo with respect to changes in particle concentrations;
4. with constant chlorophyll concentration but increasing inorganic particle load, the spectral response is enhanced and is a function of the inorganic component rather than the organic matter.

In this respect the channels of the AVHRR, covering the spectral part as shown in Fig. 2, are recording two separate parts of the spectral response of organic and inorganic material.

The resulting normalized index from this experiment is also shown in Table 2 and gives the spectral response of the suspensions integrated over the response function of the AVHRR Channels 1 and 2. As the actual measurements from satellite altitudes will include the atmospheric contribution the N1 will be negative due to the higher contribution of the atmosphere at shorter wavelengths. The data

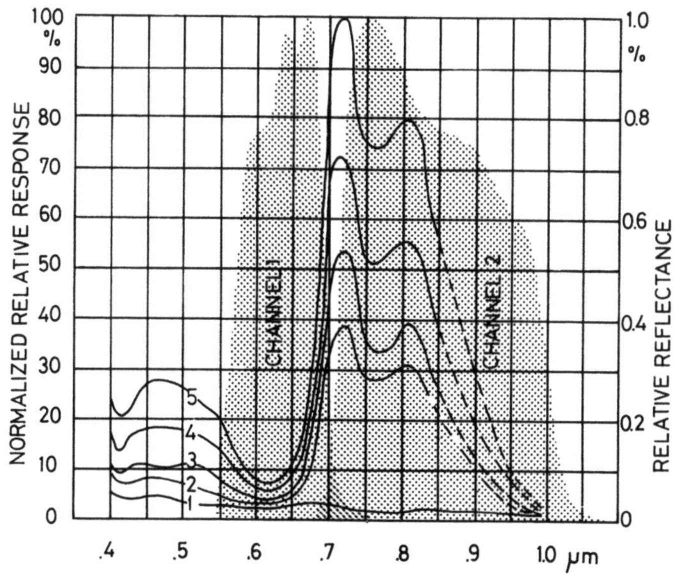


**Fig. 2.** Spectral normalized relative response of Channels 1 and 2 of the AVHRR (after Kidwell 1986) and relative reflectance of algae and clay suspensions (see Table 2)

in Table 2 also demonstrate that the rationing provides a more sensitive parameter than the single channels or the difference of both channels and also that the correlation between particle content and N1 is logarithmic in nature.

The radiometer originally measures counts which can be converted into percent albedo. Prior to launch the AVHRR instrument channels in the visible region are calibrated against a large aperture sphere equipped with 12 calibrated quartz-iodide lamps. By varying the number of bulbs which are turned on, a calibration curve from the dark level to a maximum of 12 lamps output can be obtained.

The following computations must be made in order to present the calibration in terms of percent albedo vs radiometer output and are based on data reported by Kidwell (1986): The spectral output of the sphere is integrated with the spectral response function of the AVHRR channel to yield an effective radiance for the spectral band for 12 lamps operating. This is then multiplied by the appropriate K factor to convert to n lamps. $N_L$, the effective radiance as seen by the channel in the appropriate spectral band is described by the following equation:

$$N_L = K_n \int_{\lambda^1}^{\lambda^2} C(\lambda)\, \phi(\lambda)\, d\lambda,$$

where

$K_n$ = the factor to convert to radiance for n lamps;
$C(\lambda)$ = calibrated spectral radiance of the sphere with 12 lamps on;
$\lambda$ = wavelength, in the spectral region $\lambda_1$ to $\lambda_2$;
$\phi(\lambda)$ = the measured spectral response of the channel being calibrated.

If one takes the solar irradiance at the top of the atmosphere and performs a similar calculation:

$$N_S = 1/\pi \int_{\lambda_2}^{\lambda_1} S(\lambda)\, \phi(\lambda)\, d\lambda,$$

where

$N_S$ = effective radiance of the radiometer viewing reflected sunlight;
$S(\lambda)$ = spectral irradiance viewed at the top of the atmosphere;
$\phi(\lambda)$ = spectral response function of the channel;

Then the resultant $N_S$ represents what would be "seen" from space with a 100% reflecting, diffuse surface when the solar zenith angle is zero.

Thus, the percent albedo A is:

$$A = N_L/N_S \times 100.$$

## 3 Results

The high particulate matter concentration in the North Sea makes it possible to monitor concentration gradients with the visible channels of the AVHRR.

The analysis of satellite data and the use of an index as a parameter for particulate matter in the North Sea indicate that the response of the dual channel approach is responsive to the concentration and the composition of material in suspension. Consequently, the measurements or values obtained with the AVHRR normalized index also indicate processes acting on the particles in suspension, i.e., photosynthesis, sedimentation, and resuspension of eroded material (Szekielda et al. 1983). The index, therefore, is a potential tool for monitoring processes in connection with turbulence. As tidal currents are responsible for resuspension of particles due to erosion and settling of particles to the bottom, the pattern recognized in the satellite image as well as data collected with conventional methods do not necessarily reflect the average distribution of particles. It can be assumed that according to the velocities of tidal currents in the North Sea, a change in concentration of particles may be expected as a function of current velocities. This in fact shows that besides ship observations, repetitive time series are required to better understand the dynamics of particulate matter in the North Sea.

Experiments to combine the information received by the different channels given in Fig. 3. With respect to the information content, the presentation of the AVHRR data is optional, except that the perception and interpretation of thermal data as an indicator for physical oceanographic processes and the visible data as an indicator for biological activities are difficult to comprehend for a scientist untrained in color interpretation. This includes the understanding of hue, saturation, and brightness for a three-dimensional display of data and the effects of additive color mixing when combining the signals from multispectral data. The separate presentation of the thermal data and the normalized index as a function of Channels 1 and 2 for another date is shown in Fig. 4. This display presents additional information on the color distribution which is neither obtained by the display of the single visible channels nor by the thermal infrared.

The presentation of the normalized index for the two case studies in the North Sea is presented in Fig. 5. It shows the impact of tides on erosion as well as the high particle load carried by rivers into coastal environment.

For a more quantitative approach to analyze patterns based on the distribution of particulate matter, different data processing methods have been applied along the East China Coast. Patterns recognized in the visible part of the electromagnetic spectrum along the East China coast are basically influenced by the bathymetry and the discharge of high turbidity water. In the northern part, the spectral response is mainly due to bathymetry which is also manifested in the thermal band due to the low heat storing capacity of the coastal areas. As a consequence, during summer, the near-coastal region warms to above 20 °C while cooling to almost freezing during the winter. Based on the extreme temperature difference between the coastal and offshore waters, the areas with shallow water are well outlined in the infrared part of the spectrum. On the other hand, the visible reflectance values alone make it difficult to distinguish features created by bottom topography from those created by sediment loaded water.

Another feature which has been observed several times is connected to an upwelling process offshore of the Jianggang region. Therefore, the interpretation of the visible data has to consider spectral responses from the eroded bottom sediments, concentration of suspended matter from the river outflow, and the possible effect of nutrients transported through the river system, or brought to the surface by upwelling processes. All these processes between the thermal infrared

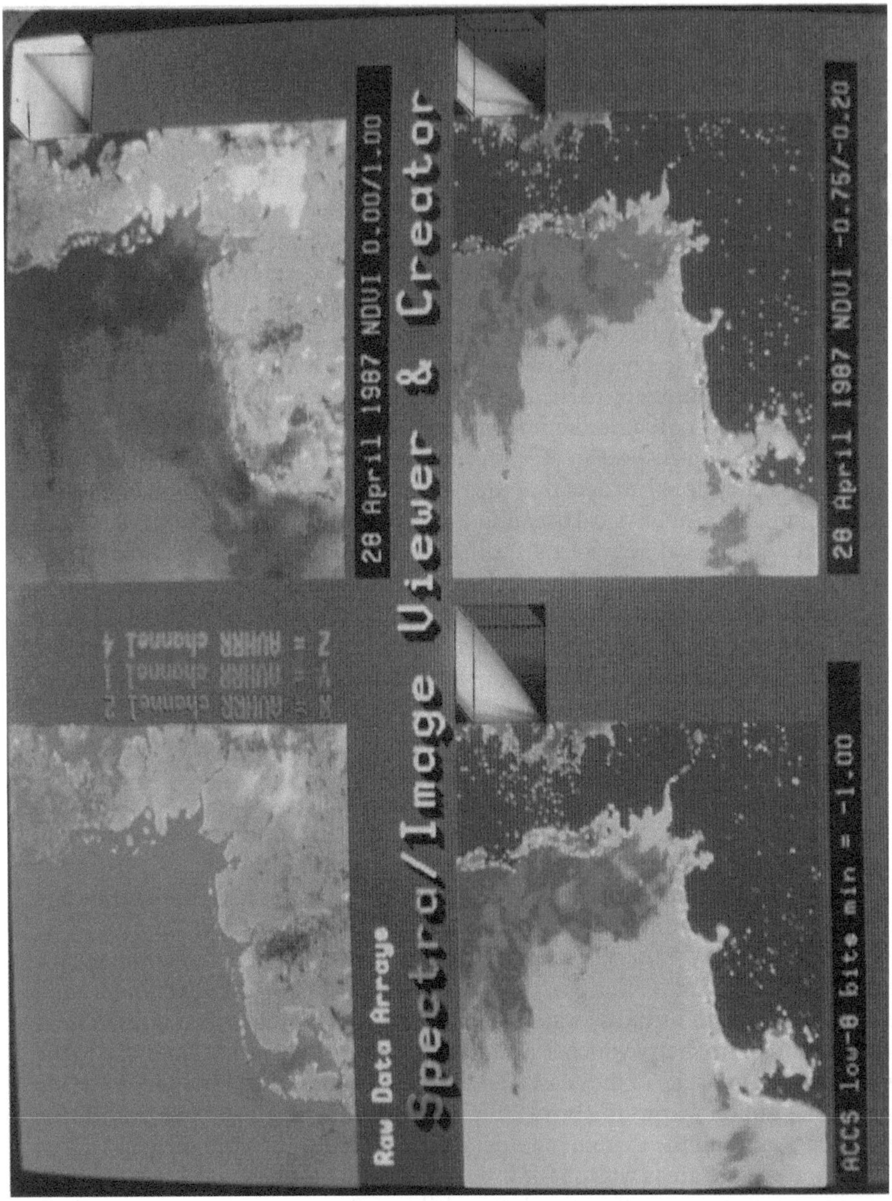
Raw Data Arrays
Spectra/Image Viewer & Creator
28 April 1987 NDVI 0.00/1.00
X = AVHRR channel 2
Y = AVHRR channel 1
Z = AVHRR channel 4
ACCS low-8 bits min = -1.00
28 April 1987 NDVI -0.75/-0.20

Fig. 3

data and the normalized index are shown in Fig. 7. Included in Fig. 8 are the clusters for the selected test sites (see Table 3; Fig. 6 Part II). This analysis shows that the normalized index in connection with the infrared observations discriminates the water masses and also has specific signatures.

The previous assumption that ratioing the two visible channels of the AVHRR leads to partial elimination of the atmospheric effects can be demonstrated with the observations of the East China coast. The original in Fig. 6 has been processed for the normalized index and is shown in Fig. 9. In the samples, the shown high concentration of suspended material, mainly from river discharge and erosion of deposits, has been responsible for thc response of the AVHRR. However, as can be shown with Fig. 10 in the upwelling area of the Somali Coast, pure plankton communities also give a clear signal.

## 4 Conclusions

The approach to interpret the red reflectance of particulate matter showed that patterns can be identified if careful selection of cloud-free observations has been done. The display of the normalized index, as well as displays of the two visible channels in combination with the thermal channel through clustering, demonstrate that the identification of water masses can be accomplished. Also, part of the atmospheric effect can be eliminated if ratios are applied in the image display. This has been demonstrated with data from the North Sea, the Changjiang and the Somali Coast. It also was found that the normalized index images indicate the changes in particle concentration as well as its composition.

**Fig. 3.** AVHRR LAC image of the North Sea using only the low 8 bits of the 10 bits on the tape. This set of images shows the raw data and three color presentations for the 28 April 1987 case. *Upper left:* Raw data 200×200 pixel arrays sampled from a 512×512 LAC image using every other pixel from a 400×400 section of the image. This is a three-dimensional RGB (red, green, and blue) image: *red*, near-IR (Channel 2; x-axis); *green*, visible (Channel 1; y-axis); *blue*, thermal-IR (Channel 4; z-axis). *Upper right:* Three-dimensional IHS (intensity, hue, and saturation) color image of the normalized difference index (NDI) for land surfaces (NDVI values between -0.03 and 1.00) which uses the thermal-IR channel as a cloud mask by reducing color saturation for colder pixels. The result is white clouds. Scene brightness is the same as it is for false color-IR images. Water is *magenta* if it is warm enough (*dark gray* if it is cold), bare soil is *red*, and increasing vegetation is *yellow, green, cyan*, and finally *blue* for the highest vegetation levels. A three-dimensional coordinate cube is plotted on or near each image to help interpret the physical meaning of the colors. The x-axis is at the bottom, the y-axis is along the left edge, and the z-axis goes back into the image. These axes are labeled on each image next to the raw data display in *panel 1*. [Intensity, brightness of visible or near-IR; hue, NDVI; saturation, thermal (cold objects become white).] *Lower left:* Two-dimensional IHS color image of the NDVI for water surfaces (NDVI values between -1.00 and 0.0). Land is *dark gray*. Clear water is *dark blue* and increasing turbidity is shown by *blue* through *cyan, green, yellow* to *red* which is the equivalent of bare soil. (Intensity, brightness of visible or near-IR; hue, NDVI; saturation, not used.) *Lower right:* Same as *panel 3* except NDVI limits have been narrowed to -0.75 to -0.20 in order to enhance information content

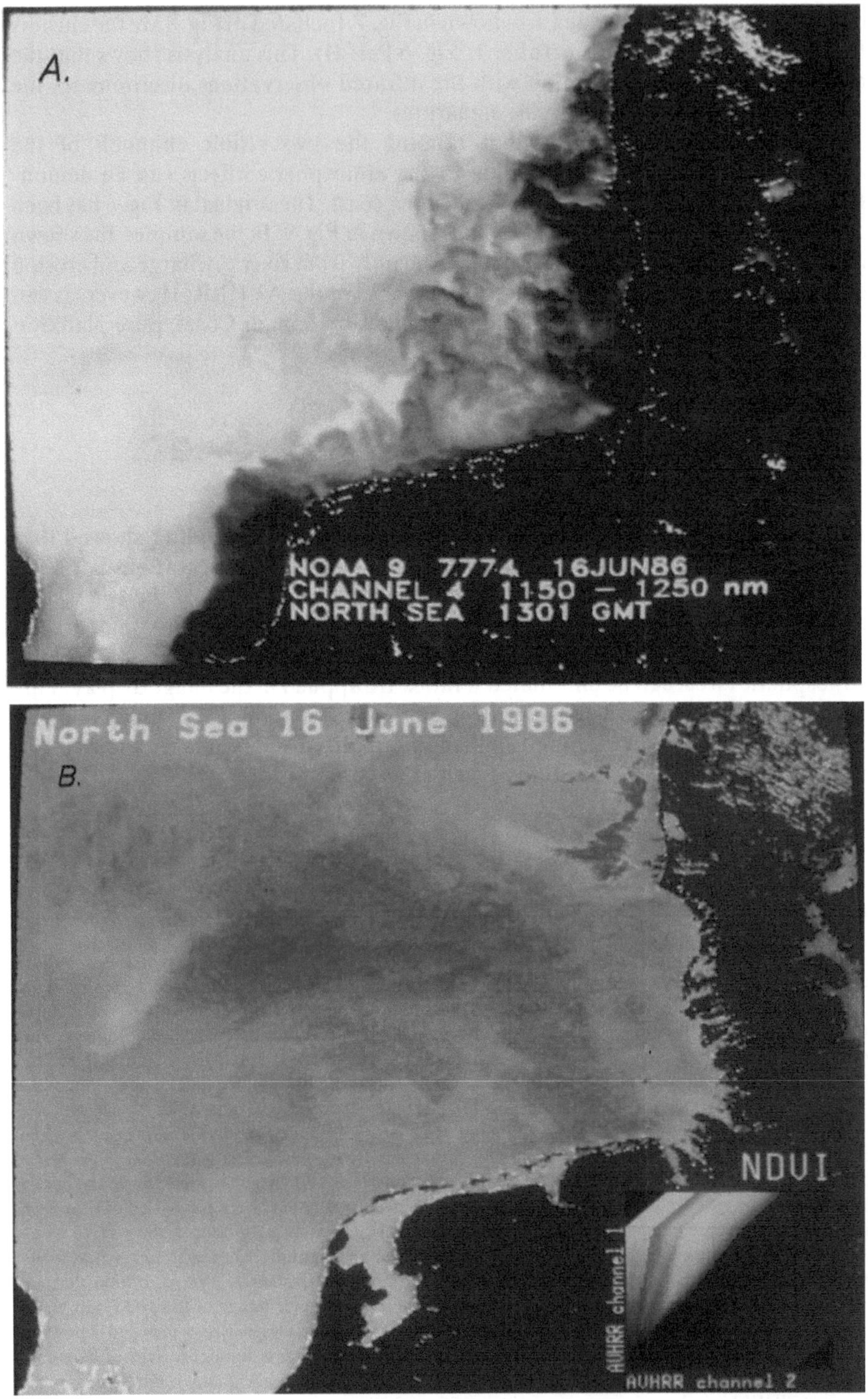
A.
NOAA 9 7774 16JUN86
CHANNEL 4 1150 – 1250 nm
NORTH SEA 1301 GMT
North Sea 16 June 1986
B.
NDVI
AVHRR channel 1
AVHRR channel 2

Fig. 4A,B

**Table 3.** Parameters for selected test sites along the E-China coast (for location see Fig. 6)

| | Albedo | | | Temperature |
|---|---|---|---|---|
| Target No. | $CH_1$[a] | $CH_2$[b] | NI[c] | range (K) |
| 1. Coastal erosion + discharge | 5.2 | 2.7 | -0.316 | 276.1-276.9 |
| 2. Outflowing from coast | 4.9 | 2.6 | -0.307 | 276.0-277.7 |
| 3. Plume not identified Yangtze or erosion | 5.5 | 3.0 | -0.294 | 277.6-278.0 |
| 4. Part of warm current Huanghai Sea | 3.7 | 2.3 | -0.233 | 279.4-280.0 |
| 5. Ejected erosion patch from coast | 4.6 | 2.3 | -0.333 | 276.3-276.8 |
| 6. Area of eutrophi-cation | 2.3 | 1.9 | -0.395 | 277.1-277.8 |
| 7. Coastal erosion | 5.0 | 3.4 | -0.190 | 275.1-275.3 |

[a] $A = 0.1071 \times \text{counts} - 3.9$.

[b] $A = 0.1051 \times \text{counts} - 3.5$.

[c] $NI = \frac{CH_2 - CH_1}{CH_2 + CH_1}$.

**Fig. 4A,B.** AVHRR data analyzed over the North Sea for 16 June 1986. **A** Infrared data from Channel 4 covering 11.5–12.5 μm. **B** Data produced from an AVHRR LAC image using only the low 8 bits of the 10 bits on the tape. This image shows one of many possible AVHRR color displays for water surfaces. Three-dimensional IHS (intensity, hue, and saturation) color image of the normalized difference vegetation index (*NDVI*) for land surfaces (NDVI values between -0.03 and +1.00) which uses the thermal-IR channel as a cloud mask. Scene brightness is the same as it is for false color-IR images. Intensity, brightest of visible or near-IR; hue, NDVI; saturation, thermal (cold objects become white)

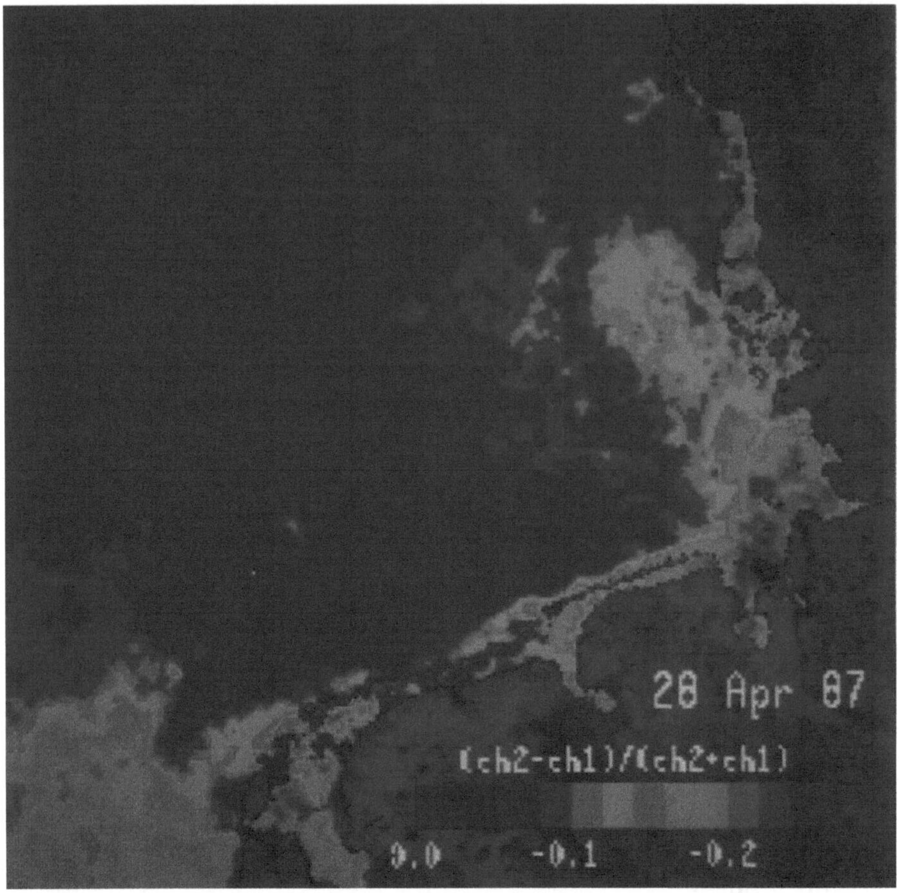


**Fig. 5.** Normalized index over the North Sea for two different dates

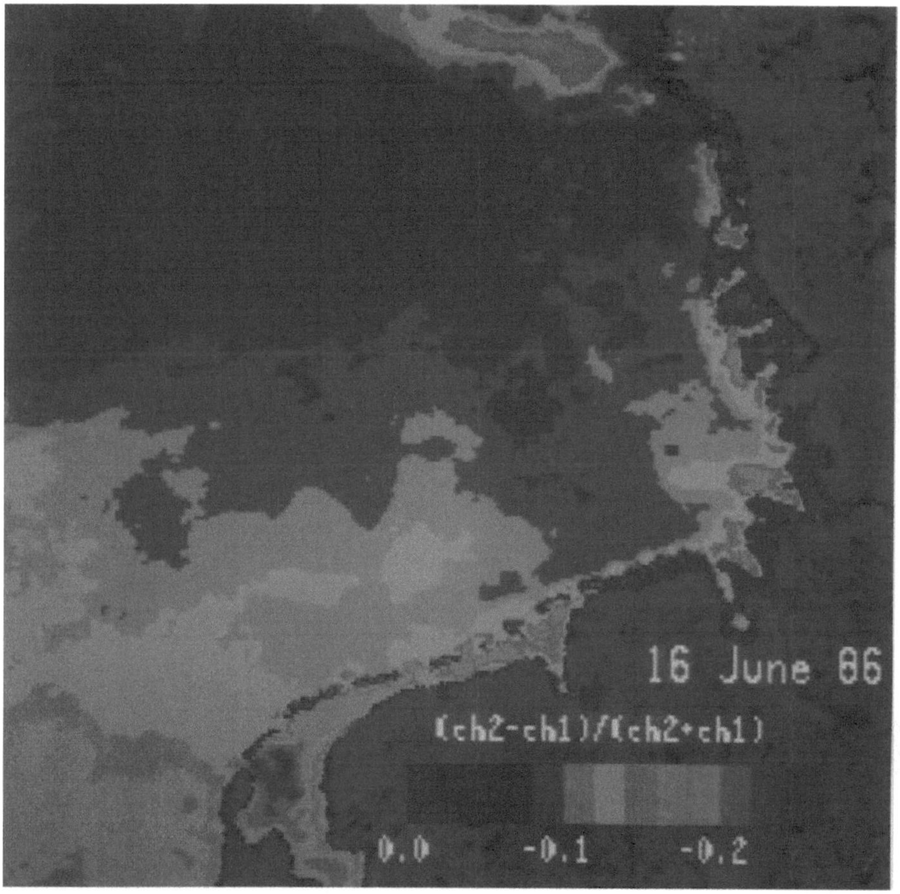


**Fig. 5, Part II**

**Fig. 6.** Distribution of suspended matter as recorded by NOAA 6. AVHRR on 9 March 1982 in Channel 1 (0.6–0.7 μm). *Above* Original recording; *below* selected targets for studies and location for test site used in Fig. 7. Description of targets *1–7* is given in Table 3

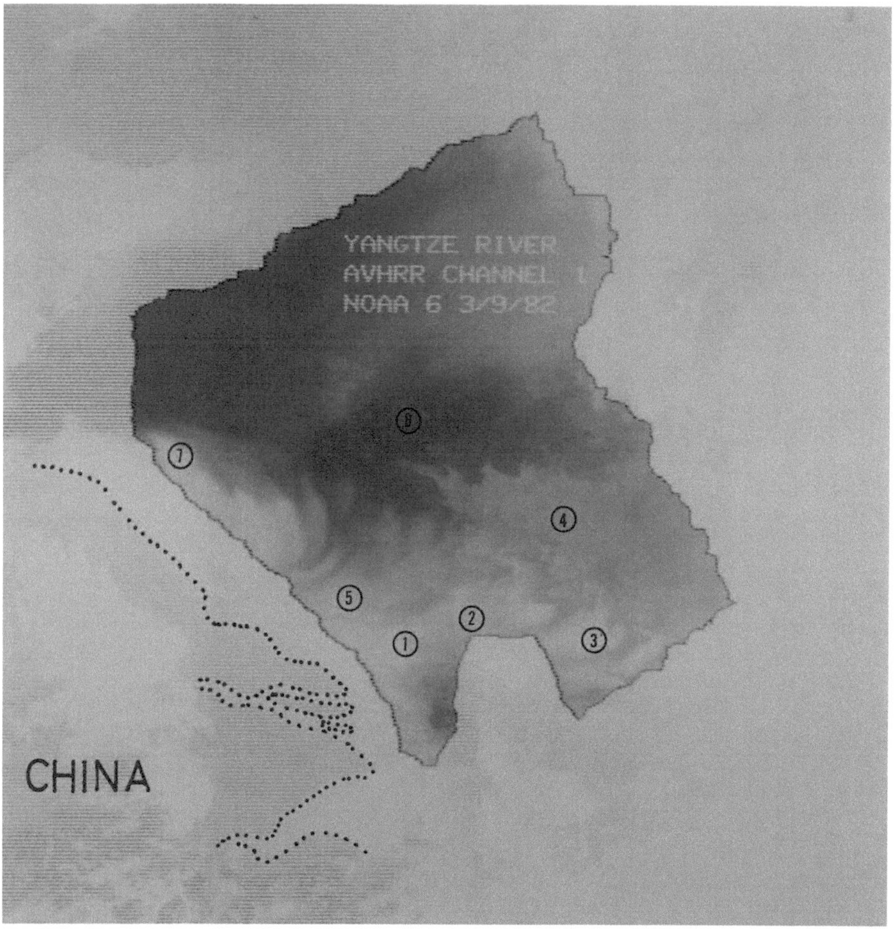


**Fig. 6, Part II**

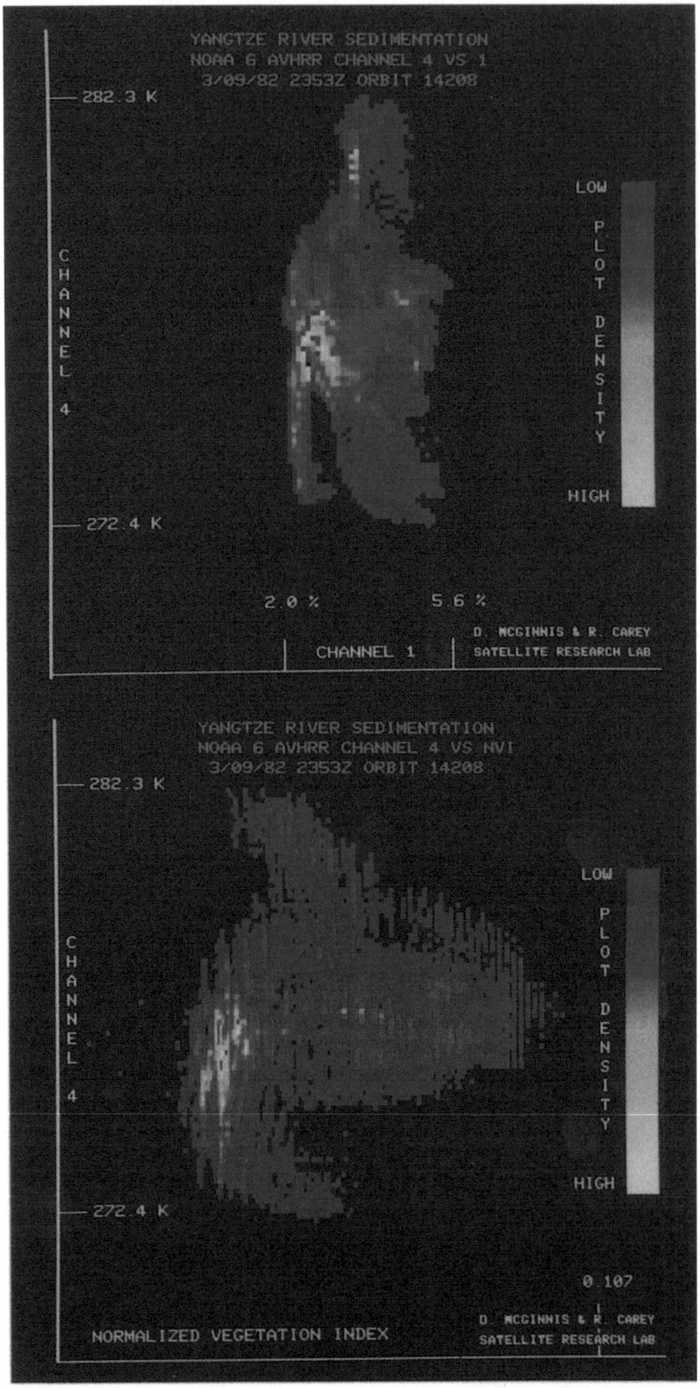


**Fig. 7.** Clusters for the area encircled in lower part of Fig. 6 Part II. *Upper cluster* shows Channel 4 versus albedo values. *Lower part* shows the cluster for Channel 4 versus NVI counts. The frequency of occurring data is shown in the inserted *plot density profile*

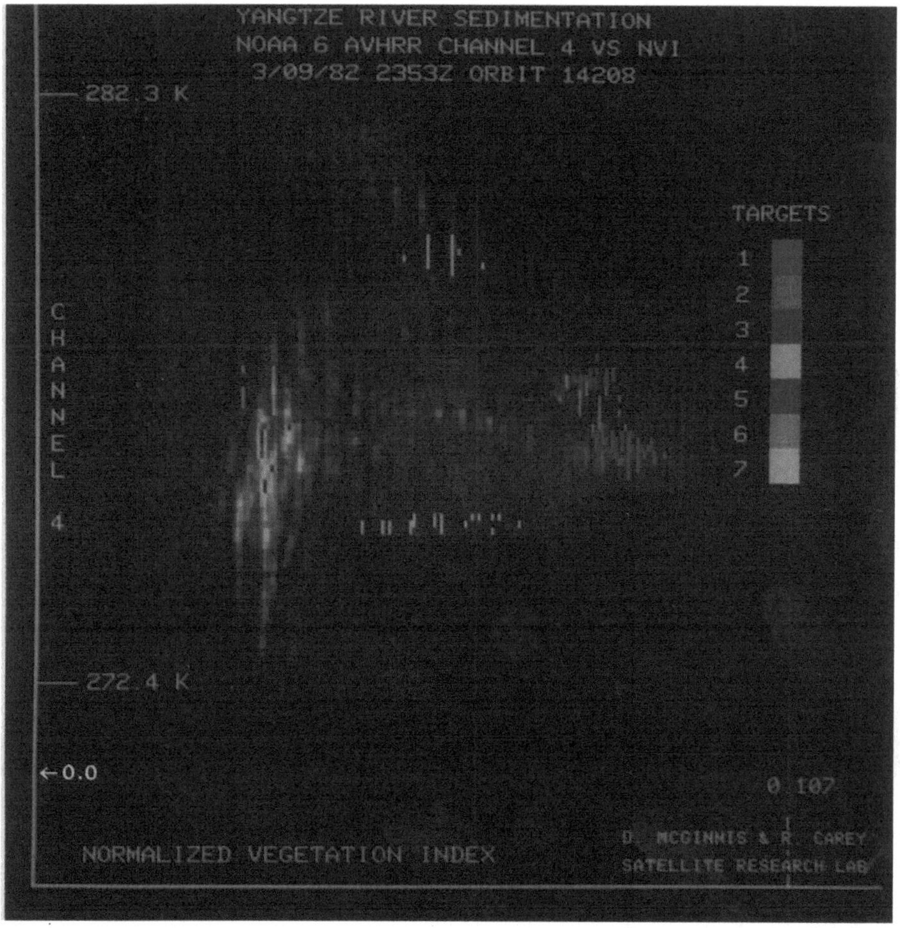


**Fig. 8.** Normalized index versus thermal infrared data of the area shown in Fig. 6 Part II and Table 3

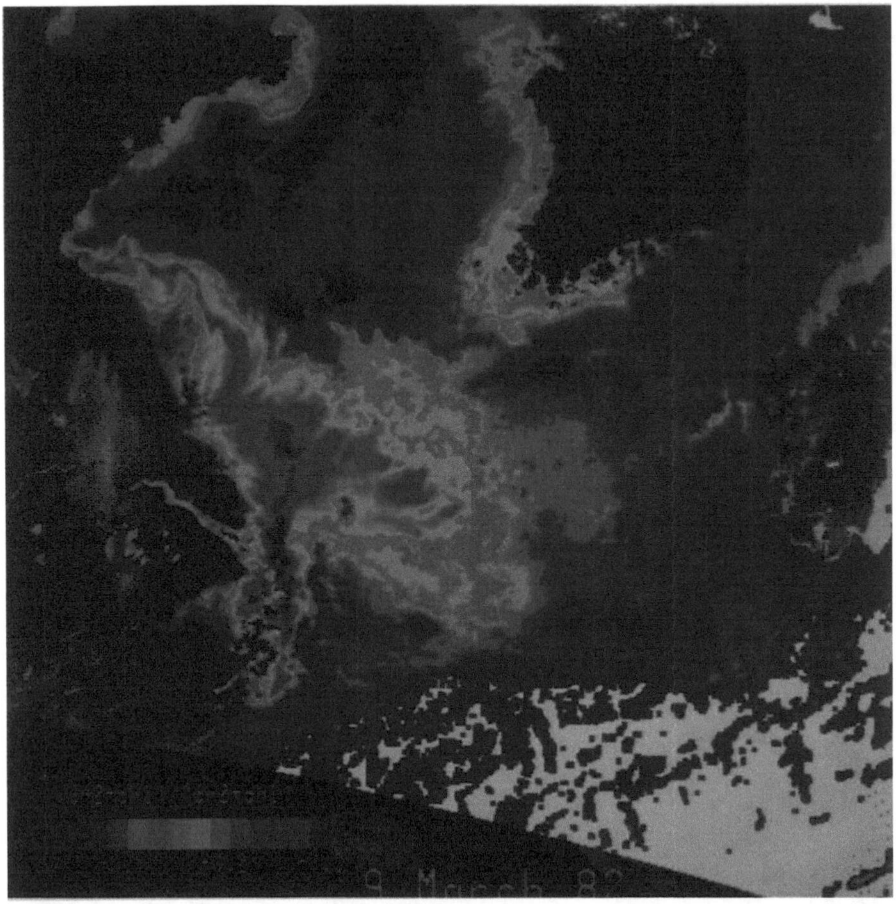

**Fig. 9.** Normalized index for the eastern coast of China

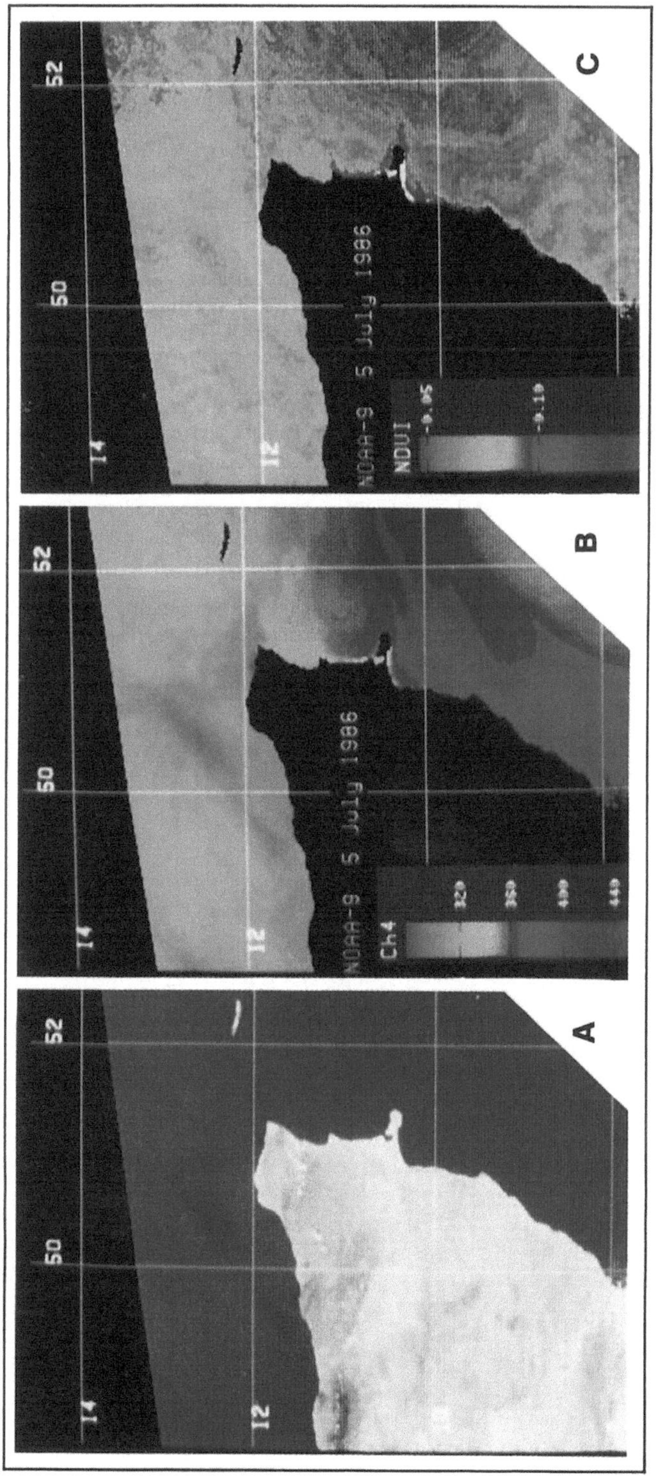


**Fig. 10A-C.** AVHRR recordings over the upwelling region of Somalia. **A** Simulated real color to demonstrate cloud-free conditions; **B** thermal infrared; **C** normalized index

# References

Gallo KP, Eidenshink JC (1988) Difference in visible and near-IR responses, and derived vegetation indices for the NOAA-9 and NOAA-10 AVHRRs, a case study. Photogram Engin Remote Sensing, 54:485–490

Gillespie AR, Kahle AB, Walker RE (1987) Colour enhancement of highly correlated images; channel ratio and chromaticity transformation techniques. Remote Sensing of Environment 22:343–365

Kidwell KB (1986) NOAA polar orbiter data user's guide. U.S. Department of Commerce, National Oceanic and Atmospheric Administration, National Environmental Satellite, Data, and Information Service, Washington, DC

Stumpf RP (1987) Application of AVHRR satellite data to the study of sediment and chlorophyll in turbid coastal water. NOAA Technical Memorandum NESDIS AISC 7, Washington, DC, 50 pp

Szekielda K-H (1982) Investigations with satellites on eutrophication of coastal regions. In: Degens ET, Kempe S, Soliman H (eds) Transport of carbon and minerals in major world rivers, Pt. 1. SCOPE/UNEP Sonderband, 52:13–37

Szekielda K-H (1988) Investigations with satellites on eutrophication of coastal regions Pt. 7. Response of the Somali Upwelling into monsoonal changes. Mitt Geol Palaont University of Hamburg, SCOPE/UNEP Sonderband Heft 66, 1–33

Szekielda K-H, McGinnis D (1985) Investigations with satellites on eutrophication of coastal regions – Pt. 4. The Changjiang River and the Huanghai Sea. SCOPE/UNEP Sonderband, 58:49–84

Szekielda K-H, McGinnis D (1987) Investigations with satellites on eutrophication of coastal regions. Pt. 4. Characterization of water masses in the Huanghai Sea. SCOPE/UNEP Sonderband, 64:93–112

Szekielda K-H, McGinnis D, Gird R (1983) Investigations with satellites on eutrophication of coastal regions. In: Degens ET, Kempe S, Soliman H (eds) Transport of carbon and minerals in major world rivers, Pt. 2, SCOPE/UNEP Sonderband, 55:55–84

# Lateral Distribution and Sources of Sediment-Associated Heavy Metals in the North Sea

G. IRION and G. MÜLLER

## Introduction

Although increasing pollution of the North Sea by fluvial and atmospheric input and by dumping activities has been known for some time, no comparable results of heavy metal contamination in sediments of the whole North Sea area are available.

The most extensive heavy metal survey carried out by Nicholson and Moore (1981) covered the English and much of the Scottish sector of the North Sea. However, trace metal analyses refer to whole sediment samples and give little information as the data were not normalized with respect to particle size variation.

In view of this problem other workers normalized sediment data by analyzing only the fine fractions. Since there is no international consensus on which fractions to use, a large diversity of grain size limits was chosen: < 90 μm (outer Thames estuary; Norton et al. 1981), < 63 μm (Dutch tidal flat areas; Kramer and van der Vlies 1983), < 20 μm (German Bight and adjoining areas; Irion and Schwedhelm 1983; coastal sediments off Lower Saxony; Steffen 1987), < 2 μm (German tidal flat areas; Schwedhelm and Irion 1983, 1985).

This dilemna was clearly recognized by Carlson (1986) in his review: "There are many published data sets on the metal content of North Sea sediments but, due to the different methodologies used, few are comparable with one another" and: "The preceding discussion demonstrates the need for caution when comparing sediment samples of differing particle size distribution. In particular, there is a need for a harmonized scheme of sediment monitoring such that comparison is possible, i.e. a standard fraction should be analyzed by a standardized chemical procedure."

The present study is aimed at filling a gap in our knowledge on the present heavy metal status of North Sea sediments. It is based on the analysis of the < 2 μm ("clay") fraction of altogether 157 surface sediments (Fig. 1) collected in a regular pattern during 1986 from different national institutions (Belgium, Denmark, F.R. Germany, the Netherlands and the United Kingdom) within an international project "North Sea Benthos Survey".

## 2 Materials and Methods

The sediments under study predominantly consist of coarse grained materials: sand, gravelly sand and – to a much lesser extent – muddy sand. With the exception of four samples all sediments contain less than 10% of clay sized material (< 2 μm); in more than half of the samples the clay content does not exceed 3%.

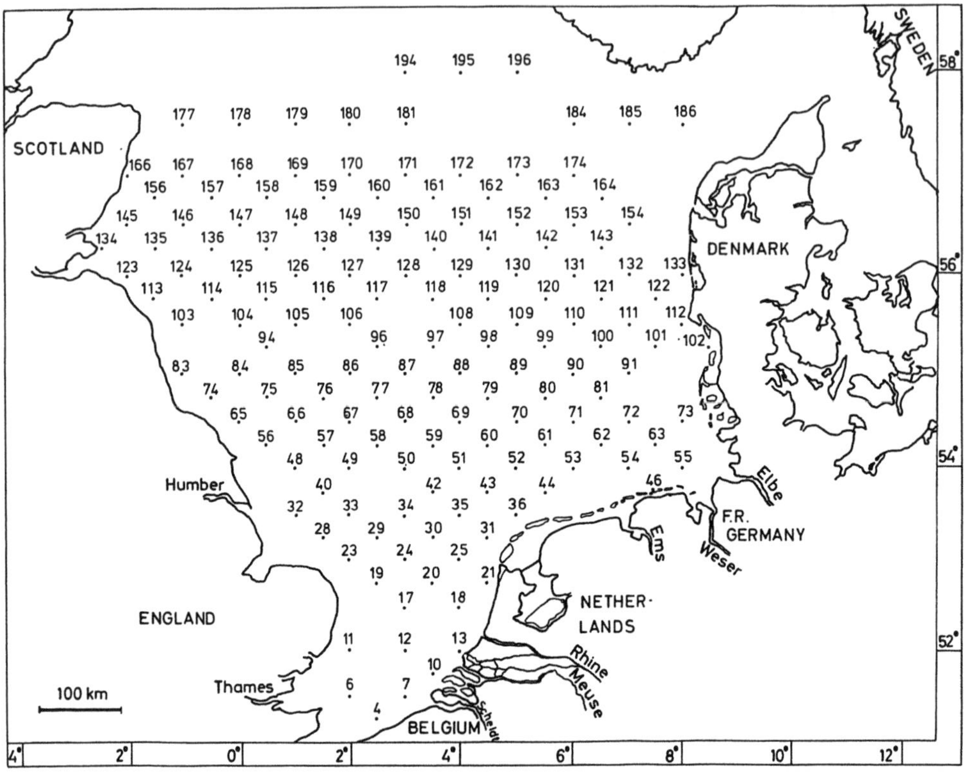


**Fig. 1.** "North Sea Benthos Survey" sampling stations studied in this report

The uppermost 10 cm of each grab sample was homogenized and a representative portion hereof washed for the removal of sea water. The clay fraction was separated quantitatively by repeated sedimentation in Atterberg cylinders (Müller 1967). After drying at 80° C the clay fraction was digested with conc. nitric acid and boiled until the organic matter was completely decomposed.

Metal determination was carried out applying standard methods of atomic spectroscopy, organic carbon was determined by wet combustion.

## 3 Results

The results of the analyses are listed in Table 1. Table 2 contains important statistical parameters. Figures 2–11 depict the lateral distribution of heavy metals and organic carbon in the study area.

**Table 1.** Heavy metal and organic carbon concentrations in the clay fraction of surficial North Sea sediments. Figures in mg/kg except for Fe and $C_{org}$ in %

| | Benthos | Zn | Pb | Cu | Cd | Fe | Mn | Cr | Ni | Co | $C_{org}$ |
|---|---|---|---|---|---|---|---|---|---|---|---|
| 1 | 4 | 152.5 | 82.1 | 21.8 | 0.70 | 2.61 | 912 | 62.0 | 31.8 | 13.4 | - |
| 2 | 6 | 160.0 | 100.0 | 28.7 | 0.73 | 2.80 | 2720 | 37.5 | 45.0 | 43.7 | - |
| 3 | 7 | 380.0 | 207.5 | 21.2 | 0.70 | 5.99 | 5514 | 68.7 | 35.0 | 42.5 | 3.11 |
| 4 | 10 | 327.5 | 178.7 | 25.0 | 0.83 | 2.62 | 5645 | 57.5 | 25.0 | 25.0 | 3.10 |
| 5 | 11 | 145.0 | 56.2 | 33.7 | 0.37 | 4.75 | 229 | 86.2 | 43.7 | 27.5 | 4.54 |
| 6 | 12 | 189.8 | 139.0 | 28.7 | 0.55 | 2.76 | 852 | 61.8 | 44.1 | 33.1 | 5.27 |
| 7 | 13 | 568.7 | 206.2 | 77.5 | 3.83 | 3.25 | 755 | 175.0 | 41.2 | 21.2 | 4.45 |
| 8 | 17 | 177.6 | 139.3 | 38.3 | 0.92 | 3.06 | 809 | 51.9 | 30.1 | 24.6 | 3.32 |
| 9 | 18 | 198.3 | 134.8 | 27.0 | 0.79 | 3.20 | 472 | 102.0 | 40.4 | 17.3 | 6.27 |
| 10 | 19 | 200.0 | 142.5 | 25.0 | 0.10 | 5.96 | 3571 | 53.7 | 36.2 | 37.5 | 2.28 |
| 11 | 20 | 149.7 | 73.5 | 29.9 | 0.26 | 3.35 | 719 | 89.8 | 65.3 | 27.2 | - |
| 12 | 21 | 758.7 | 252.5 | 53.7 | 0.75 | 3.50 | 1055 | 145.0 | 56.2 | 20.0 | 3.44 |
| 13 | 23 | 214.9 | 124.4 | 22.6 | 0.16 | 6.90 | 2446 | 74.9 | 31.1 | 45.2 | - |
| 14 | 24 | 191.2 | 171.2 | 32.5 | 0.35 | 5.54 | 3254 | 50.0 | 35.0 | 52.5 | 2.93 |
| 15 | 25 | 185.0 | 116.3 | 31.7 | 1.06 | 3.63 | 319 | 86.3 | 44.0 | 15.9 | - |
| 16 | 28 | 151.2 | 110.0 | 33.7 | 0.58 | 4.19 | 1689 | 56.2 | 45.0 | 28.7 | 2.74 |
| 17 | 29 | 261.2 | 222.5 | 32.5 | 0.48 | 5.51 | 2380 | 83.7 | 42.5 | 36.2 | 4.90 |
| 18 | 30 | 259.6 | 263.7 | 38.1 | 0.37 | 4.65 | 1969 | 78.8 | 40.8 | 29.9 | - |
| 19 | 31 | 196.4 | 131.7 | 37.9 | 0.66 | 3.26 | 2489 | 55.8 | 35.7 | 31.2 | - |
| 20 | 33 | 163.3 | 120.3 | 28.7 | 0.64 | 6.05 | 2169 | 71.6 | 34.4 | 43.0 | - |
| 21 | 34 | 185.0 | 141.2 | 28.7 | 0.64 | 3.72 | 429 | 58.7 | 35.0 | 18.7 | 3.93 |
| 22 | 35 | 220.0 | 153.7 | 33.7 | 0.75 | 4.29 | 394 | 67.5 | 41.2 | 25.0 | 3.14 |
| 23 | 36 | 215.9 | 110.2 | 36.2 | 1.56 | 3.03 | 654 | 77.0 | 46.8 | 21.1 | - |
| 24 | 39 | 163.7 | 161.2 | 40.0 | 0.97 | 2.67 | 356 | 47.5 | 40.0 | 17.5 | - |
| 25 | 40 | 147.5 | 107.5 | 26.2 | 0.31 | 5.06 | 1550 | 62.5 | 47.5 | 32.5 | 7.36 |
| 26 | 42 | 202.5 | 200.0 | 33.7 | 0.81 | 4.30 | 581 | 67.5 | 45.0 | 21.2 | 3.74 |
| 27 | 43 | 171.2 | 116.2 | 31.3 | 0.31 | 4.40 | 306 | 77.5 | 47.5 | 23.7 | 2.74 |
| 28 | 44 | 257.5 | 150.0 | 32.5 | 0.96 | 3.52 | 426 | 70.0 | 32.5 | 22.5 | 4.85 |
| 29 | 46 | 228.7 | 118.7 | 32.5 | 2.04 | 2.70 | 436 | 71.2 | 30.0 | 17.5 | - |
| 30 | 48 | 222.3 | 222.3 | 29.9 | 1.17 | 3.83 | 336 | 81.2 | 47.0 | 27.0 | 5.11 |
| 31 | 49 | 153.7 | 81.2 | 27.5 | 0.44 | 3.90 | 287 | 72.5 | 46.2 | 17.5 | 3.22 |

**Table 1.** *(Continued)*

| Benthos | | Zn | Pb | Cu | Cd | Fe | Mn | Cr | Ni | Co | $C_{org}$ |
|---|---|---|---|---|---|---|---|---|---|---|---|
| 32 | 50 | 200.3 | 225.7 | 36.8 | 0.59 | 5.44 | 8360 | 54.5 | 48.2 | 50.7 | - |
| 33 | 51 | 160.0 | 80.0 | 27.5 | 0.44 | 4.30 | 335 | 67.5 | 42.5 | 15.0 | 3.29 |
| 34 | 52 | 165.0 | 93.8 | 31.3 | 0.35 | 3.72 | 307 | 60.0 | 41.2 | 12.5 | 2.58 |
| 35 | 53 | 176.2 | 91.2 | 30.0 | 0.48 | 3.20 | 701 | 55.0 | 32.5 | 13.7 | - |
| 36 | 54 | 352.5 | 176.2 | 45.0 | 1.50 | 3.37 | 320 | 77.5 | 42.5 | 20.0 | 5.78 |
| 37 | 55 | 343.8 | 136.2 | 40.0 | 2.27 | 3.70 | 552 | 80.0 | 42.5 | 15.0 | 4.92 |
| 38 | 56 | 295.3 | 431.2 | 36.4 | 0.47 | 4.73 | 9588 | 95.2 | 44.8 | 30.8 | 5.88 |
| 39 | 57 | 185.0 | 137.5 | 27.5 | 0.97 | 4.01 | 306 | 85.0 | 45.0 | 11.2 | 4.45 |
| 40 | 58 | 185.0 | 140.0 | 31.3 | 1.29 | 3.26 | 311 | 67.5 | 42.5 | 13.7 | 5.05 |
| 41 | 59 | 162.5 | 93.8 | 28.7 | 0.53 | 4.10 | 276 | 73.7 | 42.5 | 13.7 | 3.65 |
| 42 | 60 | 157.5 | 66.2 | 27.5 | 0.20 | 4.37 | 495 | 68.7 | 40.0 | 18.7 | 3.17 |
| 43 | 61 | 171.2 | 93.8 | 30.0 | 0.44 | 4.09 | 300 | 67.5 | 38.7 | 16.2 | 2.96 |
| 44 | 62 | 230.0 | 141.2 | 35.0 | 1.00 | 3.90 | 269 | 66.2 | 36.2 | 13.7 | 3.59 |
| 45 | 63 | 323.7 | 148.7 | 31.3 | 1.28 | 4.46 | 1188 | 83.7 | 33.7 | 22.5 | 4.87 |
| 46 | 65 | 221.2 | 292.5 | 26.2 | 0.54 | 6.00 | 1389 | 82.5 | 48.7 | 27.5 | 4.02 |
| 47 | 66 | 130.0 | 81.2 | 22.5 | 0.92 | 4.30 | 317 | 83.7 | 47.5 | 20.0 | 3.83 |
| 48 | 67 | 157.5 | 86.1 | 31.5 | 0.84 | 3.74 | 321 | 79.8 | 48.3 | 23.1 | - |
| 49 | 68 | 166.2 | 120.0 | 26.2 | 0.44 | 3.56 | 289 | 80.0 | 48.7 | 17.5 | 4.84 |
| 50 | 69 | 167.5 | 110.0 | 27.5 | 0.40 | 4.39 | 382 | 72.5 | 45.0 | 17.5 | 3.37 |
| 51 | 70 | 161.2 | 107.5 | 30.0 | 0.38 | 4.19 | 313 | 68.7 | 40.0 | 18.7 | 2.84 |
| 52 | 71 | 163.7 | 87.5 | 31.3 | 0.23 | 3.79 | 306 | 71.2 | 40.0 | 17.5 | 5.55 |
| 53 | 72 | 337.5 | 175.0 | 30.0 | 1.78 | 4.57 | 1394 | 82.5 | 32.5 | 27.5 | 4.77 |
| 54 | 73 | 402.5 | 345.0 | 92.5 | 0.97 | 3.72 | 14692 | 57.5 | 37.5 | 77.5 | - |
| 55 | 74 | 238.7 | 302.5 | 35.0 | 0.21 | 4.60 | 752 | 102.5 | 50.0 | 21.2 | 6.11 |
| 56 | 75 | 190.0 | 188.7 | 31.3 | 0.19 | 3.77 | 412 | 93.8 | 47.5 | 20.0 | 5.98 |
| 57 | 76 | 93.8 | 55.9 | 18.1 | 0.16 | 2.67 | 168 | 60.9 | 21.4 | 9.9 | - |
| 58 | 77 | 132.5 | 111.2 | 22.5 | 0.39 | 2.95 | 300 | 60.0 | 32.5 | 16.2 | 4.94 |
| 59 | 78 | 157.5 | 136.2 | 25.0 | 0.33 | 4.19 | 360 | 73.7 | 43.7 | 15.0 | 4.37 |
| 60 | 79 | 161.2 | 107.5 | 28.7 | 0.33 | 4.71 | 517 | 71.2 | 40.0 | 18.7 | 2.93 |
| 61 | 80 | 187.5 | 147.5 | 32.5 | 0.47 | 4.60 | 1292 | 72.5 | 36.2 | 15.0 | 3.49 |
| 62 | 81 | 176.2 | 103.7 | 33.7 | 0.76 | 4.09 | 341 | 67.5 | 33.7 | 20.0 | 3.50 |
| 63 | 83 | 150.0 | 98.7 | 32.5 | 0.47 | 4.26 | 305 | 93.8 | 50.0 | 13.7 | 5.13 |

| | | | | | | | | | | | |
|---|---|---|---|---|---|---|---|---|---|---|---|
| 64 | 84 | 192.5 | 196.2 | 33.7 | 0.20 | 4.14 | 355 | 90.0 | 52.5 | 20.0 | 6.11 |
| 65 | 85 | 163.7 | 146.2 | 41.2 | 0.14 | 3.37 | 277 | 81.2 | 50.0 | 17.5 | 5.66 |
| 66 | 86 | 168.0 | 87.0 | 42.0 | 0.42 | 3.75 | 398 | 84.0 | 48.0 | 9.0 | - |
| 67 | 87 | 160.0 | 90.0 | 37.5 | 1.05 | 3.22 | 295 | 75.0 | 40.0 | 5.0 | 5.07 |
| 68 | 88 | 156.3 | 108.7 | 28.7 | 0.19 | 3.95 | 294 | 76.2 | 43.7 | 13.7 | 3.88 |
| 69 | 89 | 155.0 | 116.2 | 28.7 | 0.34 | 3.95 | 357 | 70.0 | 37.5 | 16.2 | - |
| 70 | 90 | 161.2 | 88.7 | 27.5 | 0.31 | 4.37 | 606 | 68.7 | 36.2 | 22.5 | - |
| 71 | 91 | 247.5 | 90.0 | 47.5 | 0.58 | 3.41 | 207 | 81.2 | 45.0 | 17.5 | 4.81 |
| 72 | 94 | 167.5 | 93.8 | 32.5 | 0.11 | 3.55 | 296 | 87.5 | 51.2 | 16.2 | - |
| 73 | 96 | 142.5 | 140.0 | 30.0 | 0.18 | 3.32 | 561 | 78.7 | 42.5 | 17.5 | 5.13 |
| 74 | 97 | 137.5 | 97.5 | 32.5 | 0.55 | 2.75 | 252 | 70.0 | 30.0 | 1.0 | - |
| 75 | 98 | 171.2 | 107.5 | 31.3 | 0.23 | 3.86 | 337 | 83.7 | 50.0 | 13.7 | 5.73 |
| 76 | 99 | 172.5 | 108.7 | 31.3 | 0.16 | 4.19 | 587 | 81.2 | 42.5 | 15.0 | 4.83 |
| 77 | 100 | 197.3 | 163.0 | 49.3 | 0.62 | 5.28 | 3194 | 81.5 | 34.3 | 36.5 | - |
| 78 | 101 | 361.2 | 142.5 | 70.0 | 1.02 | 3.76 | 1891 | 92.5 | 42.5 | 18.7 | - |
| 79 | 103 | 141.2 | 87.5 | 30.0 | 0.11 | 3.60 | 336 | 91.2 | 50.0 | 8.7 | 6.17 |
| 80 | 104 | 150.0 | 93.0 | 30.0 | 0.32 | 3.12 | 504 | 96.0 | 49.4 | 4.4 | 6.85 |
| 81 | 105 | 205.0 | 145.0 | 33.7 | 0.24 | 4.10 | 334 | 105.0 | 51.2 | 16.2 | 7.19 |
| 82 | 106 | 153.7 | 147.5 | 30.0 | 0.26 | 3.51 | 465 | 87.5 | 50.0 | 15.0 | 5.87 |
| 83 | 108 | 115.0 | 87.5 | 40.0 | 0.80 | 2.35 | 235 | 77.5 | 37.5 | 7.5 | - |
| 84 | 109 | 173.7 | 126.2 | 28.7 | 0.43 | 3.75 | 434 | 91.2 | 47.5 | 15.0 | 5.93 |
| 85 | 110 | 161.2 | 73.7 | 27.5 | 0.13 | 4.27 | 407 | 78.7 | 42.5 | 23.7 | 2.94 |
| 86 | 111 | 273.7 | 156.3 | 73.7 | 0.73 | 3.90 | 962 | 92.5 | 41.2 | 23.7 | 7.49 |
| 87 | 112 | 194.7 | 85.8 | 55.8 | 0.21 | 2.48 | 1398 | 68.1 | 30.0 | 19.1 | - |
| 88 | 113 | 147.5 | 100.0 | 30.0 | 0.10 | 3.74 | 381 | 81.2 | 50.0 | 13.7 | 5.24 |
| 89 | 114 | 160.0 | 136.2 | 27.5 | 0.11 | 3.91 | 552 | 101.2 | 51.2 | 15.0 | 6.29 |
| 90 | 115 | 171.2 | 183.7 | 30.0 | 0.17 | 4.07 | 754 | 106.2 | 53.7 | 15.0 | 7.89 |
| 91 | 116 | 147.5 | 112.5 | 26.2 | 0.36 | 3.22 | 296 | 105.0 | 55.0 | 10.0 | 8.15 |
| 92 | 117 | 150.0 | 103.7 | 26.2 | 0.17 | 3.19 | 260 | 96.2 | 51.2 | 12.5 | 6.73 |
| 93 | 118 | 155.0 | 180.0 | 23.7 | 0.21 | 3.84 | 526 | 102.5 | 43.7 | 11.2 | 6.54 |
| 94 | 119 | 132.5 | 155.0 | 22.5 | 0.25 | 3.32 | 784 | 92.5 | 36.2 | 7.5 | 6.93 |
| 95 | 120 | 151.2 | 91.2 | 25.0 | 0.07 | 3.99 | 385 | 85.0 | 38.7 | 13.7 | 3.61 |
| 96 | 121 | 220.0 | 138.7 | 45.0 | 0.33 | 4.46 | 539 | 76.2 | 37.5 | 17.5 | 4.03 |
| 97 | 122 | 317.5 | 138.7 | 56.2 | 1.19 | 3.90 | 544 | 85.0 | 41.2 | 20.0 | 8.23 |
| 98 | 123 | 137.5 | 88.7 | 30.0 | 0.21 | 3.19 | 291 | 82.5 | 43.7 | 11.2 | 6.06 |
| 99 | 124 | 150.0 | 163.7 | 28.7 | 0.17 | 3.24 | 1034 | 73.7 | 40.0 | 10.0 | 6.09 |

Table 1. *(Continued)*

| Benthos | | Zn | Pb | Cu | Cd | Fe | Mn | Cr | Ni | Co | $C_{org}$ |
|---|---|---|---|---|---|---|---|---|---|---|---|
| 100 | 125 | 126.2 | 102.5 | 23.7 | 0.17 | 2.74 | 447 | 75.0 | 40.0 | 10.0 | 6.73 |
| 101 | 126 | 138.7 | 96.2 | 26.2 | 0.19 | 2.94 | 313 | 100.0 | 48.7 | 12.5 | - |
| 102 | 127 | 156.3 | 113.7 | 26.2 | 0.29 | 3.06 | 364 | 85.0 | 47.5 | 12.5 | 8.74 |
| 103 | 128 | 162.5 | 175.0 | 25.0 | 0.17 | 3.80 | 465 | 106.2 | 46.2 | 10.0 | 8.23 |
| 104 | 129 | 156.3 | 185.0 | 20.0 | 0.31 | 3.54 | 567 | 102.5 | 36.2 | 12.5 | 8.24 |
| 105 | 130 | 119.2 | 139.7 | 27.9 | 0.41 | 4.82 | 657 | 122.9 | 24.2 | 5.6 | - |
| 106 | 131 | 158.7 | 86.2 | 28.7 | 0.11 | 4.19 | 472 | 72.5 | 36.2 | 15.0 | 2.91 |
| 107 | 132 | 188.7 | 101.2 | 90.0 | 0.33 | 3.14 | 819 | 60.0 | 33.7 | 17.5 | 4.26 |
| 108 | 133 | 241.2 | 107.5 | 37.5 | 0.61 | 3.76 | 436 | 78.7 | 40.0 | 15.0 | 4.40 |
| 109 | 134 | 142.5 | 112.5 | 31.3 | 0.25 | 3.50 | 366 | 75.0 | 45.0 | 12.5 | 5.25 |
| 110 | 135 | 208.7 | 150.0 | 33.7 | 0.17 | 2.25 | 1564 | 75.0 | 41.2 | 11.2 | 3.37 |
| 111 | 136 | 143.7 | 88.7 | 25.0 | 0.38 | 2.60 | 349 | 83.7 | 51.2 | 10.0 | 7.15 |
| 112 | 137 | 120.0 | 73.7 | 23.7 | 0.35 | 2.37 | 264 | 67.5 | 45.0 | 16.2 | 8.08 |
| 113 | 138 | 156.3 | 166.2 | 27.5 | 0.28 | 3.26 | 627 | 103.7 | 53.7 | 15.0 | 8.43 |
| 114 | 139 | 166.2 | 198.7 | 26.2 | 0.25 | 3.77 | 489 | 110.0 | 47.5 | 12.5 | 8.97 |
| 115 | 140 | 172.5 | 158.7 | 27.5 | 0.23 | 3.34 | 334 | 107.5 | 50.0 | 13.7 | 9.18 |
| 116 | 141 | 155.0 | 172.5 | 23.7 | 0.32 | 3.19 | 249 | 107.5 | 46.2 | 11.2 | 8.23 |
| 117 | 142 | 152.5 | 95.0 | 28.7 | 0.09 | 4.26 | 392 | 75.0 | 41.2 | 17.5 | 3.30 |
| 118 | 143 | 196.2 | 123.7 | 38.7 | 0.16 | 4.35 | 356 | 86.2 | 40.0 | 13.7 | 3.22 |
| 119 | 145 | 135.0 | 86.2 | 31.3 | 0.11 | 2.26 | 1189 | 62.5 | 36.2 | 10.0 | 3.98 |
| 120 | 146 | 120.0 | 75.0 | 27.5 | 0.17 | 2.27 | 241 | 82.5 | 42.5 | 10.0 | 5.87 |
| 121 | 147 | 168.7 | 166.2 | 33.7 | 0.06 | 4.49 | 800 | 141.2 | 42.5 | 7.5 | 7.65 |
| 122 | 148 | 132.5 | 83.7 | 28.7 | 0.15 | 2.97 | 306 | 90.0 | 50.0 | 8.7 | 7.42 |
| 123 | 149 | 165.0 | 171.2 | 32.5 | 0.11 | 3.67 | 342 | 125.0 | 52.5 | 7.5 | 8.31 |
| 124 | 150 | 130.0 | 132.5 | 23.7 | 0.09 | 2.94 | 386 | 102.5 | 42.5 | 11.2 | 7.34 |
| 125 | 151 | 173.7 | 151.2 | 27.5 | 0.14 | 2.96 | 360 | 98.7 | 41.2 | 12.5 | - |
| 126 | 152 | 175.0 | 143.7 | 27.5 | 0.15 | 3.82 | 359 | 108.7 | 46.2 | 11.2 | 5.53 |
| 127 | 153 | 257.5 | 181.2 | 66.2 | 0.24 | 4.87 | 4946 | 87.5 | 37.5 | 31.3 | - |
| 128 | 154 | 232.5 | 198.7 | 53.7 | 0.27 | 4.62 | 2885 | 77.5 | 33.7 | 23.7 | - |
| 129 | 156 | 177.5 | 98.7 | 28.7 | 0.24 | 2.49 | 1736 | 62.5 | 37.5 | 10.0 | 4.60 |
| 130 | 157 | 298.7 | 81.2 | 27.5 | 0.87 | 3.02 | 431 | 81.2 | 55.0 | 11.2 | 6.51 |
| 131 | 158 | 141.2 | 111.2 | 32.5 | 0.10 | 3.19 | 491 | 102.5 | 50.0 | 7.5 | 7.70 |

| | | | | | | | | | | | |
|---|---|---|---|---|---|---|---|---|---|---|---|
| 132 | 159 | 137.5 | 76.2 | 27.5 | 0.06 | 3.14 | 287 | 93.8 | 48.7 | 7.5 | 7.18 |
| 133 | 160 | 173.7 | 177.5 | 26.2 | 0.05 | 3.65 | 561 | 122.5 | 52.5 | 12.5 | 9.06 |
| 134 | 161 | 142.0 | 146.8 | 38.5 | 0.15 | 3.58 | 570 | 117.9 | 45.7 | 9.6 | - |
| 135 | 162 | 138.7 | 100.0 | 28.7 | 0.08 | 2.59 | 221 | 110.0 | 47.5 | 3.7 | 10.92 |
| 136 | 163 | 200.0 | 317.5 | 27.5 | 0.09 | 8.24 | 1932 | 157.5 | 23.7 | 13.7 | - |
| 137 | 164 | 130.0 | 130.0 | 31.3 | 0.19 | 3.80 | 5277 | 47.5 | 22.5 | 21.2 | - |
| 138 | 166 | 220.0 | 201.2 | 35.0 | 0.12 | 5.06 | 1564 | 88.7 | 41.2 | 11.2 | 5.43 |
| 139 | 167 | 246.2 | 115.0 | 30.0 | 0.14 | 2.84 | 3372 | 53.7 | 53.7 | 17.5 | 4.08 |
| 140 | 168 | 150.0 | 91.2 | 31.3 | 0.10 | 3.15 | 554 | 95.0 | 51.2 | 1.2 | 7.10 |
| 141 | 169 | 132.5 | 95.0 | 31.3 | 0.08 | 2.77 | 331 | 78.7 | 50.0 | 7.5 | 7.25 |
| 142 | 170 | 143.7 | 128.7 | 30.0 | 0.08 | 3.35 | 416 | 91.2 | 48.7 | 8.7 | 7.77 |
| 143 | 171 | 151.2 | 163.7 | 28.7 | 0.16 | 3.30 | 931 | 111.2 | 52.5 | 11.2 | - |
| 144 | 172 | 140.9 | 175.2 | 34.2 | 0.21 | 3.16 | 621 | 117.7 | 43.8 | 10.9 | - |
| 145 | 173 | 108.2 | 158.5 | 30.9 | 0.16 | 4.70 | 1225 | 104.4 | 23.2 | 13.5 | - |
| 146 | 174 | 240.0 | 295.0 | 75.0 | 0.20 | 4.65 | 2832 | 106.2 | 28.7 | 17.5 | 4.95 |
| 147 | 177 | 181.2 | 225.0 | 38.7 | 0.34 | 5.17 | 7139 | 80.0 | 55.0 | 21.2 | 4.53 |
| 148 | 178 | 123.7 | 118.7 | 32.5 | 0.45 | 2.31 | 3814 | 57.5 | 57.5 | 10.0 | 3.98 |
| 149 | 179 | 151.2 | 107.5 | 35.0 | 0.46 | 2.34 | 262 | 68.7 | 46.2 | 7.5 | 6.53 |
| 150 | 180 | 195.0 | 240.0 | 37.5 | 0.18 | 4.82 | 546 | 138.7 | 50.0 | 6.2 | 8.26 |
| 151 | 181 | 194.9 | 221.8 | 47.4 | 0.58 | 2.83 | 819 | 115.4 | 42.3 | 6.4 | - |
| 152 | 184 | 247.5 | 333.7 | 55.0 | 0.41 | 5.19 | 3952 | 87.5 | 43.7 | 20.0 | 5.87 |
| 153 | 185 | 173.7 | 156.3 | 42.5 | 0.37 | 4.16 | 914 | 78.7 | 31.3 | 12.5 | 4.85 |
| 154 | 186 | 210.0 | 172.5 | 43.7 | 0.43 | 3.51 | 1866 | 60.0 | 30.0 | 17.5 | 4.33 |
| 155 | 194 | 157.5 | 142.5 | 56.2 | 0.52 | 2.61 | 313 | 93.8 | 46.2 | 7.5 | - |
| 156 | 195 | 148.7 | 142.5 | 43.7 | 0.27 | 3.80 | 324 | 105.0 | 41.2 | 12.5 | 5.86 |
| 157 | 196 | 176.2 | 116.2 | 45.0 | 0.57 | 2.84 | 259 | 71.2 | 53.7 | 12.5 | 5.82 |

**Table 2.** Statistical parameters for heavy metals and organic carbon in the clay fraction of North Sea sediments. n = 157 (heavy metals), 119 ($C_{org}$). Fe and $C_{org}$ in %, all other figures in mg/kg

| | Mean | Standard deviation | % | Median | Minimum | Maximum |
|---|---|---|---|---|---|---|
| Zn | 190.10 | 77.97 | 41.02 | 167.50 | 93.80 | 758.70 |
| Pb | 140.72 | 59.54 | 42.31 | 131.70 | 55.90 | 431.20 |
| Cu | 34.11 | 12.08 | 35.41 | 31.30 | 18.10 | 92.50 |
| Cd | 0.47 | 0.47 | 100.0 | 0.33 | 0.05 | 3.83 |
| Fe | 3.78 | 0.93 | 24.60 | 3.74 | 2.25 | 8.24 |
| Mn | 1140.85 | 1841.15 | 161.38 | 472.00 | 168.00 | 14692.00 |
| Cr | 83.58 | 21.24 | 25.41 | 81.20 | 37.50 | 175.00 |
| Ni | 42.52 | 7.75 | 18.23 | 42.50 | 21.40 | 65.30 |
| Co | 17.59 | 10.26 | 58.39 | 15.00 | 1.00 | 77.60 |
| $C_{org}$ | 5.40 | 1.84 | 34.06 | 5.13 | 2.28 | 10.92 |

## 3.1 Zinc

Zinc concentrations vary between 93.8 and 758.7 mg/kg. The mean concentration (190.10 ± 79.97 mg/kg) is slightly higher than the median (167.50 mg/kg). Most of the samples fall into the range 150–200 mg/kg, followed by sediments with less than 150 mg/kg.

The lateral distribution pattern (Fig. 2) shows a distinct zonation: highest Zn-contents (> 300 mg/kg) occur in a narrow band of nearshore sediments between the mouth of the Scheldt and the island of Texel (Netherlands) and in a broader area off the Ems-Weser-Elbe estuaries extending northwards across the German Bight.

A zone of high concentrations (200–300 mg/kg) fringes the two maximum areas and connects them. In addition, smaller isolated areas occur off the English and Scottish coast and in the northern part of the eastern North Sea. The largest parts of the study area fall into the intermediate (150–200 mg/kg) and relatively low (< 150 mg/kg) concentration range.

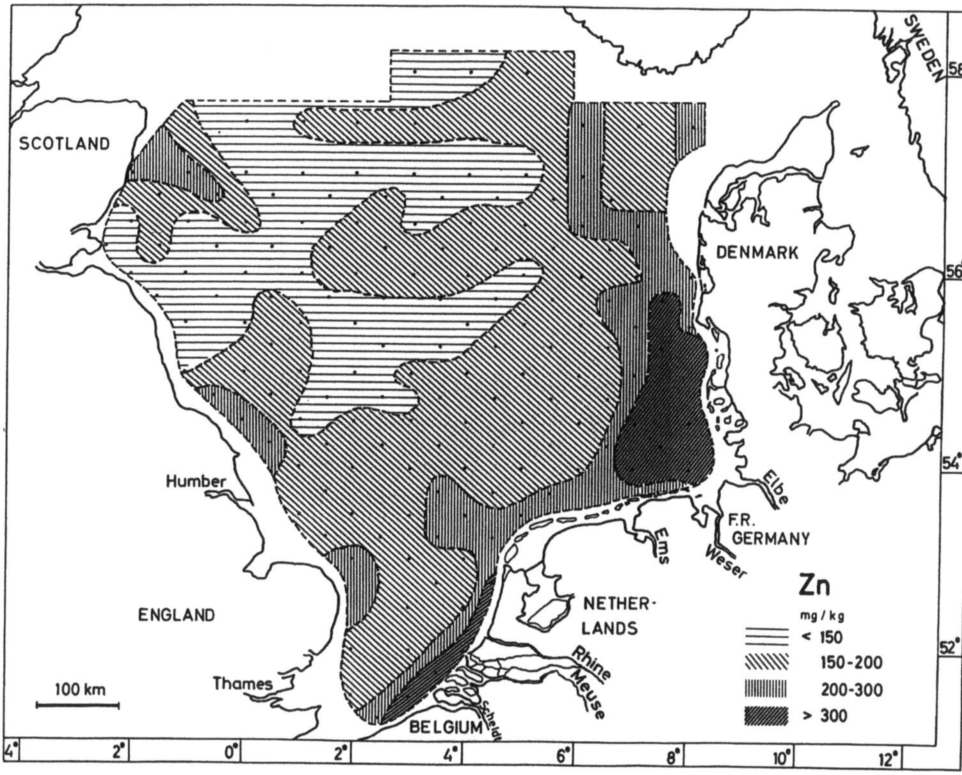


**Fig. 2.** Lateral distribution of zinc in the clay fraction of North Sea sediments

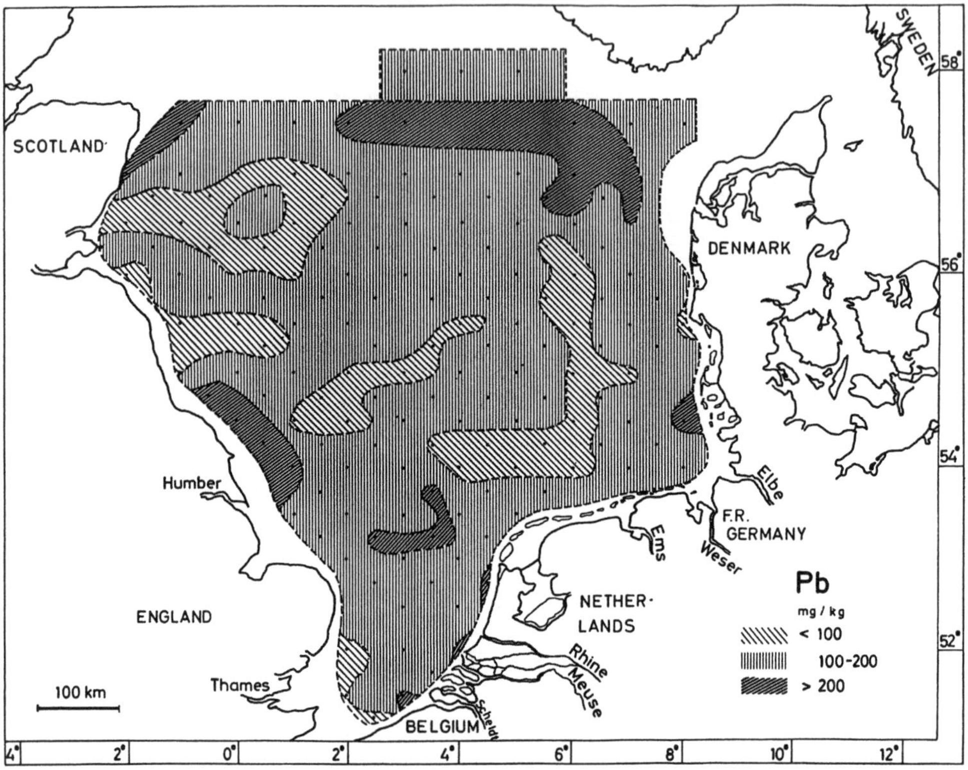


**Fig. 3.** Lateral distribution of lead in the clay fraction of North Sea sediments

### 3.2 Lead

Lead concentrations vary between 55.9 and 431.2 mg/kg. The mean concentration (140.72 ± 59.54 mg/kg) is slightly higher than the median (131.7 mg/kg).

By far the largest parts of the study area fall into the intermediate concentration range (100–200 mg/kg) which forms a framework into which irregular patches of high (> 200 mg/kg) and relatively low (< 100 mg/kg) concentration ranges are distributed more or less irregularly (Fig. 3).

Patches with low concentrations are arranged along a line from the Elbe estuary to Scotland. In addition, low concentrations are found near the entrance of the Channel.

High concentrations occur in nearshore zones north of the Humber estuary, off the mouth of the Scheldt river, between the mouth of the Rhine and the island of Texel and in the eastern part of the German Bight (with the highest Pb content of this study). The largest offshore patch extends in W-E direction in the northern part of the study area, other small patches occur in the southern North Sea.

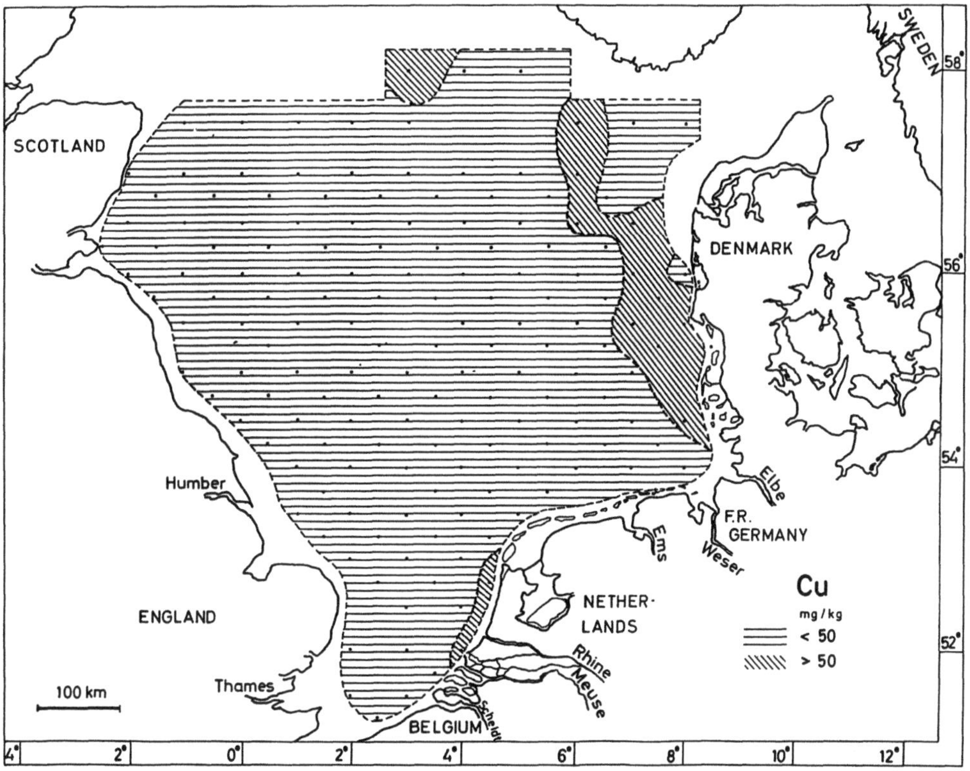


**Fig. 4.** Lateral distribution of copper in the clay fraction of North Sea sediments

### 3.3 Copper

Copper concentrations vary between 18.1 and 92.5 mg/kg. The mean concentration (34.11 ± 12.08 mg/kg) is very slightly higher than the median (31.3 mg/kg).

Most of the sediments fall into the 50–30 mg/kg concentration range followed by a group of samples with concentrations < 30 mg/kg. The lateral distribution (Fig. 4) shows an elongated area of higher (> 50 mg/kg) concentrations in the easternmost part of the North Sea which extends northwards from the mouth of the Elbe along the Danish coast.

High concentrations also occur in nearshore sediments between the mouth of the Rhine and the island of Texel.

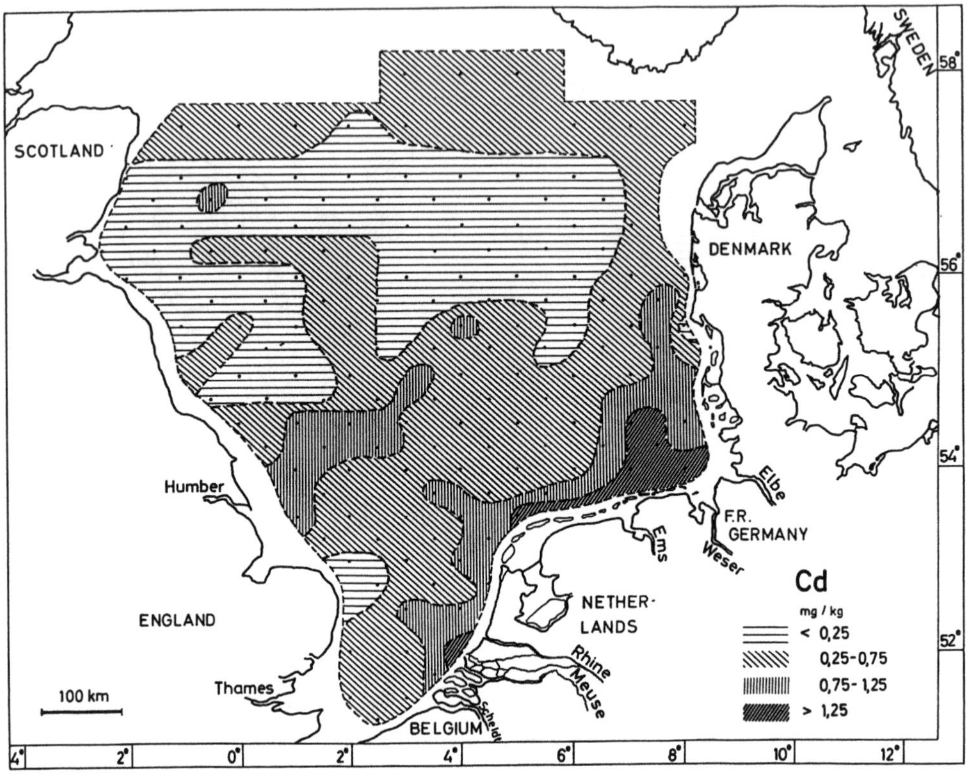


**Fig. 5.** Lateral distribution of cadmium in the clay fraction of North Sea sediments

## 3.4 Cadmium

Cadmium concentrations vary between 0.05 and 3.83 mg/kg. The mean concentration (0.47 $\pm$ 0.47 mg/kg) is considerably higher than the median (0.33 mg/kg).

Approximately equal areas of the North Sea sediments exhibit concentration ranges of 0.25–0.75 and < 0.25 mg/kg, respectively, followed by the concentration range of 0.75–1.25 mg/kg.

The lateral distribution pattern (Fig. 5) resembles in general that of the zinc distribution (Fig. 2). This is especially true when both highest (> 1.25 mg/kg) and high (0.75–1.25 mg/kg) concentration ranges areas are considered together: They cover an area along the whole Dutch and German and part of the Danish coast and include the German Bight. Another high concentration area extends northeastwards off the Humber estuary into the central part of the North Sea.

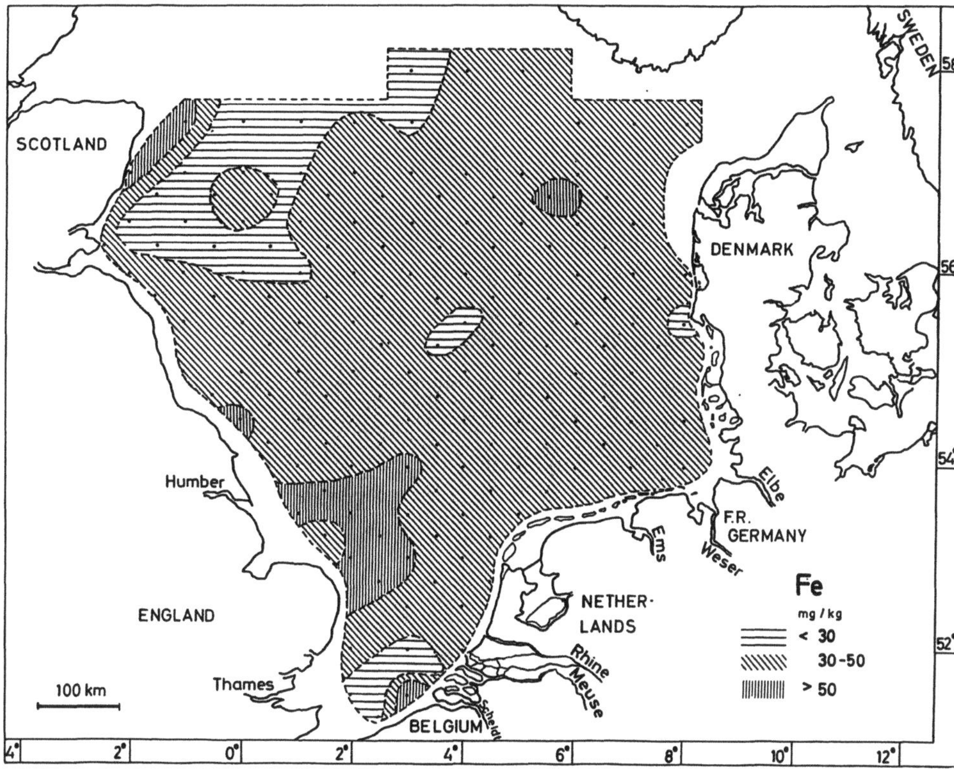


**Fig. 6.** Lateral distribution of iron in the clay fraction of North Sea sediments

## 3.5 Iron

Iron concentrations vary between 2.25 and 8.24%. The mean concentration (3.78 ± 0.93%) coincides with the median (3.74%). By far the most sediments fall into the intermediate concentration range (3–5%) which dominates in the lateral distribution pattern (Fig. 6). Areas with lower concentrations (< 3%) lie in the NW and N part of the study area and in the Southern Bight. Higher concentrations (> 5%) occupy an area off the southeastern English coast.

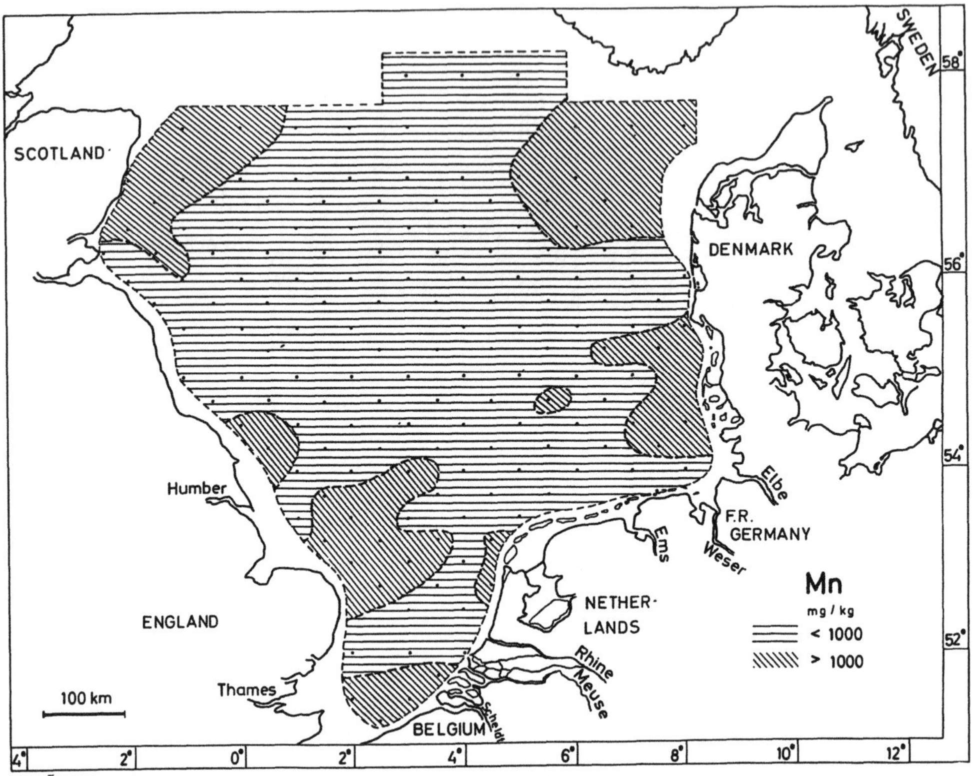


**Fig. 7.** Lateral distribution of manganese in the clay fraction of North Sea sediments

### 3.6 Manganese

Manganese concentrations vary between 168 and 14692 mg/kg. The mean concentration (1140 ± 1841 mg/kg) is about 2.4 times higher than the median (472 mg/kg).

More than two thirds of the sediments fall into the < 1000 mg/kg concentration range. The single large area occupied by this range is surrounded by altogether 7 smaller areas of higher concentrations (> 1000 mg/kg) attached to the coasts of Scotland, England, the Netherlands, Germany and Denmark (Fig. 7).

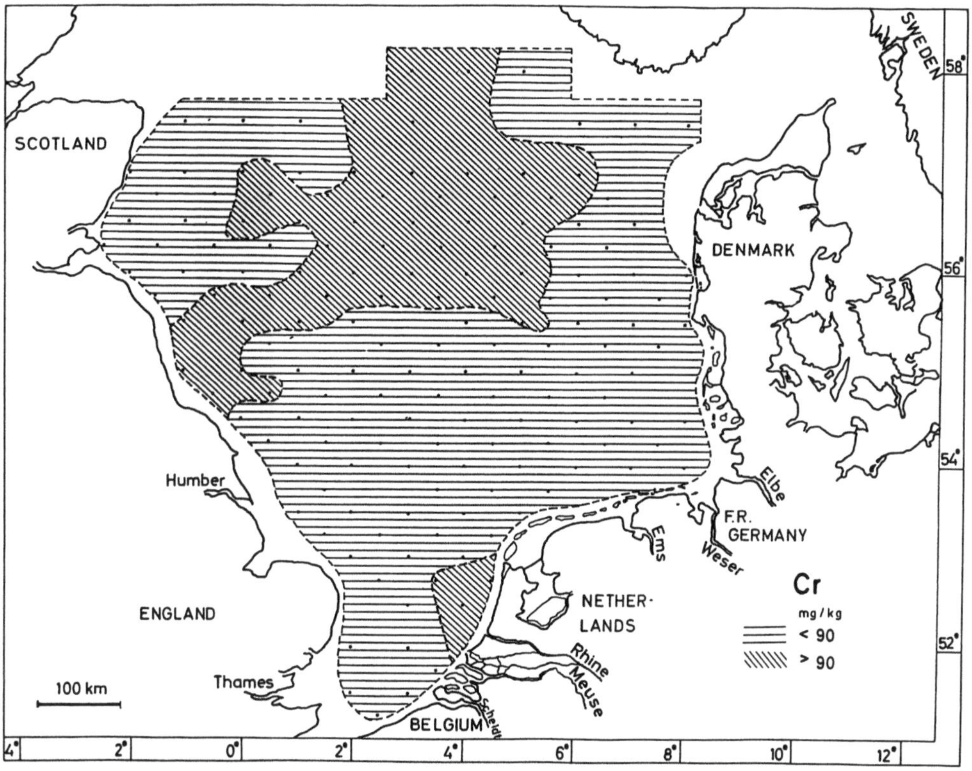


**Fig. 8.** Lateral distribution of chromium in the clay fraction of North Sea sediments

### 3.7 Chromium

Chromium concentrations vary between 37.5 and 175.0 mg/kg. The mean concentration (83.58 $\pm$ 21.24 mg/kg) is about the same as the median (81.24 mg/kg).

About two thirds of the sediments fall into the concentration range of $< 90$ mg/kg. Sediments with higher concentrations ($> 90$ mg/kg) occupy a closed area extending from the northern English coast into the central and northern North Sea, a small area occurs between the mouth of the Rhine (highest Cr concentration) and the island of Texel (Fig. 8).

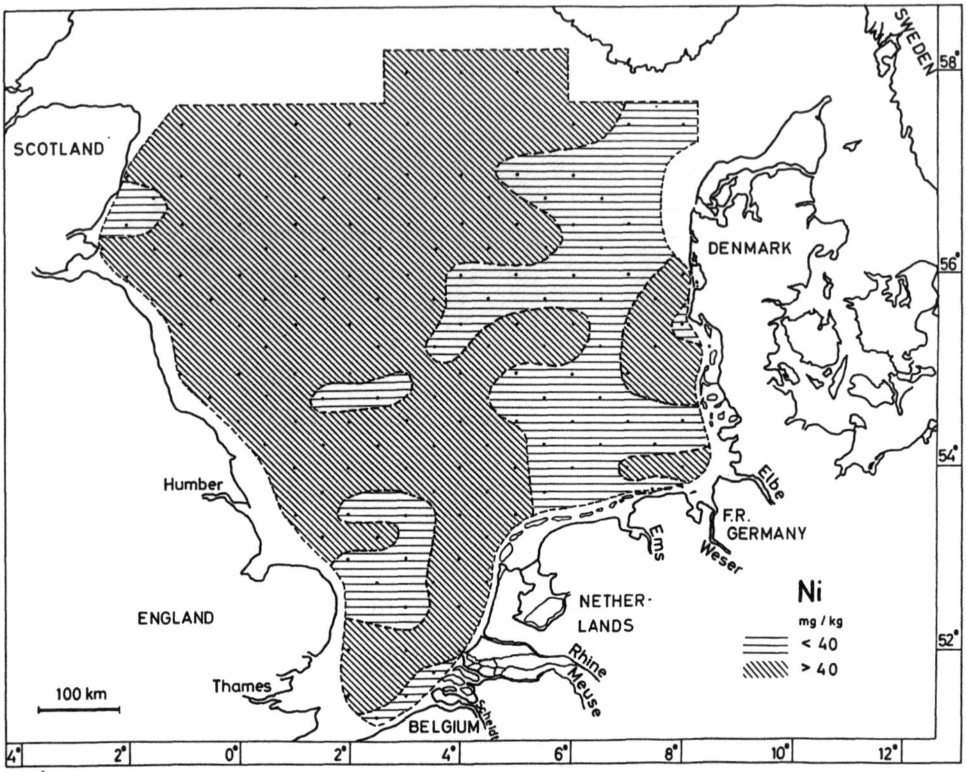


**Fig. 9.** Lateral distribution of nickel in the clay fraction of North Sea sediments

## 3.8 Nickel

Nickel concentrations vary between 21.4 and 63.5 mg/kg. The mean concentration (42.52 ± 7.75 mg/kg) is identical with the median (42.50 mg/kg).

About three quarters of the sediments fall into the > 40 mg/kg concentration range, lower concentrations are only to be found in an area extending from the West Friesian islands into the Skagerrak and within an area off the coast of East England (Fig. 9).

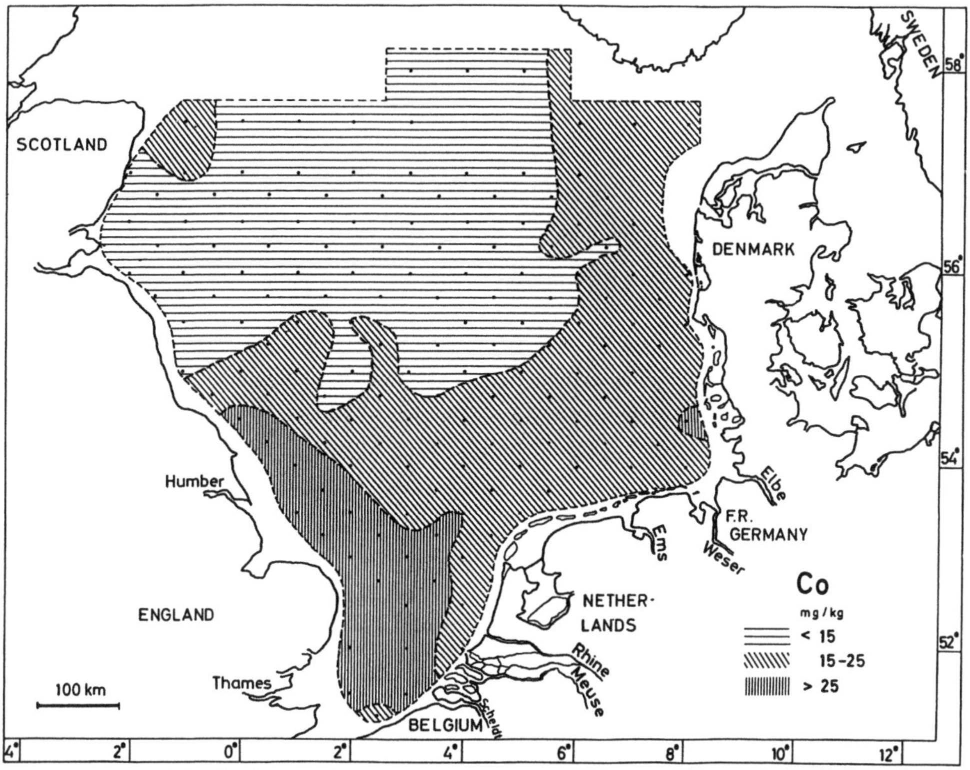


**Fig. 10.** Lateral distribution of cobalt in the clay fraction of North Sea sediments

## 3.9 Cobalt

Cobalt concentrations vary between 1.0 and 77.5 mg/kg. The mean concentration (17.59 ± 10.27 mg/kg) is slightly higher than the median (15.00 mg/kg).

Fig. 10 shows a distinct areal distribution of the different concentration ranges: relatively high concentrations (> 25 mg/kg) lie in an area off the coast of central and southern England and include large parts of the Southern Bight. Intermediate concentrations (15–25 mg/kg) mainly occupy the southern and eastern part of the North Sea and low concentrations (< 15 mg/kg) are bound to sediments in the central and northwestern part.

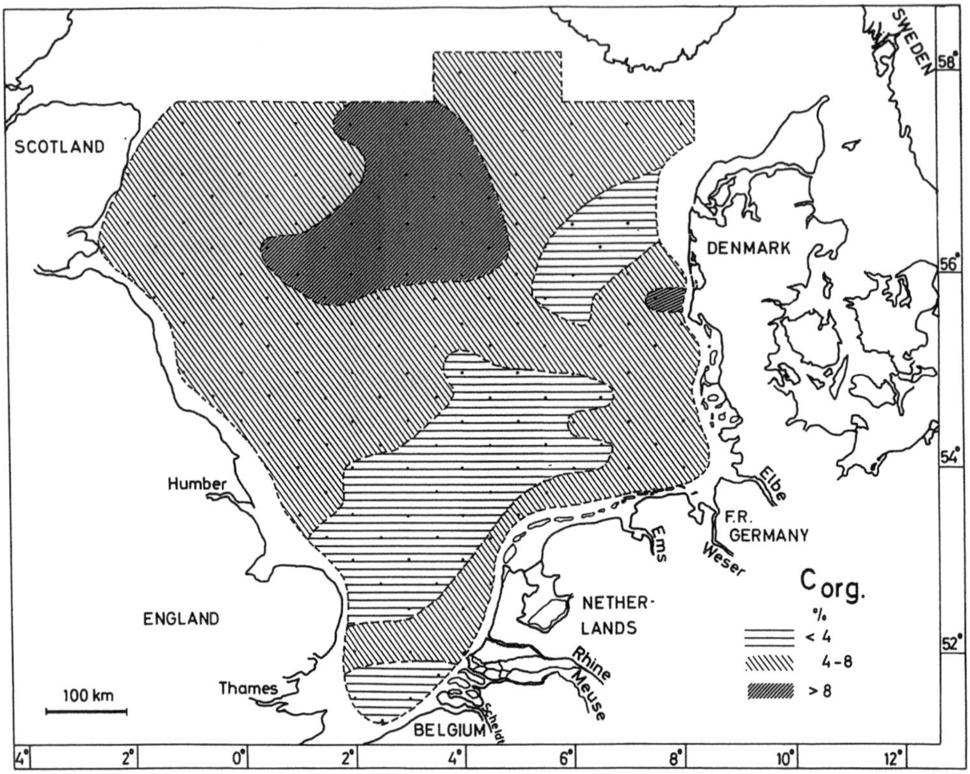


**Fig. 11.** Lateral distribution of organic carbon in the fraction of North Sea sediments

### 3.10 Organic Carbon

$C_{org}$ concentrations vary between 2.28 and 10.92%. The mean concentration (5.40 ± 1.84%) is only slightly higher than the median (5.13%).

Most sediments fall into the 4–8% concentration range. High concentrations (> 8%) nearly exclusively occur in the central part of the northern North Sea, low concentrations (< 2%) are bound to larger areas in the southern and north-eastern part.

## 4 Discussion

### 4.1 Geogenic Background Versus Anthropogenic Concentration

Heavy metals in recent sediments have two sources:

- they are "geogenic" and their concentration depends on the composition of the source rock from which the sedimentary particles are derived during weathering;

- they are "anthropogenic" and relate to the activity of man as a byproduct of industrialization and population growth during the past 100–120 years (Müller 1981).

Since about all present-day sediments are to some extent contaminated by anthropogenic heavy metals, the question of geogenic ("background") concentrations plays an important role in judging the extent of anthropogenic pollution.

The best way to establish background values for a specific area is to determine trace element concentration in sediments of that area, which were deposited in pre-industrial times.

Unfortunately, only few dated sediment cores which penetrated these "older" sediments in the North Sea have been studied so far (Table 3): Förstner and Reineck (1974) and Dominik et al. (1978) published heavy metal background concentrations in the clay fraction of a sediment core from the German Bight SE of the island of Helgoland; Müller and Irion (1984) reported results of pre-industrial heavy metal levels in the clay fraction of a sediment core collected in the deep water (513 m) of the Skagerrak, and only recently Irion (in prep.) analyzed Zn, Pb and Cu in clay sized material of sediments laid down in pre-industrial times from a core gathered 100 km NW of the island of Helgoland, German Bight.

A comparison (Table 3) shows that there is a very good agreement with copper, iron and manganese and still a relatively good agreement with cadmium, zinc, lead, nickel and chromium. Only cobalt varies considerably. When the mean pre-industrial values of the North Sea sites are compared with the "average shale" concentrations derived from a worldwide survey of clayey sediments of different geological ages (Turekian and Wedepohl 1961), no major differences can be observed with respect to Zn, Cd, Cr, Ni and Co. This is not the case, however, with Pb and Cu. Here the pre-industrial North Sea sediments exhibit a strong surplus in Pb and a strong deficit in Cu.

**Table 3.** Heavy metal concentration in the clay fraction of dated North Sea sediments deposited during pre-industrial times as compared with "average shale" (Turekian and Wedepohl 1961)

| | a) German Bight SE Helgoland n = 5 | b) Skagerrak n = 6 | c) German Bight 100 km NW Helgoland n = 3 | Average from a – c | "Average shale" |
|---|---|---|---|---|---|
| Zn | 113 | 137 | 99 | 116.3 | 95 |
| Pb | 48 | 65 | 39 | 50.7 | 20 |
| Cu | 22 | 24 | 20 | 22.0 | 45 |
| Cd | 0.3 | 0.22 | not determined | 0.26 | 0.3 |
| Fe | 3.22 | 3.52 | not determined | 3.37 | 4.72 |
| Mn | 400 | 423 | not determined | 411.5 | 850 |
| Cr | 60 | 83 | not determined | 71.5 | 90 |
| Ni | 34 | 55 | not determined | 44.5 | 68 |
| Co | 9 | 20 | not determined | 14.5 | 19 |

a) Förstner and Reineck 1974; Dominik et al. 1978.
b) Müller and Irion 1984.
c) Irion (in prep.).
Fe in %, all other figures in mg/kg.

**Table 4.** Enrichment factors for heavy metals in the clay fraction of surficial North Sea sediments if compared with pre-industrial sediments from 2 sediment cores

| | a<br>Mean conc. surficial sediments this study | b<br>Maximum conc. surficial sediments this study | c<br>Mean conc. pre-industrial North Sea sediments see Table 3 | d<br>a:b enrichment factor | e<br>b:c enrichment factor |
|---|---|---|---|---|---|
| Zn | 190.1 | 758.7 | 116.3 | 1.63 | 6.52 |
| Pb | 140.7 | 431.2 | 50.7 | 2.78 | 8.30 |
| Cu | 34.1 | 92.5 | 22.0 | 1.55 | 4.20 |
| Cd | 0.47 | 3.83 | 0.26 | 1.81 | 14.81 |
| Fe | 3.78 | 8.24 | 3.37 | 1.12 | 2.44 |
| Mn | 1140.8 | 14692.0 | 411.5 | 2.77 | 35.70 |
| Cr | 83.6 | 175.0 | 71.5 | 1.17 | 2.45 |
| Ni | 42.5 | 65.3 | 44.5 | 0.96 | 1.47 |
| Co | 17.6 | 77.5 | 14.5 | 1.21 | 5.34 |

Fe in %, all other figures in mg/kg.

Table 4 compares the mean heavy metal concentrations in the clay fraction of present-day North Sea sediments with their mean concentrations in pre-industrial time: Pb shows the largest enrichment (factor 2.78) followed by Cd (1.81), Zn (1.63), and Cu (1.55). Co (1.21), Cr (1.17) and Ni (0.96) show only weaker or no enrichment at all. Fe and Mn, both mostly of geogenic origin, will not be considered in this context, their concentration largely depends on redox conditions within the sediments.

This grouping of heavy metals has similarly been found in many other sedimentary environments which were affected by anthropogenic contamination and may be related to the ratio of the annual world consumption of an industrial metal to its mean concentration in the geosphere. The "index of the relative pollution potential" (Förstner and Müller 1974) describes the extent to which such a metal – if evenly distributed – might be enriched in soils or sediments. The indices calculated for Co, Ni and Cr are low (0.2, 1 and 2, respectively) whereas for Zn, Cd, Cu and Pb indices of 10, 25, 30 and 35 are obtained.

If maximum heavy metal concentrations found in surficial North Sea, sediments are compared with mean pre-industrial concentrations (Table 4, column e), enrichment factors grade from 14.81 (Cd) – 8.30 (Pb) – 6.52 (Zn) – 5.34 (Co) – 4.20 (Cu) – 2.45 (Cr) to 1.47 (Ni).

## 4.2 Areal Distribution and Sources of Heavy Metals in North Sea Sediments

According to their enrichment factors (Table 4) and lateral distribution patterns (Figs. 2–11) heavy metals can be grouped as follows:

*Nickel*, with the lowest (zero) enrichment factor and the lowest standard deviation shows a more or less even distribution pattern. Concentration differences depicted in Fig. 9 are only low.

*Chromium and cobalt*, both with low enrichment factors, exhibit an uneven distribution pattern with a preferential S-N-zonation. Co concentrations decrease from south to north whereas Cr concentrations are higher in the north than in the south.

*Copper, zinc and cadmium,* with intermediate enrichment factors again show an uneven distribution pattern, however, with a preferential E-W zonation. In general, concentrations decrease from east to west.

*Lead*, with a high enrichment factor exhibits an uneven, patchy distribution pattern without preferential zonation. Also *manganese* exhibits an uneven patchy distribution in which patches of higher concentrations frame a large area with relatively low concentrations.

The distribution pattern of *iron* is uneven and patchy, too. In general, concentrations in the north are lower than in the south.

For the understanding of the distribution patterns the knowledge of the input of a specific heavy metal and its pathway are important.

Table 5 lists the estimated annual input of heavy metals from different sources into the North Sea (Bericht der Bundesregierung an den Deutschen Bundestag zur Vorbereitung der 2. Internationalen Nordseeschutz-Konferenz vom 21. September 1987). There is still a great uncertainty in the estimation of the atmospheric input, maximum and minimum estimates vary considerably.

Relatively good agreement exists between calculations of the riverine input. Table 6 lists the heavy metal supply of the various rivers after van Pagee and Postma (1987). The figures clearly show that amongst all rivers the Rhine-Meuse outflow provides by far the largest heavy metal input (generally 50% or more of the total riverine load), followed by the Elbe, Weser, Humber and Scheldt rivers. Differences in the total riverine input between Tables 6 and 5 may be explained by different

**Table 5.** Estimates of heavy metal input (in t/a) to the North Sea from different sources. Where the estimate of a contaminant load has been supplied as a range, this has been expressed in the table as a max. (maximum) and a min. (minimum) value. If only one value has been supplied as an estimate of the load, this has been assigned to the max. value and a dash (–) has been assigned to the min. value

| Source | Zn | | Pb | | Cu | | Cd | | Cr | | Ni | |
|---|---|---|---|---|---|---|---|---|---|---|---|---|
| | max. | min. | max. | min. | max. | min. | max. | min. | max. | min. | max. | min. |
| River Inputs | 7370 | 7360 | 980 | 920 | 1330 | 1290 | 52 | 46 | 630 | 590 | 270 | 240 |
| Direct Discharges | 1170 | – | 170 | – | 315 | – | 20 | 20 | 490 | – | 115 | – |
| Atmospheric | 11000 | 4900 | 7400 | 2600 | 1600 | 400 | 240 | 45 | 900 | 300 | 950 | 300 |
| *Dumpings* | | | | | | | | | | | | |
| Dredgings | 8000 | – | 2000 | – | 1000 | – | 20 | – | 2500 | – | 700 | – |
| Sewage Sludge | 220 | – | 100 | – | 100 | – | 3 | – | 40 | – | 15 | – |
| Industrial Waste | 450 | – | 200 | – | 160 | – | 0.3 | – | 350 | – | 70 | – |
| Incinceration at Sea | 12 | – | 2 | – | 3 | – | 0.1 | – | 1.7 | – | 3 | – |
| Total | 28000 | 22000 | 1100 | 6000 | 4500 | 3000 | 335 | 135 | 5000 | 4200 | 2100 | 1450 |

From "Bericht der Bundesregierung an den Deutschen Bundestag zur Vorbereitung der 2. Internationalen Nordseeschutz-Konferenz vom 21. September 1987".

**Table 6.** Heavy metals input (t/a) for 1980 into the North Sea between the Strait of Dover and 56°N by rivers. After van Pagee and Postma (1987)

| Source | Zn | Pb | Cu | Cr | Cd | Hg |
|---|---|---|---|---|---|---|
| *United Kingdom* | | | | | | |
| Forth | 72.5 | 29.0 | 30.0 | 17.5 | 1.5 | 0.54 |
| Tees | 56.0 | 12.5 | 10.0 | 8.0 | 0.6 | 0.16 |
| Tyne | 199.0 | 26.5 | 26.5 | 17.5 | 1.5 | 0.43 |
| Humber | 470.0 | 132.0 | 162.0 | 100.0 | 8.6 | 2.36 |
| Wash | 175.0 | 50.0 | 44.0 | 48.0 | 1.6 | 0.43 |
| Thames | 72.0 | 16.5 | 16.0 | 15.0 | 2.2 | 1.29 |
| *Belgium* | | | | | | |
| | 30.0 | 3.0 | 4.0 | 3.0 | 0.2 | 0.12 |
| *Netherlands* | | | | | | |
| Scheldt | 504.0 | 121.0 | 72.0 | 122.0 | 12.4 | 2.5 |
| Rhine/Meuse | 4986.0 | 634.0 | 737.0 | 845.0 | 73.0 | 9.56 |
| Ems | 95.0 | 20.4 | 51.5 | 13.0 | 3.4 | 0.76 |
| *Germany, F.R.* | | | | | | |
| Weser | 1655.0 | 83.4 | 191.0 | 54.0 | 4.7 | 1.18 |
| Elbe | 2500.0 | 135.0 | 300.0 | 200.0 | 15.0 | 7.5 |
| *Denmark* | | | | | | |
| | 140.0 | 32.0 | 21.0 | 21.0 | 3.6 | – |
| Totals | 10954.5 | 1295.3 | 1665.0 | 1464.0 | 128.3 | 26.8 |

years for which calculations were made: Table 6 refers to 1980 values whereas Table 5 is based on 1986 data.

It should be noted that the figures for riverine supply in Tables 5 and 6 refer to the total heavy metal input and do not differentiate between dissolved and particulate heavy metals. For a better understanding of metal concentration in the clay fraction of the North Sea sediments a knowledge of their concentration in the clay fraction of the suspended load of the rivers draining into the North Sea would be very helpful.

From a 1985 reconnaissance survey (Müller 1985) comparable data are available for the Rhine, Ems, Weser and Elbe rivers (Table 7), which clearly show a high contamination of the Rhine and Elbe sediments with cadmium and zinc (enrichment factor $> 10$), lead and copper (enrichment factor 5–10) and – to a lesser extent – with chromium (enrichment factor 2–3). Nickel and cobalt are not significantly enriched at all.

Governed by the prevailing surface currents in the North Sea (Fig. 12), the heavy metal load of the rivers mainly affects coastal areas, especially those of the southern North Sea and the German Bight. The zinc and cadmium distribution patterns and – to a lesser extent – the distribution of copper reflect the current system in the southern and eastern part of the North Sea.

The influence of the Scottish and English rivers is much less recognizable: heavy metal inputs make up only 1/10 to 1/5 of the total riverine input and a more complex current system with longshore and offshore currents tends to transport larger quantities of the suspended load of the rivers seawards.

**Table 7.** Average heavy metal concentrations in the clay fraction of sediments (collected in 1985) of major rivers within the Federal Republic draining into the North Sea (after MÜLLER, 1985)

| | Zn | Pb | Cu | Cd | Cr | Ni | Co |
|---|---|---|---|---|---|---|---|
| Middle and Lower | | | | | | | |
| Rhine | 1125 | 190 | 178 | 5.1 | 197 | 42 | 27 |
| Ems | 727 | 87 | 108 | 1.7 | 87 | 37 | 25 |
| Weser | 611 | 91 | 76 | 2.6 | 92 | 51 | 19 |
| Elbe | 1818 | 177 | 361 | 11.8 | 325 | 111 | 29 |
| "average shale" | 95 | 20 | 45 | 0.3 | 90 | 68 | 19 |

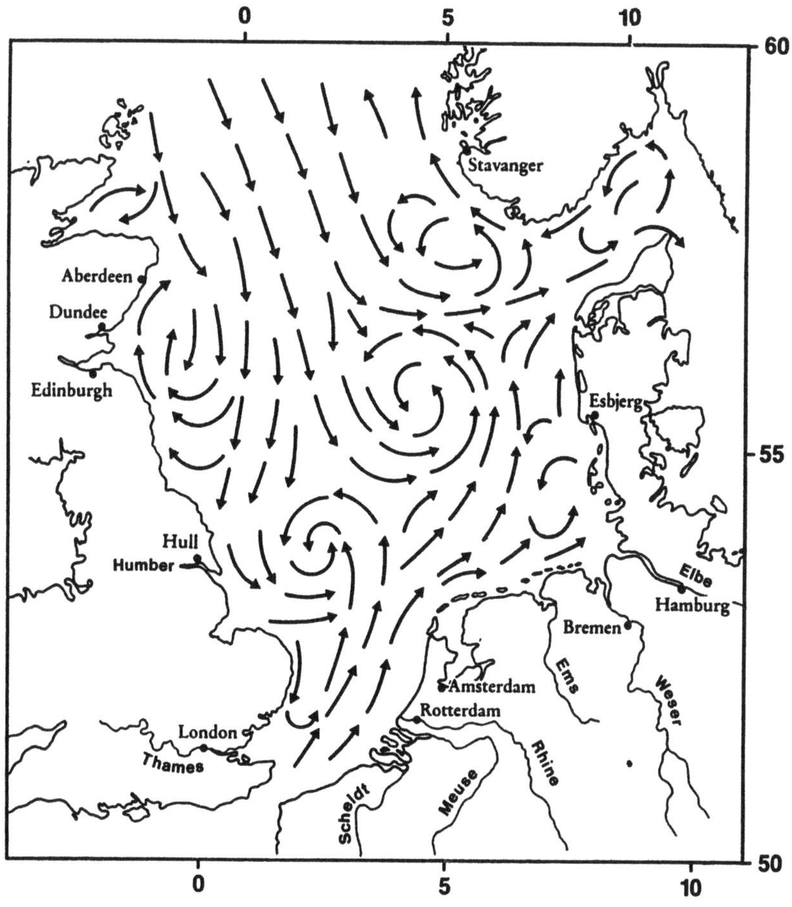


**Fig. 12.** Water circulation (model simulation with mean southwesterly wind of 3m/s speed) in the North Sea. From Müller-Navara and Mittelstaedt (1985)

Our distribution patterns for Cd and Zn are in a relatively good agreement with a model simulation of the relative contamination of the North Sea from inputs of conservative passive tracers via rivers by Hainbucher et al. (1987).

For a general consideration of the heavy metal distribution within the North Sea sediments one should keep in mind that riverine sediment supply amounts only to about 12% of the total sediment mass balance (Table 8). By far the largest sediment supply comes from the Atlantic or reaches the North Sea via the Channel.

The distribution pattern of cobalt in sediments (Fig. 10) may serve as an example how different geochemical distributive provinces contribute to the sediment supply of the North Sea: relatively high cobalt concentrations in large parts of the Southern Bight and along the coast of southern England might be related to a "Channel Province", whereas intermediate concentrations are probably related to a "Rhine-Meuse-Weser-Elbe Province" which borders an "Atlantic Province". In the case of cobalt anthropogenic sources can be neglected: as already stated in the previous chapter, cobalt has the lowest "index of relative pollution potential" and riverine and lacustrine sediments even in highly polluted areas exhibit cobalt concentrations which are identical with or only slightly above its geochemical background value.

The lead distribution in the surficial sediments of the North Sea cannot be explained by rivers as major contributors: The mean lead concentration of the North Sea sediments (Table 2) is only slightly lower than the mean content in highly polluted Rhine and Elbe sediments and even higher than in the Ems and Weser sediments (Table 7). Evidence of lead pollution in the uppermost layers of deep mid-ocean sediments of the north-east Atlantic which is due to atmospheric deposition by wet and dry removal processes has only recently been reported by Veron et al. (1987). It indicates that the global emission of lead into the atmosphere from anthropogenic sources (with leaded gasoline as major source) dominates other lead pathways to the ocean.

The correlation between different metals is generally low (Table 9). Only the coefficients for Zn:Pb, Zn:Co and Pb:Mn are slightly above the lower limit ($r = 0.5$) which frames the field of intermediate correlation ($r = 0.5$–$0.7$). High ($r = 0.7$–$0.9$) and very high ($r > 0.9$) correlations do not exist at all.

**Table 8.** Annual sediment supply into the North Sea (after Eisma, 1981) in $10^6$ t/a

| | |
|---|---|
| Atlantic | 10 |
| Channel | 10 (+?) |
| Baltic | 0.5 |
| Rivers | 4.5 |
| Atmosphere | 1.6 |
| Coastal erosion | 0.7 |
| Seafloor erosion | ≃ 5 (+?) |
| Primary production | ≃ 1 |
| Total | ≃ 34 (+?) |

**Table 9.** Spearman correlation coefficients (r) and probability of significance (p). n = 157 (between heavy metals), 119 heavy metals:$C_{org}$

| | | Pb | Cu | Cd | Fe | Mn | Cr | Ni | Co | $C_{org}$ |
|---|---|---|---|---|---|---|---|---|---|---|
| Zn | r | 0.54420 | 0.45213 | 0.45143 | 0.35119 | 0.42445 | −0.07552 | −0.17299 | 0.55073 | −0.27550 |
| | p | 0.0001 | 0.0001 | 0.0001 | 0.0001 | 0.0001 | 0.3472 | 0.0303 | 0.0001 | 0.0024 |
| Pb | r | | 0.27497 | 0.11245 | 0.34891 | 0.51671 | 0.30987 | −0.04375 | 0.24438 | 0.15341 |
| | p | | 0.0005 | 0.1608 | 0.0001 | 0.0001 | 0.0001 | 0.5864 | 0.0020 | 0.0958 |
| Cu | r | | | 0.29324 | 0.08608 | 0.19712 | −0.02061 | −0.07677 | 0.17233 | −0.15513 |
| | p | | | 0.0002 | 0.2837 | 0.0133 | 0.7978 | 0.3393 | 0.0309 | 0.0921 |
| Cd | r | | | | 0.00427 | 0.04835 | −0.34562 | −0.22198 | 0.40477 | −0.31982 |
| | p | | | | 0.9577 | 0.5476 | 0.0001 | 0.0052 | 0.0001 | 0.0004 |
| Fe | r | | | | | 0.31401 | 0.07265 | −0.15290 | 0.49360 | −0.37676 |
| | p | | | | | 0.0001 | 0.3659 | 0.0559 | 0.0001 | 0.0001 |
| Mn | r | | | | | | −0.10984 | −0.27878 | 0.42587 | −0.14023 |
| | p | | | | | | 0.1709 | 0.0004 | 0.0001 | 0.1282 |
| Cr | r | | | | | | | 0.43108 | −0.40470 | 0.67938 |
| | p | | | | | | | 0.0001 | 0.0001 | 0.0001 |
| Ni | r | | | | | | | | −0.22366 | 0.49413 |
| | p | | | | | | | | 0.0049 | 0.0001 |
| Co | r | | | | | | | | | −0.51343 |
| | p | | | | | | | | | 0.0001 |

| Correlation (r) | |
|---|---|
| < 0.2 | very low |
| 0.2–0.5 | low |
| 0.5–0.7 | intermediate |
| 0.7–0.9 | high |
| > 0.9 | very high |

| Probability (p) | |
|---|---|
| > 0.05 | not significant |
| 0.01–0.05 | significant |
| 0.001–0.01 | very significant |
| < 0.001 | highest significant |

This deficit in correlation confirms the assumption that heavy metals in North Sea sediments stem from different (natural and anthropogenic sources and were transported to their site of deposition by different means.

Between organic matter ($C_{org}$) and heavy metals a relatively high (positive) correlation coefficient (r = 0.67938) was found with Cr. No facts are known that chromium is enriched in marine organic matter (and especially in phytoplankton) which could establish a genetic relationship between concentrations of Cr and $C_{org}$. The question of the source of elevated Cr concentrations in the sediments can therefore not be answered and needs further research, especially in the field of pre-industrial heavy metal concentrations in different parts of the North Sea.

*Acknowledgements.* The authors are very thankful to the participants of the "North Sea Benthos Survey" who generously supplied the sediment material for this study. Thanks are also due to Dipl. Geologe L. Haamann for his analytical assistance.

## References

Carlson H (ed) (1986) Quality status of the North Sea: Dtsch Hydrogr Z Erg-HB 16:424 pp

Dominik J, Förstner U, Mangini A, Reineck HE (1978) Pb-237 and Cs-127 chronology of heavy metal pollution from a sediment core from the German Bight (North Sea). Senckenb Marit 10:213–227

Eisma D (1981) Supply and deposition of suspended matter in the North Sea. Spec Publ Int Ass Sediment 5:415–428

Förstner U, Müller G (1974) Schwermetalle in Flüssen und Seen als Ausdruck der Umweltverschmutzung. Springer, Berlin Heidelberg New York, 275 pp

Förstner U, Reineck HE (1974) Die Anreicherung von Spurenelementen in den rezenten Sedimenten eines Profilkernes aus der Deutschen Bucht Senckenberg Marit 6:175–184

Hainbucher D, Pohlmann D, Backhaus J (1987) Transport of conservative passive tracers in the North Sea: First trends of a circulation and transport model. Continent Shelf Res 7:1161–1179

Irion G, Schwedhelm E (1983) Heavy metals in surface sediments of the German Bight and adjoining areas. Proc Int Conf "Heavy Metals in the Environment", Heidelberg 1983, pp 888–891, CEP Consultants Ltd, Edinburgh, UK

Kramer CMJ, van der Vlies LM (1983) Heavy metals in sediments of the Dutch Wadden Sea. Proc Int Conf "Heavy Metals in the Environment", Heidelberg 1983, pp 892–895; CEP Consultants Ltd, Edinburgh, UK

Müller G (1967) Methods in sedimentary petrology. Schweizerbart, Stuttgart and Hafner, New York/London, 283 pp

Müller G (1981) Heavy metals and other pollutants in the environment: a chronology based on the analysis of dated sediments. Proc Int Conf "Heavy Metals in the Environment", Amsterdam 1981 pp 12–17, CEP Consultants Ltd, Edinburgh, UK

Müller G (1985) Heavy metal concentration in sediments of major rivers within the Federal Republic of Germany; 1972 and 1985. Proc Int Conf "Heavy Metals in the Environment", Athens 1985: pp 110–112; CEP Consultants Ltd, Edinburgh, UK

Müller G, Irion G (1984) Chronology of heavy metal contamination in sediments from Skagerrak (North Sea). Mitt Geol-Pal Inst Univ, Hamburg 56:413–421

Müller-Navara S, Mittelstaedt E (1985) Schadstoffausbreitung und Schadstoffbelastung in der Nordsee. Eine Modellstudie. Dtsch Hydrogr Inst Hamburg, 38:1–50

Nicholson RA, Moore PJ (1981) The distribution of heavy metals in the superficial sediments of the North Sea Rapp P-V Réun Cons Int Explor Mer 181:35–48

Norton MG, Eagle RA, Nunny RS, Rolfe MS, Hardiman PA, Hampson BL (1981) The field assessment of effects of dumping wastes at sea. 8: Sewage sludge dumping in the outer Thames Estuary. Fish Res Techn Rep, Lowestoft 62: 62 pp

Pagee JA van, Postma L (1987) North Sea pollution: The use of modelling techniques for impact assessment of waste inputs. Proc 2nd North Sea 86, Rotterdam, pp 97–113

Schwedhelm E, Irion G (1983) Heavy metal distribution in tidal flat sediments of the German part of the North Sea. Proc Int Conf "Heavy Metals in the Environment", Heidelberg 1983, pp 1037–1040; CEP Consultants Ltd, Edinburgh, UK

Schwedhelm E, Irion G (1985) Schwermetalle und Nährelemente in den Sedimenten der deutschen Nordseewatten. Cour Forsch-Inst Senckenberg 73: 119 pp

Steffen D (1987) Schwermetalle in niedersächsischen Küstensedimenten. In: Niedersächs Umweltministerium (ed): Umweltvorsorge Nordsee Belastungen, Gütesituation und Maßnahmen, pp 185–203

Turekian KK, Wedepohl KH (1961) Distribution of elements in some major units of the earth's crust. Bull Geol Soc Am 72:175–192

Veron A, Lambert CE, Isley A, Linet P, Grousset F (1987) Evidence of recent lead pollution in deep north-east Atlantic sediments. Nature (London) 326:278–281

Chapter 4

# Isotopes in Biogeochemistry

# Carbon and Hydrogen Isotope Variations in Marine Sediment Gases

E. Faber, W. J. Stahl, and M. J. Whiticar

## 1 Introduction

The application of stable carbon and hydrogen isotopes is useful in studying the formation processes of hydrocarbon gases generated from organic and/or inorganic precursor materials by bacterial and thermal processes. Due to isotope fractionations between the sources and the products, hydrocarbon gases have isotope signatures characterizing their formation pathways, and the types and the rate of transformation (maturity) of the precursor material.

Due to the various papers dealing with isotope data of organic compounds and hydrocarbon gases and the limited space for this contribution this paper is restricted to hydrocarbon gases and associated products (e.g. $CO_2$).

The recent activity in the field of marine sediment hydrocarbon gases has been initiated by the investigations, firstly of diagenetic processes in unconsolidated sediments (Claypool and Kaplan 1974; Martens 1976; Reeburgh 1976; Sackett et al. 1979; Kvenvolden 1983) and secondly of catagenic processes and the application towards petroleum exploration (McIver 1967; Tissot and Welte 1978; Hunt 1979).

These investigations monitored gas concentrations and distributions rather than specifically apply or develop isotope techniques but they opened this wide field for other investigations to get an insight into generation and consumption mechanisms.

Pioneering and initial isotope research was started by Craig (1953), who was followed by other researchers (Galimov 1973; Galimov and Ivlev 1973; Stahl 1968, 1977; Colombo et al. 1965; Sackett and Menendez 1972).

More recent work describing hydrocarbon gas formation during early diagenetic sedimentary stages was published by Bernard (1979), Brooks et al. (1979), Kvenvolden and Redden (1980), Rice and Claypool (1981) and Whiticar et al. (1986). Various information is available on the mechanisms of bacterial gas formation and consumption.

The isotope data of thermal hydrocarbon gases generated at greater depth and higher temperatures at the catagenic stage have been investigated by Stahl (1975), Stahl et al. (1977), Stahl (1977), Fuex (1977), Schoell (1980, 1983), Rice (1983), Rigby and Smith (1981).

The hydrocarbon gases are affected by processes such as oxidation or mixing occurring after the gas generation (post genetic alterations, secondary effects) which change the gas quantity, the molecular gas ratio and the isotope signature. Especially methane oxidation has been extensively investigated; some of the results are reported by Hathaway and Degens (1969), Doose (1980), Devol (1983), Barker and Fritz (1981), Coleman et al. (1981), Alperin and Reeburgh (1984) and Whiticar and Faber (1986). Discussions were held on a possible methane $^{12}C$-enrichment due to diffusion processes yielding isotope effects in migrating hydrocarbon gases (Colombo et al. 1968; May et al. 1968; Fuex 1980; Craig 1968).

The application of carbon and hydrogen isotope investigations to the problems of prospecting and exploration for oil and gas was initiated by results correlating isotope data of thermal hydrocarbons to the maturity stage of the source materials (Stahl and Carey 1975, James 1983, Faber 1987). These data are basically important, not only to differentiate between hydrocarbons of bacterial or thermal origin, but also to determine the type of thermal products (oil, condensate or gas) which has been generated from the source rock together with the light hydrocarbons isotopically analyzed. The initial application was related to reservoir gases (sufficient gas quantities for analyses) but after technical developments of the laboratory apparatus, the isotope technique could also be applied to the gases available only in very small gas quantities ($< 10\,\mu l$ methane, Faber et al. 1986) from well (Menedez 1973; Reitsema et al. 1981; Stahl et al. 1977) and also surface samples (Horvitz 1978; Stahl et al. 1981; Faber and Stahl 1984).

## 2 Hydrocarbon Gas Locations

Hydrocarbon gases are situated in various positions within the sediment fabric. Sediment interstices provide the most common location for gases in significant quantities. These gases may be either in the dissolved, hydrated or free gas forms. The latter form is generally restricted to (a) biogenic gases in shallow water environments with higher organic carbon contents, or to (b) gas accumulations (e.g. reservoir gases). Gas hydrates are restricted to sediments saturated with gases, within a specific temperature/pressure range.

Sorbed gases are bound to the surfaces of mineral grains and/or organic matter. The quantity of hydrocarbon gases sorbed is frequently dependent on lithology, and subsurface petroleum potential. These gases generally have a thermogenic character (Faber 1987) a phenomenon which is poorly understood.

Hydrocarbon gases entrapped in sedimentary inclusions are found in trace amounts. For this reason, isotope data on these included gases, as well as on those entrapped in the mineral grains is rare.

## 3 Hydrocarbon Gas Classification

Interpretative schemes are available using molecular and stable carbon and hydrogen isotope compositions to genetically characterize hydrocarbon gases. These schemes are applicable to gases in both diagenetic and catagenic sediment settings. Perhaps the most frequently used combination is that of methane carbon isotope ratio ($\delta^{13}C_{CH4}$) and the light hydrocarbon molecular ratio $C_1/(C_2 + C_3)$ as shown in Fig. 1. Primary gas types (biogenic and thermogenic) occupy distinctly

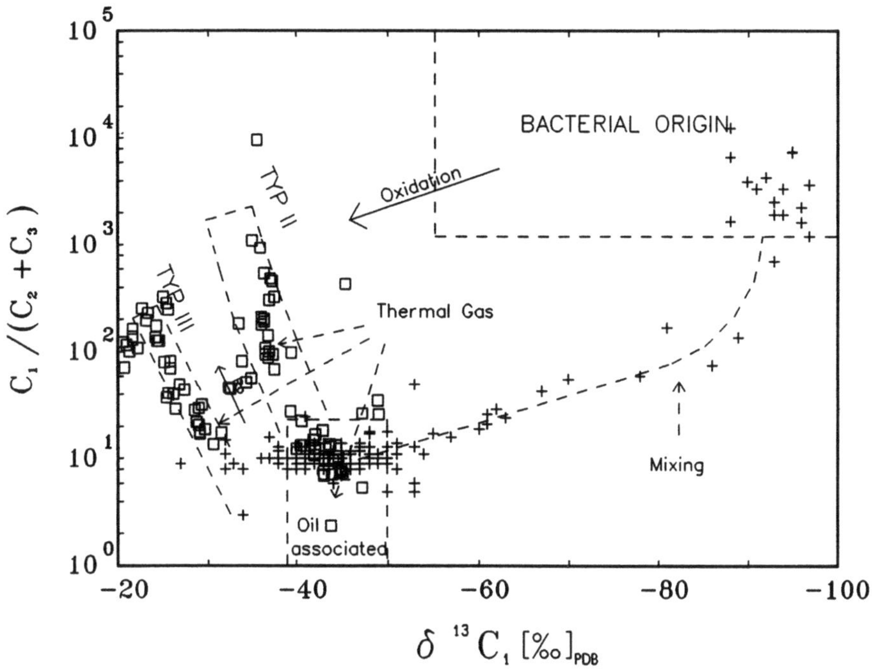


**Fig. 1.** Methane carbon isotope data and gas ratios of various hydrocarbon gases. Data used in Figs. 1 to 4 are from: Botard et al. 1972; Faber 1987; Friedman and Hardcastle 1973; Welhan and Lupton 1987; Whiticar et al. 1985; Woltemate 1982; Woltemate et al. 1984

different ranges. Mixing of gases can also be detected and mathematically described as in simple cases shown by the non-linear mixing line.

The combination of methane carbon and hydrogen isotope ratios ($\delta^{13}C_{CH4}$, $\delta D_{CH4}$, Fig. 2) also informs on the genetic origin of the thermal and bacterial methane gases, but in addition allows to separate the bacterial methane into gases formed by $CO_2$-reduction and acetate fermentation reactions (Whiticar et al. 1986).

Fig. 3 compares the carbon isotope ratios of methane and carbon dioxide ($\delta^{13}C_{CH4}$, $\delta^{13}C_{CO2}$). The bacterial methane is mostly enriched in $^{12}C$ while the $^{13}C$ of the thermal gases increases with the source rock maturity. Compared to the methane isotope variations, the $^{13}C/^{12}C$-ratios of the carbon dioxide are relatively constant.

A quantitatively relationship of methane and ethane carbon isotope data (based on reservoir gases) with source rock maturity is given in Fig. 4. It is inferred, that the geochemical data of unmixed, thermal gases from one source rock of a specific maturity are plotting on or near to the correlation line, indicating the maturity stage of the source. Mixed gases plot off the line; in simple cases (mixing of only two different gases) the mixing relationships can be determined (Faber 1987; Berner and Faber 1987).

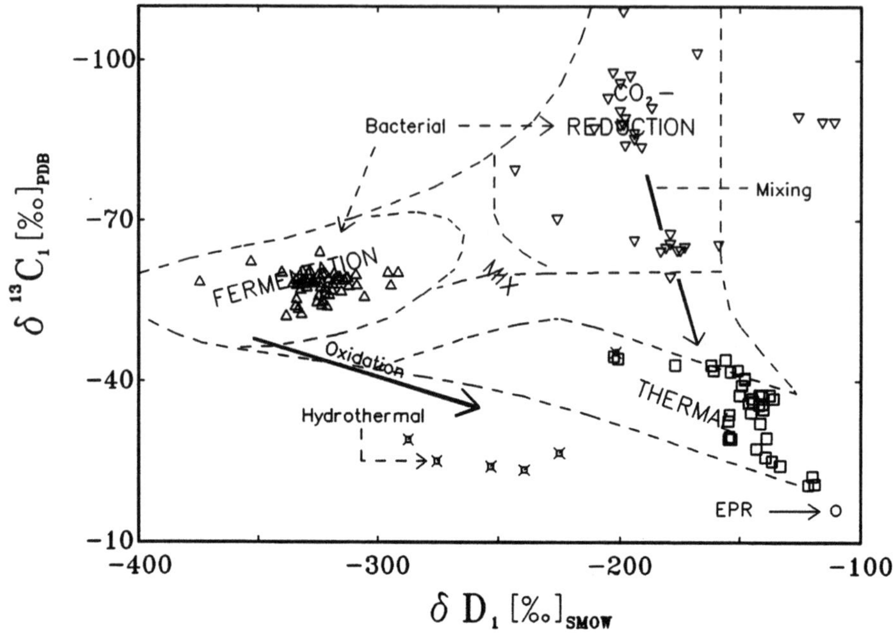


**Fig. 2.** Methane carbon and hydrogen isotope ratios of deep and shallow gases (after Whiticar et al., 1986)

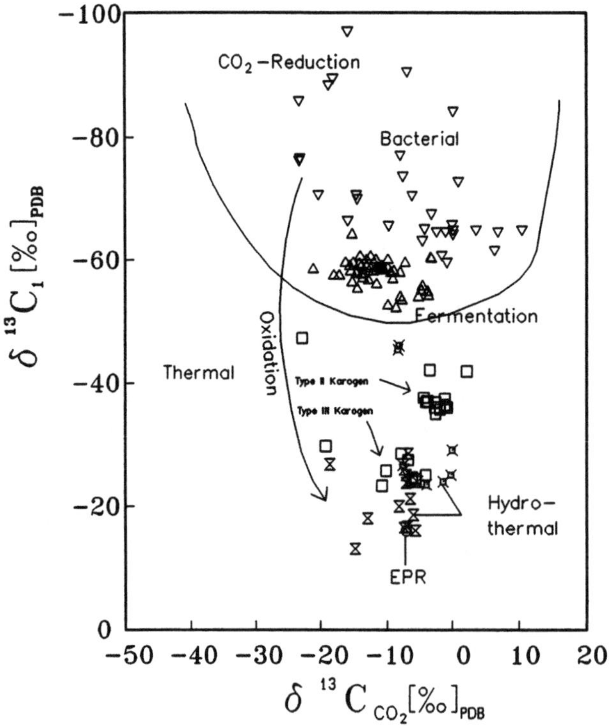


**Fig. 3.** Carbon isotope ratios of methane and carbon dioxide (after Whiticar and Faber, 1986)

**Fig. 4a,b** Methane (**Fig. 4a**; propane, **Fig. 4b**) and ethane carbon isotope ratios of natural gases in relation to source rock maturity and type (after Faber, 1987)

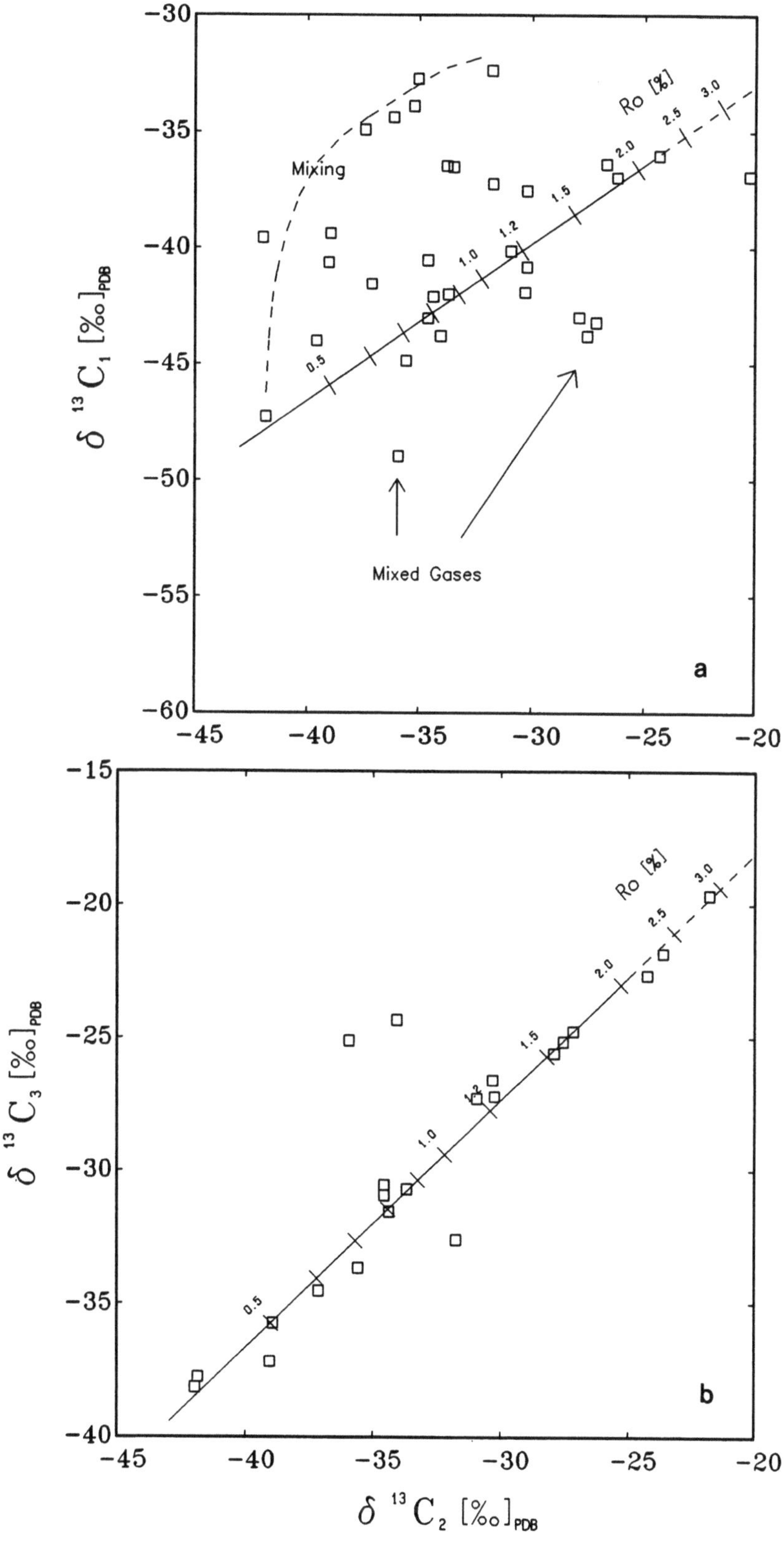

Mixing
Ro [%]
0.5
1.0
1.2
1.5
2.0
2.5
3.0
Mixed Gases
a
b
δ ¹³C₁ [‰]PDB
δ ¹³C₃ [‰]PDB
δ ¹³C₂ [‰]PDB


Fig. 4a,b

## 4 Examples

Bacterial hydrocarbon gases are dominated by methane ($C_1/(C_2 + C_3) > 500$) which is rich in $^{12}C$ (-100‰ $< \delta^{13}C_{CH4} <$ -55‰ as indicated in Fig. 1. The formation of this methane by bacterial fermentation or carbon dioxide reduction (Whiticar et al. 1986) is indicated by the carbon and hydrogen isotope data Fig. 2). The $\delta D_{CH4}$ of the fermentation gas generally is more negative while the $CO_2$-reduction gas is more positive than -250‰. A similar, but not as clear distinction is given by the $\delta^{13}C_{CH4}$ which is often more positive then -60‰ for the fermentation methane and vice versa for the $CO_2$-reduction gas. The carbon isotope ratio of the coexisting carbon dioxide ranges between -20‰ $< \delta^{13}C_{CO2} <$ 10‰. It is not as diagnostic as is the $\delta^{13}C_{CH4}$ and the $\delta D_{CH4}$.

Thermal hydrocarbons (including reservoir gases) are, relative to the bacterial gases, enriched in the $C_{2+}$-components (Fig. 1) and the methane $^{13}C$ and deuterium (Figs. 1 and 2). The kerogene type of the organic source material (type II; III) (coal) and the association to oil in reservoirs is indicated (Fig. 1). The combination of methane, ethane and propane carbon isotope ratios is useful to estimate quantitatively the source rock maturity (Fig. 4). If data points plot near the regression line in Fig. 4 a "primary", i.e. not a mixed gas, is inferred allowing one to determine the maturity (vitrinite reflectance) from the carbon isotopes.

Some of the various data of hydrothermal gases, of gases from hydrothermal waters and gases of probably abiogenic origin from the East Pacific Rise (Welhan and Lupton 1987) are included in Figs. 2 and 3. The carbon dioxide and methane $\delta^{13}C$-values are similar to the data of the gases previously discussed. However, the methane carbon isotope ratios are similar to those of the coal gases or are more positive. Therefore, these gases can be distinguished from the thermal or bacterial hydrocarbons.

Often, the post genetic alterations such as bacterial oxidation ore mixing of different gases mask or change the original isotope signature. As indicated in Fig. 2, the mixing line of two different gases concerns the $\delta^{13}C$-$\delta D$ relationship (Schoell 1983), but is non-linear concerning the $C_1/(C_2 + C_3) - \delta^{13}C_{CH4}$ diagram of Fig. 1 (Bernard 1979; Faber and Stahl 1984) and the $\delta^{13}C$-$\delta^{13}C$ relations of Fig. 4a (Faber 1987, Berner and Faber 1987). Mixing proportions in simple (two mixing members) can easily be calculated.

During hydrocarbon gas oxidation in sediments mainly the methane concentration and the $^{12}C$-methane content decrease (Faber and Stahl 1984; Whiticar and Faber 1986). A corresponding oxidation trend is indicated in Fig. 1. The oxidation simultaneously increases the deuterium content in the methane resulting in a linear oxidation trend in Fig. 2 (Coleman et al. 1981). Although both gas and isotope ratios are changed, the mixing and oxidation processes are generally detectable and often the genetic origin of these gases can be determined.

# 5 Outlook

Hydrocarbon gas classification has reached a level of sophistication that permits not only the reliable characterization of gas types, but also the identification of gases which have undergone alteration such as bacterial oxidation, or mixing. Isotopes of gases in extremely low concentrations can also be measured with the currently available analytical technologies. These classification schemes provide geochemists with the interpretative confidence to track hydrocarbon gas formation during metabolic and chemical processes acting on organic matter in marine sediments.

The reliable classification of hydrocarbon gases in marine sediments, rather than being a goal in itself, represents the prerequisite to investigate the fluxes of gases in various environments and their interactions with the geo- and atmospheres.

# References


Alperin MJ, Reeburgh WS (1984) Geochemical observations supporting anaerobic methane oxidation. In: Crawfort RL, Hanson RS (eds) Microbial Growth on C-1 Compounds. Am Soc Microbiol, Washington, pp 282–289

Barker JF, Fritz P (1981) Carbon isotope fractionation during microbial methane oxidation. Nature (Lond) 293:289–291

Batard F, Baubron JC, Bosch B, Marce A, Risler JJ (1982) Isotopic evidence of gases of a deep origin in French thermomineral waters. J Hydrol 56:1–21

Bernard BB (1979) Methane in marine sediments. Deep-Sea Res 26:429–443

Berner U, Faber E (1988) Maturity related mixing model for methane, ethane and propane, based on carbon isotopes. Org. Geochem. Vol 13, Nos 1–3, pp. 67–72

Brooks JM, Bright TJ, Bernard BB, Schwab CR (1979) Chemical aspects of a brine pool at the East Flower Garden bank, Northwestern Gulf of Mexico. Limnol Oceanogr 24:4

Claypool GE, Kaplan IR (1974) The origin and distribution of methane in marine sediments. In: Kaplan IR (ed) Natural gases in marine sediments. New York, Plenum, pp 99–139

Coleman DD, Risatti JB, Schoell M (1981) Fractionation of carbon and hydrogen isotopes by methane-oxidizing bacteria. Geochim Cosmochim Acta 45:1033–1037

Colombo U, Gazzarini F, Gonfiantini R, Tongiorni E (1965) Carbon isotopic composition of individual hydrocarbons from Italian natural gases. Nature (Lond) 205:1303–1304

Colombo U, Gazzarrini F, Gonfiantini R, Tongiorgi E, Caflisch L (1968) Carbon isotope study of hydrocarbons in Italian natural gases. Adv Org Geochem, Oxford, pp 499–516

Craig H (1953) The geochemistry of stable carbon isotopes. Geochim, Cosmochim, A 3:53–92

Craig H (1968) Isotope separation by carrier-diffusion. Science 159:93–96

Degens ET (1968) Geochemistry of sediments. Enke, Stuttgart, 282 pp

Devol AH (1983) Methane oxidation rates in the anaerobic sediments of Saanich Inlet. Limnol Oceanogr 28:738–742

Doose PR (1980) The bacterial production of methane in marine sediments. PhD Thesis, UCLA 240 pp

Faber E (1987) Zur Isotopengeochemie gasförmiger Kohlewasserstoffe. Erdöl, Erdgas und Kohle, 103. Jahrgang, Heft 5, Mai 1987

Faber E, Stahl WJ (1984) Geochemical surface exploration for hydrocarbons in North Sea. AAPG Bull 68, 3:363–386

Faber E, Dumke I, Ott A, Poggenburg J (1986) DGMK 298 – Weiterentwicklung von Isotopenverfahren für die Kohlenwasserstoff – Exploration. Ber Dtsch Ges Mineralölwiss Kohlechem eV, Hamburg

Friedman I, Hardcastle K (1973) Interstitial water studies. Leg 15 – Isotopic composition of water. In: Heezen BC, Mac Gregor ID (eds) Initial Reports Deep Sea Drilling Project. Vol 20. US Govt Printing Office, Washington, pp 901–903

Fuex AN (1977) The use of stable carbon isotopes in hydrocarbon exploration. J Geochem Expl 7:155–188
Fuex AN (1980) Experimental evidence against an appreciable isotopic fractionation of methane during migration. In: Douglas AG, Maxwell GR (eds) Advances in Organic Geochemistry 1979. Oxford, New York, Pergamon Press, pp 725–732
Galimov EM (1973) Carbon isotopes in oil-gas geology. Nedra, Moscow (in Russian)
Galimov EM, Ivlev AA (1973) Thermodynamic isotope effects in organic compounds. I. Carbon isotope effects in straight-chain alkanes. Russian J Phys Chem 47(11):1564–1566
Gunter BD, Musgrave BC (1971) New evidence on the origin of methane in hydrothermal gases. Geochim Cosmochim Acta 85:113–118
Hathaway JC, Degens ET (1969) Methane-derived marine carbonates of Pleistocene age. Science 165:690–692
Horvitz L (1978) Near-surface evidence of hydrocarbon movement from depth. In: Problems of petroleum migration: AAPG Studies in Geology 10:241–269
Hunt JM (1979) Petroleum geochemistry and geology. WH Freeman & Co, 617 pp
James AT (1983) Correlation of natural gas by use of carbon isotopic distribution between hydrocarbon components. AAPG Bull 67, 7:1176–1191
Kvenvolden KA (1983) Geochemistry of natural-gas hydrates in oceanic sediment. In: Bjoroy M (ed) Advances in Organic Geochemistry 1981. Wiley Heyden 1983, pp 442–430
Kvenvolden KA, Field ME (1981) Thermogenic hydrocarbons in unconsolidated sediment of El River Basin, offshore northern California. AAPG Bull 65:1642–1646
Kvenvolden KA, Redden GD (1980) Hydrocarbon gas in sediment from the shelf, slope, and basin, of the Bering Sea. Geochim Cosmochim Acta 44:1145–1150
Lyon G (1974) Interstitial water studies Leg 15 – Chemical and isotopic composition of gases from Cariaco Trench sediments. In: Heezen BC, MacGregor ID (eds) Initial Reports Deep Sea Drilling Project. Vol 20, pp 773–774. US Govt Printing Office, Washington
Martens CS (1976) Control of methane sediment-water transport by macroinfaunal irrigation in Cape Lookout Bight, North Caroline. Science 192:998–1000
May FW, Freund EP, Dostal KP (1968) Modellversuche über Isotopenfraktionierung von Erdgaskomponenten während der Migration. Z Angew Geol 14:376–380
McIver RD (1967) Composition of kerogen – clue to its role in the origin of petroleum. In: 7th World Petroleum Congress, Proc, Mexico City, Elsevier, London 2:26–36
Menendez R (1973) Composition isotopic du carbon dans le gaz provenant de sondages d' aquitaine. Bull Cent Rech Pau – SNPA 7:69–81
Nakai M, Yoshida A, Ando N (1974) Isotopic studies on oil and natural gas fields in Japan. Chikyakaya (Geochem) 7/8, 1:87–98
Reeburgh WS (1976) Methane consumption in Carious Trench waters and sediments. Earth Planet Sci Lett 28:337–344
Reitsema RH, Kaltenback AJ, Lindberg FA (1981) Source and migration of light hydrocarbons indicated by carbon isotopic ratios. AAPG Bull 65:1636–1542
Rice DD (1983) Relation of natural gas composition to thermal maturity and source rock typ in San Juan Basin, Northwestern New Mexico and Southwestern Colorado. AAPG Bull 67, 8:1199–1218
Rice DD, Claypool GE (1981) Origins of and conditions for shallow accumulations of natural gas. Twenty-Seventh Annu Field Conf-Wyoming Geol Assoc Guide Book, pp 267–271
Rigby D, Smith JW (1981) An isotopic study of gases and hydrocarbons in the cooper basin. APEA J 21, 1:222–229
Sackett WM, Menendez R (1972) Study of the hydrocarbons and kerogens in the Aquitan Basin, Southwest France. In: Gaertner HRV, Wehner H (eds) Adv in Org Geochem, pp 523–533
Sackett WM, Broks JM, Bernard BB, Schwab CR, Chung H, Parker RA (1979) A carbon inventory for Orca basin brines and sediments. Earth Planet Sci Lett 44:73–81
Schoell M (1980) The hydrogen and carbon isotopic composition of methane from natural gases of various origins. Geochim Cosmochim Acta 44:649–661
Schoell M (1983) Genetic characterisation of natural gases. Am Assoc Petrol Geol Bull 67, 12:2225–2238
Stahl WJ (1968) Klärung der Genese nordwestdeutscher Erdgase durch $^{13}C/^{12}C$-Isotopenanalysen. Naturwissenschaften 55:296
Stahl WJ (1975) Carbon isotope fractionations in natural gases. Nature 251 (5471):134–135

Stahl WJ (1977) Carbon and nitrogen isotopes in hydrocarbon research and exploration. Chem Geol 20:121–149
Stahl WJ, Carey BD Jr (1975) Source-rock identification by isotope analyses of natural gases from fields in the Val Verde and the Delaware Basins, West Texas. Chemical Geology 16:257–267
Stahl WJ, Faber E, Schmitt M, Carey BD (1977) Carbon isotopes in oil and gas exploration- examples of applications. Nuclear techniques and mineral resources 1977. Int Atomic Energy Agency, IAEA-SM-216/61
Stahl WJ, Faber E, Carey BD, Kirksey DL (1981) Near-surface evidence of migration of natural gas from deep reservoirs and source rocks. AAPG Bull 65, 9:1543–1550
Tissot BP, Welte DH (1978) Petroleum formation and occurrence. A new approach to oil exploration. Springer, Berlin Heidelberg New York, 527 pp
Welhan JA, Lupton JE (1987) Light hydrocarbon gases in Guaymas basin hydrothermal fluids: thermogenic versus abiogenic origin: AAPG Bull 71, 2:215–223
Whiticar MJ, Faber E (1986) Methane oxidation in sediment and water column environments- isotopic evidence. Proc "12th Int Meeting On Organic Geochemistry; Sept 16–20, Jülich"
Whiticar MJ, Suess E, Wehner H (1985) Thermogenic hydrocarbons in surface sediments of the Bransfield Strait, Antarctic Peninsula. Nature (Lond) 314:87–90
Whiticar MJ, Faber E, Schoell M (1986) Biogenic methane formation in marine and freshwater environments: CO2 reduction vs acetate formation – isotope evidence. Geochim Cosmochim Acta 50:693–709
Woltemate I (1982) Isotopische Untersuchungen zur bakteriellen Gasbildung in einem Süßwassersee. Diplomarbeit, Clauthal Universität, 90 pp
Woltemate I, Whiticar MJ, Schoell M (1984) Carbon and hydrogen isotopic composition of bacterial methane in a shallow freshwater lake. Limnol Oceanogr 29:985–992

# Stable Carbon Isotope Composition of Pelagic and Benthic Organic Matter in the North Sea and Adjacent Estuaries

R. W. P. M. Laane, E. Turkstra, and W. G. Mook

## 1 Introduction

Estuarine particulate (organic) matter is a mixture of material which mainly originates from two sources; allochthonous sources (rivers and the sea) and from the in-situ primary production.

In the Dutch estuaries the most important source of suspended particulate organic carbon (POC) is the North Sea (Fig. 1). For instance, the annual budget for the Ems-Dollard showed that $902 \times 10^3$ tons of particulate matter, of which $34 \times 10^3$ tons of POC, originate from the North Sea, compared to a river input of $130 \times 10^3$ and $12 \times 10^3$ tons, respectively (Laane and Ruardy 1988). For the other estuaries the marine input is also the major component (Eisma et al. 1982).

For studies on sediment budgets, foodweb, micropollutant and diagenetic processes it is necessary to understand the distribution and fate of riverine and marine POC in estuaries and in the adjacent coastal sea. Up to now, the most fruitful method of discriminating between different sources is by their POC isotope ratio: riverine POC has a lower $\delta^{13}C$ value (average -27‰ for the Rhine at location R, average -28‰ for the Meuse at location M, Fig. 1) than marine POC (average $\delta^{13}C \simeq -21$‰) (Fry and Sherr 1984; Sackett 1986). Gradients in $\delta^{13}C$ of POC (-15.2‰ to -28‰) and surface sedimentary organic carbon (SOC) (-14.0‰ to -27.4‰) in different estuaries have been described by Tan and Strain (1979); Salomons and Mook (1981); Eisma et al. (1982); Sherr (1982); Fry (1984); Simenstad and Wissmar (1985); Hedges et al. (1986) and Jouanneau (1987).

Complicating factors which must be taken into account before applying a simple mixing model to calculate the contribution of fluvial and marine POC at a certain location in an estuary are summarized as follows:

1. The fractionation of carbon isotopes during photosynthesis by land plants differs. Plants fixing their carbon by the C3-pathway show average $\delta^{13}C$ values of -26‰, considerably lower than plants fixing their carbon by the C4-pathway, averaging at $\delta^{13}C = -13$‰ (Deines 1980). The human associated foodweb and the derived detritus are primarily C3.
2. Marine plants show a temperature dependence of the fractionation for carbon isotopes during the photosynthesis. Sackett et al. (1965) and Fontugne and Duplessy (1981) found that plankton in cold water has lower $\delta^{13}C$ (-27‰) than plankton in warm waters. Seagrass, marsh vegetation and other benthic marine plants may have less negative $\delta^{13}C$ values than phytoplankton.
3. During diagenetic processes $\delta^{13}C$ of the organic matter is inclined to shift to lower values, due to the loss of relatively heavy compound such as carbohydrates (Deines 1980).

In this communication the distribution of the stable carbon isotope composition of POC and of SOC (Sedimentary Organic Carbon) is described in four different Dutch estuaries and the adjacent coastal sea. Taken into account the results on $\delta^{13}C$ during a tidal cycle and between different seasons and in decomposition experiments, the origin and transformation of organic matter of and into sediments is described. To obtain an overall picture, data from other papers describing the $\delta^{13}C$ in the Ems-Dollard, the Rhine, the Scheldt and the coastal zone (Anonymous 1975; Salomons and Mook 1981; Eisma et al. 1982, 1985; Marquenie et al. 1985; Van Zoest, unpubl. results) have been included.

## 2 Material and Methods

Samples were collected during different cruises. During cruise 1 in October 1984, surface-water samples were collected by pumping at stations 1–32 (Fig. 1). Sediment samples were collected with a van Veen grab from which only the upper few cm of sediment were taken for further analysis. The sediment was divided by wet sieving into a coarse ($> 63\ \mu m$) and a fine fraction ($< 63\ \mu m$). In October 1984 at station 27, surface-water samples were collected during one tidal cycle. During this cruise particulate matter was separated by centrifuging. On the other 4 cruises, water samples were collected by pumping from a depth of 1.5 meters at stations 22, 27, 28, and 29 (March-June 1985). Particulate matter was collected after immediate suction filtration through a Whatman GF/F filter.

The unfiltered water samples were incubated in the dark at field temperature and stirred continuously with a magnetic stirrer. Within 15 min after sampling and after 13 days, the content of the bottles was thoroughly mixed and subsampled. Each 150 ml was filtered over precombusted glass-fiber filters (Whatman GF/F) and stored at -20°C until analysis. The filters and bottles had been precombusted for 4.5 h at 480°C.

Methods applied for determination of temperature, salinity, and concentration of particulate organic matter have been described by Van Es and Laane (1982) and Laane (1982).

Dried suspended matter and sediment samples were first treated with dilute hydrochloric acid to remove carbonates. After washing and drying, the residues were combusted to $CO_2$ and analyzed for $^{13}C/^{12}C$ as described by Salomons and Mook (1981). The results on the carbon isotope ratios are reported as relative deviations from the VPDB standard. The more negative a $\delta^{13}C$ value, the more the organic matter is depleted in $^{13}C$.

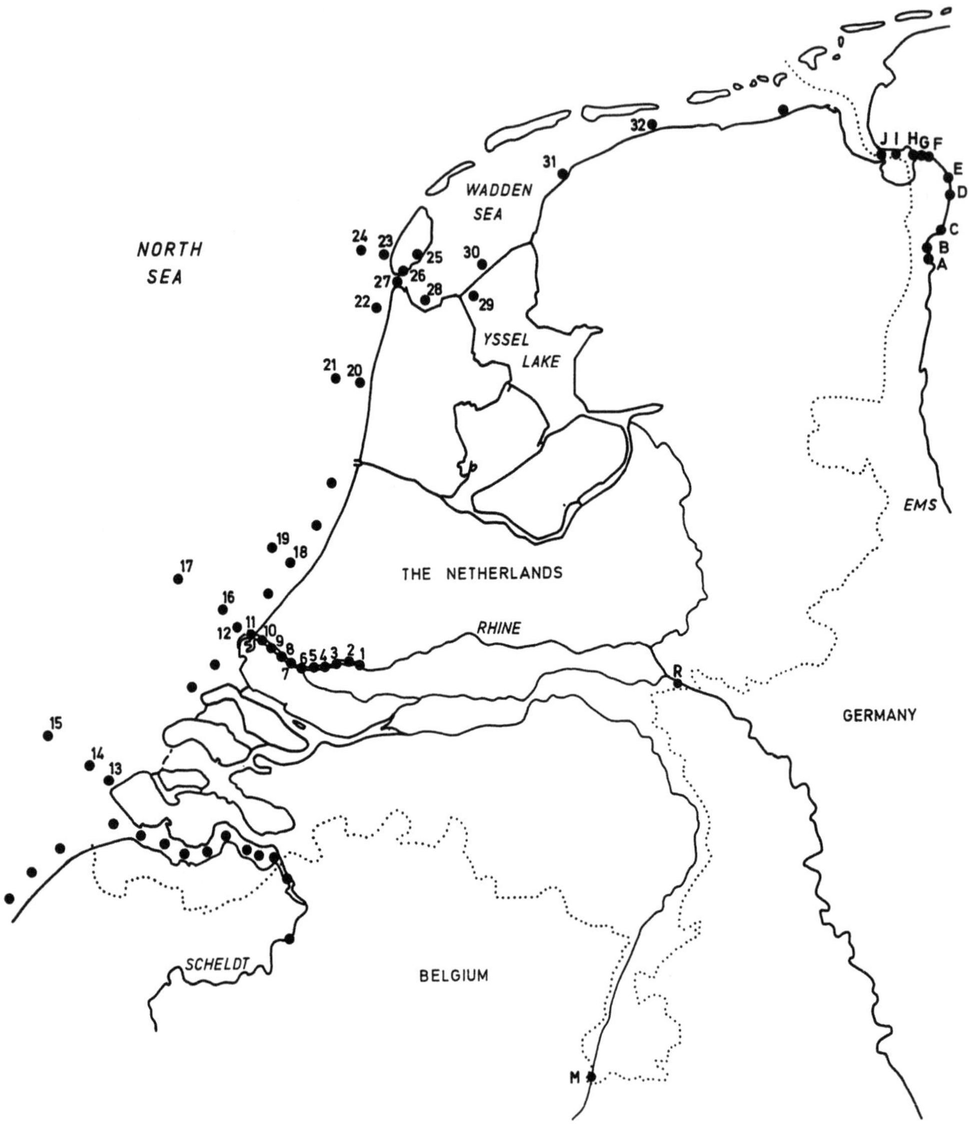


**Fig. 1.** Map of the North Sea and Dutch estuaries with sampling locations

# 3 Results

## 3.1 Estuaries

In the fresh-water compartment of the Rhine (Fig. 1, stations 1-8), $\delta^{13}C$ of POC is rather constant (Fig. 2). In the tidal compartment it increases to -17.9‰. In the first estuarine compartment station (8-9), the organic matter has a higher $\delta^{13}C$ value than the marine value of -23‰ (station 12). Here, the (sedimentary) SOC is more negative (station 10-11).

In the fresh-water compartment, $\delta^{13}C$ of SOC is less negative than that of POC. This is also observed in the river Scheldt at low salinity (Fig. 3). The opposite is found in the Ems-Dollard at low salinities and in the estuarine compartments of the Scheldt and Rhine. Here, the sediments have lower $\delta^{13}C$ than suspended matter (Figs. 2 and 3). These results are in agreement with those found by others.

Salomons and Mook (1981) observed a $\delta^{13}C$ value of -25.4 ± 0.4‰ for SOC in the sediments at station 1 for different months during the years 1958-1973. They also measured $\delta^{13}C$ of the stable POC in a vertical profile at station 8 at four different depths, showing an average value of -27.0 ± 0.2‰.

## 3.2 North Sea and Wadden Sea

$\delta^{13}C$ for SOC ranges from -20.8‰ to -25.4‰ (Fig. 3). However, relatively high values are found at station 13 (-15.9‰) and at station 16 (-16.9‰). $\delta^{13}C$ of the coarse fraction at these stations is even as high as -13.8‰ (Figs. 1 and 3).

$\delta^{13}C$ of POC varies from -15.2‰ to -25.8‰ (Fig. 3), the coarse fraction at station 18 having a $\delta^{13}C$ value of -12.5‰.

At the North Sea and the Wadden Sea stations POC has lower $\delta^{13}C$ values than the SOC.

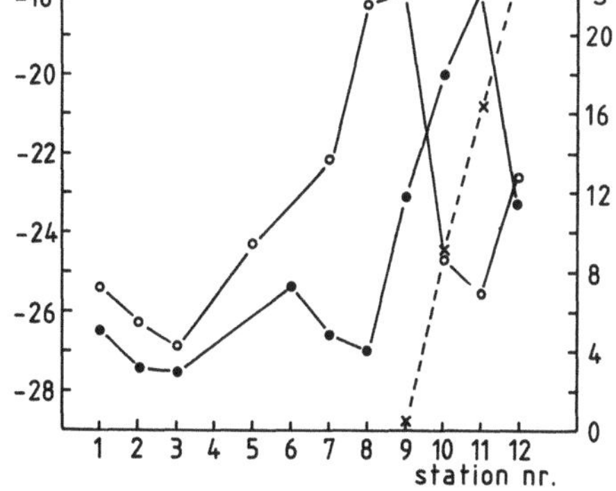


**Fig. 2.** Distribution of the $\delta^{13}C$ in suspended (POC) (●), sedimentary (SOC) (○) organic matter and salinity (×) in the Rhine estuary in October 1984. Horizontal scale: station numbers

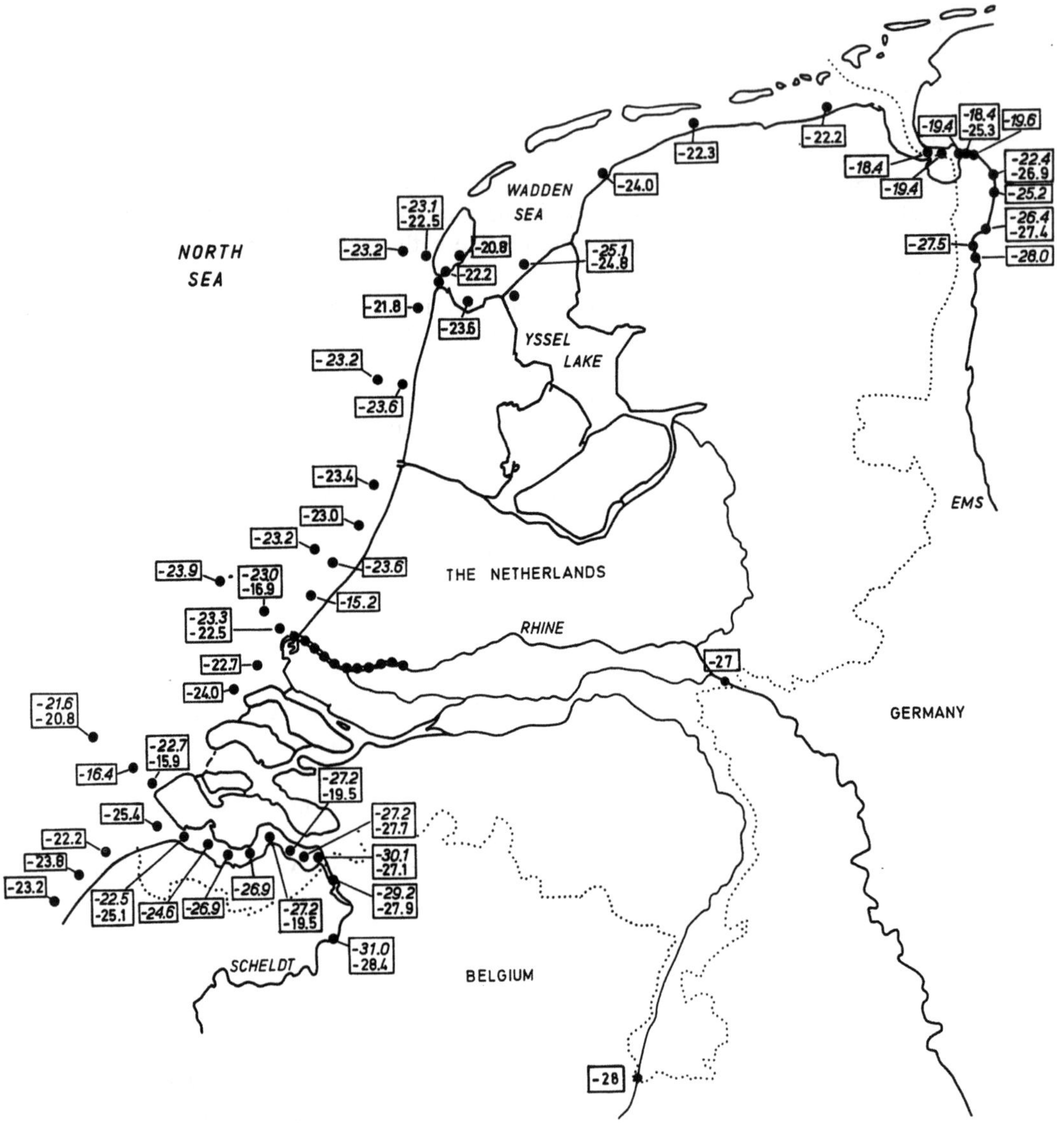


**Fig. 3.** Distribution of the $\delta^{13}C$ values of suspended (values in *italics*) and sedimentary organic matter in the Scheldt, Rhine and Ems-Dollard estuary, the coastal zone of the North Sea and the Wadden Sea

## 3.3 Tidal and Seasonal Variation

During a tidal cycle in October 1984, $\delta^{13}C$ of POC varied between -23.8‰ (LW) and -22.6‰ (HW) in a tidal inlet of the Wadden Sea (station 27).

$\delta^{13}C$ of POC at stations 22, 27, 28 and 29 is plotted against sea-water temperature in Fig. 4. At all stations $\delta^{13}C$ increases with increasing temperature. The slope of the relations is quite similar to that found by Fontugne and Duplessey (1981) for marine phytoplankton. However, the absolute values are different. Small deviations, as for instance the decrease at station 28 at 9 °C, can be explained by the fact that the samples were not taken at the same tidal phase. During cruise C and D an increase is found in $\delta^{13}C$ in POC of the Yssellake (station 29) and during the last cruise at station 27. Also a sharp increase is found at station 28 between 9 and 14 °C.

## 3.4 Decomposition Experiments

The changes in $\delta^{13}C$ of POC during decomposition are given in Fig. 5A-D for three different stations (22, 28 and 29). During the first cruise in March (Fig. 5A) $\delta^{13}C$ of POC changed only slightly during decomposition. During decomposition experiments in the other months the isotope value of POC of the Yssel Lake does not change significantly. POC from stations 28 and 29 became lower $\delta^{13}C$ during

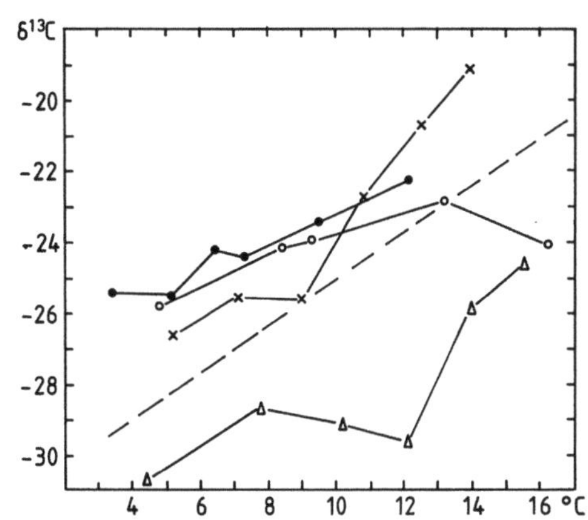


**Fig. 4.** Relation between the $\delta^{13}C$ of POC and temperature for station 22(●), 27 (○), 28(×) and 29(△) in 1985. The relation found by Fontugne and Duplessey (1981) is given as the *dotted line*

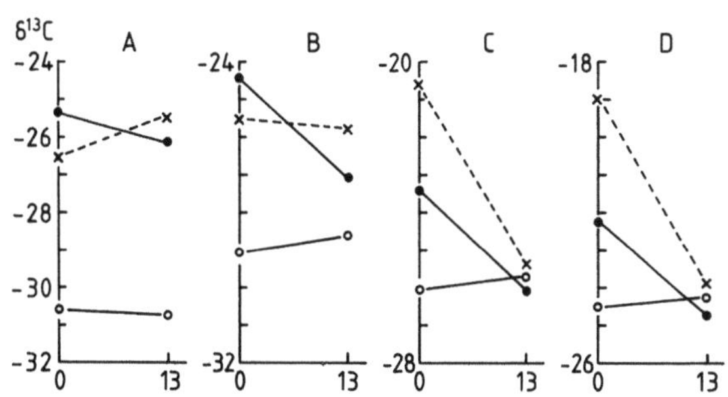


**Fig. 5A-D.** Changes in the $\delta^{13}C$ of particulate organic matter after 13 days of dark incubation for station 22(●), 28(×) and 29(○), at four different dates: **A** 23 April; **B** 7 May; **C** 23 May; **D** 6 June, 1985

decomposition experiments, resulting in a maximum difference of 4.9‰ at station 28. During May and June (Fig. 5C and D), the ultimate isotope values for the 3 different stations tend to go to similar $\delta^{13}C$ values around -25.5‰ and -24.3‰, respectively.

During all incubation experiments, the concentration of POC decreases after 13 days. Especially in the May and June experiments, up to 90% of particulate organic matter was decomposed (Laane and Kloosterhuis unpubl.).

### 3.5 Relation of $\delta^{13}C$ with the Concentration of Organic Matter

For the different estuaries and the Dutch coastal zone the relation between $\delta^{13}C$ and the percentage of SOC and POC is given in Figs. 6A and 6B, respectively, showing no clear relation. Only patterns can be recognized for the different areas. In the sediments of the Scheldt and Ems-Dollard estuary nearly the same pattern is found. In the Rhine estuary the isotope values are comparable with those found in Ems-Dollard and the Scheldt estuaries, where the organic carbon content is relatively low.

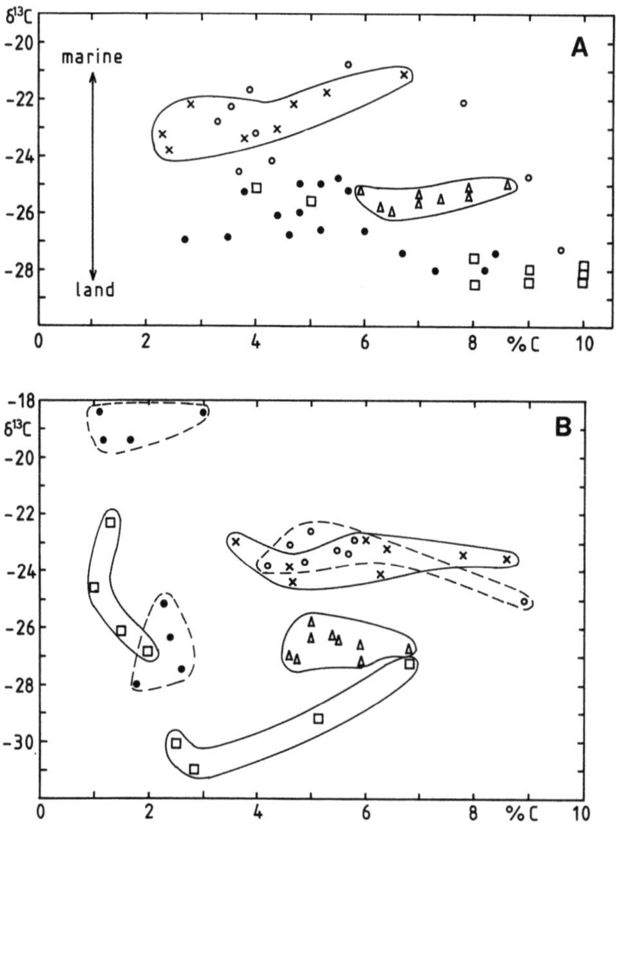


**Fig. 6A,B.** Relation between the $\delta^{13}C$ of sedimentary (**A**) and suspended (**B**) organic carbon in the Ems-Dollard (●), Wadden Sea (○), the North Sea (×), the Rhine (△) and the Scheldt (□)

The organic matter in the Wadden Sea and coastal zone sediments of the North Sea has less negative $\delta^{13}C$ than that in the 3 estuaries. Relatively low $\delta^{13}C$ values are found in the areas with predominant sedimentation of the Wadden Sea (stations 25, 26, and 28).

POC in the estuaries has lower $\delta^{13}C$ than that in the Wadden Sea, the North Sea coast and the Scheldt estuary (Fig. 6B). Two patterns are found in the Ems-Dollard and Scheldt estuaries: one in the fresh-water compartment and the other in the marine part of the estuary. Tan and Strain (1979, 1983) also observed no clear correlation between the percentage of organic matter and the $\delta^{13}C$ in the sediments of the St. Lawrence estuary.

# 4 Discussion

## 4.1 Rivers Rhine, Meuse, Scheldt, and Ems

The fresh-water POC and SOC have the lowest $\delta^{13}C$ values. This is in agreement with data from other rivers and is explained by the difference in $\delta^{13}C$ between land and marine plant material. Relatively low values may be effected by organic pollution of the river (-28‰) (Scheldt) or by a high load of humic substances from peat erosion (Ems estuary; Laane 1981; Eisma et al. 1981).

## 4.2 North Sea

Our values for POC and SOC in the North Sea are in agreement with marine results described by other authors (review by Sackett 1986), apart from some exceptionally high $\delta^{13}C$ values (between -15‰ and -17‰). This is due to the presence of seagrass and debris of marsh vegetation especially in the coarse fraction (station 13).

## 4.3 The Estuaries

The behavior of $\delta^{13}C$ of POC and SOC in the Ems-Dollard estuary is shown in Fig. 7 (data from Eisma et al. 1985 and from Salomons and Mook 1981). The same picture is found in the Scheldt estuary (cf. Fig. 3). In these two estuaries a simple mixing model, as suggested by Eisma et al. (1985) can be applied to calculate the contribution of marine and fluvial organic matter in the suspended matter and in the sediments at any location. From these data it can be concluded that in the Scheldt and in the Ems-Dollard estuary marine organic matter enters the estuary up to the first fresh-water compartment. This is in agreement with the results of Favejee (1960); Salomons (1975) and Rudert and Muller (1981), who also found an upstream marine import of suspended matter.

In the Rhine estuary (Fig. 2) this picture is more complicated. An irregular pattern with relatively high and low $\delta^{13}C$ values of particulate and sedimentary

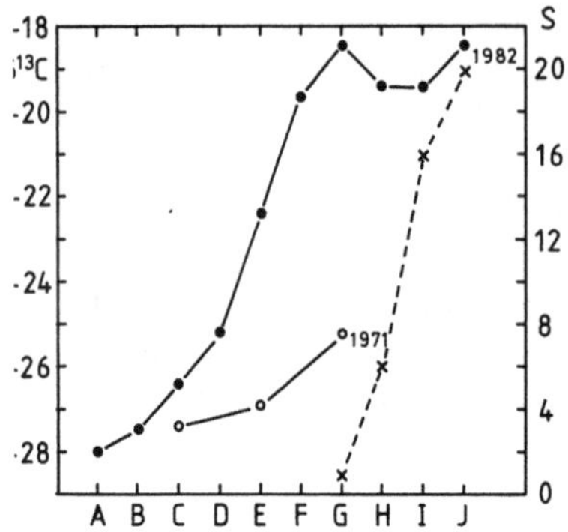


**Fig. 7.** Distribution of the $\delta^{13}C$ for suspended (●) and sedimentary organic carbon (○) and salinity (×) in the Ems-Dollard. *Horizontal scale* station numbers. (Data from Salomons and Mook 1981 and Eisma et al. 1985)

organic matter is found between the two end members. Other anthropogenic sources must be taken into account to explain this pattern.

The sedimentary carbon at stations 10 and 11 in the Rhine has about the same $\delta^{13}C$ values as found in the riverine part (station 1-3). Results on $\delta^{13}C$ of the carbonate fraction of the sediments also suggest a sedimentation of riverine sediments at stations 10 and 11 (Turkstra unpubl.).

The relatively low $\delta^{13}C$ of POC at station 11 must be explained by the input of waste associated with the human foodweb (Sackett 1986). This material is transported upstream and probably sedimentates (together with the marine sediments) around the stations 8 and 9, causing high $\delta^{13}C$ of the organic fraction. This sedimentary organic matter is mixed with fluvial sediments in the fresh-water compartment of the estuary (between stations 3 and 8) due to the salt water intrusion near the bottom of the estuary. However, this is not reflected in $\delta^{13}C$ of POC is only shown seawards of station 9.

## 4.4 Wadden Sea

An increase in water temperature resulted in a decrease of the $\delta^{13}C$ of POC in the Wadden Sea and Lake Yssel (Fig. 4). The deviation from the expected inverse linear relation (Fontugne and Deplessey 1981), as found for stations 27 and 29, could probably be explained by the presence of different species of algae (Cadée and Hegeman 1979).

The sharp increase of $\delta^{13}C$ at station 28 is probably caused by resuspension of benthic diatoms (de Jonge and van den Berg 1987) which are known to show a higher $\delta^{13}C$ value than pelagic algae (Sackett 1986). Probably due to differences in phytoplankton species and in $\delta^{13}C$ of the dissolved inorganic carbon at these stations (Mook 1970), the results do not fit the absolute values of the relation between temperature and $\delta^{13}C$ as found by Fontugne and Duplessey (1981). However, the slopes are the same for different stations, indicating the same temperature dependency of the mechanism of isotope fractionation in algae.

### 4.5 Lake Yssel

The decomposition experiments reveal that the $\delta^{13}C$ of POC in Lake Yssel (station 29) nearly remains the same during decomposition. This is in agreement with the results of Gearing et al. (1984) and Schwinghammer et al. (1983), who found no $\delta^{13}C$ change, or a slight depletion in $^{13}C$ during extensive aerobic decomposition of planktonic samples.

However, in estuarine and marine samples the organic matter $\delta^{13}C$ becomes more negative during decomposition (Fig. 5). These results are in agreement with those of Degens (1968a,b, 1969), who observed a decreasing $\delta^{13}C$ during dark respiration of phytoplankton cultures. This can partly be explained as follows. The observed $\delta^{13}C$ of POC is an average of different $\delta^{13}C$ values of the biochemical constituents. Degens et al. (1968a) and Eadie and Jeffrey (1973) showed that lipids have low $\delta^{13}C$ compared to carbohydrates. Proteins and carbohydrates are more easily decomposed than lipids (Laane et al. 1988), resulting in an overall decrease of $\delta^{13}C$ of the residual fraction. This process can not explain the relatively high organic matter content of sediments (Dean et al. 1986). These authors suggest the contribution of additional bacterial lipids with low $\delta^{13}C$ to the sediments. The opposite, however, is found in the food chain between different trophic levels. Fry et al. (1984) found an increase in $\delta^{13}C$ of the same order of magnitude as the $^{13}C$ enrichment during decomposition, about 4‰. These fractionation processes of the POC influences the ultimate isotope composition of sedimentary organic matter.

## References


Anonymous (1975) Voorkomen en gedrag van zware metalen in het slib uit de Schelde. Verslag R 994, Waterloopkundig Laboratorium en Institut voor Bodemvruchtbaarheid, p 17

Cadée GC, Hegeman J (1979) Phytoplankton primary production, chlorophyll and composition in a tidal inlet of the Western Wadden Sea (Marsdiep). Neth J Sea Res 13:224–241

Dean WA, Arthur MA, Claypool GE (1986) Depletion of $^{13}C$ in Cretaceous marine organic matter: source, diagenetic, or environmental signal? Mar Geol 70:119–157

Deines P (1980) The isotopic composition of reduced organic carbon. In: Fritz P, Fontes JC (eds) Handbook of environmental isotope geochemistry. Elsevier, Amsterdam, pp 329–406

Degens ET (1969) Biogeochemistry of stable carbon isotopes. In: Eglinton G, Murphy MTJ (eds) Organic geochemistry, methods and results. Springer, Berlin Heidelberg New York, pp 304–331

Degens ET, Guillard RRL, Sackett WM, Hellebust JA (1968a) Metabolic fractionation of carbon isotopes in marine plankton-1. Temperature and respiration experiments. Deep-Sea Res 15:1–9

Degens ET, Behrendt M, Gotthardt B, Reppman E (1968b) Metabolic fractionation of carbon isotopes in marine plankton-2. Data on samples collected off the coast of Peru and Equador. Deep-Sea Res 15:11–20

Eadie BJ, Jeffrey LM (1973) $\delta^{13}C$ analyses of oceanic particulate organic matter. Mar Chem 1:199–209

Eisma D, Cadée GC, Laane RWPM (1982) Supply of suspended matter and particulate and dissolved organic carbon from the Rhine to the coastal North Sea. Mitt Geol-Palaontol Inst Univ Hamb 52:483–506

Eisma D, Bernard P, Boon JJ, Grieken R van, Kalf J, Mook WG (1985) Loss of particulate organic matter in estuaries as exemplified by the Ems and Gironde estuaries. Mitt Geol-Palaontol Inst Univ Hamb 58:397–412

Es FB van, Laane RWPM (1982) The utility of organic matter in the Ems-Dollard estuary. Neth J Sea Res 16:300–314

Fontugne MR, Duplessey JC (1981) Organic carbon isotope fractionation by marine plankton in the temperature range -1 to 31°C. Oceanol Acta 4:85–90

Fry B (1984) $^{13}C/^{12}C$ ratios and the trophic importance of algae in Florida Syringodium filiforme seagrass meadows. Mar Biol 79:11–19

Fry B, Sherr EB (1984) $\delta^{13}C$ measurements as indicators of carbon flow in marine and fresh water ecosystems. Contrib Mar Sci 27:13–47

Fry B, Anderson RK, Entzelroth L, Bird JL, Parker PL (1984) $^{13}C$ enrichment and oceanic food web structure in the Northwestern gulf of Mexico. Contrib Mar Sci 27:49–63

Gearing JN, Gearing PJ, Rudnick DT, Requejo AG, Hutchins MJ (1984) Isotopic variability of organic carbon in a phytoplankton-based temperature estuary. Geochim Cosmochim Acta 48:1089–1098

Hedges JI, Clark WA, Quay PD, Richey JE, Devol AH, Santos UdeM (1986) Composition and fluxes of particulate organic material in the Amazon River. Limnol Oceanogr 31:717–738

Jouanneau JM (1987) The contribution of $^{14}C$ dating to a better understanding of the POM behaviour in estuaries. Mar Chem 21:189–197

Laane RWPM (1982) Chemical characteristics of the organic matter in the Ems-Dollard estuary. Thesis, University of Groningen, Groningen

Laane RWPM, Etcheber H, Relexans JC (1987) The nutritive value of particulate organic matter in estuaries and its ecological implication for macrobenthos. In: Degens ET, Kempe S, Gan W-B (eds) Transport of carbon and minerals in world rivers Pt 5 Mitt Geol-Paläontol Inst Univ Hamb 64:71–91

Laane RWPM, Ruardy P (1988) Modelling estuarine carbon fluxes. Mitt Geol-Palaontol Inst Univ Hamb 66:239–265

Marquenie JM, Simmers JW, Birnbaum E (1985) An evaluation of dredging in the western Scheldt (The Netherlands) through bioassays. TNO Report R 85/075, p 56

Mook WG (1970) Stable carbon and oxygen isotopes of natural waters in the Netherlands. Proc IAEA Conf Isotope Hydrology 1970, Vienna, pp 163–190

Rudert M, Muller G (1981) Mineralogy and provenance of suspended solids in estuarine and nearshore areas of the southern North Sea. Senckenb Marit 13:57–64

Sackett WM (1986) Uses of stable carbon isotope composition of organic carbon in sedimentological studies on tropical marine systems. Sci Total Environ 58:139–149

Sackett WM, Eckelman WR, Bender ML, Be AWH (1965) Temperature dependence of carbon isotope composition in marine plankton and sediments. Science 148:235–237

Salomons W (1975) Chemical and isotopic composition of carbonates in recent sediments and soils from Western Europe. J Sediment Petrol 45:440–449

Salomons W, Mook WG (1981) Field observations of the carbon isotopic composition of particulate organic carbon in the southern North Sea and the adjacent estuaries. Mar Geol 41:M11–M20

Schwinghamer P, Tan FC, Gordon DC Jr (1983) Stable carbon isotope studies on the Pecks mudflat ecosystem in the Cumberland basin. Can J Fish Aquat Sci 40:262–272

Sherr BE (1982) Carbon isotope composition of organic seston and sediments in a Georgia salt marsh estuary. Geochim Cosmochim Acta 46:1227–1232

Simenstad CA, Wissmar RC (1985) $\delta^{13}C$ evidence of the origins and fate of organic carbon in estuaries and nearshore food webs. Mar Ecol Progr Ser 22:141–152

Tan FC, Strain PM (1979) Organic carbon isotope ratios in recent sediments in the St. Lawrence Estuary and the Gulf of St. Lawrence. Estuarine Coastal Mar Sci 8:213–225

Tan FC, Strain PM (1983) Sources, sinks and distribution of organic carbon in the St. Lawrence estuary, Canada. Geochim Cosmochim Acta 47:125–132

# Sulphur Bacteria and Sulphur Isotope Fractionation in a Meromictic Lake near Toronto, Canada

M.D. Dickman and H.G. Thode

## 1 Introduction

Purple and green phototrophic bacteria often develop at the chemocline of meromictic lakes (Parkin and Brock 1981). Green phototrophic bacteria are obligate anaerobes, while many purple phototrophic bacteria are tolerant of minute amounts of dissolved oxygen (Pfennig 1967). For this reason the purples are typically found layered above the greens at the chemocline (Guerrero and Abella 1985). Both types of these phototrophic bacteria require light and a suitable electron donor such as hydrogen sulphide. The relative abundance of purple and green phototrophic bacteria at the chemocline of meromictic lakes is largely governed by the quality and quantity of light received at these depths. This hypothesis was independently tested by Truper and Genovese (1968) and Parkin and Brock (1981).

The purpose of this study was to follow the bacterial reduction of sulphate to hydrogen sulphide and to determine how the vertical stratification of phototrophic bacteria in Crawford Lake was related to its midwater sulphur cycle and sulphur isotope chemistry.

Crawford Lake (Fig. 1) is located on the top of the Niagara Escarpment 290 m above sea level about 70 km north of Toronto on Silurian Guelph-Anabel Dolomite. Large white cedars (*Thuja occidentalis* L.) ring the lake along its high cliffs preventing high winds from reaching the lake's surface (Boyko 1978). As a result, the lake, which is 24 m deep and has an area of 2.5 ha, has never mixed below 15.5 m during the period of our study, 1974–1987.

## 2 Methods

### Water Sampling

Dissolved oxygen (% saturation), temperature (°C), alkalinity (mg/l as $CaCO_3$), specific conductivity ($\mu$S/cm), secchi transparency (m), pH and sulphate concentration (mg/l) were measured according to methods described by Dickman and Thode (1985).

Additional surface lake water samples were collected in half-litre nalgene bottles or in two 5-l dark glass bottles and stored at 4 °C for sulphate content and isotope ratio measurements.

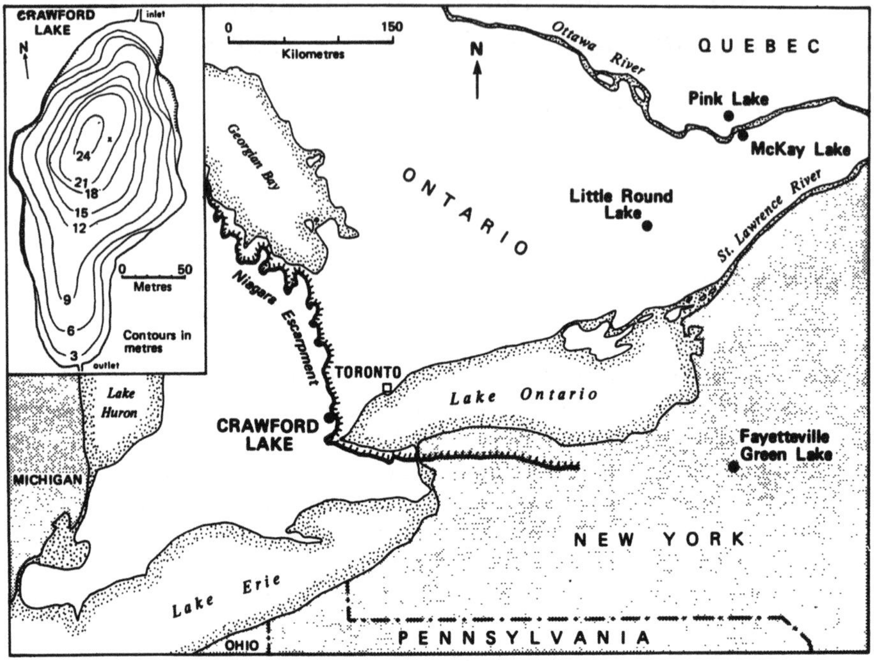


**Fig. 1.** Location map of five meromictic lakes (3 in Ontario and 1 in New York and Quebec) with an inset of the morphometry of Crawford Lake

## Sulphur Bacteria Collection and Water Sampling for Sulphur Content and Isotope Ratio Measurements

Phototrophic bacteria were collected at 0.1-m intervals from the top of the chemocline to a depth of less than 0.01% surface light intensity (circa 16 m) using a pump sampler suspended on a hose and a metred line.

Large (2–5 l) dark brown glass bottles were filled (using the submerged pump sampler) and allowed to overflow until their volumes were twice displaced. The bottles were sealed to prevent oxygen from entering and then transported to the laboratory. These samples, stored at 4 °C, were used for sulphate, sulphide contents and isotope ratio measurements.

## Pigment Analysis

In the laboratory, the water samples were filtered through a 0.45 μm Gelman glass fiber filter. Pigments were extracted in aqueous 90% acetone and analyzed according to the procedures of Sanger and Gorham (1972) and Severn (1982). The identification of bacteriochlorophylls in whole cell suspensions and acetone extracts was carried out on the basis of characteristic absorption spectra as described by Rybak (1986).

### Analytical Procedures for Sulphur Content and Isotope Ratio Measurements of Aqueous Sulphate

Standard Methods were used to determine the sulphur contents and isotope ratios of the lake water sulphate (Thode et al. 1961) after removal of any $H_2S$ present, see below. The sulphate was (1) precipitated as $BaSO_4$, (2) reduced to $H_2S$ using reducing mixture (HI, $H_3PO_2$, HCl) (Thode et al. 1961), (3) converted in steps to CdS, $Ag_2S$ and finally to $SO_2$ gas for isotope analysis. The sulphur content was determined gravimetrically either as $BaSO_4$ or as $Ag_2S$.

### Analytical Procedures for Sulphur Content and Isotope Ratios

*Lake Water Sulphide.* Standard macro methods were used to determine the sulphur content and isotope ratios of the $H_2S$ present in the chemocline and bottom waters. The $H_2S$ was flushed out of the acidified water samples with a stream of $O_2$ free nitrogen and absorbed in a cadmium acetate trap. The CdS that formed was then converted to $Ag_2S$ and finally to $SO_2$ gas for isotope analysis. The sulphur content as $H_2S$ was determined gravimetrically as $Ag_2S$.

*Other Forms of Sulphur.* The trace amounts of elemental sulphur, organic sulphur, thiosulphate etc. could not be analyzed isotopically. Previous studies by Parkin and Brock (1981) on the role of phototrophic bacteria in a meromictic lake in Wisconsin USA indicated that $H_2S$ was completely oxidized to sulphate and that intermediate sulphur compounds were, for the most part, absent.

*Sediment Sulphur.* The total sulphur in the sediment samples (dried in vacuo at 60° C) was extracted by heating with Eschka mixture MgO and $Na_2CO_3$ in a crucible at ~ 800 °C. The sulphate that formed was (1) dissolved in water, (2) precipitated as $BaSO_4$, (3) converted in steps to $SO_2$ gas for isotope analysis as described above (Thode et al. 1961). The total sulphur content was determined gravimetrically as $BaSO_4$ or $Ag_2S$.

*Mass Spectrometry.* Isotope analyses of $SO_2$ gas samples were performed using a high precision isotope ratio mass spectrometer (Thode et al. 1961). Sulphur isotope ratios were expressed in the $\delta^{34}S$ notation where

$$\delta^{34}S‰ = \left[ \frac{(^{34}S/^{32}S)_{sample}}{(^{34}S/^{32}S)_{standard}} - 1 \right] 1000.$$

Standard used is sulphur from Troilite in the Canyon Diablo meteorite.

## 3 Results and Discussion

Absorption spectra of acetone extracts of the photosynthetic bacteria between 14.5 and 16.5 m in Crawford Lake were made in order to determine the relative ratios of the pigments contained by the phototrophic bacteria at each depth. At depths near 15.5 m the bacteria displayed an absorption peak at 654 nm, which is the absorption peak for Bchl d (Parkin and Brock 1981). In addition, the 15.5-m samples absorbed at 440 nm, which also corresponds to a Bchl d absorption peak. Thus we concluded that green sulphur bacteria (Chlorobiacea) predominated in the 15.5-m depths. Bchl a which absorbs at 772 nm was not detected at elevated levels at these depths (Fig. 2A).

Nearer the surface (15 m) the acetone cell extracts absorbed most strongly at 467 nm (Fig. 2B). This is the characteristic wavelength absorbed by pigments of the family Chromataceae (purple phototrophic bacteria).

The light intensity at this depth under 10 cm of snow and 14 cm of ice (1 Feb. 1987, Table 1) was only 0.03% of the surface light intensity or approximately 12 microeinsteins/$cm^2$/s.

Absorption spectra of intact cells are shown in Fig. 2C. The Crawford Lake bacteria displayed absorbance peaks at 715 and 452 nm due to Bchl d absorption of the brown forms of the Chlorobiaceae. Low absorbance by the cells was observed in the 600 nm to 650 nm range. The Crawford Lake bacteria also exhibited a peak at 730 nm due largely to the green forms of the Chlorobiaceae (Pfennig 1967). In spring, a broad peak from 570 to 450 nm developed (not figured). This was due to the combined absorbance of both Bchl and carotenoids. The vertical distribution of bacteria and algae in Crawford Lake varies through time (e.g. Fig. 2A). However,

**Table 1.** Crawford Lake – 1 February 1987[a]

| Depth (m) | Conductivity (μS/cm) | pH | Dissolved oxygen (mg/l) | Temperature (°C) | Light (% Surface light intensity) |
|---|---|---|---|---|---|
| 0[a] | 600 | 8.0 | 8.4 | 4.1 | 100 |
| 1 | 620 | 8.0 | 8.1 | 4.1 | 70 |
| 2 | 640 | 8.0 | 7.8 | 4.1 | 45 |
| 3 | 647 | 7.9 | 7.3 | 4.1 | 20 |
| 4 | 648 | 7.6 | 6.9 | 4.1 | 12 |
| 5 | 653 | 7.4 | 6.6 | 4.1 | 7.7 |
| 6 | 655 | 7.3 | 5.8 | 4.2 | 5.2 |
| 7 | 660 | 7.2 | 5.0 | 4.3 | 3.7 |
| 8 | 667 | 7.2 | 4.8 | 4.3 | 2.5 |
| 9 | 677 | 7.2 | 4.0 | 4.3 | 1.8 |
| 10 | 681 | 7.2 | 3.2 | 4.3 | 1.2 |
| 11 | 689 | 7.1 | 2.8 | 4.4 | 0.87 |
| 12 | 704 | 7.0 | 1.0 | 4.5 | 0.73 |
| 13 | 709 | 6.9 | 0.1 | 4.5 | 0.38 |
| 14 | 723 | 6.6 | 0.0 | 4.5 | 0.25 |
| 15 | 760 | 6.6 | 0.0 | 4.7 | 0.03 |
| 15.5 | 1270 | 6.6 | 0.0 | 4.9 | 0.003 |
| 16 | 1540 | 6.6 | 0.0 | 5.0 | 0.001 |
| 17 | 1900 | 6.6 | 0.0 | 5.2 | 0.000 |
| 18 | 2290 | 6.6 | 0.0 | 5.4 | 0.000 |
| 19 | 2670 | 6.6 | 0.0 | 5.7 | 0.000 |
| 20 | 2990 | 6.6 | 0.0 | 5.8 | 0.000 |

[a] Top of ice was taken as 0 m, ice was double layered 4–6 cm of snow on top of 3–4 cm of ice underlain by 2–3 cm of water below which was 6–7 cm of ice

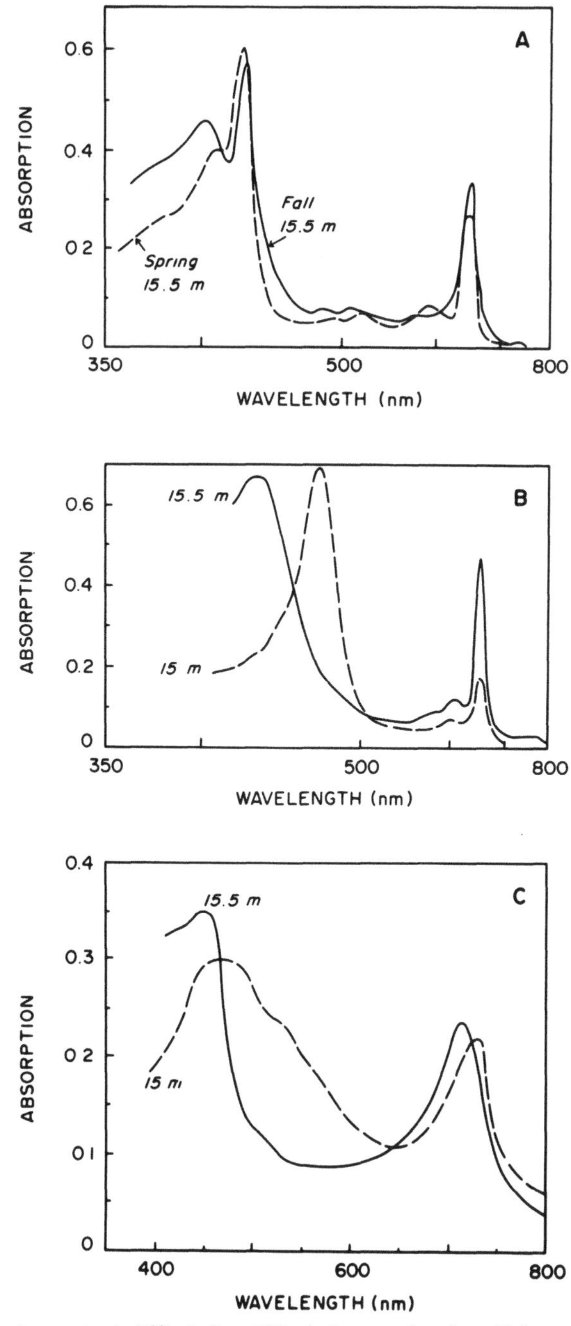


**Fig. 2A-C.** Absorption spectra for **A** acetone extracted filtrate from 200 ml of water taken from 15.5 m in Crawford Lake in May (spring) and August (summer) of 1987 (Log Scale), **B** acetone extracts of the filtrate from 200 ml of water taken from 15.0 and 15.5 m in Crawford Lake in June (Log Scale), **C** whole cell acetone-free extracts of filtrate from 200 ml of water taken from 15.0 and 15.5 m in Crawford Lake in June 1987 (Arithmetic Scale)

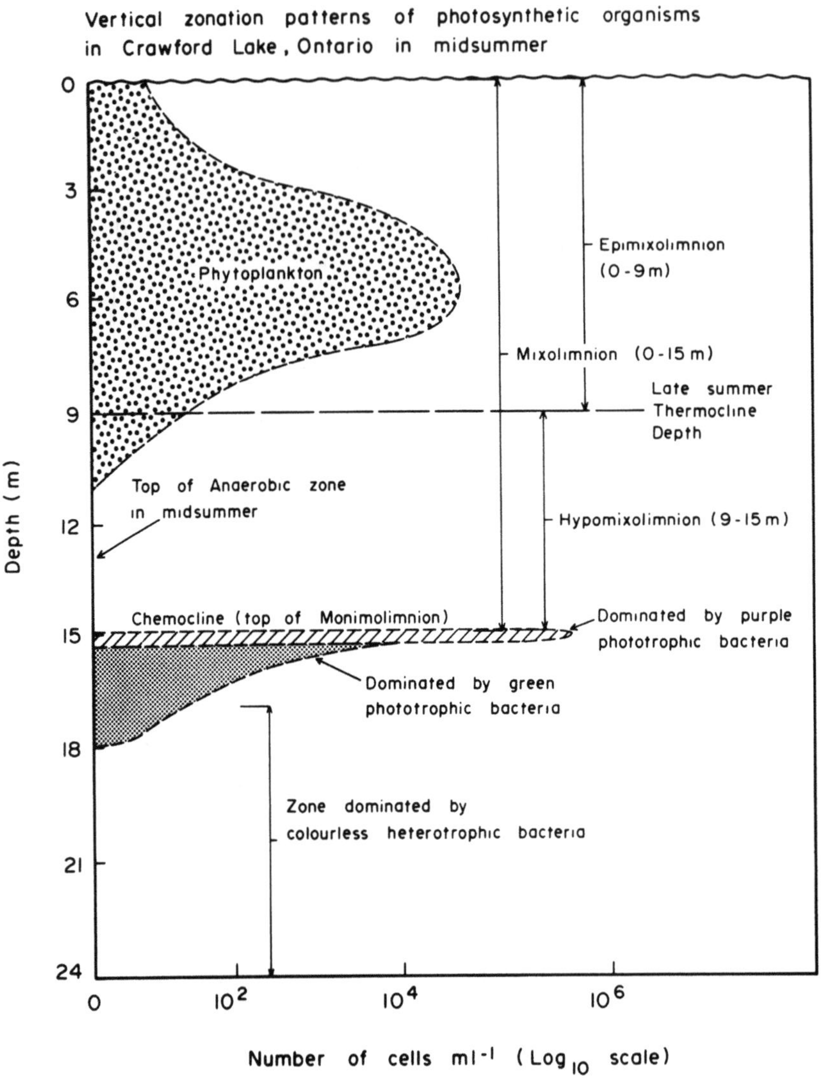


**Fig. 3.** Schematic representation of the vertical zonation of algae and bacteria in Crawford Lake during late summer

their average mid-summer distributions can be characterized schematically (Fig. 3).

The chemocline of Crawford Lake was not static. In late November, following the disappearance of the thermocline, dissolved oxygen was found as deep as 15.2 m. By mid-summer, dissolved oxygen was not observed below 12 m (Fig. 4). This dynamic process was also reflected in seasonal fluctuations of the pycnocline (density gradient). The density of the water is a function of its temperature (Fig. 3B) and its salinity. Salinity, in turn, is a function of specific conductivity (Fig. 4).

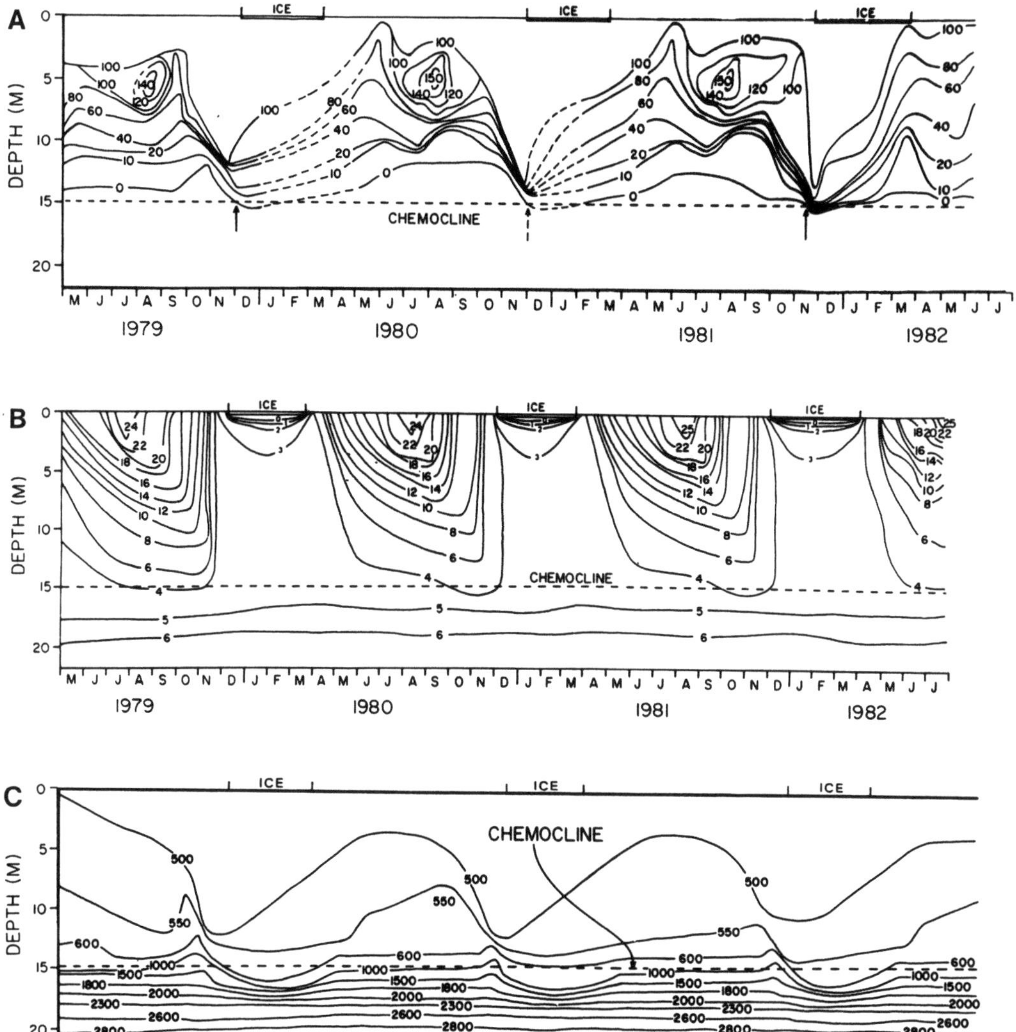


**Fig. 4A-C.** Smoothed depth-time contours (isopleths) for **A** dissolved oxygen as percent saturation, **B** temperature (°C), and **C** specific conductivity ($\mu S\ cm^{-2}$) for the period May 1979-July 1982. Samples were taken at each metre at biweekly intervals in spring, summer and fall and at monthly intervals in winter

The ventilation of the 15-m-deep chemocline in late fall was reflected by changes in dissolved oxygen, temperature and specific conductivity. The vertical arrows in Fig. 4A represent the point in time when oxygen from the overlying mixolimnion made contact with the anaerobic bacteria colonizing the lake's chemocline. This "partial ventilation" of the chemocline results in a mass mortality of phototrophic bacteria (Dickman 1985). Apart from this brief period of "chemocline ventilation" in late fall, the Crawford Lake chemocline phototrophic bacterial layer (15–16 m) operated in a relatively closed system. This fact is important for an understanding of the sulphur isotope fractionation process which is discussed in the following section.

## Sulphur Isotope Ratios and Water Column Profiles

The sulphur contents and isotope ratios ($\delta^{34}S$ values) determined for the sulphates and sulphides ($\Sigma H_2S$) extracted from the Crawford Lake water column samples collected from various depths over a period of several years are plotted in Figs. 5 and 6.

Similar data obtained from measurements made on a single series of water samples collected in late winter with the lake still under ice cover are given in Table 2. The water column profiles for the $O_2$ content (mg/l), the conductivity ($\mu S/cm$)

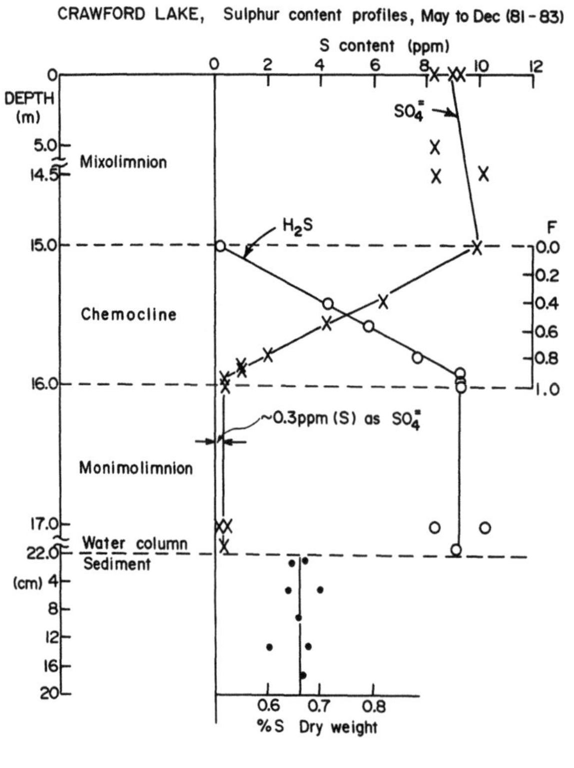


**Fig. 5.** Sulphur content $\{SO_4^{2-}(S)$ and $H_2S(S)\}$ depth profiles for the Crawford lake water column. Samples collected from various depths (May to Dec. 1981-83). $F$ is the fraction of $SO_4^{2-}$ reduced

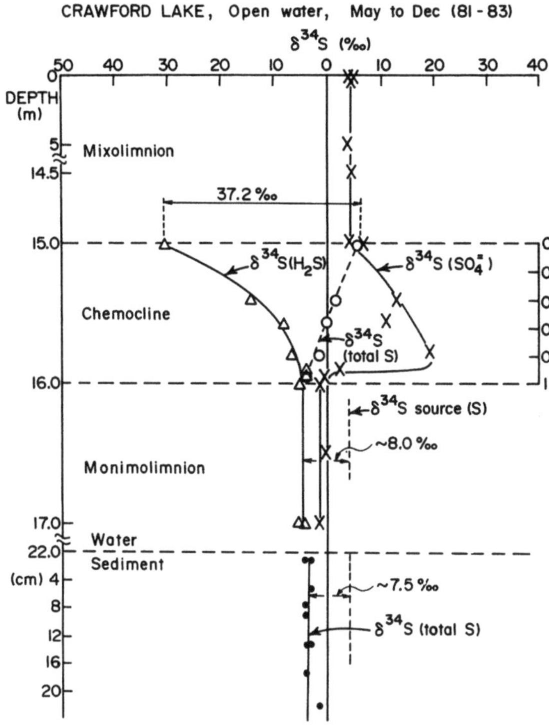


**Fig. 6.** Sulphur isotope ratio ($\delta^{34}S_{SO_4^{2-}}$ and $\delta^{34}S_{H_2S}$) depth profiles for Crawford lake water column (May to Dec. 1981-83). Samples are the same as those in Fig. 1. *f* is the fraction of $SO_4^{2-}$ reduced

**Table 2.** Sulphur and sulphur isotope ratio profiles in the Crawford Lake chemocline under ice cover, March 21, 1987

| Depth (m) | S Content ppm $\Sigma H_2S$(S) | $SO_4^{2-}$ (S) | Total (S) | $\delta^{34}S$‰ $\Sigma H_2S$(S) | $SO_4^{2-}$ (S) | Total (S) | Extent of reaction f |
|---|---|---|---|---|---|---|---|
| 14.5 | - | 9.28 | 9.28 | - | 3.67 | 3.67 | - |
| 14.7 | ~ = 0 | 9.64 | 9.64 | - | 3.06 | 3.06 | - |
| 14.9 | 0.14 | 10.09 | 10.23 | -49.6 | 4.03 | 3.30 | 0.14 |
| 15.1 | 1.34 | 8.52 | 9.86 | -39.2 | 10.60 | 3.83 | 0.138 |
| 15.3 | 3.74 | 6.27 | 10.01 | -26.6 | 21.52 | 3.54 | 0.374 |
| 15.6 | 6.44 | 3.62 | 10.06 | -16.2 | 31.52 | 0.97 | 0.640 |
| 16.0 | 9.62 | 0.65 | 10.27 | - 6.9 | 39.16 | -3.95 | 0.937 |

and the temperature (°C), as well as for the sulphur contents and isotope ratios for this latter series of samples are plotted in Figs. 7 and 8.

*Water Column Profiles.* In Crawford Lake the chemocline is characterized by a steep gradient in specific conductivity which separates the mixolimnion from the permanently stagnant deeper monimolimnion (Fig. 3). This sharply defined layer is anoxic the year round below 15.3 m, but anoxic conditions extend up to 12 m during the summer. This temporal progression in the size of the anoxic layer in Crawford Lake is shown in Fig. 4A.

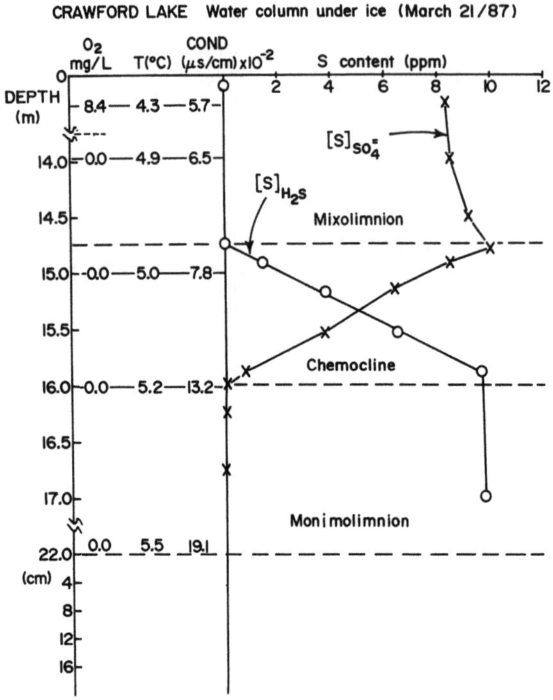


**Fig. 7.** Sulphur content $\{SO_4^{2-}(S)$ and $H_2S(S)\}$ depth profiles for Crawford lake water column (under ice) March 21, 1987. Also included are depth profiles for $O_2$ content (mg/l), conductivity (mμs/cm) and temperature (°C)

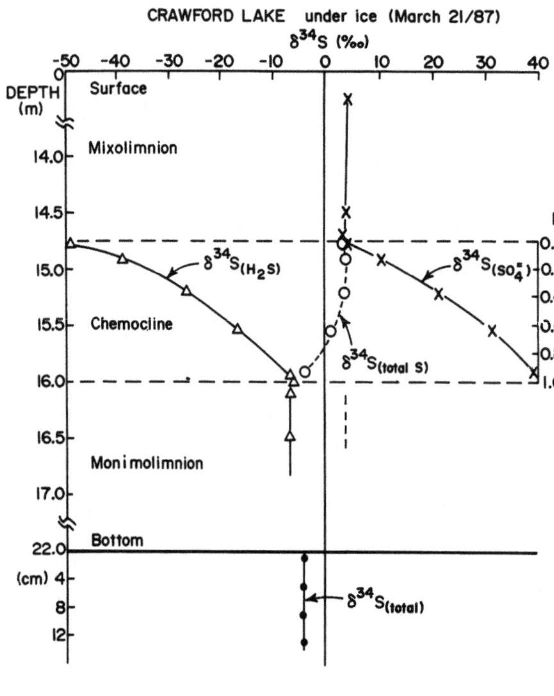


**Fig. 8.** Sulphur isotope ratio ($\delta^{34}S_{SO_4^{2-}}$ and $\delta^{34}S_{H_2S}$) depth profiles for Crawford lake water column (under ice) March 21, 1987. Samples are the same as those in Fig. 3. *F* is the fraction of $SO_4^{2-}$ reduced

*Sulphur Content Profiles.* The sulphur content and isotope ratio profiles for the Crawford Lake water column are quite remarkable (Figs. 5, 6, 7, 8). The surface waters (mixolimnion) down to 14.7 to 15 m contain essentially only $SO_4^{2-}$ (S) with a $\delta^{34}S_{SO_4^{2-}}$ value of ~ 4.0‰ close to the mean values reported for bulk precipitation in the area and similar to that of sulphur in leaves and soil collected around the lake (Nriagu et al. 1987; Thode et al. 1987). There is no isotope signature evidence of a deep ground water source or evaporite source of sulphate.

In the anoxic chemocline between 15 and 16 m, the $SO_4^{2-}$ is biologically reduced to $\Sigma H_2S$, the extent of reaction (f) reaching 100% at ~ 16 m. Thus the bottom waters (monimolimnion) below 16 m contain essentially only $\Sigma H_2S$-(S) (Figs. 5 and 7). However, the total sulphur content ($SO_4^{2-}$(S) + $\Sigma H_2S$(S)) is fairly constant throughout the water column; the mean values obtained for samples from the three zones over a period of years are 9.7 ± 1.6 PPM; 9.8 ± 0.8 PPM and 9.4 ± 0.6 PPM respectively top to bottom. Thus the sulphate in the biologically active chemocline is completely reduced in a largely closed system within a 1- to 1.2-m layer, the boundaries of the zone being sharply defined over a long period of time.

The traces of $SO_4^{2-}$(S) of up to 0.33 PPM, indicated in the bottom waters collected during summer and fall (Fig. 5), and not indicated in samples collected during ice cover (Fig. 7), are probably due at least in part to $H_2S$ oxidation during sample preparation. Greater precautions were taken to prevent $O_2$ contamination during the processing of the latter samples. Further, the presence of methanogens in the monimolimnion would play a role in establishing a minimum $SO_4^{2-}$ concentration during the summer and fall. Lovely and Klug (1986) have shown that in mixed cultures of methanogens and "sulphate reducers", that at low $SO_4^{2-}$ concentrations (1 to 2 PPM), the methanogens take over, utilizing all the nutrient (acetate). Under these conditions, the "sulphate reducers" stop growing and further $SO_4^{2-}$ reduction is no longer possible.

*Sulphur Isotope Ratio ($\delta^{34}S$) Profiles.* In view of the uncertainties in the measurements of sample depths and slight variations in the depth of the chemocline from season to season, the $\delta^{34}S_{SO_4^{2-}}$ and $\delta^{34}S$ $\Sigma H_2S$ values are plotted in Figs. 6 and 8 against f, the fraction or extent of $SO_4^{2-}$ reduction instead of depth within the chemocline; f = 0 corresponding to the depth at which a trace of $H_2S$ first appears ~ 15 m. It is seen from the data of Table 2 that there is essentially a linear relationship between water depth and extent of reaction (f) for each sample.

The large differences in $\delta^{34}S$ values between the $SO_4^{2-}$ and coexisting product $\Sigma H_2S$ in the chemocline indicates a large sulphur isotope fractionation effect in the bacterial reduction of $SO_4^{2-}$ to $H_2S$ in the Crawford Lake environment (Figs. 6 and 8).

*Theoretical Isotope Effect.* The isotope effect in per mil is defined as $(k_1/k_2-1)\,1000$, where $k_1$ and $k_2$ are the specific rate constants for the reduction of the light and heavy isotopic species respectively. In a Rayleigh-type closed system process, the isotope effect may be taken as the difference in $\delta$ values between the initial $SO_4^{2-}$ and the first $H_2S$ product to form or for small fractions of reaction. It is clear from Figs. 6 (summer) and 8 (winter) that isotope effects of 37.2‰ and 53.5‰ respectively are indicated.

Also the isotope effect may be calculated for each depth or fraction of reaction (Table 3) using the Rayleigh equation

$$\frac{k_1}{k_2} = \frac{\log(1-f)}{\log(1-rf)} \quad \text{(closed system)}$$

where f = fraction of reaction

and

$$r = \frac{(^{34}S/^{32}S)\ H_2 \text{ at } f = f}{(^{34}S/^{32}S)\ SO_4^{2-} \text{ (mean value for } f = 0 \text{ and } f = f).}$$

These high sulphur isotope effects (~ 50‰) obtained even at the lowest $SO_4^{2-}$ concentrations near the bottom of the chemocline were totally unexpected in a freshwater lake environment. They are equal to or slightly exceed the maximum values (~ 50‰) obtained in laboratory experiments carried out under extreme conditions, (high $SO_4^{2-}$ concentrations and very low metabolic rates or nutrient deficient systems) (Ford 1957; Kaplan and Rittenberg 1964). These extreme conditions have been found to prevail in some marine environments such as the Black Sea water column and sediments and in some lakes with high $SO_4^{2-}$ concentrations, such as Green Lake in N.Y. State and Lake Sakovo in the Soviet Union (Vinogradov et al. 1962; Torgersen et al. 1981).

*Sulphur Isotope Fractionation Model.* The bacterial reaction of $SO_4^{2-}$ to $H_2S$ takes place through a sequence of enzyme catalyzed steps involving the breaking of-S-O-bonds (Postgate 1968; Peck 1974; Kemp and Thode 1968). According to the steady-state model of C.E. Rees (1973) for sulphur isotope fractionation, the overall sulphur isotope effect becomes a summing up of the net isotope effect for each step as the rate of the back reactions approaches the rate for the forward reactions. Thus maximum (~50‰) and minimum (~0) values of the isotope effect are predicted for

**Table 3.** Sulphur isotope effects in the bacterial reduction of sulphate in Crawford Lake chemocline

| Depth (m) | Extent of reaction f | $\delta^{34}S$‰ $H_2S$ (S) | $SO_4^{2-}$ S | $\Delta\delta^{34}S$ ‰ | $\left(\frac{k_1}{k_2} - 1\right)$ 1000‰[a] |
|---|---|---|---|---|---|
| 14.7 | - | - | 3.4 | - | - |
| 14.9 | 0.14 | -49.6 | 4.0 | 53.6 | 53.4 |
| 15.1 | 1.38 | -39.2 | 10.6 | 49.8 | 49.8 |
| 15.3 | 3.74 | -26.6 | 21.5 | 48.1 | 50.3 |
| 15.6 | | -16.2 | 31.5 | 47.7 | 60.0 |
| 16.0 | | - 6.9 | 39.2 | 46.0 | - |

[a]Calculated assuming Rayleigh Process (Closed System)

Kinetic isotope effect defined as $(\frac{k_1}{k_2}-1) \times 1000‰$ where $k_1$ and $k_2$ are specific rate constants for the removal of the light and heavy isotopic species of $SO_4^{2-}$ respectively.

For a Rayleigh process close system

$$\frac{k_1}{k_2} = \frac{\log(1\text{-}f)}{\log(1\text{-}rf)}$$

where f is fraction of reaction

$$\text{and } r = \frac{^{34}S/^{32}S \text{ in } H_2S \text{ at } f = f}{^{34}S/^{32}S \text{ in } SO_4^{2-} \text{ (mean value for } f = 0 \text{ and } f = f)}$$

very slow and very rapid rates of $H_2S$ production (metabolic rates) respectively. This model assumes adequate $SO_4^{2-}$ supply or zero order kinetics with respect to $SO_4^{2-}$ concentrations. Sulphur isotope effects reported for bacterial reduction of $SO_4^{2-}$ in freshwater lakes are low (Matrosov et al. 1975; Chambers and Trudinger 1978). It has generally been assumed that in these soft water lakes the rate of $H_2S$ production becomes dependent on the rate of $SO_4^{2-}$ uptake by sulphate-reducing bacteria. Since there is only a small inverse isotope effect (~ 0–3‰) associated with this rate of uptake, the overall effect in the bacterial reduction of $SO_4^{2-}$ to $H_2S$ will approach ~ zero as this initial step becomes rate determining. This was found to be the case in laboratory experiments at $10^{-5}$ molar $SO_4^{2-}$ concentrations (Harrison and Thode 1958).

By contrast, in the natural Crawford Lake environment, under ice cover, it is seen that almost maximum isotope effects (Fig. 8) are realized, even at the very lowest $SO_4^{2-}$ concentrations near the bottom of the chemocline, where $SO_4^{2-}$ reduction is almost complete (Table 2). Therefore, according to the sulphur isotope fractionation model, conditions in the Crawford lake chemocline (15–16 m) under ice cover are such that the rate of bacterial reduction of $SO_4^{2-}$ to $H_2S$ is metabolic rate limiting and not $SO_4^{2-}$ concentration limiting, even down to 1 to 2 PPM $SO_4^{2-}$ concentration levels. The extremely low metabolic rates implied, possibly due to a nutrient deficiency, would then account for the high sulphur isotope fractionation values found.

*Apparent Isotope Effect.* Crawford Lake is somewhat unique in that the sulphate in the chemocline (15 to 16 m) is completely reduced biogenically in a large closed system and the theoretical isotope effect can be measured and calculated. In the case of most freshwater lakes it is only possible to measure an apparent isotope effect such as the $\delta^{34}S$ difference between the source sulphate and the reduced sulphur in the anoxic bottom waters or sediments.

This apparent isotope effect will approximate the theoretical value for systems completely open to $SO_4^{2-}$, where $SO_4^{2-}$ concentrations are high and the supply is rapid relative to reduction rates. However, in soft water lakes with low $SO_4^{2-}$ concentrations the apparent isotope effect will be very much lower than the theoretical value, approaching zero as sulphate reduction becomes complete in a closed system.

In the case of Crawford Lake the apparent isotope effect is seen to be ~ 8‰ (see Figs. 6 and 8) and not zero as it would be for a completely closed system reduction. This isotopic imbalance indicates a loss of $^{34}S$ enriched $SO_4^{2-}$ from the chemocline, since only this sulphur has $\delta^{34}S$ values greater than that of the source sulphur (+4‰). Some small ventilation of this $SO_4^{2-}$ from the top of the chemocline (Fig. 3) could account for this isotopic imbalance.

The high sulphur isotope fractionation effect found for Crawford Lake (53.5‰) suggests the possibility of other low $SO_4^{2-}$ (< 27 ppm) meromictic lakes with equally high values.

*$\delta^{34}S$ in Crawford Lake Sediments.* Only the total sulphur was extracted and analyzed isotopically from the Crawford Lake sediment. There was visible evidence of some FeS and $FeS_2$ in the sediments. However, as in the case of other freshwater lakes, most of the sulphur was bound up in organic matter. The two main

sources of this reduced sulphur are: (1) the $H_2S$ from the bacterial reduction of $SO_4^{2-}$ and (2) that derived from organic matter. Initially the organic matter would have a $\delta^{34}S$ value of ~ 4.0‰, the same as that of the source sulphate from which it was derived, since there is no isotope effect in the plant metabolism of $SO_4^{2-}$ (Ishii 1953; Kaplan and Rittenberg 1964). The $H_2S$ in the monimolimnion of Crawford Lake is seen to have a $\delta^{34}S$ value of ~ -4.5‰. The $\delta^{34}S$ value for the total sulphur in all sediment samples is close to -3.5‰ (see Figs. 6,8). This suggests that most of the sulphur in the sediments has come from the high concentrations of $H_2S$ in the monimolimnion, the $H_sS$ being introduced and exchanged with organic material through biomitigated reactions.

The most interesting feature of the $\delta^{34}S_{Total\ (S)}$ sediment profiles is the constancy of the $\delta$ values down to 24 cm in depth, indicating little change in the stratified nature of the Crawford Lake water column in well over 100 years.

*The Biological Sulphur Cycle in Crawford Lake.* The biological sulphur cycle in Crawford Lake is confined largely to the anoxic chemocline (15 to 16 m), where in the presence of both $SO_4^{2-}$ and $\Sigma H_2S$, both kinds of anaerobes, sulphate reducers (*D. desulfuricans*) and phototrophic oxidizing bacteria (*Chromatium* and *Clorobium*) can exist. Whereas the "sulphate reducers" provide the $H_2S$ required by the phototrophic bacteria, the phototrophic bacteria could provide the necessary nutrient for the "sulphate reducers" (e.g. acetate). However, the phototrophic bacteria will tend to migrate to the top of the chemocline where $H_2S$ first appears and where the light intensity is highest, (see Fig. 3) since the light intensity falls off rapidly with depth through the chemocline reaching almost zero at 16 m.

The dense populations of phototrophic bacteria present in the Crawford Lake chemocline (15 to 16 m) will, with adequate light intensity, oxidize the $H_2S$ produced by "sulphate reducers" to S°, $S_2O_3^{2-}$ and finally through to $SO_4^{2-}$. As in the case of meromictic Knaack Lake in Wisconsin USA, (Parkin and Brock 1981), the oxidation carried through to $SO_4^{2-}$ with little or no accumulation of intermediates.

In the biological sulphur cycle in the Crawford Lake chemocline (15 to 16 m) there is a net reduction of $SO_4^{2-}$ to $\Sigma H_2S$, the extent of reaction reaching 100% within ~ 1 m depth. The Rayleigh $\delta^{34}S$ profile curves for $SO_4^{2-}$ and $H_2S$ through this sharply defined layer will be essentially the curves for bacterial reduction of $SO_4^{2-}$ since there is little or no isotope effect in the phototrophic bacterial oxidation of $H_2S$ to $SO_4^{2-}$. Fry et al. (1985) have reported a small inverse isotope effect for oxidation by chromatium of ~ 3‰. The isotope effects calculated from these curves should therefore be essentially those for the bacterial reduction of $SO_4^{2-}$ to $H_2S$.

The absence of significant amounts of $H_2S$ in the oxygen free zone just above the chemocline of Crawford Lake (see Fig. 4) (14.9 m) suggests a thriving population of phototrophic bacteria, with rates of biogenic oxidation of $H_2S$ about equal to rates of $SO_4^{2-}$ reduction by the "sulphate reducers".

There is also isotopic evidence of phototrophic bacterial oxidation of $\Sigma H_2S$ to $SO_4^{2-}$ in the Crawford Lake lower chemocline (15.5 to 16 m) during the summer and fall (see Fig. 6). The Rayleigh type of $\delta^{34}S$ depth profiles obtained for the summer and fall samples (Fig. 6) are somewhat perturbed compared to those obtained for samples collected under the ice (Fig. 8). In particular the $\delta^{34}S_{SO_4^{2-}}$ values reach a minimum and then fall off rapidly with depth reaching values below that of the source $SO_4^{2-}$ (~ +4‰), as the residual $SO_4^{2-}$ concentration approaches zero. This could only result from some net bacterial oxidation of $H_2S$ to $SO_4^{2-}$.

Although the light intensity at the lower levels in the lake chemocline (15.5 to 16 m) are low even in the summer time, ~ 5 times that under ice cover, there is nevertheless sufficient activity to change the $\delta^{34}S_{SO_4^{2-}}$ values where residual $SO_4^{2-}$ concentrations are low.

## 4 Conclusions

The predominant form of dissolved sulphur in water is sulphate. Nearly all assimilation of sulphur is as sulphate, but during decomposition of organic matter, sulphur is released largely as hydrogen sulphide (Hutchinson 1957). This process of sulphide generation typically occurs in the anaerobic sediments of holomictic lakes. However, in meromictic lakes, this process occurs at the chemocline where phototrophic bacteria may consume the sulphides as rapidly as they are produced. As one descends through the chemocline in Crawford Lake, light intensity rapidly falls and sulphide uptake by phototrophic bacteria falls off as well. This permits sulphide concentrations to increase even though the rate of sulphide generation is negligible below the photic zone.

The rather specific requirements of the colourless sulphate-reducing bacteria, and the photosynthetic sulphide-oxidizing bacteria results in the development of a massive population of these two bacterial groups at 14.9 to 15.9 m in Crawford Lake. There was no evidence of any phototrophic activity below a depth of 16 m in Crawford Lake. Therefore, phototrophic bacteria which were collected below 16 m must be inactive as would be expected due to the virtual absence of light at 16 m.

Light levels at 15 m in Crawford Lake were always low. During winter, under ice and snow, there was usually less than 0.01% of surface light intensities at 15 m. This fact resulted in rather striking differences in the winter and summer phototrophic bacteria species composition and the sulphur isotope profiles at 15 to 16 m. The Rayleigh type $\delta^{34}S$ summer profiles differed substantially from those collected in winter. This may well reflect the fact that during winter, shade tolerant phototrophic bacteria replace the less shade tolerant summer forms. At low (winter) light intensities the rate of photosynthesis would necessarily decline. This, in turn, would result in a reduction of the rate of production of metabolites produced by the phototrophic bacteria and then consumed by the nearby sulphate reducing bacteria. This fact is reflected in the observed difference between our summer and winter Rayleigh type $^{34}S$ profiles.

In the batch Rayleigh type reduction process (observed in Crawford Lake between ventilation periods) maximum sulphur isotope effects (~ 50‰) were evident even at the lower $SO_4^{2-}$ concentrations where $SO_4^{2-}$ reduction was almost complete.

It may therefore be concluded (1) that in Crawford Lake the rate of diffusion or rate of $SO_4^{2-}$ supply from the mixolimnion down into the chemocline must be slow relative to the rate of bacterial reduction. This is a necessary requirement of a closed system process, and (2) the overall rate of $SO_4^{2-}$ reduction to $H_2S$ must be limited by the metabolic rate and not by the rate of $SO_4^{2-}$ uptake by the bacteria, a requirement of the bacterial reduction model, for the realization of maximum isotope effects (Rees 1973).

*Acknowledgments.* The authors are grateful to NSERC (The Natural Sciences and Engineering Council of Canada) for funding this research.

## References


Boyko M (1978) European impact on the vegetation around Crawford Lake in Southern Ontario. M Sc Thesis, University of Toronto, Ontario

Chambers LA, Trudinger PA (1978) Microbiological fractionation of stable sulfur isotopes. A review critique. Geomicrobiol J 1:249-293

Dickman MD (1985) Seasonal succession of microlamina formation in a meromictic lake displaying varved sediments. Sedimentology 32:109-118

Dickman MD, Thode HG (1985) The rate of lake acidification in four lakes north of Lake Superior and its relationship to downcore sulphur isotope ratios. Water Air Soil Pollut 26:233-253

Ford RW (1957) Sulphur isotope effects in chemical and biological processes. PhD Thesis, McMaster University, Hamilton, Ontario, Canada

Fry B, Gest H, Hayes JM (1984) Isotope effects associated with the anaerobic oxidation of sulfite and thiosulphate by photosynthetic bacterium, *Chromatium, Vinosum*. FEMS Microbiol Lett 27:227-232

Guerrero RE, Abella R (1985) Phototrophic sulphur bacteria in two Spanish lakes: Vertical distribution and limiting factors. Limnol Oceanogr 30:919-931

Harrison AG, Thode HG (1958) Mechanism of bacterial reduction from isotope fractionation studies. Faraday Soc Trans 54:84-92

Hutchinson GE (1957) A treatise on limnology I. Wiley, New York, p 1015

Ishii MV (1953) Fractionation of sulphur isotopes in the plant metabolism of sulphates. M.Sc. Thesis, McMaster University, Hamilton, Ontario, Canada

Kaplan IR, Rittenberg SC (1964) Microbiological fractionation of sulphur isotopes. J Gen Microbiol 34:195-212

Kemp ALW, Thode HG (1968) The mechanism of the bacterial reduction of sulphate and of sulphide from isotope fractionation studies. Geochim Cosmochim Acta 32:71-91

Lovely DR, Klug MJ (1986) Model for the distribution of sulphate reduction and methanogenesis in freshwater sediments. Geochim Cosmochim Acta 50:11-18

Matrosov AG, Chebotarev YeN, Kudryavtseva AJ, Zyukun AM, Ivanov MV (1975) Sulphur isotope composition in freshwater lakes containing $H_2S$. Trans Geokhimiya 6:943-947

Nriagu JO, Holdway DA, Coker RD (1987) Biogenic sulphur and the acidity of rainfall in remote areas of Canada. Science 237:1189-1192

Parkin TB, Brock TD (1981) The role of phototrophic bacteria in the sulfur cycle of a meromictic lake. Limnol Oceanogr 26(5):880-890

Peck HD (1974) The evolutionary significance of inorganic sulfur metabolism. In: Carlile, Shekel (eds) Symp Soc Gen Microbiol 24:241-262

Pfennig N (1967) Photosynthetic bacteria. Ann Rev 21:464-480

Postgate JR (1968) The sulfur cycle. In: Nickless G (ed) Inorganic Sulfur Chemistry. Elsevier, Amsterdam, pp 259-279

Rees CE (1973) A steady state model for sulphur isotope fractionation in bacterial reduction processes. Geochim Cosmochim Acta 37:1141-1162

Rybak M (1986) Chlorobium Chlorophyll as an indicator of organic pollution in a paleolimnological investigation. Acta Hydrobiol 14:255-261

Sanger JE, Gorham E (1972) Stratigraphy of fossil pigments as a guide to the postglacial history on Kirchner Marsh, Minnesota. Limnol Oceanogr 17:840-854

Severn SR (1982) Factors influencing the seasonal changes in primary productivity of the phototrophic bacteria of Crawford Lake, Ontario. M.Sc. Thesis, Brock University Biological Sciences Dept, St. Catharines, Ontario, Canada, p 204
Thode HG, Monster J, Dunford HB (1961) Sulphur isotope geochemistry. Geochim Cosmochim Acta 25:150–174
Thode HG, Dickman MD, Rao SS (1987) Effects of acid precipitation on sediment downcore profiles of diatoms, bacterial densities and sulphur isotope ratios in lakes north of Lake Superior. Arch Hydrobiol Suppl 74:397–422
Torgersen T, Hammond DE, Clarke WB, Peng TH (1981) Fayetteville, Green Lake, New York: $^{3}H$-$^{3}He$ water mass ages and secondary chemical structure. Limnol Oceanogr 26:110–122
Truper HG, Genovese S (1968) Characterization of photosynthetic sulphur bacteria causing red water in Lake Faro (Massince, Sicily). Limnol Oceanogr 13:225–232
Vinogradov AP, Grinenko VA, Ustinov VI (1962) Isotope composition of sulphur compounds in the Black Sea. Geokhimiya 10:973–997

# Carbonate Crusts in the Red Sea: Their Composition and Isotope Geochemistry

P. Stoffers and R. Botz

## 1 Introduction

Deep-sea sediments in the Red Sea often contain lithified carbonate layers as was first described by Natterer (1898). Herman (1965) ascribed these layers to paleoceanographic conditions where increasing temperatures at the end of the last glaciation led to $CO_2$ loss in the water and subsequent calcium carbonate formation. Gevirtz and Friedman (1966) found aragonite to be the most important carbonate mineral precipitating as cement during times of high salinity and temperature at the sea bottom. Milliman et al. (1969) estimated that more than half of the deep-sea carbonates in the Red Sea precipitated inorganically from the sea water. Furthermore they suggested that aragonite-cemented layers were formed from highly saline waters during glacially lowered sea level stands. Magnesian calcite, which is also found in the carbonate crusts, is believed to be an inversion product of aragonite and/or to have been precipitated directly from normal Red Sea water.

Isotope analyses have proved to be of great value in clarifying the carbonate formation processes. The $\delta^{13}C$ values of carbonates depend mainly on the source of the dissolved $CO_2$ whereas the $\delta^{18}O$ values depend on both the $^{18}O$ content of the water and the temperature during precipitation.

The aim of the present work was to study the isotopic composition of carbonate crusts from the Red Sea and, thus, try to increase our knowledge concerning their genesis.

## 2 Methods Applied

Figure 1 shows the Red Sea area and the sites where the carbonate crusts were sampled. Table 1 lists the stations of the Sonne 29 cruise 1984. All crusts were described macroscopically. As they are usually inhomogenous-looking, representative subsamples were taken and investigated separately. For analysis, we ground large amounts (several g) of each crust or subsample, respectively, to ensure representative results. There was no way of mechanically and/or chemically separating aragonite from magnesian calcite. However, we successfully separated dolomite from siderite both of which occurred in crust 369. The detailed macroscopic description of the individual crusts (and subsamples) is given by Mistacidou (1986). In general, five major types of crusts could be identified by means of macroscopic and microscopic descriptions. These are (1) pteropod-rich crusts with fibrous cement; (2) laminated carbonatic crusts with foraminifera; (3) goethite-rich crusts; (4) volcanic glass-rich crusts; (5) crusts containing coarse biogenic debris.

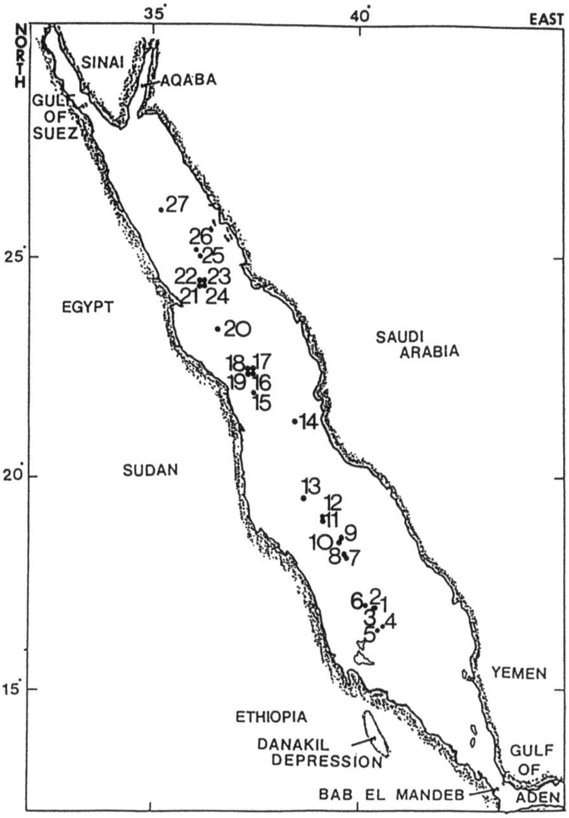


**Fig. 1.** The Red Sea and the sample sites of the RV SONNE-29

**Table 1.** Station protocol of the RV SONNE 29-cruise

| No in Fig. 1 | Station[a]/Sample | Longitude (E) | Latitude (N) |
|---|---|---|---|
| 1 | 213 DC | 40°.388 | 17°.013 |
| 2 | 214 DC a | 40°.387 | 17°.007 |
| | 214 DC b | | |
| | 214 DC c | | |
| 3 | 217 DC a | 40°.384 | 17°.004 |
| | 217 DC b | | |
| | 217 DC c | | |
| 4 | 220 DC a | 40°.510 | 16°.498 |
| | 220 DC b | | |
| | 220 DC c | | |
| | 220 DC d | | |
| 5 | 221 DC a | 40°.480 | 16°.443 |
| | 221 DC b | | |
| 6 | 238 DC | 40°.306 | 17°.115 |
| 7 | 246 DC | 39°.580 | 18°.094 |
| 8 | 247 DC a | 39°.596 | 18°.089 |
| | 247 DC b | | |

**Table 1.** *(Continued)*

| No in Fig. 1 | Station[a]/Sample | Longitude (E) | Latitude (N) |
|---|---|---|---|
| 9 | 255 DC a | 39°.437 | 18°.337 |
| | 255 DC b | | |
| 10 | 259 DC | 39°.407 | 18°.310 |
| 11 | 271 DC a | 39°.154 | 19°.039 |
| | 271 DC b | | |
| | 271 DC c | | |
| 12 | 272 DC a | 39°.153 | 19°.028 |
| | 272 DC b | | |
| | 272 DC c | | |
| | 272 DC d | | |
| | 272 DC e | | |
| 13 | 278 DC a | 38°.466 | 19°.345 |
| | 278 DC b | | |
| 14 | 293 KG | 38°.399 | 21°.243 |
| 15 | 304 DC a | 37°.517 | 22°.033 |
| | 304 DC b | | |
| | 304 DC c | | |
| | 304 DC d | | |
| 16 | 310 KG a | 37°.427 | 22°.423 |
| | 310 KG b | | |
| 17 | 311 KL a | 37°.360 | 22°.481 |
| | 311 KL b | | |
| 18 | 318 FG | 37°.347 | 22°.523 |
| 19 | 331 DC a | 37°.360 | 22°.390 |
| | 331 DC b | | |
| 20 | 341 FG | 36°.494 | 23°.500 |
| 21 | 354 DC a | 36°.162 | 24°.432 |
| | 354 DC b | | |
| 22 | 357 FG | 36°.168 | 24°.435 |
| 23 | 363 DC | 36°.167 | 24°.430 |
| 24 | 364 FG | 36°.166 | 24°.435 |
| 25 | 367 DC a | 36°.161 | 25°.013 |
| | 367 DC b | | |
| | 367 DC c | | |
| | 367 DC d | | |
| 26 | 369 DC a | 36°.013 | 25°.220 |
| | 369 DC b | | |
| | 369 DC c | | |
| 27 | 381 DC | 35°.207 | 26°.132 |

[a]KL = Kasten corer; FG = TV grab.

The carbonate content of the crusts was measured gasometrically (Müller and Gastner 1971). The carbonate powder was analyzed by XRD (Cu K $\alpha$ radiation) for mineralogical analysis. As an internal standard NaCl was added.

The element concentrations of Ca, Mg, Sr, Fe, Mn, Zn, and Cu were determined by atomic absorption spectroscopy (acid digestion).

Isotopic analysis was carried out on 40 samples applying the standard method (McCrea 1950). The $\delta$-values are given as per mil deviations from the PDB-standard.

Nine crusts were dated by means of radiometric carbon-14 analysis.

# 3 Results

The carbonate content of the crusts covers a wide range from 5% to 99% (Table 2). The results of the mineralogical analyses are shown in Table 2 and Fig. 2. Generally, the carbonate crusts in the Red Sea consist of mixtures of variable amounts of low and high Mg-calcite (with up to 15 mol% Mg) and aragonite. Only three crusts (311, 331, 369) contain significant amounts of dolomite ($Ca_{58}$; $Mg_{42}$). One subsample of crust No. 369 consists of pure siderite. The geochemical results also reflect this classification (Fig. 3). Aragonite samples plot in the "Ca-corner" of the triangle, whereas samples containing higher amounts of Mg-calcite or dolomite group close to its center. The siderite sample (369) has a Fe-concentration of 17.6% and, thus, plots in the "Fe-corner" of Fig. 3. However, in general, the iron content of the samples is due to the presence of pyrite and goethite rather than Fe-carbonates. In addition, samples with high Fe concentrations may also have high Zn contents indicating hydrothermal influence (Table 3).

The isotopic composition of the carbonate crusts is given in Table 2. The $\delta^{13}C$ values range from − 6‰ to + 4‰. The $\delta^{18}O$ values vary between − 1.3‰ and + 7.1‰. As Fig. 4 shows, three groups are distinguishable by the δ-values. Group Two differs from Group One by higher $\delta^{18}O$ values, whereas carbonate crusts from Group Three have negative $\delta^{13}C$ values.

The results of the radiocarbon dating are given in Table 4. The ages of the crusts range from 4,670 to 26,000 years.

**Table 2.** Carbonate content, carbonate mineralogy and isotopic composition of the bulk carbonate crusts investigated

| Sample no | Carbonate content % | Carbonate mineralogy (%) Calcite | Mg-Calcite | Aragonite | Dolomite | Siderite | $MgCO_3$ mol% | $\delta^{13}C$ ‰ | $\delta^{18}O$ ‰ |
|---|---|---|---|---|---|---|---|---|---|
| 213 DC | 83 | 3 | 5 | 92 | | | 13 | +3.9 | +5.8 |
| 214 DC[a] | 19 | 27 | 73 | | | | 14 | | |
| 214 DC[b] | < 10 | 51 | 49 | | | | 9 | | |
| 214 DC[c] | 27 | 40 | 23 | 37 | | | 7 | +1.4 | +0.3 |
| 217 DC[a] | 95 | 6 | 15 | 79 | | | 11 | +3.8 | +6.3 |
| 217 DC[b] | 29 | 24 | 76 | | | | 10 | | |
| 217 DC[c] | 40 | 32 | 68 | | | | 11 | +2.3 | +2.2 |
| 220 DC[a] | 79 | 12 | 30 | 58 | | | 12 | +3.4 | +4.5 |
| 220 DC[b] | 73 | 18 | 53 | 29 | | | 11 | +3.2 | +3.2 |
| 220 DC[c] | 71 | 16 | 28 | 56 | | | 9 | +3.0 | +5.3 |
| 220 DC[d] | 68 | 24 | 36 | 40 | | | 9 | +2.9 | +4.6 |
| 221 DC[a] | 98 | 6 | 15 | 79 | | | 11 | +3.7 | +6.0 |
| 221 DC[b] | 86 | 9 | 16 | 75 | | | 9 | +3.7 | +5.5 |
| 238 DC | 82 | 33 | 56 | 11 | | | 9 | +3.0 | +2.1 |
| 246 DC | 91 | 9 | 20 | 71 | | | 12 | +3.9 | +5.0 |
| 247 DC[a] | 85 | 29 | 60 | 11 | | | 10 | +3.1 | +1.9 |
| 247 DC[b] | 95 | 30 | 61 | 9 | | | 6 | +3.3 | +2.3 |
| 255 DC[a] | 95 | 35 | 57 | 8 | | | 6 | +3.3 | +2.2 |
| 255 DC[b] | 94 | 24 | 68 | 8 | | | 7 | | |
| 259 DC | 89 | 36 | 57 | 7 | | | 10 | +3.1 | +2.1 |
| 271 DC[a] | 30 | 35 | 41 | 24 | | | 10 | +2.6 | +2.4 |
| 271 DC[b] | 50 | 23 | 77 | | | | 9 | +3.6 | +4.0 |

**Table 2.** *(Continued)*

| Sample no | Carbonate content % | Carbonate mineralogy (%) Calcite | Mg-Calcite | Aragonite | Dolomite | Siderite | $MgCO_3$ mol% | $\delta^{13}C$ % | $\delta^{18}o$ ‰ |
|---|---|---|---|---|---|---|---|---|---|
| 271 DC[c] | 98 | 5 | 73 | 22 | | | 9 | +3.7 | +7.1 |
| 272 DC[a] | < 10 | 61 | 39 | | | | 8 | | |
| 272 DC[b] | 94 | 27 | 65 | 8 | | | 13 | | |
| 272 DC[c] | 90 | 29 | 64 | 7 | | | 5 | +3.4 | +2.5 |
| 272 DC[d] | 89 | 30 | 62 | 8 | | | 9 | | |
| 272 DC[e] | 88 | 26 | 64 | 10 | | | 9 | | |
| 278 DC[a] | 71 | 31 | 69 | | | | 11 | | |
| 278 DC[b] | 86 | 25 | 75 | | | | 10 | +3.5 | +2.3 |
| 293 KG | 95 | 23 | 64 | 13 | | | 9 | +3.4 | +2.4 |
| 304 DC[a] | 97 | 45 | 55 | | | | 8 | | |
| 304 DC[b] | 97 | 48 | 52 | | | | 7 | +2.8 | +1.7 |
| 304 DC[c] | 95 | 38 | 62 | | | | 7 | | |
| 304 DC[d] | 82 | | 100 | | | | 6 | +3.0 | +3.0 |
| 310 KG[a] | 99 | 20 | 80 | | | | 8 | +3.7 | +2.6 |
| 310 KG[b] | 94 | 29 | 66 | 5 | | | 9 | +1.4 | |
| 311 KL[a] | 95 | 48 | 15 | 3 | 34 | | 10 | +3.1 | +2.4 |
| 311 KL[b] | 10 | 16 | 22 | 55 | 7 | | 10 | +3.8 | +4.6 |
| 318 FG | 85 | 10 | 22 | 68 | | | 11 | +4.0 | +4.8 |
| 331 DC[a] | 97 | 3 | 6 | 90 | 1 | | 12 | | +6.5 |
| 331 DC[b] | 91 | 4 | 9 | 85 | 2 | | 13 | +3.5 | |
| 341 FG | 94 | 6 | 19 | 75 | | | 14 | −2.7 | +5.8 |
| 354 DC[a] | < 10 | | | | | 100 | | | +5.5 |
| 354 DC[b] | < 10 | | | | | | | +2.8 | |
| 357 FG | 47 | 39 | 49 | 12 | | | 6 | −0.9 | +1.9 |
| 363 DC | 78 | 65 | 35 | | | | | +3.5 | +3.7 |
| 364 FG | 94 | 5 | 9 | 86 | | | 9 | +2.3 | +6.6 |
| 367 DC[a] | 88 | 56 | 50 | 4 | | | 7 | +2.1 | +2.0 |
| 367 DC[b] | 88 | 46 | 54 | | | | 7 | | +2.5 |
| 367 DC[c] | 97 | 29 | 62 | 9 | | | 7 | +2.8 | |
| 367 DC[d] | 81 | 45 | 55 | | | | 8 | −6.0 | +2.6 |
| 369 DC[a] | < 10 | | | | | 100 | | −4.4 | −1.3 |
| 369 DC[b] | 90 | | | | 100 | | | +3.2 | −0.9 |
| 369 DC[c] | 95 | 13 | 25 | 62 | | | 10 | | +5.6 |

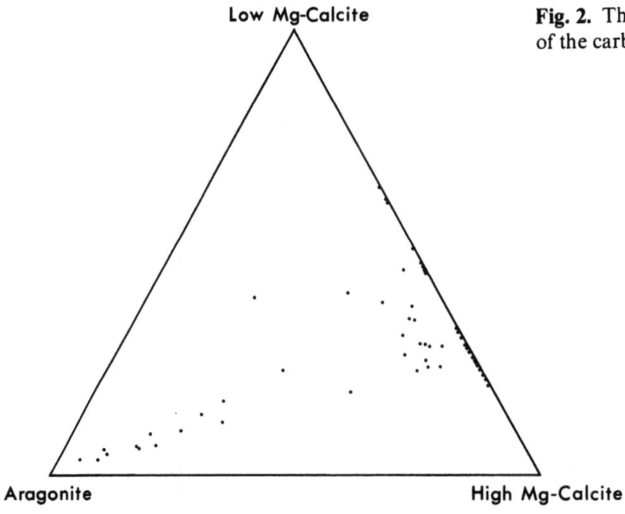


**Fig. 2.** The mineralogical composition of the carbonate crusts

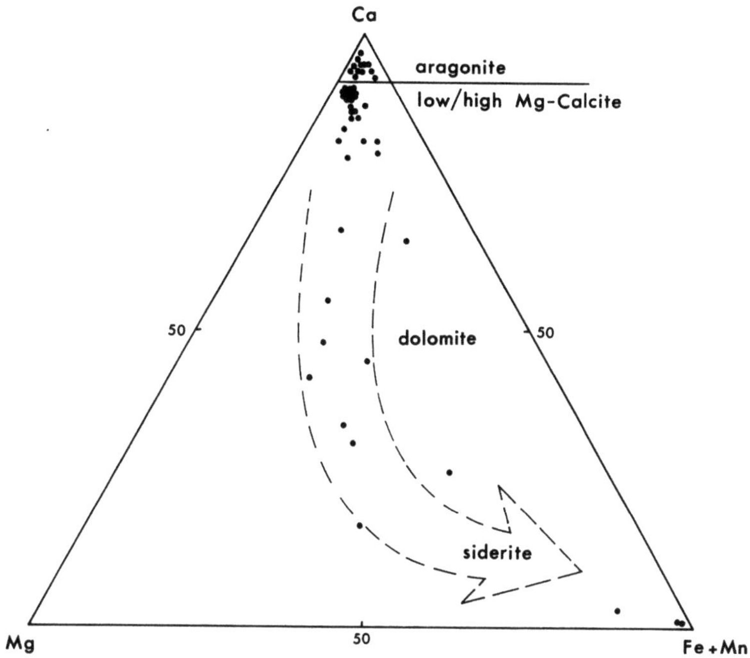


**Fig. 3.** The Ca-Mg-Fe + Mn triangle including the crusts investigated

**Table 3.** The elemental composition of the carbonate crusts

| Sample | Ca % | Mg % | Sr % | Fe % | Mn % | Cu ppm | Zn ppm |
|---|---|---|---|---|---|---|---|
| 213 DC | 30.13 | 0.61 | 0.72 | 0.83 | 0.14 | 24 | 89 |
| 214 DC a | 4.26 | 10.08 | 0.05 | 8.12 | 1.94 | 97 | 220 |
| 214 DC b | 5.14 | 5.51 | 0.04 | 4.5 | 0.09 | 73 | 77 |
| 214 DC c | 8.48 | 7.37 | 0.07 | 3.99 | 0.06 | 54 | 143 |
| 217 DC a | 32.18 | 1.17 | 0.69 | 0.42 | 0.07 | 15 | 64 |
| 217 DC b | 7.44 | 8.42 | 0.05 | 5.94 | 1.77 | 105 | 66 |
| 217 DC c | 9.09 | 6.03 | 0.05 | 3.40 | 0.2 | 71 | 85 |
| 220 DC a | 24.72 | 2.43 | 0.41 | 0.83 | 2.75 | 23 | 132 |
| 220 DC b | 23.27 | 3.56 | 0.18 | 1.31 | 0.19 | 11 | 86 |
| 220 DC c | 24.28 | 1.64 | 0.39 | 0.89 | 0.85 | 14 | 61 |
| 220 DC d | 22.97 | 2 | 0.26 | 2.72 | 0.34 | 55 | 90 |
| 221 DC a | 33.0 | 1.08 | 0.69 | 0.41 | 0.03 | 14 | 58 |
| 221 DC b | 30.15 | 1.24 | 0.65 | 0.62 | 0.05 | 23 | 54 |
| 238 DC | 25.43 | 3.21 | 0.09 | 0.84 | 0.71 | 42 | 383 |
| 246 DC | 31.44 | 1.63 | 0.53 | 0.43 | 0.1 | 31 | 166 |
| 247 DC a | 26.95 | 2.29 | 0.08 | 0.86 | 0.19 | 16 | 104 |
| 247 DC b | 30.68 | 2.56 | 0.09 | 0.59 | 0.18 | 21 | 112 |
| 255 DC a | 30.26 | 2.36 | 0.07 | 0.61 | 0.23 | 21 | 119 |
| 255 DC b | 30.46 | 2.71 | 0.09 | 0.43 | 1.31 | 20 | 90 |
| 259 DC a | 30.01 | 2.31 | 0.08 | 0.61 | 0.15 | 22 | 106 |
| 271 DC a | 8.57 | 5.11 | 0.07 | 5.06 | 0.46 | 78 | 237 |

**Table 3.** *(Continued)*

| Sample | Ca % | Mg % | Sr % | Fe % | Mn % | Cu ppm | Zn ppm |
|---|---|---|---|---|---|---|---|
| 271 DC b | 15.48 | 4.62 | 0.05 | 2.96 | 0.09 | 59 | 85 |
| 271 DC c | 32.23 | 3.05 | 0.22 | 0.37 | 0.21 | 23 | 206 |
| 272 DC a | 3.91 | 3.6 | 0.34 | 7.37 | 0.29 | 132 | 62 |
| 272 DC b | 30.02 | 2.53 | 0.07 | 0.82 | 0.29 | 26 | 156 |
| 272 DC c | 21.27 | 1.89 | 0.06 | 0.75 | 0.24 | 34 | 149 |
| 272 DC d | 29.12 | 2.41 | 0.07 | 0.82 | 0.24 | 26 | 137 |
| 272 DC e | 29.46 | 2.57 | 0.08 | 0.87 | 0.92 | 34 | 156 |
| 278 DC a | 24.11 | 2.67 | 0.08 | 2.06 | 0.59 | 32 | 144 |
| 278 DC b | 25.97 | 2.51 | 0.07 | 1.27 | 0.36 | 15 | 181 |
| 293 KG | 28.52 | 2.62 | 0.10 | 1.04 | 0.15 | 32 | 289 |
| 304 DC a | 29.19 | 2.21 | 0.08 | 0.90 | 0.14 | 29 | 201 |
| 304 DC b | 30.26 | 1.93 | 0.09 | 0.85 | 0.19 | 60 | 117 |
| 304 DC c | 28.32 | 2.32 | 0.08 | 0.73 | 0.21 | 39 | 111 |
| 304 DC d | 26.85 | 2.89 | 0.05 | 1.40 | 0.06 | 38 | 78 |
| 310 KG a | 30.97 | 2.79 | 0.07 | 0.49 | 0.23 | 13 | 152 |
| 310 KG b | 31.48 | 2.76 | 0.09 | | 0.29 | 23 | 153 |
| 311 KL a | 25.14 | 4.07 | 0.07 | 1.17 | 1.63 | 58 | 69 |
| 311 KL b | 19.14 | 3.21 | 0.29 | 3.47 | 3.7 | 68 | 388 |
| 318 FG | 31.8 | 1.26 | 0.54 | 0.7 | 0.14 | 33 | 243 |
| 331 DC a | 32.29 | 0.53 | 0.81 | 0.55 | 0.84 | 40 | 68 |
| 331 DC b | 32.7 | 0.6 | 0.79 | 0.52 | 1.42 | 39 | 140 |
| 341 FG | 31.35 | 1.14 | 0.65 | 0.51 | 0.02 | 12 | 52 |
| 354 DC a | 0.27 | 0.38 | 0.02 | 44.32 | 0.1 | 5 | 644 |
| 354 DC b | 0.05 | 0.06 | 0.02 | 7.10 | 0.01 | 6 | 5556 |
| 357 FG | 33.85 | 1.73 | 0.78 | 0.36 | 0.01 | 20 | 48 |
| 363 DC | 22.01 | 2.4 | 0.08 | | 0.43 | 12 | 102 |
| 364 FG | 33.85 | 0.74 | 0.78 | 0.36 | 0.01 | 20 | 48 |
| 367 DC a | 28.52 | 2.03 | 0.08 | 2.93 | 0.09 | 23 | 124 |
| 367 DC b | 27.21 | 1.83 | 0.07 | 0.85 | 0.16 | 28 | 100 |
| 367 DC c | 31.86 | 2.84 | 0.09 | 0.37 | 0.04 | 18 | 36 |
| 367 DC d | 28.97 | 2.29 | 0.08 | 0.84 | 0.04 | 20 | 70 |
| 369 DC a | 0.56 | 2.29 | 0.03 | 17.57 | 1.63 | 30 | 76 |
| 369 DC b | 14.47 | 7.54 | 0.03 | 3.72 | 0.70 | 35 | 243 |
| 369 DC c | 33.16 | 1.27 | 0.54 | 0.33 | 0.04 | 21 | 41 |

**Table 4.** Radiocarbon ages of the crusts

| $^{14}C$ Sample | $^{14}C$-age B.P. |
|---|---|
| 220 DC c | 14.000 ± 130 |
| 255 DC | 3.900 ± 50 |
| 238 DC | 6.050 ± 50 |
| 259 DC | 4.670 ± 65 |
| 311 KL b | 10.800 ± 170 |
| 354 DC | 19.500 ± 400 |
| 354 DC b | 23.700 ± 650 |
| 363 DC | 21.300 ± 525 |
| 369 DC b | 26.000 ± 1700 |

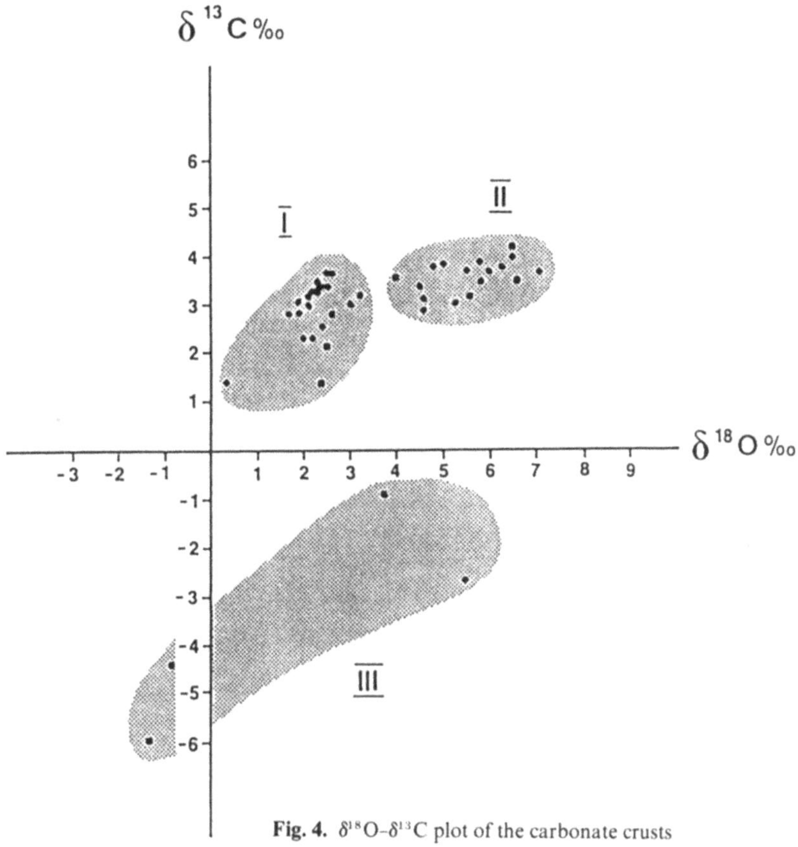


**Fig. 4.** $\delta^{18}O$–$\delta^{13}C$ plot of the carbonate crusts

# 4 Discussion

The mineralogical composition of the carbonate crusts from the Red Sea is shown in Fig. 2. As can be seen in this figure, many crusts consist of pure low and high magnesian calcite. On the other hand, only two crusts (213, 331) contain more than 90% aragonite. Aragonite occurs in two different forms within the crusts. First, there are abundant pteropodes, and second, fibrous aragonitic cement may bind individual grains together. It appears that aragonite-rich crusts occur preferentially in the southern and central part of the Red Sea.

According to Milliman et al. (1969), low and high magnesian calcite-containing crusts probably precipitated from normal Red Sea water, whereas aragonite formed during low sea level stands with associated highly saline conditions. These variations in the environmental conditions should be preserved in the isotopic composition of the carbonate crusts.

Figure 4 shows three groups of carbonate crusts distinguishable by isotopic analysis. According to Keeling (1958), atmospheric $CO_2$ has a $\delta^{13}C$ value of $-7‰$. The fractionation between solid $CaCO_3$ and gaseous $CO_2$ is 10.2‰ at 20 °C (Emrich et al. 1970). This results in a $\delta^{13}C$ value of 3.2‰ for carbonates which precipitated in isotopic equilibrium with atmospheric $CO_2$. Carbonates from Groups One and Two formed thus in atmospheric equilibrium as indicated by their $\delta^{13}C$ values near 3‰. However, the $\delta^{18}O$ values of carbonates from both groups differ significantly (Fig. 4).

Present Red Sea water has a $\delta^{18}O$ value of + 1.8‰ (Schoell and Faber 1978). Thus carbonate crusts from the first group with $\delta^{18}O$ values of + 2 to + 3‰ precipitated from normal Red Sea water at normal sedimentary temperatures (20 °C). Carbonates from the second group, however, have highly positive $\delta^{18}O$ values in the range of + 4 to + 7‰ suggesting carbonate formation from highly saline (evaporitically modified) sea water. Figure 5 shows that the $\delta^{18}O$ values increase with increasing aragonite content. As the oxygen isotopic fractionation of aragonite relative to calcite is only 0.6‰ (Rubinson and Clayton 1969) these positive $\delta^{18}O$ values of aragonitic crusts from group Two very well confirm the idea of Milliman et al. (1969) that aragonite formed from highly saline sea water. The carbonate crusts from the third group have negative $\delta^{13}C$ values. This indicates that an additional source of $^{12}C$-rich $CO_2$ must have contributed to carbonate formation. Although we do not have many data on this group, the mineralogy of the third-group carbonates appears to be dominated by dolomite and siderite. These minerals are probably of diagenetic origin rather than primary precipitates from surface waters.

Although the isotopically light carbonate crusts were not sampled within hot brine areas (Fig. 1), it could theoretically be that these carbonates formed from deep brines occurring in various areas of the Red Sea. The carbon isotopic composition of the dissolved inorganic carbon in the Atlantis II Deep was investigated by Schoell and Stahl (1972). The most negtive $\delta^{13}C$ value (–7‰) was measured for a hot (59 °C) brine. However, the positive $\delta^{18}O$ values of crusts 354 and 363 ($\delta^{18}O$ = + 5.5‰ and + 3.7‰, respectively) suggest that the carbonates did not precipitate at a similarly high temperature (the $\delta^{18}O$ values of brine waters were found to range from + 1‰ to + 2.6‰; Schoell and Faber 1978). The situation is not so clear for

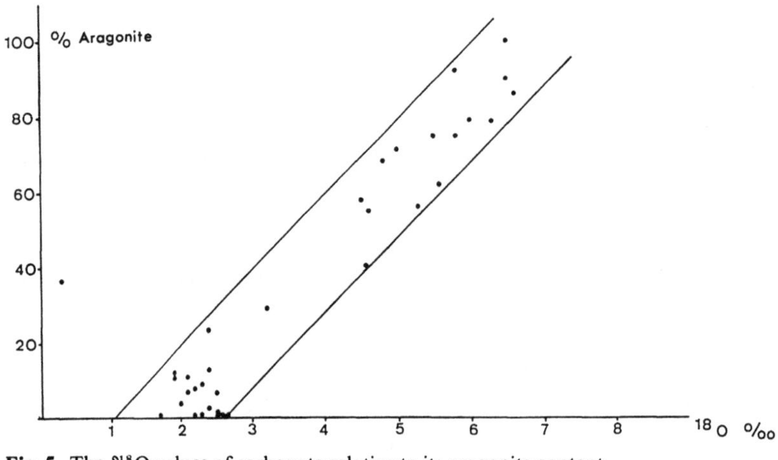


**Fig. 5.** The $\delta^{18}O$ values of carbonate relative to its aragonite content

crust 369. Carbonates from this crust have slightly negative $\delta^{18}O$ values (-1.3‰ and -0.9‰, respectively), suggesting formation at somewhat higher temperature. As pyrite is closely associated with the carbonate phase, it is likely that sulfate reduction occurred within the sediments. During this process organic matter is degraded anaerobically by microbial activity. It is known that diagenetic carbonates which formed during sulfate reduction are enriched in $^{12}C$ (Irwin 1980; Kelts and McKenzie 1980; Pisciotto and Mahoney 1981). Considering both the origin of the crusts in areas off hot brine deeps and their mineralogical composition, it is likely that sulfate reduction during the anaerobic degradation of sedimentary organic matter led to the formation of biogenic ($^{12}C$-rich) carbon dioxide which was mixed with the "primary" $CO_2$ dissolved in the seawater.

If the radiocarbon analyses give reliable information about the true ages of these diagenetic crusts (e.g., if no "old" carbon was added by, for example, hydrothermal activity) then it appears that crusts which lithified during early diagenesis are older than the others (Table 4). Their ages are around 20,000 to 26,000 years in contrast to crusts containing only primary components which were dated to be around 4,000 to 14,000 years old.

## 5 Conclusions

Primary carbonate crusts in the Red Sea consist of low and high Mg-calcite and aragonite. Aragonite precipitated from highly evaporated seawater. This is shown by the positive $\delta^{18}O$ values up to 7.1‰ rel. PDB. Only a few crusts contain diagenetic minerals such as dolomite and siderite. These carbonates are enriched in $^{12}C$ ($\delta^{13}C$ to $-6$‰). It is assumed that these minerals formed within the sediments during anaerobic sulfate reduction.

## References

Emrich K, Ehhalt DH, Vogel JC (1970) Carbon isotope fractionation during the precipitation of calcium carbonate. Earth Planet Sci Lett 8:363–371

Gevirtz JL, Friedman GM (1966) Deep-sea carbonate sediments of the Red Sea and their implications on marine lithification. J Sediment Petrol 36:143–152

Herman YR (1965) Etudes des sédiments Quaternaires de la Mer Rouge. PhD Thésis, Université Paris. Masson, Paris, pp 341–415

Irwin H (1980) Early diagenetic carbonate precipitation and pore fluid migration in the Kimmeridge clay of Dorset, England. Sedimentology 27:577–591

Keeling CD (1958) The concentration and isotopic abundance of carbon dioxide in rural areas. Geochim Cosmochim Acta 13:322–334

Kelts KR, McKenzie JA (1980) Formation of deep-sea dolomite in anoxic diatomaceous oozes. 26th Intern Geol Congr, Paris, France (Abstr)

McCrea JM (1950) On the isotopic chemistry of carbonates and a paleotemperature scale. J Chem Phys 18:849–857

Milliman JD, Ross DA, Teh-Lung K (1969) Precipitation and lithification of deep-sea carbonates in the Red Sea. J Sediment Petrol 39:724–736

Mistacidou E (1986) Sedimentologische, mineralogische, geochemische und radiometrische Altersbestimmungen anhand von Karbonatkrusten aus dem Roten Meer im Zusammenhang zu den Sauerstoff- und Kohlenstoffisotopen-Analysen. Diplomarbeit, Univ Heidelberg 56 pp
Müller G, Gastner M (1971) The "Karbonatbombe", a simple device for the determination of the carbonate content in sediments, soils and other materials. Neues Jahrb Miner Monatsh 1971:466–469
Natterer K (1898) Expedition SM Schiff "Pola" in das Rote Meer, nördliche Hälfte (Oktober 1895 – Mai 1896). Denk Akad Wiss Wien, Math-Naturw 65:445–572
Pisciotto KA, Mahoney JJ (1981) Isotopic survey of diagenetic carbonates, Deep Sea Drilling Project Leg 63. In: Yeats RS, Haq BU et al. (eds) Initial Reports DSDP 63, pp 595–609
Rubinson M, Clayton RN (1969) Carbon-13 fractionation between aragonite and calcite. Geochim Cosmochim Acta 33:997
Schoell M, Faber E (1978) New isotopic evidence for the origin of the Red Sea brines. Nature 275:436–438
Schoell M, Stahl W (1972) The carbon isotopic composition and the concentration of the dissolved inorganic carbon in the Atlantis II deep brines/ Red Sea. Earth Planet Sci Lett 15:206–211

Chapter 5

# Inorganic Geochemistry

# Calcium Carbonate Supersaturation and the Formation of in situ Calcified Stromatolites

S. Kempe and J. Kaźmierczak

## 1 Introduction

Stromatolites are products of benthic microbial communities composed predominantly of photosynthetic microorganisms, mainly cyanobacteria (blue-green algae), with occasional admixture of eukaryotic algae. The microbial communities (mats) may either act as traps for current-transported sediment particles, or they may induce in situ precipitation of minerals, mostly Ca and Mg carbonates. The in situ calcified, laminated microbial mats are classified as *stromatolites* whereas those characterized by clotted internal structures are termed *thrombolites*. Both structures can occur together in the same microbial sedimentary deposit and their clear-cut separation is often difficult (for review see: Kennard and James 1986; Burne and Moore 1987). The lamination in stromatolites may reflect diurnal, tidal, synodic, seasonal, annual or irregular growth increments which may produce by accretion a great variety of macroscopic bodies grouped sometimes into large reefoid structures (e.g., Hofmann 1973; Monty 1973; Walter 1976).

Both the trapping and the mineralizing algal mats occur today, the former almost exclusively in marine environments, the latter in lagoonal, paralacustrine and lacustrine environments. This has not, however, been so in the geologic past: throughout most of the Precambrian stromatolites – the only large and widespread marine biological structures of that time – formed predominantly by in situ mineralization of cyanobacteria-like microbiota (e.g., Monty 1973; Serebryakov and Semikhatov 1974; Awramik 1982). It is still debated why stromatolites are so rare in modern seas and why marine cyanobacterial mats diminished their ability to calcify with the end of the Cretaceous. The present paper attempts to answer Monty's (1972, p. 747) anxious question concerning the absence of in situ calcification in extant marine cyanobacteria.

## 2 Geologic Record of in situ Calcified Stromatolites

In the following synopsis, we limit ourselves to those in situ calcified stromatolites which clearly originated from submerged marine environments. We will, however, also consider calcareous stromatolitic structures which could be the result of early post-mortem calcification caused by the activity of heterotrophic or sulfate reducing bacteria and/or fungi, as observed in recent cyanobacterial mats (e.g., Krumbein and Cohen 1977; Lyons et al. 1984).

The Archean stromatolite record is scarce and only a few sites are unquestionably marine (Walter 1983; Hofmann et al. 1985). Some of these stromatolites are composed of dolomitic limestone (e.g., the 2.64 Ga old Bulawayan stromatolites; see Schopf et al. 1971) or of alternating siliceous and dolomitic layers (e.g., the > 2.58 Ga old Whalen Group stromatolites; see Hofmann and Snyder 1985). It is difficult to assess, however, if these carbonates are primary or secondary precipitates or derive from detrital particles. Since the carbonates occurring in the Archean stromatolites are dominated by dolostones with the incidental appearance of aragonite (e.g., in the > 2.6-Ga-old Steeprock Group; Walter 1983) the Mg/Ca ratio of their environment must have been high. From approximately 2.3 Ga on, calcareous stromatolites became more and more abundant (e.g., Serebryakov and Semikhatov 1974; Hofmann 1977; Awramik 1984) reaching their acme in the late Proterozoic (Riphean) as highly diversified structures (e.g., Krylov 1963; Hofmann 1969; Walter 1976). Though it is still a matter of considerable controversy whether these stromatolites were formed in the result of trapping and binding of fine calcareous particles or rather by in situ calcification (Walter 1972; Monty 1973; Gebelein 1974; Semikhatov et al. 1979), the latter possibility is gaining more acceptance particularly among Soviet specialists (e.g., Korolyuk 1963; Krylov et al. 1971; Serebryakov and Semikhatov 1974).

Mineralogy and time distribution of in situ calcified stromatolites fit the model of an early alkaline, silica-rich and $Ca^{2+}$-poor ocean suggested by Kempe and Degens (1985) and Kempe et al. (1987), according to which alkalinities were supposedly very high during the Archean, keeping the Ca-ion concentration well below $10^{-4}$ mol/kg. Thus, though the ocean was highly supersaturated with respect to calcite and dolomite, the low abundance of alkaline earth ions limited the amount of calcareous stromatolites generated. Near the end of the Proterozoic decreasing alkalinity and pH supposedly caused a rather rapid increase in $Ca^{2+}$ concentration in the ocean. High degrees of calcite and dolomite supersaturation would still have prevailed, but now larger amounts of carbonates could be removed per kg seawater than before, and may have favored the widespread and fast growth of calcareous stromatolites.

The late Proterozoic intensive formation of calcareous stromatolites significantly declined during Vendian and early Cambrian. This phenomenon is commonly attributed to the appearance of grazing gastropods and small acoelomate animals (Garrett 1970; Awramik 1971; Walter and Heys 1985), to substratum competition by encrusting eukaryotic organisms (Monty 1973; Pratt 1982), or to changing sedimentological conditions caused by the evolution of skeletal metazoans (Pratt 1982).

The common belief in a dramatic decline of stromatolites close the transition Precambrian/Cambrian has been reconsidered recently by Pratt (1982): his tabulation of calcareous stromatolites, including thrombolites and cryptalgal limestones, shows that in situ calcified microbialites (sensu Burne and Moore 1987) are common components of many Phanerozoic carbonate deposits. It should be noted that Pratt's data do not include the still controversial stromatoporoid stromatolites (Kazmierczak 1976; Kazmierczak and Krumbein 1983) which are dominating structures in the early Paleozoic platformal carbonates. Modern counterparts of cyanobacterial stromatoporoids have been newly discovered in a sea-linked volcanic lake of Indonesia (Kazmierczak and Kempe in preparation). Nevertheless, well-defined, independently growing, calcareous stromatolites became less obvious and of rather local importance in marine Phanerozoic environments. They are particularly limited during late Permian and early Triassic.

Since mid-Cretaceous in situ calcified cyanobacterial communities became rare in normal marine environments and apparently disappeared in the lower Cenozoic from the marine record (Monty 1973, 1977; Pentecost and Riding 1986) but not from peri-marine and fluviolacustrine settings. These observations are not contradicted by the recent discoveries of modern subtidal calcareous stromatolites in normal salinity seawater from the Bahamas (Dravis 1983; Dill et al. 1986) as they accrete through trapping of calcareous sediment particles cemented by early diagenetic Mg-calcite or aragonite. Riding (1982) has associated this exclusion of stromatolites from the marine realm with changes in the Mg/Ca ratio of seawater while Kazmierczak et al. (1985) suggest changes of $Ca^{2+}$ concentration in epicontinental seas. We suggest that the cessation of marine cynobacterial calcification may also be the result of changes in calcite supersaturation in seawater.

## 3 Recent Stromatolites

Recent stromatolites have been reported from a wide range of continental and marine settings. The typical modern marine cyanobacterial stromatolite is a sediment-trapping structure as e.g. documented for the Bermudas (Gebelein 1969), the Bahamas (Neumann et al. 1970; Dravis 1983; Dill et al. 1986) and Shark Bay, Australia (Logan 1961; Logan et al. 1974). Its mineral content is not determined by the mat or its chemical environment, but by the local supply of fine-grained sediment and by skeletal mineralogy of the marine organisms which produce it. Mostly aragonite and calcite have thus to be expected. Diagenetic cementation of Mg-calcite or aragonite consolidates the primary structure. In situ calcified mats have been reported from hypersaline and highly alkaline continental settings such as Mono Lake/Utah U.S.A. (Whiticar pers. commun.), from hypersaline marine-derived lagoonal waters (e.g., Laguna Mormona, Baja California, Mexico: Horodyski and vonder Haar 1975; Marion Lake, South Australia: von der Borch et al. 1977), from mesohaline alkaline lakes, from many hard water lakes (e.g., from alpine lakes, Schneider et al. 1983; Schröder et al. 1983) and from karst water bodies (e.g., a sinkhole in dolomitic limestone; Gomes 1985). Even though the suite of carbonate minerals in mineralizing mats is rather large: aragonite is reported from Laguna Mormona and Marion Lake, hydromagnesite is reported from Coorong Lagoon, South Australia (Walter et al. 1973), high-Mg-calcite is reported from lake Tanganyika (Cohen and Thouin 1987), the main mineral is undoubtedly low-Mg-calcite in these mats. In the following we describe four stromatolite sites – in the order of decreasing salinity – in more detail because they will later serve as case studies for the geochemical discussion.

### 3.1 Satonda Island Crater Lake

Satonda is a small volcanic island 2 km offshore Sumbawa/Indonesia north of the Tambora Volcano. Its double crater is filled with seawater forming a lake 1.2 × 0.9 km wide and 69 m deep. At 13 sites along the otherwise sandy lake shore massive calcareous reefs grow outward from rocky points. In 1986, during the R/V SONNE 45B cruise (chief-scientist E.T. Degens) the authors spend 10 days in a field camp to investigate the reefs and the chemistry of the lake (Kempe and Kazmierczak in press).

The reefs are 1 to 2 m thick and occur to a water depth of at least 23 m. Their upper parts form strongly corroded subcircular heads and emerge about half a meter above the water level at the end of the dry season (Fig. 1). The reef walls are very steep, partly overhanging and composed of brittle, cavernous limestone consisting of aragonite and low-Mg-calcite, in part silicified. Filaments of siphonocladalean algae (Chlorophyta) densely overgrow the underwater surfaces down to a depth of 4 to 8 m. These filaments can calcify only near the water surface and in small evaporative lagoons. The reef growth proceeds by alternating layers of in vivo calcifying pleurocapsalean cyanobacteria and red algae (*Peyssonnelia* sp., *Lithoporella* sp.), often separated by accumulations of gastropod fecal pellets settled in cyanobacterial micrite (Fig. 2 A-C). The red algae occur only within the first 1 cm of the reef and the rest is almost entirely constructed of calcified coccoid cyanobacteria. Down to a depth of about 8–10 m, the living reef surface has a cauliflower-like appearance, while at a depth > ca. 10 m cystous structures resembling corn flakes dominate (composed mostly of *Peyssonnelia* sp. thalli). Reef growth proceeds on rocky substratum, on boulders and on wood but not along the sandy beaches. The internal structure of the reef proved to be quite complicated. Lamination can occur, but irregular, clotty growth is the rule. Cavities may be lined with diagenetic calcite cements or may be filled with siliciclastic sediment. In thin sections and under the scanning electron microscope coccoid cyanobacteria can be seen as the main calcifying agents (Fig. 2 D,E). Usually, only the outer sheaths surrounding the coccoid aggregates are calcified. They are particularly well preserved in the silicified sections of the reef. Shells and fecal pellets, calcified or silicified during early diagenesis, produced by a large population of the only gastropod species living in the crater lake at present (*Cerithium* sp.) can contribute significantly to the total mass of the reef. The macrobiota of the lake is extremely species-poor and consists, besides the gastropods, of a monospecific population of monaxonid sponges (*Suberites* sp.) overgrowing the reef surface and intertwinning with the green algae, and of fish a few centimeters long that live close to the reef wall. Also observed were rare amphipods dwelling at the reef surface. A dense population of an oligochaete is living in the black sandy mud along the lake shore.

### 3.2 Walker Lake

Lake Walker in western Utah/USA is a relatively shallow (30 m), closed lake some 27 km long and 8 km wide. Extensive use of water from tributary rivers for agriculture has lowered the lake level considerably during the last 100 years causing

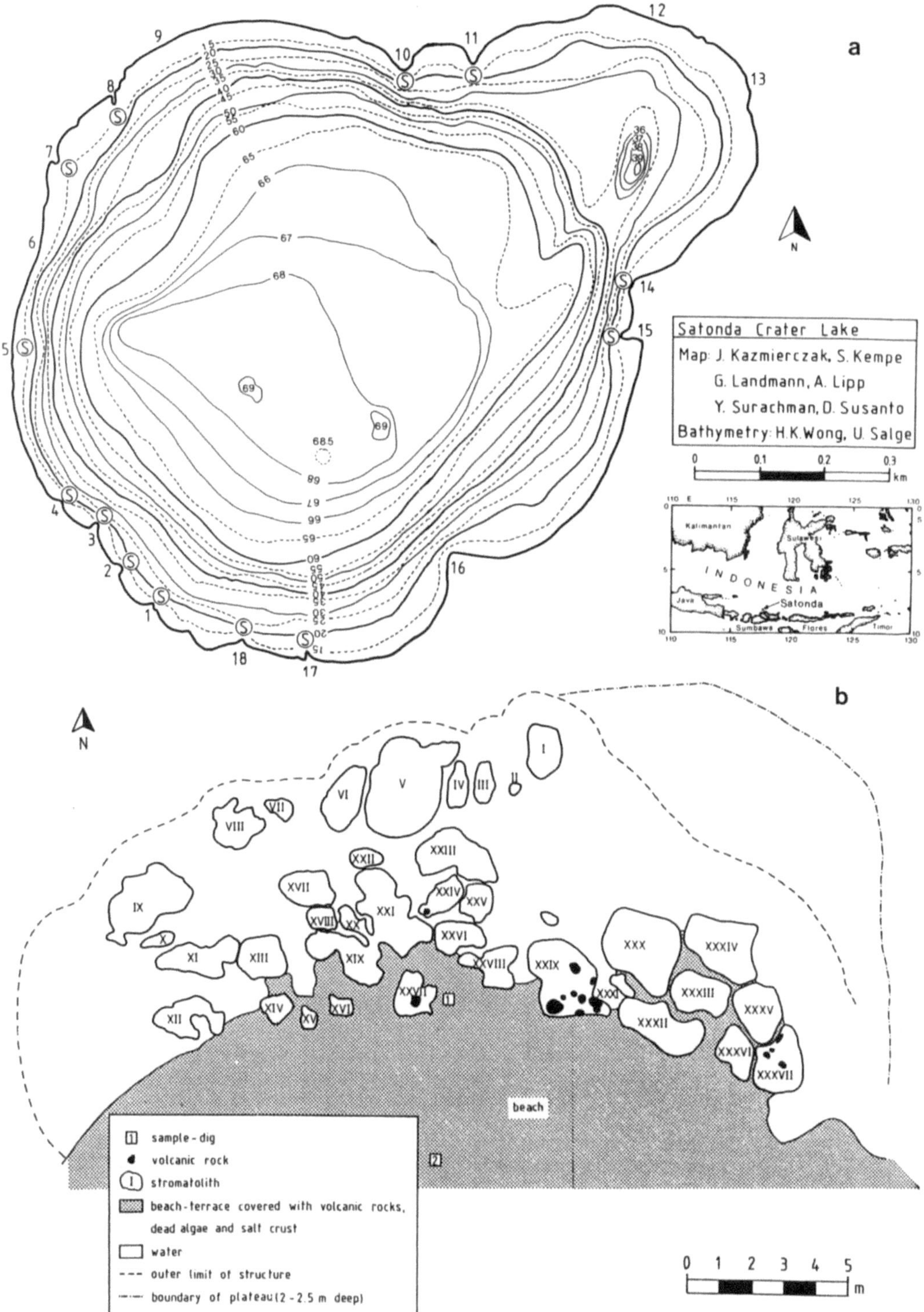


**Fig. 1a,b.** (Figure 1c and legend see page 260)

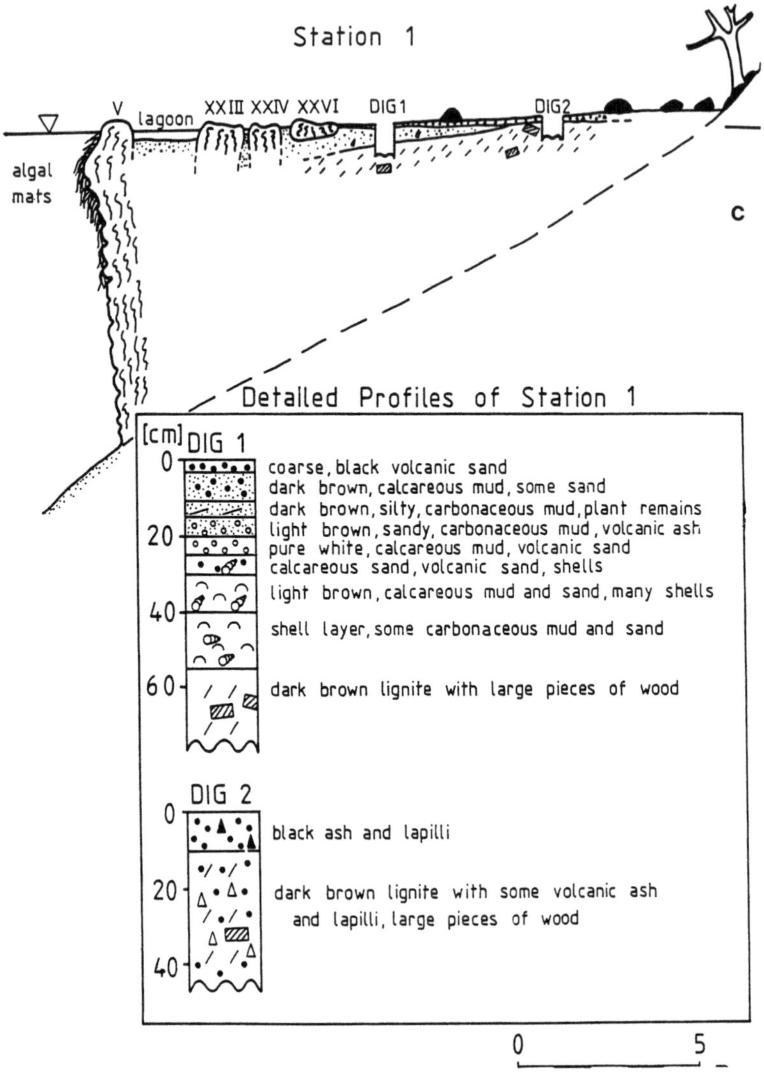


**Fig. 1c**

**Fig. 1. a** Bathymetric map of Satonda Crater Lake and locations of stromatolitic $CaCO_3$-reefs (*circled S*). Numbers denote stations at shore line. **b** Map Station 1, stromatolitic structures as apparent above water level in October, 1986. **c** Vertical section of Station 1 reef as observed by diving

**Fig. 2A-E.** Cyanobacterial/red algal calcareous reefs from the seawater filled Satonda Island Crater Lake/Indonesia: **A** Vertical cut through specimen from the reef surface; *arrow* indicates the boundary between the upper part of the reef composed of calcified thalli of red algae (*Peyssonnelia* sp., *Lithoporella* sp.) alternating with calcified coccoid cyanobacteria, and the lower part constructed by cyanobacterial micrite with numerous shells and fecal pellets of gastropods (*Cerithium* sp.), and tiny serpulid tubes; sample 29, Station 17, depth 23 m (bar = 2 cm). **B** Enlarged vertical thin section of the upper part of the specimen shown in **A**, demonstrating details of the red algae thalli (*Peyssonnelia* sp.). and the clotty (thrombolitic) character of the cyanobacterial micrite (bar = 100 μm). **C** Enlarged vertical thin section of the lower part of the specimen shown in **A**, illustrating the clotty character of the

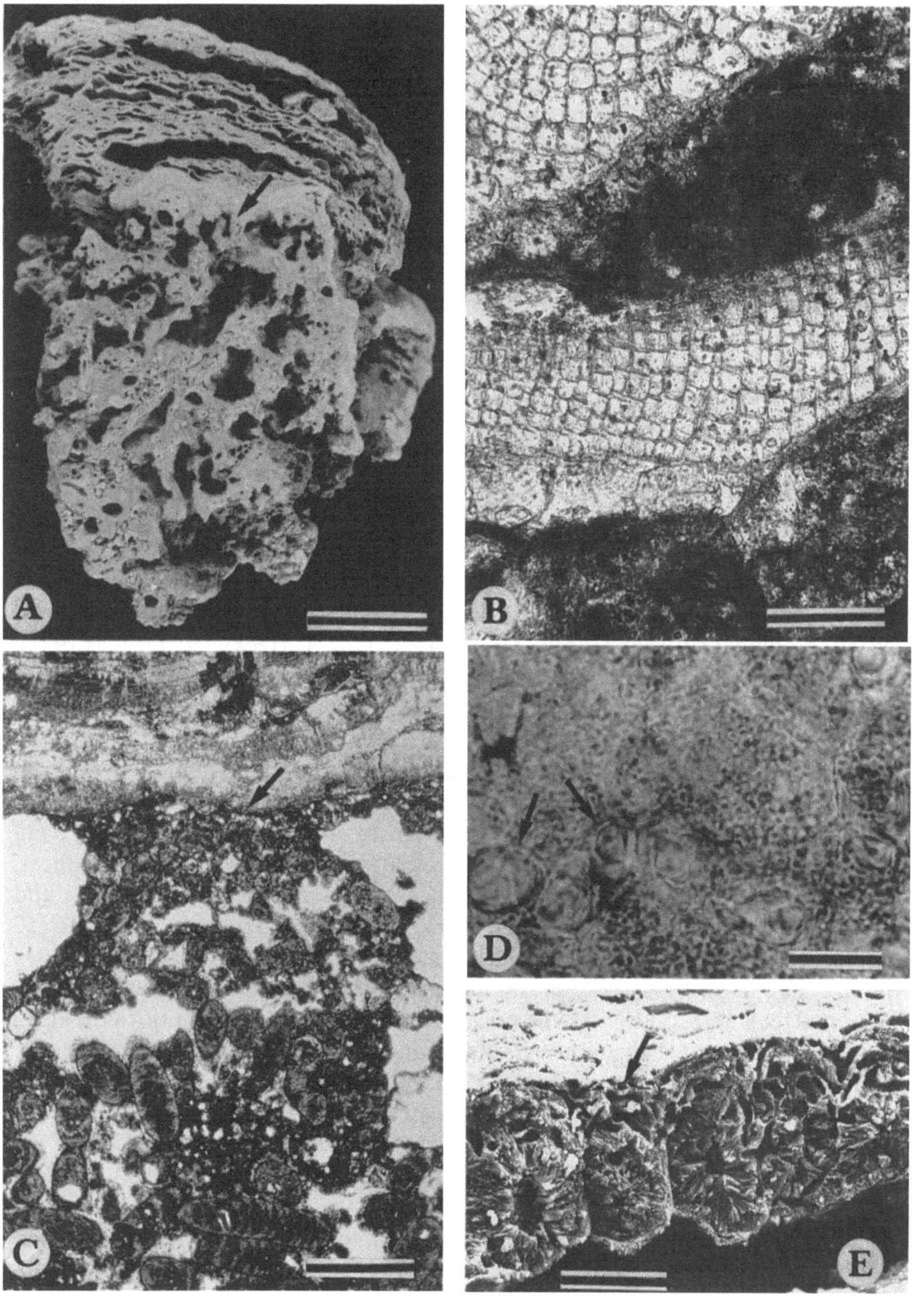


cyanobacterial micrite and the numerous gastropod fecal pellets; *arrow* indicates the same level as the arrow in **A** (bar = 500 μm). **D** Tangential thin section of the living reef surface; visible are larger cells (some marked with *arrows*) of the somewhat deeper laying thallus of the red alga (*Lithoporella* sp.) covered by dense aggregates of pleurocapsalean cyanobacteria; sample 25, Station 17, depth 17 m (bar = 20 μm). **E** SEM picture of vertically fractured reef surface showing a row of heavily calcified cells of the red algae *Lithoporella* sp. covered by a thin layer of calcified sheaths of pleurocapsalean cyanobacteria (marked by *arrow*) sample 25, Station 17, depth 17 m (bar = 20 μm) its salt content to rise from 2.5 g/l (1884) to 8.6 g/l in 1966. Osborne et al. (1982) report on stromatolites in shallow and well mixed waters along its coast where pebbles, boulders or exposed bedrock provide hard grounds. The observed algal mats precipitate low-Mg-calcite and form small, laminated generalized, cabbage-head or columnar stromatolites, which are up to 4 cm high and sometimes club-shaped stromatolites. Epilithic green algae including filamentous *Cladophora glomerata*, and *Ulothrix* cf. *aequalis* thrive on rocks and stromatolites. Mats are formed by green alga *Gongrosira* and by the oscillatoriacean cyanobacteria *Schizothrix*. In the calcifying lower stratum of the mats other trichomeous blue-green algae dominate: *Amphithix janthina, Calothrix, Homeothrix, Spirulina, Anabaena* and *Lyngbya*. Coccoid species like *Entophysalis, Anacestis* and *Dermacapsa* occur as well. Diatoms may also be present and the composition of the mats is seasonally variable.

### 3.3 Lake Tanganyika

In April/March 1970 Egon T. Degens organized and headed the first thorough investigation of the hydrochemistry, sedimentology and tectonic structure of Lake Tanganyika by a multidisciplinary group of scientists from the Woods Hole Oceanographic Institution (WHOI) (Degens et al. 1971; Hecky and Degens 1973). After this pioneering effort, exciting discoveries were made in the lake: Cohen and Thouin reported on recent microbial calcareous structures from the lake for the first time in 1987. At the northeastern shores of Lake Tanganyika they found seven successive carbonate facies from the water level to a depth of 40 to 50 m (Fig. 3). Thrombolitic reefs occur in the lake between 15 m and at least 50 m depth. In the depth range between 30 and 40 m, up to 3 m high thrombolite heads were discovered. They carry a 2 to 3 mm thick microbial mat at their upward surfaces. These mats are heavily grazed by two gastropods (*Lavigeria* cf. *L. nassa* and *Paramelania* cf. *P. damoni*) which show population densities of up to 50 individuals/m$^2$. Cohen and Thouin (1987) suggest that this grazing pressure causes the structure to be highly porous and nonlaminated. They therefore classify these microbial reefs as thrombolites and not as stromatolites. No description of the algae constituting the mat is given though, but cyanobacteria (*Anabaena* sp.) constitute an appreciable part of the phytoplankton in the lake and can reach the highest biomass of all phytoplankton and of all seasons during the early stagnation period (Hecky and Kling 1981). Stromatolite growth seems to be possible well below the euphotic layer ($>$ 1% light) which was found to be 28 m deep by Hecky and Fee (1981).

It is interesting to note that the thrombolites (and all other shore-based carbonates) consist of high-Mg-calcites. High-Mg-calcites also account for the bulk of carbonates of the abyssal sediments in the lake which therefore may originate from resuspended shore sediments (Stoffers and Hecky 1978). In addition, three thin layers of aragonite occur in the sediments of the northern lake basin which probably were swept down from Lake Kivu with the Rusizi River (Degens pers. commun.).

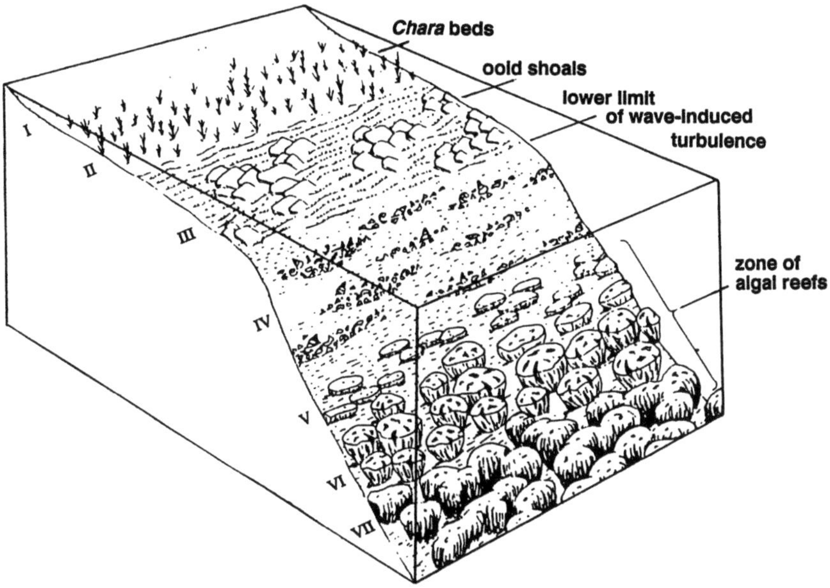


**Fig. 3.** Scheme of thrombolites and succession of carbonate facies with depth in Lake Tanganyika at the Burundi coast 115 km south of the Rusizi mouth (Cohen and Thouin 1987). Nearshore (**I**) clastic sands and (**II**) bioturbated *Chara*-bearing calcareous silts occur. Rippled ooid shoals and cemented oolite ridges follow below 2 m (**III**). Below the influence of turbulence induced by waves (approx. 4 m depth) shelly lime muds or coquinas (areas without fine particles) were found (**IV**). Microbial heads dominate below 20 m: they are thin and isolated between 20 and 25 m (**V**), become more erect below 25 m (**VI**) and form sills and ledge systems below 30 m (**VII**). In deeper waters thrombolites can grow directly on the Precambrian basement

## 3.4 Andros Island

Recent in situ calcifying stromatolites occur in inland fresh water ponds on Andros Island/Great Bahamas Bank. Monty (1972) described their morphology and taxonomy in detail. The ponds are very shallow and dry out almost completely at the end of the winter in March. The algal mats cover the pond bottoms and form four zones according to water depth and duration of inundation. Only thin cyanobacterial mats develop in the upper shore zone before the water recedes. In the eulitoral zone smooth, a few cm thick, slightly domed and polygonal, subcircular or amoeboid mats 10 to 15 cm in diameter develop. They show laminated structure with alternating micritic carbonates and soft, highly organic layers. Toward the deeper part of these ponds, pinnacles and short columns a few centimeters high develop on the elliptical domes. Even deeper, the cyanobacterial mats have to compete with reeds and other higher plants, which they encrust up to the water level. At the final stage of evaporation whitings occur in the residual ponds, covering mats and higher plants alike.

The only carbonate mineral precipitated is low-Mg-calcite. This also shows that the material is not marine in origin, because then it would be composed of aragonite, the most common mineral in carbonates of the Bahamas. It is interesting to note that the stromatolites are not lithified, but remain earthy and soft in texture. The domes are made of vertical filaments of *Scytonema*, calcifying below their growing tips, and of thin, horizontal films of *Schizothrix calcicola* which calcify daily. This gives the mat a rather durable structure. Development of gas bubbles may extent the upper soft layer which is also inhabited by aggregates of *Entophysalis*. Upon desiccation, growth can continue inside the mat, but then no calcification occurs. Gas bubbles can lift the top of domes or peel off mats, thus causing discontinuities in the laminae or even inverted layering. The calcareous mud is not very deep, 30 cm on average, and rests on the Pleistocene marine carbonates of the Great Bahama Bank.

## 4 Precipitation of Carbonate Minerals

Dissolution of carbonate minerals is governed by thermodynamics. Once the ion-activity product of the alkaline earth ion in question ($Ca^{2+}$ or $Mg^{2+}$) and the carbonate ion ($CO_3^{2-}$) becomes less than the equilibrium constant of the respective mineral, dissolution can start:

$$\text{dissolution} \quad 1 < \frac{(Ca^{2+})(CO_3^{2-})}{K_{Calcite}} < 1 \quad \text{supersaturation.} \tag{1}$$

In order to facilitate discussion of saturation with respect to a certain mineral the saturation index (SI) is used (Langmuir, 1971):

$$\log \frac{(Ca^{2+})(CO_3^{2-})}{K_{Calcite}} = SI_{Calcite}. \tag{2}$$

The SI becomes zero (log of 1) at exact saturation, negative at undersaturation and positive at conditions of supersaturation.

Depending on the velocity of the transport of the reaction partners to and from the mineral surface by diffusion or turbulent advection, dissolution proceeds as long as an undersaturation with respect to the mineral exists (Dreybrodt 1981).

Precipitation of carbonate minerals, however, does not start as soon as the SI rises above zero. Rather, natural carbonate solutions can become highly supersaturated before precipitating carbonate minerals spontaneously. In fact, spontaneous calcite precipitation occurs only in very restricted environments: in caves (formation of flowstones, e.g., Dreybrodt 1980), during rapid evaporation of thin water films (caliche formation) and upon mixing of fresh waters with highly alkaline waters (formation of whitings, i.e. aragonite muds in soda lakes, such as Lake Van, Müller et al. 1972; Kempe 1977). It appears that in all other cases where carbonate minerals precipitate, biogenic processes – especially photosynthesis – mediate the precipitation.

Even though the ocean surface waters are highly supersaturated with regard to all major carbonate minerals ($SI_{Calcite}$ = 0.3 to 0.5), the majority of calcareous marine sediments are precipitated enzymatically. This is true for calcareous algae such as coccolithophoridae, for protozoans such as foraminifera, and for multicellular animals such as echinoderms, molluscs, corals and vertebrates (Degens 1979). Enzymatic carbonate precipitation, which does not involve an isotopic fractionation, can be described by:

$$Ca^{2+} + 2\,HCO_3^- \xrightarrow{\text{no isotopic fractionation}} Ca(HCO_3)_2 ===> CaCO_3 + H_2CO_3. \quad (3)$$

While thermodynamic precipitation produces carbonates about 5 per mil heavier in $^{13}C$ than the parent solution:

$$Ca^{2+} + 2\,HCO_3^- \xrightarrow{\text{isotopic fractionation}} CaCO_3 + H_2O + CO_2. \quad (4)$$

Calculation of saturation indices is not as straightforward, as Eq. (1) suggests, however: one needs to calculate activities from measured total concentrations of (in the case of $SI_{Calcite}$) $Ca^{2+}$ and $CO_3^{2-}$. These are depending on the activity coefficient, which in turn is a function of total ionic strength. Also, only activities of the free ion species are needed and concentrations of neutral ion pairs such as $CaCO_3^0$ or charged ion pairs such as $CaHCO_3^+$ have to be taken into account. Furthermore, all the equilibrium constants vary with temperature. To solve the problem, iterative, computerized electrolyte models have been developed. Wigley (1971, 1973a,b) gives such a procedure for carbonate fresh water (program see Kempe 1975) and Wigley and Plummer (1976) presented a program (WATMIX) which allows calculation of SIs from waters of higher salinities, notably seawater. For more detailed discussion of the carbonate system see Kelts and Hsü (1978), Kempe (1982) or Pytkowicz (1983).

In general, the saturation state of a solution with regard to carbonate minerals is altered by the extraction or addition of $CO_2$. $CO_2$ extraction continues by diffusion as long as the water has a higher $pCO_2$ than the ambient air, which has a $pCO_2$ of about 340 ppmv (Degens et al. 1984). Diffusion across the interface is greatly enhanced by turbulence in rapidly running creeks or in waterfalls. This leads to the formation of travertine dams (Kempe and Emeis 1985; Emeis et al. 1987; Herman and Lorah 1987). Diffusional loss of $CO_2$ may be enhanced by photosynthetical uptake of $CO_2$ further amplifying supersaturation. Seasonal extraction of $CO_2$ by phytoplankton and macrophytes can cause high supersaturation in temperate hard water lakes which in turn triggers precipitation of fine grains of calcite called Seekreide (Kelts and Hsü 1978; Schneider et al. 1983; Schröder et al. 1983; Kempe et al. 1985). Respiration on the other hand adds $CO_2$ to the water, thus reducing supersaturation.

## 5 Case Studies

Since cyanobacterial and other microbial mats precipitate $CaCO_3$ in a biologically induced, i.e., nonenzymatic manner (for review see Golubic 1973; Krumbein 1979; Pentecost and Riding 1986), i.e., as a result of interactions between the activity of the organisms and its surrounding environment (Simkiss 1986), they must do so in $CaCO_3$-supersaturated waters. However, it is as yet unclear what ranges of supersaturation occur at stromatolite growth sites and how they differ from the average marine supersaturation. If we could answer these questions, we may gain an understanding of what is the paleoecological and paleoenvironmental advantage of calcifying cyanobacteria compared to other calcifying organisms. Thus, we will present in this study saturation calculations for a variety of calcareous stromatolite-bearing sites in order to compare them with "average" marine or fresh water conditions.

Sites which are now dry are of no use for such a comparison. Furthermore, only those sites can be used for which ample geochemical data have been published. In most cases stromatolites have been described morphologically, taxonomically or mineralogically, but not in terms of hydrochemistry. In some cases, hydrochemical data are available, but important parameters like alkalinity or pH have not been reported. Thus only a few sites have complete enough hydrochemical data to be discussed.

### 5.1 Satonda Island Crater Lake

The chemical composition of Satonda Lake waters is given in Table 1 together with a seawater sample for comparison. A vertical profile of some of the parameters is plotted in Fig. 4. Results show, that the water body has three layers, an oxygenated upper layer, 22.8 m thick, and somewhat less saline than seawater, and two anaerobic bottom layers of higher salinity than seawater. The main chemical features are the significantly higher alkalinity and pH of the surface layer as compared to seawater and its lower Ca and Mg concentrations. At depth, pH decreases across the extremely sharp interface at 22.8 m, indicating an increase in $pCO_2$. Calculation with WATMIX shows, that the surface layer is almost in equilibrium with the atmosphere (340 ppmv), but that a pressure of 240,000 ppmv is reached at depth. The carbonate alkalinity (total alkalinity corrected for borate) rises from 3.4 meq/kg (seawater 2.0), to almost 50 meq/kg at the bottom of the lake while Ca concentrations rise only slightly. Measurements of $\delta\,^{13}C$ for dissolved inorganic carbon in the lake gave values of -8.4, -11.4 and -21.4 per mill PDB for samples from the three layers at 10, 40 and 60 m depths. This suggests, that organic detritus (plants = -25 $\delta\,^{13}C$) falling into the lakes is progressively respired at depth, releasing isotopically light $CO_2$. The increase in $pCO_2$ causes weathering of the silicates in the lake bottom (and of the about 80 cm of ash the island received during the 1815 Tambora cataclysm) which in turn increases the alkalinity. Degassing of this water at the surface causes the increase in pH and, due to the high alkalinity, a supersaturation with respect to carbonate minerals. SIs for calcite and dolomite reach values above 0.8 and 2.8, respectively, in surface waters. At depth these values decrease rapidly, indicating undersaturation below the chemocline. Due to the high supersaturation in the surface layer, $CaCO_3$ has been extracted by growing stromatolites, thus explaining the low Ca-values. Extraction is done nonenzymatically, as is illustrated by the $\delta\,^{13}C$-values of the reefs which range between -2.1 and -5.8 with a mean of -3.5 ± 1.0‰. Compared to dissolved carbon of the surface layer of -8.4‰ carbonates display the typical fractionation of nonenzymatic calcite precipitation of about +5‰. In detail, the history of the lake water body seems to

**Table 1.** Composition of Satonda Crater Lake water

| Sample No | Depth m | Temp. °C | pH | Eh mV | $O_2$ mg/l | Sal. ‰ | σ T | Na ( | K | Mg meq/kg | Ca | Carb.-Alk ) | Mg/Ca | $pCO_2$ ppmv | Sat.-Index Calc. | Dolo. |
|---|---|---|---|---|---|---|---|---|---|---|---|---|---|---|---|---|
| Satonda Crater Lake, surface waters (0.1 m) | | | | | | | | | | | | | | | | |
| 1 tuff beach | | 33.1 | 8.45 | 357 | 7.33 | 30.78 | 17.51 | 421.1 | 19.13 | 81.16 | 10.41 | 3.66 | 7.80 | 363 | 0.88 | 2.91 |
| 4 reef front | | 31.8 | 8.50 | 403 | 10.31 | 30.69 | 17.91 | 428.8 | 11.37 | 79.09 | 10.94 | 3.49 | 7.23 | 297 | 0.90 | 2.93 |
| 9 lagoon end | | 39.0 | 8.38 | 324 | 7.89 | 35.18 | 18.50 | 491.7 | 12.26 | 90.86 | 12.95 | 3.20 | 7.01 | 369 | 0.88 | 2.88 |
| Satonda Crater Lake, profile | | | | | | | | | | | | | | | | |
| 20 | 2 | 29.9 | 8.43 | 411 | 6.67 | 30.87 | 18.70 | 425.2 | 11.12 | 86.43 | 10.55 | 3.39 | 8.19 | 351 | 0.81 | 2.78 |
| 21 | 10 | 29.9 | 8.42 | 333 | 6.69 | 30.87 | 18.73 | 427.0 | 11.14 | 84.35 | 10.84 | 3.47 | 7.78 | 372 | 0.82 | 2.79 |
| 22 | 20 | 28.6 | 8.33 | 339 | 3.36 | 30.61 | 18.93 | 423.1 | 11.04 | 83.97 | 10.55 | 3.41 | 7.96 | 476 | 0.73 | 2.60 |
| 23 | 22 | 28.3 | 8.29 | 330 | 2.80 | 30.69 | 19.09 | 424.5 | 11.02 | 83.88 | 10.80 | 3.50 | 7.77 | 546 | 0.72 | 2.57 |
| 24 | 24 | 28.7 | 7.31 | - 66 | 0.00 | 34.73 | 21.98 | 478.8 | 12.36 | 96.14 | 12.62 | 5.61 | 7.61 | 10100 | 0.09 | 1.31 |
| 25 | 26 | 29.7 | 7.13 | -138 | $H_2S$ | 36.82 | 23.22 | 510.8 | 13.16 | 99.18 | 12.89 | 6.42 | 7.69 | 17700 | -0.01 | 1.12 |
| 26 | 30 | 29.8 | 7.27 | -123 | $H_2S$ | 36.82 | 23.18 | 509.8 | 13.06 | 100.22 | 12.94 | 6.26 | 7.75 | 12500 | 0.12 | 1.38 |
| 27 | 40 | 29.0 | 7.33 | -136 | $H_2S$ | 36.82 | 23.45 | 510.0 | 13.11 | 98.94 | 13.96 | 6.37 | 7.09 | 10900 | 0.21 | 1.52 |
| 28 | 50 | 29.0 | 7.12 | -143 | $H_2S$ | 36.91 | 23.52 | 511.5 | 13.11 | 99.41 | 13.57 | 7.51 | 7.36 | 21100 | 0.07 | 1.24 |
| 30 | 55 | 29.1 | 6.90 | -210 | $H_2S$ | 39.02 | 25.08 | 534.3 | 14.93 | 111.33 | 13.50 | 32.97 | 8.25 | 154000 | 0.48 | 2.11 |
| 29 | 60 | 29.2 | 6.92 | -192 | $H_2S$ | 40.60 | 26.23 | 561.5 | 14.94 | 110.65 | 14.17 | 37.71 | 7.81 | 167000 | 0.58 | 2.28 |
| 31 | 62 | 28.9 | 6.87 | -224 | $H_2S$ | 41.06 | 26.69 | 566.3 | 14.99 | 114.66 | 13.34 | 48.19 | 8.59 | 239000 | 0.61 | 2.36 |
| For comparison: seawater, Satonda Bay | | | | | | | | | | | | | | | | |
| 0 | 0.2 | 29.5 | 8.27 | - | - | 34.37 | 21.44 | 460.8 | 9.46 | 102.23 | 21.20 | 1.99 | 4.82 | 309 | 0.73 | 2.39 |

Samples above the chemocline were recovered 8.00 to 10.00 h in the morning. Temp., pH, Eh: measured immediately after sample recovery.
Salinity: calculated from conductivity measured immediately after sample recovery.
Na: calculated as charge balance of cations, total cation charge derived from ideal seawater composition as given by Millero, 1974.
K, Mg, Ca: measured by AAS as mg/l and recalculated for meq/kg for density at 20°C.
Alkalinity: Total carbonate alkalinity, calculated from total alkalinity titrations done a few hours after sample recovery and corrected for borate alkalinity at in-situ pH and temperature.
Mg/Ca: mol ratio.
$pCO_2$: in-situ $CO_2$-partial pressure as calculated by WATMIX (Wigley and Plummer 1976).
Saturation Index: as calculated by WATMIX, the saturation index is the log of the ratio of the ion-activity products to the respective solubility products at in-situ temperatures. 0.0 denotes saturation, positive values indicate supersaturation and negative values undersaturation.

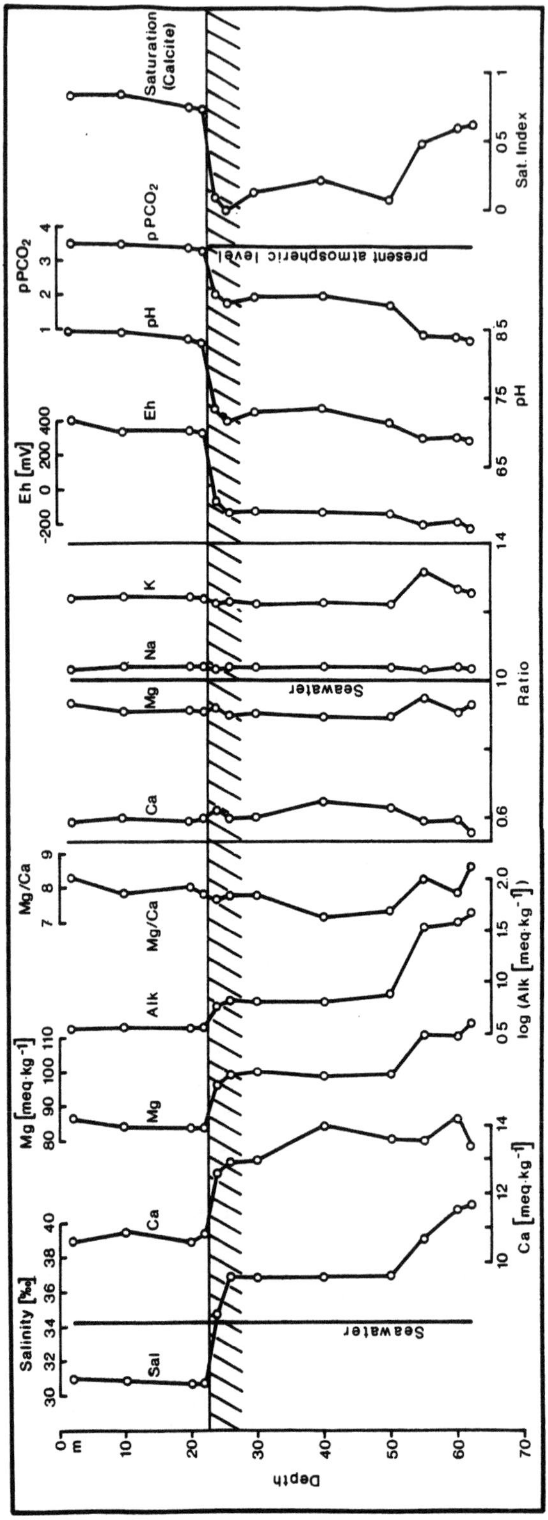


**Fig. 4.** Vertical structure of hydrochemistry in Satonda Crater Lake/Indonesia (horizontal hatched band marks dense manganese oxide layer, below which no light penetrates). $pPCO_2$ denotes negative log of $PCO_2$

be complicated (for more discussion see Kempe and Kazmierczak, in press) but $^{14}C$ dates suggest that the lake was invaded by seawater about 4000 years ago (possibly following the collapse of part of the crater rim) and that reef growth in the quickly alkalized water started soon after. In addition to the samples of the vertical profile, Table 1 lists three samples from the vicinity of the reef at station 1. The highest salinity and temperature was measured in a small lagoon between reef tops where evaporation concentrated the lake water significantly. The highest oxygen value and the lowest $pCO_2$ were measured in front of the reef, just above the blooming green algae. These samples show that physical and biological processes can cause significant deviations of surface waters within the topography of a reef. Measurements of the micro-environments of the mats may show even larger deviations. The calcite supersaturation stays at around SI 0.8 to 0.9 in all samples, i.e., significantly higher than in normal seawater.

## 5.2 Walker Lake

Osborne et al. (1982) give data on the composition of Walker Lake water (Table 2). The salinity of the lake (main salt is NaCl) is five times less than seawater, but it contains ($HCO_3^-$ + $CO_3^{2-}$) ≫ ($Mg^{2+}$ + $Ca^{2+}$) (i.e., 43.0 versus 10.4 meq/l). Lake Walker is thus of the soda lake type, containing appreciable amounts of alkaline ions ($Na^+$ + $K^+$) balanced by carbonates. The pH-values reported were 9.0 in the morning and 9.3 in the early afternoon. The Mg/Ca ratio is about six times higher than in Satonda Crater Lake, and absolute Ca-concentrations are much lower.

Calculation of saturation indices with the WATMIX model yields high supersaturations for calcite and extremely high values for dolomite both in the morning (0.70 and 3.2) and in the afternoon (0.85 and 3.5) (Table 2). Due to photosynthesis, the $pCO_2$ decreases during the day and saturations increase. If reported pH-values are correct, then the lake is not in equilibrium with the atmosphere but experiences higher respirative than photosynthetic activity. This, however, may be an effect of seasons, since in autumn large amounts of algae die and are remineralized in the lake (Osborne et al. 1982).

Since the occurrence of stromatolites is relatively limited, we may assume that supersaturation in the Lake was established only relatively recently, possibly within the last few decades. This is in accordance with the reported increase of salinity in the lake since the last century. Former and higher lake stands of Lake Walker are also marked by stromatolites (Osborne et al. 1982). Thus, high alkalinities must have prevailed in former lakes as well. Primarily the Na-balanced alkalinity derives from weathering of fresh volcanic rocks in the drainage basin of Lake Walker, secondarily older evaporative deposits may be leached at present.

**Table 2.** Composition of Lake Walker waters

| Temp. | pH | Tot.ions | Na | K | Mg | Ca | Carb.-Alk | $SO_4$ | Cl | Mg/Ca | $pCO_2$ | Sat.-Index | |
|---|---|---|---|---|---|---|---|---|---|---|---|---|---|
| °C | | mg/l | ( | | | meq/l | | | ) | | ppmv | Calc. | Dolo. |
| 20 | 9.0 | 8 610 | 132 | 4.1 | 10.2 | 0.21 | 43.08 | 40.2 | 57.0 | 48.6 | 1400 | 0.70 | 3.23 |
| 15 | 9.3 | " | " | " | " | " | " | " | " | " | 570 | 0.85 | 3.54 |

### 5.3 Lake Tanganyika

In case of Lake Tanganyika extensive geochemical data collected during the 1970 WHOI expedition has been published by Degens et al. (1971), Hecky and Degens (1973) and Degens and Ittekkot (1983). The water samples, however, were taken mid-lake from a ship far to the south of the thrombolite sites which are at the Burundi coast. Also no detailed pH values were published, it was just stated that they range from 8.5 to 9.2 (Degens et al. 1971).

Luckily Cohen and Thouin (1987) published concentrations of Ca, Mg, Na, K, and alkalinity measured on samples taken along the shore where the thrombolites occur which compare favorably well with the overall compositional average of the 1970 samples (Table 3). However, they did not measure $SO_4^{2-}$ nor $Cl^-$, concentration data of which have to be taken from Talling and Talling (1965). Again, no in situ pH and temperature measurements were published.

Temperature can easily be guessed from the 1970 data to be around 28 °C. Also the pH value (stated to be mildly alkaline in Cohen and Thouin 1987) can be derived from the data, simply by assuming that the $CO_2$ pressure of the surface waters of Lake Tanganyika is in equilibrium with the atmosphere and then calculating the parameters of the carbonate system with the numerical electrolyte model. Results indicate that at a $pCO_2$ of 340 ppmv, the water would have a pH of 8.95. Altering the $pCO_2$ and temperature slighty within their presumed seasonal ranges will change the pH value by a few hundredth pH units only.

The corresponding calcite and dolomite saturation indices are unexpectedly high (Table 3), i.e., 0.96 and 1.44. Thus we have to conclude (contrary to Degens et al. 1971), that Lake Tanganyika surface waters are highly supersaturated with respect to all carbonate minerals, certainly much higher than average marine surface waters. The reason of this high supersaturation is the rather high alkalinity of Lake Tanganyika of 6.6. meq/l. Even though the total dissolved ions add up to only 576 mg/l (i.e., fresh water), the lake belongs to the soda lake type: the sum of the alkaline earth ions is 4.35 meq/l compared to an alkalinity of 6.6 meq/l. This means that part of the alkalinity is balanced by alkaline ions as was also the case in Walker Lake. If the water of Lake Tanganyika would be subject to prolonged evaporation, a very alkaline brine could be formed (this process has been investigated thermodynamically by Garrels and Mackenzie 1967 and Hardie and Eugster 1970), similar to those observed in some of the terminal lakes along the East-African rift (Eugster and Hardie 1978). The hydrochemical signature of Lake Tanganyika is derived from the composition of Lake Kivu, which feeds Lake Tanganyika via the Rusizi River. Lake Kivu itself is influenced by volcanic hydrothermal activity (Degens 1973; Degens et al. 1973; Degens and Stoffers 1976) which causes reaction of volcanic $CO_2$ with fresh silicates, a process liberating alkaline earth and alkaline ions from fresh volcanic silicates under high $CO_2$ pressures. When these waters outgas their excess $CO_2$ at the surface of Lake Kivu, some of the alkaline earth ions are precipitated and solutions of Na and K and excess alkalinity stay behind and give Lake Tanganyika its high alkalinity.

The WHOI data show that Lake Tanganyika has a pronounced thermocline between 70 and 100 m depth and that anoxia begins below 100 m in the north and below 150 m depth in the south. This implies an increase in $pCO_2$ in the layers below

**Table 3.** Composition of Lake Tanganyika waters

| Temp. °C | pH | Tot. ions mg/l | Na ( | K | Mg | Ca meq/l | $SO_4$ | Cl | Alk ) | $\Sigma CO_2$ mmol/l | Mg/Ca | $pCO_2$ ppmv | Sat.-Index Calc. | Sat.-Index Dolo. |
|---|---|---|---|---|---|---|---|---|---|---|---|---|---|---|
| 28[a] | 8.95[b] | 575.6 | 2.40[c] | 0.73[3] | 3.73[c] | 0.62[c] | 0.10[d] | 0.75[d] | 6.60[c] | | 5.9 | 342[b] | 0.96[b] | 1.44[b] |
| | | | ±0.10[c] | ±0.03[3] | ±0.16[c] | ±0.11[c] | – | – | ±0.10[c] | | | | | |
| | 8.5–9.2[e] | 551 | 2.83[e] | 0.79[e] | 3.45[e] | 0.35[e] | 0.07[e] | 0.62[e] | | 6.24[e] | 9.9 | | | |

[a] Temperature assumed according to Degens and Hecky 1973.
[b] Values calculated (see text).
[c] Data of Cohen and Thouin (1987) for samples taken in 1985, standard deviations based on 26 samples for major ions and 69 of alkalinity along the Burundi shore of Lake Tanganyika.
[d] Data of Talling and Talling (1965) for samples taken in 1961.
[e] Data of Degens et al. (1971) for 54 samples taken in 1970 from various depths of the lake.

the thermocline due to respiration of sinking organic material. Since no pH values are available, it is not possible to calculate these $pCO_2$ values nor the decrease of supersaturation with depth at this time. However, it will not take much $CO_2$ to undo the high supersaturation found at the surface. If one slowly increases the $pCO_2$ for a water of the composition found at a depth of 1000 m in the southern Tanganyika basin to only 1500 ppmv, a pH of 8.3 and a $SI_{Calcite}$ of 0.0 will result. Such a $pCO_2$ (five times the surface $pCO_2$) is not uncommon in stagnating water bodies and may result in eutrophied lakes within a few months during seasonal stratification easily (Kempe 1982). Thus, the high supersaturation of Lake Tanganyika is probably only a surface layer characteristic, limiting authochtonous carbonate precipitation to depths shallower than roughly 80 m. In fact, at depth carbonates probably dissolve as witnessed by relative low concentrations of carbonates or even their absence in the abyssal sediments. About 4000 years ago significant conservation of high-Mg-calcite in sediments began, connected with a major positive shift in the $\delta^{13}C$ of the organic matter in the sediment (Degens et al. 1971). This shift is most likely connected with the sudden increase of water tributary from Lake Kivu, establishing higher salinities in Lake Tanganyika and introducing supersaturation at the surface. Since that time stromatolites may have started to grow along the shore of Lake Tanganyika.

## 5.4 Andros Island

Monty (1972) published water analyses from three stromatolite ponds on Andros Island/Bahamas. Samples were taken in January, i.e., at relatively high waters in the ponds. Original values and the calculated $pCO_2$ and saturation indices are given in Table 4. The waters belong to the calcium-bicarbonate type with low values of other ions, save for some aerosol-derived NaCl. Such water is typical for areas composed of carbonate rocks. Total mineralization is only 180 to 230 mg/l. Since no temperature measurements were given, 20°C is assumed.

None of the waters indicates supersaturation of carbonate minerals at first glance (Table 4). However, Monty did not give any information as to when during the day nor where exactly the samples were taken. Using his pH-values one derives at a $pCO_2$ much higher than atmospheric $pCO_2$, which indicates that either the samples were taken early in the morning when respirative activity in the lake was able to produce an excess of $CO_2$ during the night, or that the pH-measurements were in error. Erroneous analytical results are also indicated by a too large anion equivalent sum compared to total cations. Using the given concentrations nevertheless, one can outgas the water numerically. This is done for Wilmo Lake (second row Table 4). Calcite supersaturation then becomes 0.37. If one further assumes that evaporation will concentrate the water when the ponds dry up by half (i.e., concentrations are doubled) and keeps the water in equilibrium with the atmospheric $pCO_2$ (Table 4, fourth line), then very high supersaturations both with respect to calcite and dolomite are obtained. Also the photosynthetic withdrawal of $CO_2$ will cause supersaturation inside the mat some time after the sun rose and calcification may well begin in the early afternoon due to supersaturation of the mat microenvironment.

**Table 4.** Composition of Andros Island pond waters

| Location | Temp. °C | pH | Tot. ions mg/l | Na ( | K | Mg | Ca meq/l | Carb. -Alk | $SO_4$ | Cl ) | Mg/Ca | $pCO_2$ ppmv | Sat. -Index Calc. | Dolo. |
|---|---|---|---|---|---|---|---|---|---|---|---|---|---|---|
| Wilmo Lake | 20 | 8.0[a] | 231 | 1.25[a] | 0.07[a] | 0.49[a] | 1.49[a] | 1.98[a] | - | 1.61[a] | 0.33 | 1000 | -0.06 | -0.25 |
| degassing | 20 | 8.45 | ” | ” | ” | ” | ” | ” | | ” | ” | 340 | 0.37 | 0.18 |
| evaporation | 20 | 8.0 | 462 | 2.5 | 0.13 | 0.99 | 2.98 | 3.86 | - | 3.22 | 0.33 | 1900 | 0.50 | 0.31 |
| degassing | 20 | 8.7 | ” | ” | ” | ” | ” | ” | - | ” | ” | 350 | 1.14 | 0.95 |
| Worry Lake | 20 | 8.0[a] | 195 | 1.10[a] | 0.05[a] | 0.43[a] | 1.28[a] | 1.55[a] | - | 1.55[a] | 0.33 | 810 | -0.22 | -0.40 |
| Unnamed Lake | 20 | 8.1[a] | 177 | 0.90[a] | 0.05[a] | 0.30[a] | 1.30[a] | 1.52[a] | - | 1.35[a] | 0.23 | 630 | -0.12 | -0.38 |

Temperature assumed
[a] Data of Monty 1972.

The relative $Ca^{2+}$ and $HCO_3^-$ concentrations of Andros pond waters are surprising. They cannot result just from rainwater collecting in depressions. This is because rainwater can – at an atmospheric $pCO_2$ of 340 ppmv – dissolve only about 40 mg$CaCO_3$/l. Thus, exchange with the underlying karstic fresh water lens must occur which holds higher $CaCO_3$-concentrations due to additional supply of $CO_2$ from soil air with seepage water. Alternatively pond waters are interflow waters, i.e., rainwater, which has travelled through voids of the shallow soils. Definitely more data is needed to fully understand the seasonal changes of Bahama pond hydrochemistry. Nevertheless, even with the few data at hand, one can assume that Andros pond stromatolites also form under conditions of high calcite supersaturation.

## 6 Conclusions

We have seen that cyanobacteria mats can precipitate various carbonate minerals over wide ranges of depth, light intensity, salinity, alkalinity, Ca-concentrations, pH and temperature. The common stress in all four discussed cases seems to be an extremely high supersaturation with regard to carbonate minerals. Thus, we tentatively conclude that in situ calcifying stromatolites are caused by high supersaturation of the ambient medium. Today's ocean seems to have a too low supersaturation to allow for stromatolites of the mineralizing type to grow fast enough to be competitive with the biomineralizing organisms.

The enzymatic, i.e., biologically controlled, biocalcification typical for eukaryots has recently been viewed as being originally a product of the $Ca^{2+}$ regulation system which maintains the extremely low $Ca^{2+}$ concentrations ($10^{-7}$ mol) in the cytosol of cells (Simkiss 1977; Kazmierczak et al. 1985; Degens et al. 1986). Cyanobacteria as well as many other prokaryotes can, however, bind calcium on or within their cell walls and external layers (sheaths, glycocalyx). They may even provide $CaCO_3$ nucleation sites on their outer sheaths (Pentecost 1985), and thus keep the inside of their protective envelopes at less critical calcium concentrations. It is known that increased pH values enhance significantly the migration of calcium ions into cells (Roos and Boron 1981). Hence, synergistic action of alkaline conditions and relatively high $Ca^{2+}$ concentrations in the surrounding of the cells (expressed as high $SI_{calcite}$) may stimulate an intensive production of the $Ca^{2+}$-affine extracellular substances by the cyanobacterial cells (e.g., Foerster 1964) and ultimately $CaCO_3$ deposition.

Since we observe that present stromatolites grow in calcite-supersaturated waters and since we observe that in the late Precambrian in situ calcifying stromatolites occurred widely, we may, by analogy, plausibly conclude that oceans at that time experienced a high supersaturation with regard to carbonate minerals. Kempe and Degens (1985) and Kempe et al. (1987) suggested, based on thermodynamic, kinetic and mass balance arguments, that the early ocean must have been alkaline but low in calcium. This is in accordance with the findings of this study, which shows that high alkalinities are coupled with low calcium concentrations. Also, there is no need to assume that ancient stromatolites were formed by other organisms than their contemporary counterparts, i.e., mainly by cyanobacteria (Walter 1976, 1983), or that cyanobacteria had a principally different metabolism in the Precambrian.

The variable post-Precambrian occurrence in marine environments of in situ calcified cyanobacterial stromatolitic communities and particularly their dramatic post-Cretaceous collapse is, to our opinion, much better explicable by long-term changes in $SI_{calcite}$ than by changes in the Mg/Ca ratio alone as suggested by Riding (1982). The presence of a vigorously calcifying cyanobacterial community in Satonda Crater Lake filled with a moderately alkaline seawater of much higher $SI_{calcite}$ than in "normal" seawater shows that relatively small variations in seawater chemistry may have large ecological effects. In a sense, Satonda Lake water is a recreated model of late Precambrian ocean chemistry.

*Dedication and Acknowledgment.* This paper is gratefully dedicated to Egon T. Degens on occasion of his 60th anniversary. The text clearly bears witness to the long-standing and fruitful influence which Degens and his scientific work has on the authors.

The authors acknowledge continued support by the German Ministry of Research and Technology (Förderkennzeichen 03 R 372 7) and by the Polish Academy of Sciences (Project CPBP 04.03). Discussion with Antoni Hoffman (Warszawa), Jürgen Schneider (Göttingen) and technical assistance by Zbigniew Strak was of great help. This paper is a contribution to the IGCP Project 261 "Stromatolites".

## References

Awramik SM (1971) Precambrian columnar stromatolite diversity: Reflection of metazoan appearance. Science 174:825–827

Awramik SM (1982) The pre-Phanerozoic fossil record. In: Holland HD, Schidlowski M (eds) Mineral deposits and the evolution of the biosphere. Springer, Berlin Heidelberg New York, pp 67–82

Awramik SM (1984) Ancient stromatolites and microbial mats. In: Cohen Y, Castenholz RW, Halvorson HO (eds) Microbial mats: Stromatolites. Alan R Liss, New York, pp 1–22

Borch CC von der, Bolton B, Warren JK (1977) Environmental setting and microstructure of subfossil lithified stromatolites associated with evaporites, Marion Lake, South Australia. Sedimentology 24:693–708

Burne RV, Moore LS (1987) Microbialites: Organosedimentary deposits of benthic microbial communities. Palaios 2:241–254

Cohen AS, Thouin C (1987) Nearshore carbonate deposits in Lake Tanganyika. Geology 15:414–418

Degens ET (1973) Accounting for the salts in the sea. Nature 243:504–507

Degens ET (1979) Why do organisms calcify? Chem Geol 25:257–269

Degens ET, Ittekkot V (1983) Dissolved organic matter in Lake Tanganyika and Lake Baikal -a brief survey-. In:Degens ET, Kempe S, Soliman H (eds) Transport of carbon and minerals in major world rivers, part 2. Mitt Geol-Paläont Inst Univ Hamburg, SCOPE/UNEP Sonderbd 55:129–143

Degens ET, Stoffers P (1976) Stratified waters as a key to the past. Nature 263:22–27

Degens ET, Herzen RP von, Wong HK (1971) Lake Tanganyika: Water chemistry, sediments, geological structure. Naturwissenschaften 58:229–240

Degens ET, Herzen RP von, Wong HK, Deuser WG, Jannasch HW (1973) Lake Kivu: Structure, chemistry and biology of an East African rift lake. Geol Rundsch 62:245–277

Degens ET, Kempe S, Spitzy A (1984) Carbon dioxide: a biogeochemical portrait. In: Hutzinger O (ed) The handbook of environmental chemistry, vol 1/C. Springer, Berlin Heidelberg New York, pp 127–215

Degens ET, Kazmierczak J, Ittekkot V (1986) Cellular response to $Ca^{2+}$ stress and its geological implications. Acta Palaeontol Pol 30:115–135 (for 1985)

Dill RF, Shinn EA, Jones AT, Kelly K, Steinen RP (1986) Giant subtidal stromatolites forming in normal salinity waters. Nature 324:55–58
Dravis JJ (1983) Hardened subtidal stromatolites, Bahamas. Science 219:385–386
Dreybrodt W (1980) Deposition of calcite from thin films of natural calcareous solutions and the growth of speleothems. Chem Geol 29:89–105
Dreybrodt W (1981) Kinetics of the dissolution of calcite and its application to karstification. Chem Geol 31:245–269
Emeis K, Richnow H-H, Kempe S (1987) Travertine formation in Plitvice National Park/Yugoslavia: Chemical vs biological control. Sedimentology 34:595–609
Eugster HP, Hardie LA (1978) Saline lakes. In: Lerman A (ed) Lakes – chemistry, geology, physics. Springer, Berlin Heidelberg New York, pp 237–293
Foerster JW (1964) The use of calcium and magnesium ions to stimulate sheaths formation in *Oscillatoria limosa* (Roth), Agardh, CA. Trans Am Microsc Soc 83:420–427
Garrels RM, Mackenzie FT (1967) Origin of the chemical composition of some springs and lakes. In: Equilibrium concepts in natural waters. Am Chem Soc, Advances in Chemistry 67:222–242
Garrett P (1970) Phanerozoic stromatolites: Non-competitive ecologic restriction by grazing and burrowing animals. Science 169:171–173
Gebelein CD (1969) Distribution, morphology and accretion rate of recent subtidal algal stromatólites, Bermuda, J Sediment Petrol 39:49–69
Gebelein CD (1974) Biologic control of stromatolite microstructure: implications for Precambrian time stratigraphy. Am J Sci 274:575–598
Golubic S (1973) The relationship between blue-green algae and carbonate deposits. In: Carr NG, Whitton BA (eds) The biology of blue-green algae. Blackwell, Oxford, pp 434–472
Gomes NA de NC (1985) Modern stromatolites in a karst structure from the Malmani subgroup, Transvaal sequence, South Africa. Trans Geol Soc S Afr 88:1–9
Hardie LA, Eugster PH (1970) The evolution of closed-basin brines. Miner Soc Am, Spec Publ 3:273–290
Hecky RE, Degens ET (1973) Late Pleistocene-Holocene Chemical Stratigraphy and Paleolimnology of the Rift Valley Lakes of Central Africa. Woods Hole Oceanogr Inst Tech Rep 73-28:114
Hecky RE, Fee EJ (1981) Primary production and rates of algal growth in Lake Tanganyika. Limnol Oceanogr 26:532–547
Hecky RE, Kling HJ (1981) The phytoplankton and protozooplankton of the euphotic zone of Lake Tanganyika: Species composition, biomass, chlorophyll content and spatiotemporal distribution. Limnol Oceanogr 26:548–564
Herman JS, Lorah MM (1987) $CO_2$ outgassing and calcite precipitation in Falling Spring Creek, Virginia, U.S.A.. Chem Geol 62:251–262
Hofman HJ (1969) Stromatolites from the Proterozoic Animikie and Sibley Groups, Ontario. Geol Surv Can Pap pp 68–69, 77
Hofmann HJ (1973) Stromatolites: Characteristics and utility. Earth-Sci Rev 9:339–373
Hofmann HJ (1977) On Aphebian stromatolites and Riphean stromatolite stratigraphy. Precambrian Res 5:175–205
Hofmann HJ, Snyder GL (1985) Archean stromatolites from Hartville Uplift, eastern Wyoming. Geol Soc Am Bull 96:842–849
Hofmann HJ, Thurnston PC, Wallace H (1985) Archean stromatolite from Uchi Greenstone Belt, northwestern Ontario. In: Ayres LD, Thurnston PC, Card KD, Weber W (eds) Evolution of Archean supracrustal sequences. Geol Assoc Can Spec Pap 28:125–132
Horodyski RJ, Haar SP Von der (1975) Recent calcareous stromatolites from Laguna Mormona (Baja California) Mexico. J Sediment Petrol 45:894–906
Kazmierczak J (1976) Cyanophycean nature of stromatoporoids. Nature 264:49–51
Kazmierczak J, Krumbein WE (1983) Identification of calcified coccoid cyanobacteria forming stromatoporoid stromatolites. Lethaia 16:207–215
Kazmierczak J, Ittekkot V, Degens ET (1985) Biocalcification through time: environmental challenge and cellular response. Paläontol Z 59:15–33
Kelts K, Hsü KJ (1978) Freshwater carbonate sedimentation. In: Lerman A (ed) Lakes – chemistry, geology, physics. Springer, Berlin Heidelberg New York, pp 295–323
Kempe S (1975) A computer program for hydrothermal problems in karstic water. Ann Spéléol 30:699–702

Kempe S (1977) Hydrographie, Warven-Chronologie und organische Geochemie des Van Sees, Ost-Türkei. Thesis, Mitt Geol-Paläont Inst Univ Hamburg 47:125–228

Kempe S (1982) Long-term records of $CO_2$-pressure fluctuations in fresh waters. Habilitationsschrift, In: Degens ET (ed) Transport of carbon and minerals in major world rivers, Part 1. Mitt Geol-Paläont Inst Univ Hamburg, SCOPE/UNEP Sonderbd 52:91–332

Kempe S, Degens ET (1985) An early soda ocean? Chem Geol 53:95–108

Kempe S, Emeis K (1985) Carbonate chemistry and the formation of Plitvice Lakes. In: Degens ET, Kempe S, Herrera R (eds) Transport of carbon and minerals in major world rivers, Part 3. Mitt Geol-Paläont Inst Univ Hamburg, SCOPE/UNEP Sonderbd 58:351–383

Kempe S, Kaźmierczak J (1990) Chemistry and stromatolites of the sea-linked Satruda Crater Lake, Indonesia: a recent model for the Precambrian sea? Chem Geol (in press)

Kempe S, Kaźmierczak J, Degens ET (1989) The soda ocean concept and its bearing on biotic and crustal evolution. In: Crick R E (ed) Origin, evolution and modern aspects of biomineralization in plants and animals, Plenum Press, New York, pp 29-43

Kennard JM, James NP (1986) Thrombolites and stromatolites: Two distinct types of microbial structures. Palaios 1:492–503

Korolyuk IK (1963) Stromatolites of the Upper Precambrian. In: Keller BM (ed) Upper Precambrian stratigraphy of USSR, Pt 2. Gosgeoltekhizdat, Moscow, pp 479–498 (in Russian)

Krumbein WE (1979) Calcification by bacteria and algae. In: Trudinger PA, Swain DJ (eds) Biogeochemical cycling of mineral-forming elements. Elsevier, Amsterdam, pp 47–68

Krumbein WE, Cohen Y (1977) Primary production, mat formation and lithification: contribution of oxygenic and facultative anoxygenic cyanobacteria. In: Flügel E (ed) Fossil algae. Springer, Berlin Heidelberg New York, pp 37–56

Krylov IN (1963) Columnar branching stromatolites of Riphean beds of the southern Urals and their significance for the stratigraphy of the Upper Precambrian. Trudy Geol Inst Akad Nauk SSSR 69:133 (in Russian)

Krylov IN, Shapalova IG, Kolosov PN, Fedonkin AM (1971) Riphean deposits of the lower part of Lena River. Sov Geol 7:85–95 (in Russian)

Langmuir D (1971) The geochemistry of some carbonate ground waters in Central Pennsylvania. Geochim Cosmochim Acta 35:1023–1045

Logan BW (1961) Cryptozoon and associated stromatolites from the Recent, Shark Bay, Western Australia. J Geol 69:517–533

Logan BW, Read F, Hagan GM, Hoffmann D, Brown RG, Woods PJ, Gebelein CD (eds) (1974) Evolution and diagenesis of Quaternary carbonate sequences, Shark Bay, Western Australia. Am Assoc Petrol Geol Mem 22:358

Lyons WB, Long DT, Hines ME, Gaudette HE, Armstrong PB (1984) Calcification of cyanobacterial mats in Solar Lake, Sinai. Geology 12:623–626

Monty CLV (1972) Recent algal stromatolitic deposits, Andros Island, Bahamas, preliminary report. Geol Rundsch 61:742–783

Monty CLV (1973) Precambrian background and Phanerozoic history of stromatolitic communities, an overview. Ann Soc Geol Belg 96:585–624

Müller G, Irion G, Foerstner U (1972) Formation and diagenesis of inorganic Ca-Mg-carbonates in the lacustrine environment. Naturwissenschaften 59:158–164

Neumann AC, Gebelein CD, Scoffin TP (1970) The composition, structure and erodability of subtidal mats, Abaco, Bahamas. J Sediment Petrol 40:274–297

Osborne RH, Licari GR, Link MH (1982) Modern lacustrine stromatolites, Walker Lake, Nevada. Sediment Geol 32:39–61

Pentecost A (1985) Association of cyanobacteria with tufa deposits: identity, enumeration and nature of the sheath material revealed by histochemistry. Geomicrobiol J 4:285–298

Pentecost A, Riding R (1986) Calcification in cyanobacteria. In: Leadbeater BSC, Riding R (eds) Biomineralization in lower plants and animals. Syst Assoc Spec 30:73–90, Oxford Univ Press, Oxford

Pratt BR (1982) Stromatolite decline – a reconsideration. Geology 10:512–515

Pytkowicz RM (1983) Equilibria, nonequilibria, and natural waters, Vol I, II. Wiley, New York, 353 pp

Riding R (1982) Cyanophyte calcification and changes in ocean chemistry. Nature 299:814–815

Roos A, Boron WF (1981) Intracellular pH. Physiol Rev 61:296–434

Schneider J, Schröder HG, Campion-Alsumard T Le (1983) Algal micro-reefs – coated grains from freshwater environments. In: Peryt TM (ed) Coated grains. Springer, Berlin Heidelberg New York, pp 284–298

Schopf JW, Oehler DZ, Horodyski RJ, Kvenfolden KA (1971) Biogenecity and significance of the oldest known stromatolites. J Paleontol 45:477–485
Schröder HG, Windolph H, Schneider J (1983) Bilanzierung der biogenen Karbonatproduktion eines oligotrophen Sees (Attersee, Salzkammergut-Österreich). Arch Hydrobiol 97:356–372
Semikhatov MA, Gebelein CD, Cloud P, Awramik SM, Benmore WC (1979) Stromatolite morphogenesis – progress and problems. Can J Earth Sci 16:992–1015
Serebryakov SN, Semikhatov MA (1974) Riphean and Recent stromatolites: a comparison. Am J Sci 274:556–574
Simkiss K (1977) Biomineralization and detoxification. Calcif Tissue Res 24:199–200
Simkiss K (1986) The process of biomineralization in lower plants and animals – an overview. In: Leadbeater BSC, Riding R (eds) Biomineralization in lower plants and animals. Syst Assoc Spec 30:19–36, Oxford Univ Press, Oxford
Stoffers P, Hecky RE (1978) Late Pleistocene-Holocene evolution of the Kivu-Tanganyika Basin. Spec Publs Int Ass Sediment 2:43–55
Talling JF, Talling IB (1965) The chemical composition of African lake waters. Int Rev Hydrobiol 50:421–463
Walter MR (1972) Stromatolites and the biostratigraphy of the Australian Precambrian and Cambrian. Palaeontol Assoc Lond Spec Pap Palaeontol 11:190
Walter MR (ed) (1976) Stromatolites. Elsevier, Amsterdam, 790 pp
Walter MR (1983) Archean stromatolites: evidence of the Earth's earliest benthos. In: Schopf JW (ed) Earth's earliest biosphere. Princeton Univ Press, Princeton, pp 187–213
Walter MR, Heys GR (1985) Links between the rise of the metazoa and the decline of stromatolites. Precambrian Res 29:149–174
Walter MR, Golubic S, Preiss WV (1973) Recent stromatolites from hydromagnesite and aragonite depositing lakes near the Coorong Lagoon, South Australia. J Sediment Petrol 43:1021–1030
Wigley TML (1971) Ion pairing and water quality measurements. Can J Earth Sci 8:468–476
Wigley TML (1973a) Chemical solution of the system calcite-gypsum-water. Can J Earth Sci 10:306–315
Wigley TML (1973b) The incongruent solution of dolomite. Geochim Cosmochim Acta 37:1397–1404
Wigley TML, Plummer N (1976) Mixing of carbonate waters. Geochim Cosmochim Acta 40:989–995

# Pleistocene/Upper Pliocene Sapropels in the Tyrrhenian Sea

K.-C. Emeis and ODP Leg 107 Scientific Party

## 1 Introduction

A most significant finding of the ODP Leg 107 drilling campaign was the recovery of at least 56 distinct sapropel intervals in upper Pliocene to Pleistocene sediments of six sites drilled in the Tyrrhenian Sea (Fig. 1). Except for 3 reports of disturbed organic-rich sediment – recovered in Core 201 of the Swedish Deep-Sea Expedition (Olausson 1961), in Core 2R-1, 107 cm of Site 373 (Leg 13 DSDP; Ryan

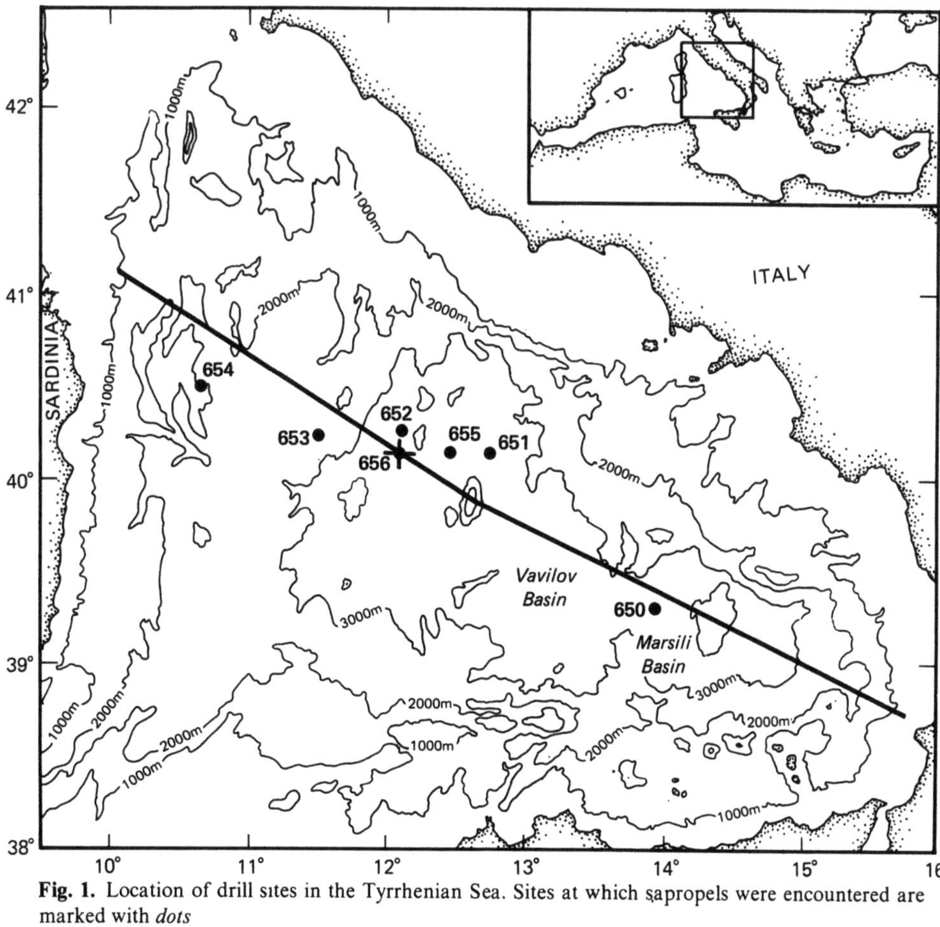


**Fig. 1.** Location of drill sites in the Tyrrhenian Sea. Sites at which sapropels were encountered are marked with *dots*

Hsü et al. 1973) and at Site 373, Core 1-2, 0-5 cm of DSDP Leg 42A (Kidd et al. 1978) – sapropels had previously only been described from the eastern Mediterranean (Olausson 1961; Kidd et al. 1978; Cita and Grignani 1982; Thunnell et al. 1984; Anastasakis and Stanley 1984; ten Haven 1986; and many others) and the Black Sea (Degens and Ross 1972; 1974). Scientific deep-sea drilling in the Tyrrhenian Sea during DSDP Legs 13 and 42A apparently missed most of these deposits due to spot coring and rotary drilling techniques; high sedimentation rates may have precluded recovery by conventional gravity coring devices.

The recovery of multiple layers of sapropels and sapropelic sediments in the Tyrrhenian Sea demonstrates that oceanographic conditions conducive to sapropel formation were not confined to the Black Sea and eastern Mediterranean, but occurred sporadically and possibly simultaneously in the entire Mediterranean during the Pliocene and Pleistocene. In the light of this finding, previous models of sapropel genesis may need reconsideration. In this paper, we present some initial data on the Tyrrhenian sapropels and suggest some implications of their massive occurrence in the western Mediterranean realm. We end by outlining possible causes for deposition of sapropels in an attempt to revive the interest in sapropels and their paleoceanographic significance.

## 2 Occurrence and Description of Sapropels in Leg 107 Cores

Table 1 lists those sapropel and sapropelic intervals recovered during Leg 107 that were verified by measuring the organic carbon content. Figure 1 shows the site locations of Leg 107. According to the classification of Kidd et al. (1978), sapropelic sediments contain 0.5 to 2% $C_{org}$, while sapropels contain in excess of 2% $C_{org}$. In Table 1, we did not include organic carbon-rich sediments of Messinian and older ages, because the term sapropel *sensu stricto* should in our opinion be used only in connection with organic-rich sediments deposited in an open marine sedimentary environment. This environment was established in the Tyrrhenian Sea at the base of the Pliocene section after transgression on to the Messinian strata.

A preliminary analysis of ages for the sapropels and sapropelic sediments recovered during Leg 107 is given in Table 1 and is depicted in the lithologic columns of Fig. 2. These ages are based on sedimentation rate estimates extrapolated from shipboard and shorebased stratigraphic work (Müller, in press; Channell, in press). Attempts to date two of the sapropels (107-655A-2H-4, and -3H-4) by radiometric techniques failed because their age exceeds the limits of the $Th^{230}/U^{234}$ method used (Mangini, pers. comm).

We may assume that published accounts of eastern Mediterranean sapropels and their assigned ages are biased towards sapropel intervals of late Pleistocene age, because their occurrence is mainly documented by gravity-coring with limited penetration. Upper Pleistocene sapropels corresponding to the $S_1$-$S_{12}$ sequence of Ryan (1972) appear to be missing in the Tyrrhenian Sea, and at our present state of knowledge we cannot conceive a way to conclusively correlate the Tyrrhenian occurrences to older sapropels described from the eastern Mediterranean by Kidd et al. (1978).

**Table 1.** Occurrence of sapropels and sapropelic sediments/ ODP Leg 107

| Hole | Core | Section | Interval* | Depth** | $C_{org}$ % | Age*** |
|---|---|---|---|---|---|---|
| 650 | 3 | 2 | 71- 72 | 14.9 | 0.94 | 0.027 |
| | 5 | 4 | 80- 81 | 38.1 | 1.45 | 0.07 |
| | 5 | 6 | 78- 79 | 41.1 | 0.74 | 0.076 |
| | 17 | 1 | 99-101 | 152.3 | 0.78 | 0.28 |
| | 18 | 1 | 117-119 | 161.7 | 0.79 | 0.30 |
| | 19 | 1 | 14- 15 | 170.3 | 0.58 | 0.31 |
| | 21 | 2 | 23- 25 | 191.2 | 1.18 | 0.35 |
| | 21 | 2 | 93- 96 | 191.9 | 0.86 | 0.35 |
| | 22 | 1 | 112-115 | 200.3 | 0.79 | 0.366 |
| | 47 | cc | 28- 30 | 431.0 | 0.74 | 0.85 |
| | 50 | 4 | 138-141 | 456.3 | 1.91 | 0.98 |

Base NN21 (0.275 Ma) at 147.88 mbsf, Brunhes/Matuyama (0.73 Ma) at 408.66 mbsf Base NN19 at 579.15 mbsf.

| Hole | Core | Section | Interval* | Depth** | $C_{org}$ % | Age*** |
|---|---|---|---|---|---|---|
| 651 | 18 | 5 | 134-138 | 164.0 | 1.30 | 0.386 |
| | 27 | 2 | 8- 10 | 243.6 | 0.56 | 0.574 |
| | 27 | 2 | 68- 70 | 244.2 | 1.97 | 0.58 |
| | 28 | cc | 1- 2 | 260.7 | 1.96 | 0.62 |
| | 34 | 1 | 53- 55 | 309.8 | 1.68 | 0.74 |
| | 35 | 1 | 6- 8 | 319.0 | 2.88 | 1.00 |
| | 36 | 1 | 54- 56 | 329.1 | 4.16 | 1.28 |
| | 36 | 6 | 82- 85 | 336.9 | 1.78 | 1.50 |
| | 37 | 2 | 107-110 | 340.8 | 2.01 | 1.61 |
| | 37 | 3 | 36- 38 | 341.6 | 1.74 | 1.63 |
| | 37 | 4 | 140-142 | 345.6 | 2.94 | 1.74 |

Brunhes/Matuyama (0.73 Ma) at 308.48 mbsf, top of Olduvai (1.63 Ma) at 341.55 mbsf.

| Hole | Core | Section | Interval* | Depth** | $C_{org}$ % | Age*** |
|---|---|---|---|---|---|---|
| 652 | 3 | 2 | 1- 2 | 18.5 | 1.91 | 0.374 |
| | 4 | 5 | 114-130 | 34.1 | 3.36 | 0.69 |
| | 4 | 6 | 17- 20 | 34.4 | 2.31 | 0.70 |
| | 6 | 1 | 45- 47 | 46.3 | 4.18 | 1.03 |
| | 7 | 4 | 30- 32 | 60.2 | 0.96 | 1.43 |
| | 8 | 1 | 113-115 | 66.0 | 3.30 | 1.60 |
| | 9 | 2 | 37- 39 | 76.4 | 2.77 | 1.90 |
| | 9 | 2 | 96- 99 | 77.0 | 3.11 | 1.91 |

Brunhes/Matuyama (0.73 Ma) at 36.05 mbsf, base NN19 (1.63 Ma) at 67.1 mbsf.

| Hole | Core | Section | Interval* | Depth** | $C_{org}$ % | Age*** |
|---|---|---|---|---|---|---|
| 653B | 1 | 4 | 39- 46 | 4.9 | 1.12 | 0.10 |
| | 5 | 5 | 142-1448 | 44.0 | 0.61 | 0.69 |
| | 6 | 5 | 58- 66 | 52.6 | 0.57 | 0.89 |
| | 7 | 1 | 75- 76 | 56.1 | 2.67 | 0.97 |
| | 7 | 2 | 71- 73 | 57.2 | 2.30 | 1.00 |
| | 7 | 3 | 143-149 | 59.4 | 0.57 | 1.05 |
| | 7 | 4 | 29- 31 | 60.2 | 4.20 | 1.07 |
| | 7 | 5 | 63- 65 | 62.0 | 2.62 | 1.11 |
| | 7 | 6 | 113-114 | 64.0 | 0.74 | 1.15 |
| | 8 | 1 | 4- 6 | 64.9 | 2.70 | 1.18 |
| | 8 | 1 | 8- 10 | 65.0 | 3.39 | 1.18 |

**Table 1.** *(Continued)*

| Hole | Core | Section | Interval* | Depth** | $C_{org}$% | Age*** |
|---|---|---|---|---|---|---|
| | 8 | 2 | 50– 56 | 66.9 | 0.65 | 1.22 |
| | 9 | 2 | 120–126 | 77.1 | 0.69 | 1.46 |
| | 9 | 6 | 4– 7 | 81.9 | 1.73 | 1.57 |

Biostratigraphic boundaries of Hole 653A were used. Base NN21 (0.275 Ma) at 13.74, base of NN20 (0.474 Ma) at 34.8 mbsf, base of NN19 (1.63 Ma) at 84.4 mbsf.

| | | | | | | |
|---|---|---|---|---|---|---|
| 654 | 5 | cc | 5– 7 | 40.2 | 0.93 | 0.89 |
| | 6 | 4 | 133–135 | 46.1 | 1.88 | 1.02 |
| | 6 | 4 | 142–144 | 46.2 | 3.68 | 1.03 |
| | 6 | 5 | 15– 17 | 46.5 | 1.35 | 1.03 |
| | 8 | 2 | 49– 50 | 61.1 | 0.97 | 1.37 |
| | 8 | 4 | 42– 43 | 64.2 | 1.70 | 1.44 |
| | 10 | 1 | 78– 80 | 79.8 | 0.66 | 1.87 |
| | 13 | 1 | 100–103 | 109.0 | 0.61 | 2.87 |
| | 19 | 3 | 23– 26 | 169.0 | 0.68 | 4.40 |
| | 21 | 5 | 20– 22 | 190.8 | 0.75 | 4.87 |

Base NN20 (0.474 Ma) at 21.9 mbsf, base of NN19 (1.63 Ma) at 72.6 mbsf, Base NN16 (3.5 Ma) at 127.4 mbsf, top and base of NN14 (3.7 and 4.6 Ma) at 151.1 and 178.3 mbsf, respectively.

| | | | | | | |
|---|---|---|---|---|---|---|
| 655 | 2 | 4 | 63– 78 | 8.2 | 3.51 | 1.01 |
| | 3 | 1 | 52– 76 | 13.3 | 5.69 | 1.50 |
| | 3 | 4 | 67– 75 | 17.45 | 3.60 | 2.03 |

Base of NN20 (0.474 Ma) at 2.7 mbsf, base of NN19 (1.63 Ma) at 14.6 mbsf, base of NN17 (2.5 Ma) at 20.7 mbsf.

*Intervals (in cm) of sample on which $C_{org}$ was determined; **Depth below seafloor in meters; ***age in million years computed from datum levels of Channell et al. (in press) and Müller (in press). Underlined are sapropels *sensu stricto*, i.e. $C_{org}$ exceeds 2% by weight.

In spite of the limitation of our age determinations for the occurrences described here, it appears that sapropels in the Tyrrhenian Sea were deposited in certain time intervals. According to our age estimates (Fig. 2), most sapropels *s.s.* were deposited in three periods between 0.2 and 0.4, 1.0 and 1.2, and 1.4 to 1.6 Ma, respectively (Figure 3 and Table 1). It is intriguing that these periods of sapropel accumulation appear to coincide with high sedimentation rates in both the basins and on the Sardinian margin and with tectonic phases recognized on land. The occurrence of sapropels around the Pliocene/Pleistocene boundary is noteworthy, because multiple "sapropel" layers were described from the biostratigraphic stratotype of the Vrica section in Calabria (Selli et al. 1977).

Figures 4 to 7 depict examples of typical sapropels encountered in Holes 655A and 652A. Figure 4 shows a sequence from Hole 655A-2H-4, 50–80 cm (8.2 m below seafloor): sandwiched between light grey (2.5Y7/2 in the U.S.G.S. Rock Color Chart) marly nannofossil ooze, the sapropel interval of 5 cm thickness is characterized by laminated greenish-grey (5GY5/1) calcareous mud to black (5Y2.5/1), finely laminated and pyrite-bearing mud. Organic carbon concentrations increase

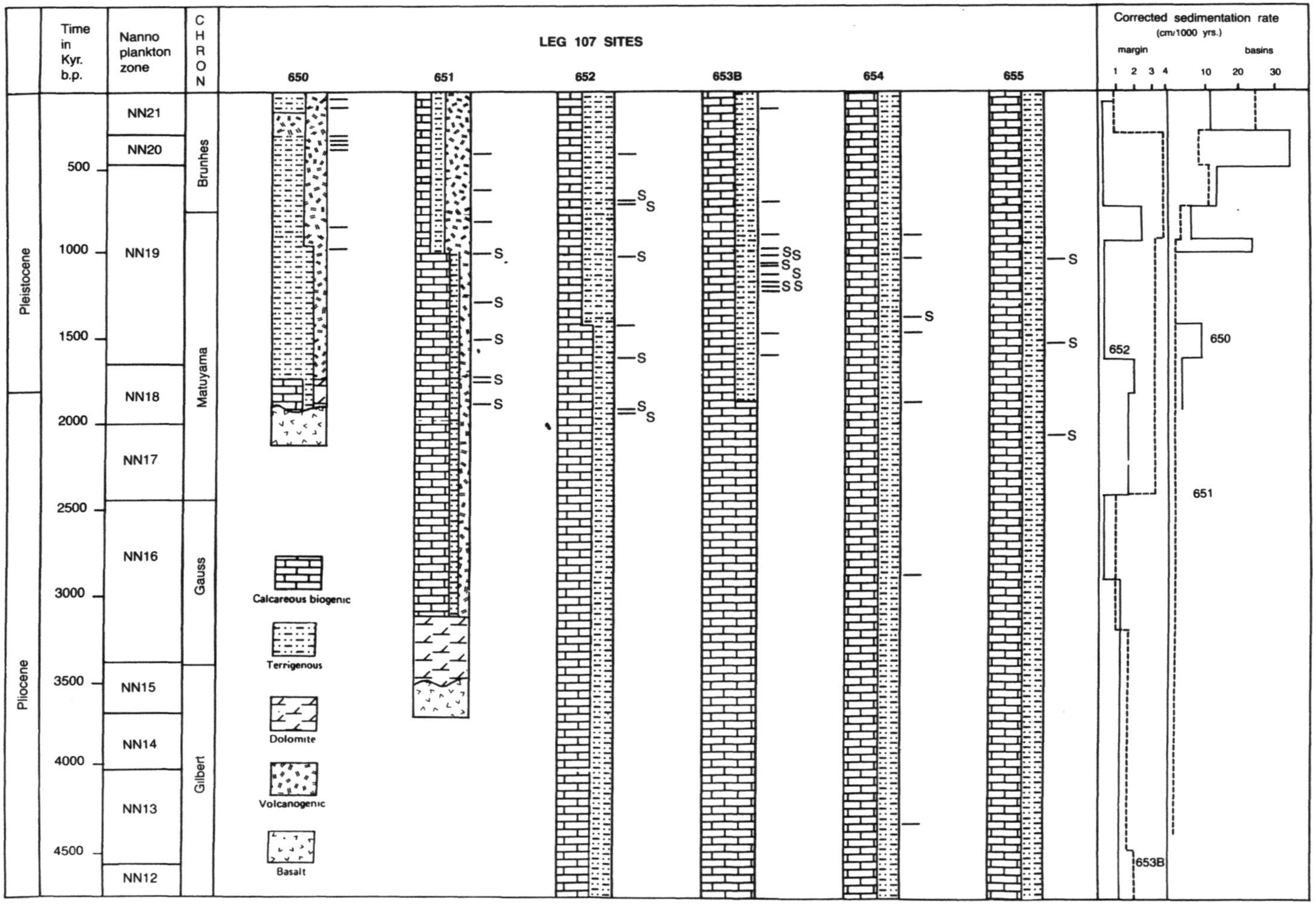


**Fig. 2.** Synthesis diagram of lithologies and preliminary age estimates of sapropels (denoted *s* on the right of a black line) and sapropelic intervals (*black lines*) for six holes drilled in a transect across the Tyrrhenian Sea. On the right, sedimentation rates corrected for compaction are given for holes in pelagic basins (650 and 651) with extremely high sedimentation rates, and for sites on the margin of Sardinia

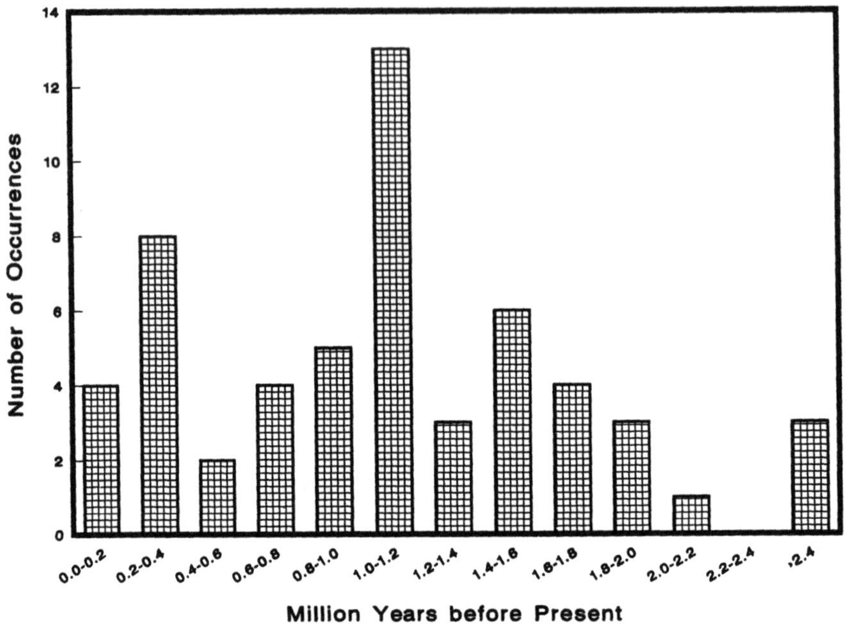


**Fig. 3.** Histogram of sapropel occurrences in cores from ODP Leg 107 in the Tyrrhenian Sea. Occurrences are plotted in increments of 0.2 m.y. from the data in Table 1

from less than 0.4% in the underlying nannofossil ooze to 5.3% at the immediate top of the black sapropel, and drop to less than 0.4% in the overlying calcareous mud. A well-defined green and brown to black redox-front extends over 6 cm into the overlying lithology. This chemical boundary coincides with a marked increase in bioturbation. In the interval three centimeters below and extending to the first centimeter of the actual sapropel, the number of foraminiferal tests increases dramatically.

Carbonate values in the sapropel drop from 51% in the underlying foraminifer-nannofossil ooze to less than 15% in the sapropel proper; the low value coincides with the highest $C_{org}$ value of 5.3%. Only gradually the $CaCO_3$ content increases to 30%. In Fig. 4, faint laminations in the upper part of the sapropel (around 65 cm) indicate that benthic activity was greatly impeded and that bioturbation did not destroy primary sedimentary features. The zone between the top of the sapropel at 64 cm and the redox front at 59 cm is a remarkable piece of evidence for depositional conditions. Organic carbon values in this interval are low (around 0.2%, see Table 2), and carbonate values are rather low as well. The preservation of primary sedimentary structures and laminations in this zone implies that the sediments were deposited under anoxic conditions as well, when benthic stirring and bioturbation was impaired. A second interpretation, which follows the model of Wilson et al. (1986) for the formation of metal-rich oxidation fronts in pelagic sediments, is that the laminations between the redox front and the sapropel mark previous pauses during the progression of the front towards the sediment-water interface. Some remaining disseminated metal still leaves the

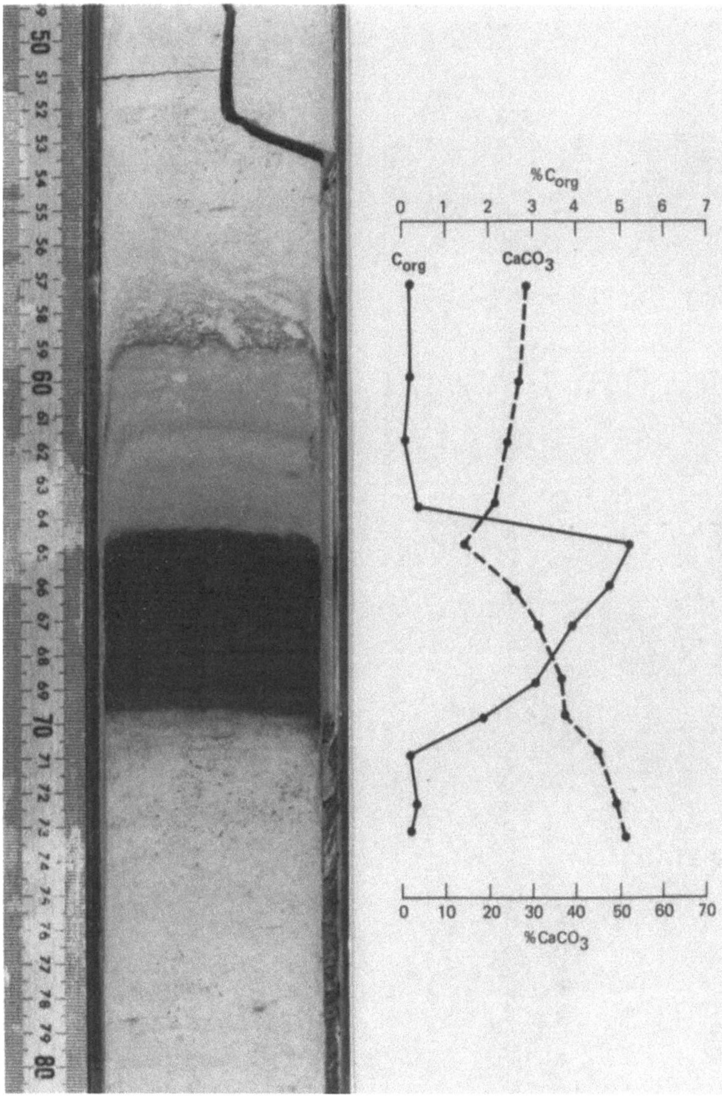


**Fig. 4.** Sapropel sequence 655-2H-4, 50 to 80 cm (8.1 to 8.4 mbsf). Note the abundance of foraminiferal tests at the base of the sapropel and the redox-front extending 10 cm into overlying sediments. The sapropel displays microlaminations in the most organic carbon-rich top centimetre. On the right, concentration of calcium carbonate and organic carbon are plotted at the sampling site in the sapropel. Note also the abrupt drop in organic carbon at the top of the sapropel and the continuously low $CaCO_3$ values in the redox-zone. Bioturbation is evident around the redox front, but did not destroy the laminated mud overlying the sapropel. See text for discussion

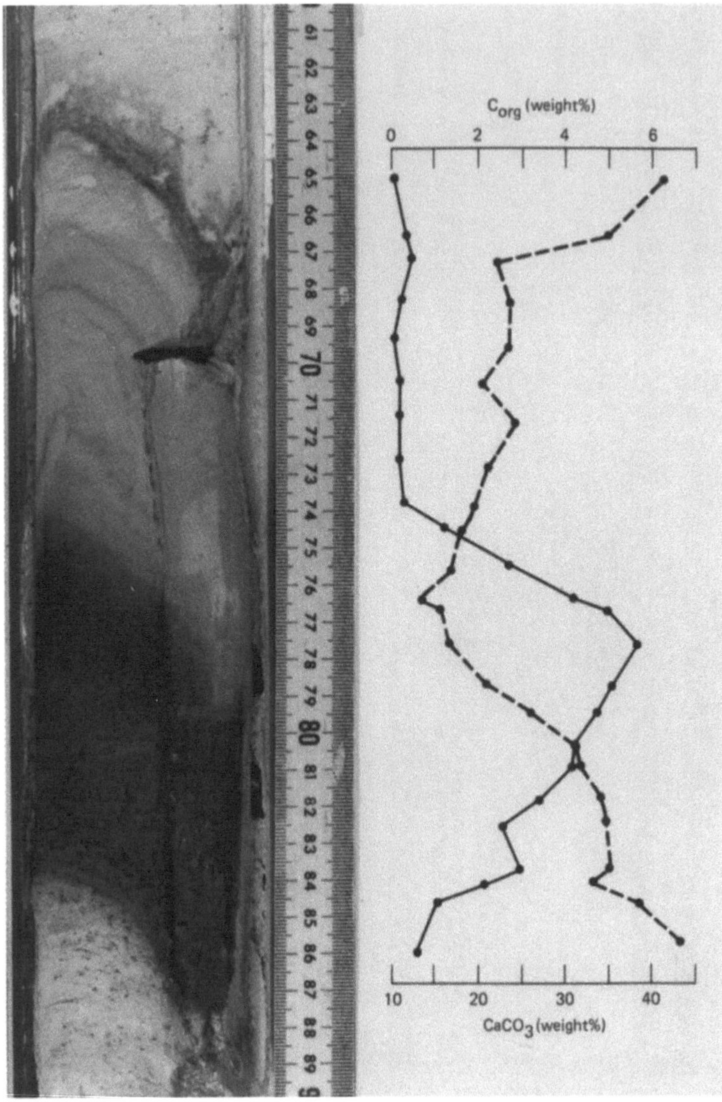


**Fig. 5.** Sapropel sequence recovered in Core 655A-3H-1, at 60 to 90 cm. Even though the sequence is slightly disturbed by drilling, the facies patterns as outlined for sapropel 107-655A-2H-4, 60–80 cm is evident in this sample. Note the abundance of foraminifers at the transition from nannofossil marl to sapropel and the well-developed redox aura extending into the overlying sediments. Again, $CaCO_3$ and $C_{org}$ data are plotted for individual samples of a high-resolution transect across the sequence. Bioturbation did not extend to the depth of the sapropel, even though traces of bioturbation are evident at the top of the redox front. Carbonate concentration declines within the sapropel and the redox zone, but reaches values equivalent to those of the underlying sediments beyond the redox front

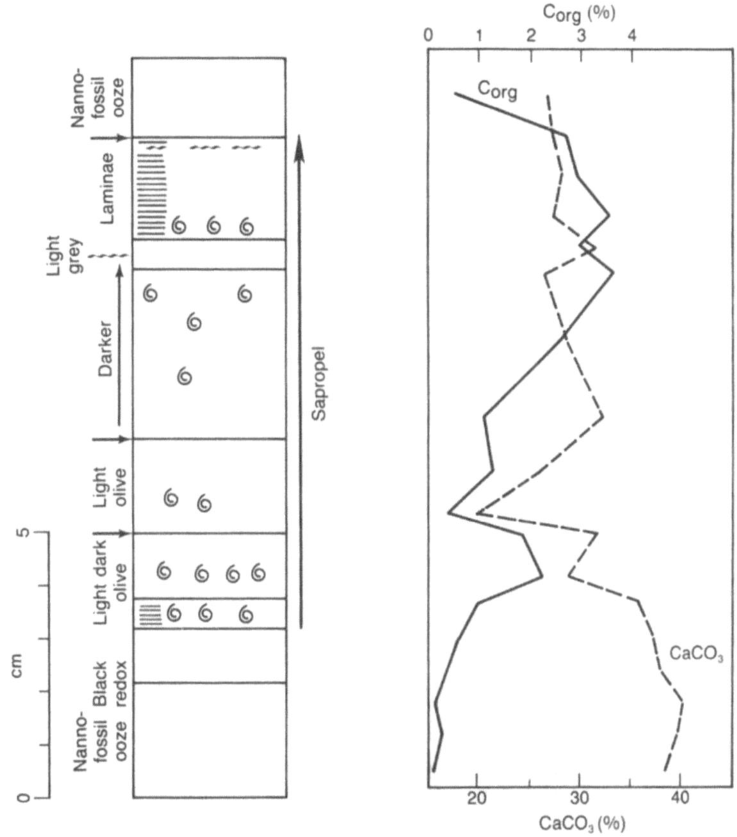


**Fig. 6.** Line drawing of sapropel sequence 655A-3H-4, 63 to 78 cm. This sapropel is different from the ones discussed before in that the increase and decrease in organic carbon at the transition to the surrounding nannofossil marl and ooze is gradual, as is the decline of $CaCO_3$ values. A further peculiarity of this sapropel is the presence of a redox front extending downsection from the organic-rich layer, as opposed to upsection in the other sapropels. Furthermore, foraminifers are not found concentrated at the base of the sapropel, bnt occur dispersed throughout the organic-rich layer, and faunal analysis is not supportive of anoxic conditions during deposition of this particular sapropel

**Fig. 7.** Sapropel sequence 652A-6R-1, 30–60 cm. In spite of similarities in organic matter concentration (4.18%) and type, this sapropel lacks the redox front observed in the previous examples. The contact to overlying mud is gradual, and seems to be affected by bioturbation

**Table 2.** Chemical Evaluation of Sapropel Samples/Hole 655A

| Core | Sec. | Sample # | $T_{max}$ | S1 | S2 | S3 | P1 | S2/S3 | $C_{org}$ | HI | OI | N | C/N | $CaCO_3$ |
|---|---|---|---|---|---|---|---|---|---|---|---|---|---|---|
| 2H | 4 | 1 | 518 | 0.02 | 0.17 | 2.36 | 0.11 | 0.07 | 0.18 | 94 | 1331 | 0.05 | 3.6 | 51.0 |
| 2H | 4 | 2 | 530 | 0.00 | 0.18 | 2.08 | 0.00 | 0.08 | 0.31 | 58 | 671 | 0.06 | 5.2 | 49.4 |
| 2H | 4 | 3 | 523 | 0.02 | 0.22 | 2.28 | 0.08 | 0.09 | 0.16 | 138 | 1425 | 0.07 | 2.3 | 45.2 |
| 2H | 4 | 4 | 424 | 0.17 | 2.65 | 4.04 | 0.06 | 0.65 | 1.83 | 145 | 221 | 0.17 | 10.8 | 37.6 |
| 2H | 4 | 5 | 415 | 0.49 | 6.76 | 4.84 | 0.07 | 1.39 | 3.05 | 222 | 159 | 0.28 | 10.9 | 37.7 |
| 2H | 4 | 6 | 424 | 0.58 | 8.97 | 5.87 | 0.06 | 1.52 | 3.88 | 231 | 151 | * | * | 31.5 |
| 2H | 4 | 7 | 426 | 0.62 | 10.36 | 6.89 | 0.06 | 1.50 | 4.75 | 218 | 145 | 0.40 | 11.9 | 25.4 |
| 2H | 4 | 8 | 418 | 0.82 | 13.61 | 6.40 | 0.06 | 2.12 | 5.30 | 257 | 121 | 0.73 | 7.2 | 13.9 |
| 2H | 4 | 10 | 418 | 0.05 | 0.65 | 2.74 | 0.07 | 0.23 | 0.34 | 191 | 806 | 0.15 | 2.3 | 20.8 |
| 2H | 4 | 11 | 478 | 0.03 | 0.32 | 2.75 | 0.09 | 0.11 | 0.02 | * | * | 0.13 | * | 23.9 |
| 2H | 4 | 12 | 510 | 0.01 | 0.16 | 2.86 | 0.06 | 0.05 | 0.21 | 76 | 1362 | 0.13 | 1.6 | 26.8 |
| 2H | 4 | 13 | 468 | 0.01 | 0.16 | 3.80 | 0.06 | 0.04 | 0.20 | 80 | * | 0.16 | 1.3 | 28.3 |
| 3H | 1 | 1 | 465 | 0.02 | 0.13 | 2.99 | 0.14 | 0.04 | 0.63 | 20 | 474 | 0.02 | 31.0 | 44.0 |
| 3H | 1 | 2 | 410 | 0.05 | 0.48 | 3.08 | 0.10 | 0.15 | 1.09 | 44 | 282 | 0.04 | 27.3 | 38.3 |
| 3H | 1 | 3 | 418 | 0.17 | 1.79 | 4.14 | 0.09 | 0.43 | 2.17 | 82 | 191 | 0.06 | 36.2 | 33.0 |
| 3H | 1 | 4 | 421 | 0.45 | 6.19 | 5.49 | 0.07 | 1.12 | 3.00 | 206 | 183 | 0.15 | 20.0 | 35.1 |
| 3H | 1 | 6 | 417 | 0.50 | 6.37 | 5.74 | 0.07 | 1.10 | 2.60 | 245 | 221 | 0.16 | 16.3 | 34.7 |
| 3H | 1 | 7 | 423 | 0.58 | 7.43 | 5.79 | 0.07 | 1.28 | 3.43 | 217 | 169 | 0.18 | 19.1 | 34.2 |
| 3H | 1 | 8 | 423 | 0.70 | 10.32 | 6.45 | 0.06 | 1.60 | 4.25 | 243 | 152 | 0.22 | 19.3 | 31.4 |
| 3H | 1 | 9 | 428 | 0.79 | 12.02 | 6.72 | 0.06 | 1.78 | 4.79 | 245 | 141 | 0.24 | 18.3 | 25.9 |
| 3H | 1 | 10 | 424 | 0.73 | 11.77 | 7.27 | 0.06 | 1.61 | 5.06 | 238 | 133 | 0.27 | 18.7 | 21.0 |
| 3H | 1 | 11 | 422 | 1.05 | 14.44 | 7.63 | 0.07 | 1.89 | 5.69 | 207 | 128 | 0.29 | 19.6 | 16.7 |
| 3H | 1 | 12 | 423 | 0.89 | 12.89 | 7.10 | 0.06 | 1.81 | 5.05 | 286 | 151 | 0.29 | 17.4 | 15.7 |
| 3H | 1 | 15 | 428 | 0.37 | 5.54 | 5.44 | 0.06 | 1.01 | 2.73 | 301 | 166 | 0.23 | 18.6 | 17.3 |
| 3H | 1 | 17 | 433 | 0.17 | 2.07 | 3.90 | 0.08 | 0.53 | 1.33 | 156 | 293 | 0.11 | 12.1 | 18.3 |
| 3H | 1 | 18 | 410 | 0.06 | 0.38 | 3.24 | 0.14 | 0.11 | 0.34 | 112 | 953 | 0.11 | 3.1 | 19.7 |
| 3H | 1 | 19 | 453 | 0.03 | 0.20 | 3.05 | 0.14 | 0.06 | 0.25 | 80 | 1220 | 0.04 | 6.3 | 21.3 |
| 3H | 1 | 20 | 502 | 0.01 | 0.16 | 3.82 | 0.06 | 0.04 | 0.22 | 73 | 1736 | 0.02 | 11.0 | 24.1 |
| 3H | 1 | 21 | 365 | 0.04 | 0.17 | 3.41 | 0.20 | 0.04 | 0.27 | 62 | 1262 | 0.07 | 3.9 | 20.6 |
| 3H | 1 | 22 | 438 | 0.04 | 0.29 | 3.94 | 0.12 | 0.07 | 0.11 | * | * | 0.09 | 1.2 | 23.8 |
| 3H | 1 | 23 | 464 | 0.02 | 0.15 | 4.30 | 0.12 | 0.03 | 0.27 | 56 | 1592 | 0.03 | 9.0 | 23.7 |
| 3H | 1 | 24 | 503 | 0.02 | 0.17 | 4.30 | 0.11 | 0.03 | 0.51 | 33 | 843 | 0.10 | 5.1 | 22.1 |
| 3H | 1 | 26 | 416 | 0.02 | 0.16 | 5.19 | 0.11 | 0.03 | 0.42 | 38 | 1236 | 0.01 | 42.0 | 34.9 |
| 3H | 1 | 28 | 503 | 0.03 | 0.13 | 4.97 | 0.19 | 0.02 | 0.13 | * | * | 0.05 | 2.6 | 40.2 |
| 3H | 4 | 1 | 399 | 0.01 | 0.18 | 3.55 | 0.06 | 0.05 | 0.08 | * | * | 0.02 | 4.0 | 38.6 |
| 3H | 4 | 2 | 485 | 0.02 | 0.34 | 3.68 | 0.06 | 0.09 | 0.21 | 162 | 1752 | 0.03 | 7.0 | 39.7 |
| 3H | 4 | 3 | 530 | 0.00 | 0.10 | 3.23 | 0.00 | 0.03 | 0.10 | 100 | * | 0.02 | 5.0 | 40.3 |
| 3H | 4 | 4 | 469 | 0.02 | 0.22 | 3.20 | 0.08 | 0.06 | 0.32 | 69 | 1000 | 0.07 | 4.6 | 37.9 |
| 3H | 4 | 5 | 419 | 0.04 | 0.47 | 3.32 | 0.08 | 0.14 | 0.58 | 81 | 572 | 0.08 | 7.3 | 37.1 |
| 3H | 4 | 6 | 427 | 0.11 | 1.23 | 3.80 | 0.08 | 0.32 | 0.97 | 262 | 809 | 0.08 | 5.9 | 35.8 |
| 3H | 4 | 7 | 431 | 0.15 | 3.29 | 3.67 | 0.04 | 0.89 | 2.20 | 150 | 167 | 0.09 | 24.4 | 28.9 |
| 3H | 4 | 8 | 431 | 0.16 | 3.36 | 4.33 | 0.05 | 0.77 | 1.79 | 188 | 242 | 0.09 | 19.9 | 31.4 |
| 3H | 4 | 9 | 401 | 0.09 | 0.48 | 3.63 | 0.16 | 0.13 | 0.37 | 130 | 885 | 0.11 | 3.4 | 19.5 |
| 3H | 4 | 10 | 431 | 0.14 | 1.90 | 4.35 | 0.07 | 0.43 | 1.20 | 158 | 363 | 0.07 | 17.1 | 25.7 |
| 3H | 4 | 12 | 424 | 0.32 | 5.51 | 4.71 | 0.05 | 1.16 | 1.05 | 524 | 448 | 0.13 | 8.1 | 32.1 |
| 3H | 4 | 13 | 424 | 0.37 | 5.89 | 5.58 | 0.06 | 1.05 | 2.53 | 233 | 239 | 0.13 | 19.5 | 28.4 |
| 3H | 4 | 14 | 426 | 0.42 | 9.52 | 5.74 | 0.04 | 1.65 | 3.60 | 264 | 159 | 0.17 | 21.2 | 26.5 |
| 3H | 4 | 15 | 431 | 0.38 | 7.77 | 5.33 | 0.05 | 1.45 | 2.96 | 262 | 180 | 0.15 | 19.7 | 31.5 |
| 3H | 4 | 16 | 431 | 0.39 | 9.26 | 5.44 | 0.04 | 1.70 | 3.51 | 264 | 155 | 0.17 | 20.6 | 27.1 |
| 3H | 4 | 17 | 433 | 0.32 | 7.74 | 5.23 | 0.04 | 1.47 | 2.91 | 266 | 180 | 0.15 | 19.4 | 28.1 |
| 3H | 4 | 18 | 432 | 0.35 | 6.64 | 5.39 | 0.05 | 1.23 | 2.68 | 248 | 217 | 0.19 | 14.1 | 27.4 |
| 3H | 4 | 19 | 448 | 0.05 | 0.29 | 3.75 | 0.15 | 0.07 | 0.47 | 62 | 798 | 0.05 | 9.4 | 25.2 |

impression of lamination. If we consider the laminations to be primary features, then the prominent redox front at 59 cm, which presumably is composed of disseminated manganese and iron oxo-hydroxides, gives several additional clues concerning the conditions during sapropel deposition and re-establishment of oxic conditions at the sediment/water interface. Iron is virtually insoluble as $Fe^{3+}$, as is the oxidized form of manganese, $Mn^{4+}$. The formation of a redox front implies that conditions in the sediment were reducing, and that dissolved $Fe^{2+}$ and $Mn^{2+}$ have migrated along concentration gradients to the sediment/water interface (see Wilson et al. 1986). This is an indiction for a concentration gradient during the impregnation of the redox front, which marks the time when a steady state was attained with regard to supply of oxidants ($NO_3^-$ and $O_2$) and reductants (organic matter, $Fe^{2+}$ and $Mn^{2+}$). During this impregnation, the bottom water thus must have either been oxic, or sedimentation rate and supply of organic matter must have greatly deviated from steady state.

We believe that two models are able to explain the combinations of sapropel, redox front, and lamination. Both follow the reasoning that under conditions of bottom water oxia, the dissolved metals would reprecipitate at a redox boundary in the sediment by oxygen and nitrate diffusing or being vented by benthos into the sediment. This zone may only be a few millimeters thick, and at steady-state sedimentation rates and bottom-water oxygen levels above 0.2 mL/L, the precipitation zone would follow the sediment/water interface, would be disturbed by burrowing organisms and would be diffuse. Our first model suggests that the lamination between redox front and sapropel are primary features, and that the much lower organic carbon values in the redox zone are not caused by diagenetic carbon "burn-down". Preservation of lamination in the mud zone and the sapropel proper underneath the redox front in the case of the sapropel in Figs. 4 and 5 indicates then that benthos was absent during deposition of the mud overlying the sapropel. We can assume from this, and from the pattern of bioturbation above, in, and under the redox front (Fig. 4), that the redox front has not migrated away from the sapropel during multiple iterations of dissolution and reprecipitation of the metals in the redox front. We believe that the redox front marks the time, when bottom-water conditions returned to normal, oxygenated conditions, thus creating concentration gradients for the reduced metals and for molecular oxygen. This observation, if true, is a remarkable clue to sapropel formation, because it implies clearly that anoxia alone is not the cause for the extraordinary concentration of organic carbon – by preservation – in the sapropels. Anoxia at the site of sapropel deposition probably lasted well through the deposition of the sapropels proper and continued during deposition of the organic-lean muds between sapropel and the redox fronts, but did not result in accumulations of organic carbon above low background levels. We thus have to assume that very high production of organic carbon has the main part in deposition of some of the sapropels encountered here.

Our second model follows the concept of a "burn-down" of organic matter in an oxidation front after disequilibration of a sedimentary system as described by Wilson et al. (1986). A progressive oxidation front may be initiated by changes in sedimentation rate, in organic matter burial, in bottom water oxygenation or other changes. In this model, oxidants diffuse into the sediment from the bottom water, while reductants diffuse upwards. When the flux of oxidants ($NO_3^-$ and $O_2$) equals the flux of reductants, the progression of the front ceases and steady-state conditions have been reached. Because this mechanism occurs below the influence of bioturbation, it may account for lamination in sediments that are not indicative of missing bioturbation. We are fortunate in that bioturbation is nicely preserved in the redox front depicted in Sapropel 655A-2H-4, thus discrediting the latter hypothesis considerably.

The sapropel at 655A-3H-1, 52–76 cm (Fig. 5) shows the same pattern of nannofossil ooze grading over forminifer-nannofossil ooze to black, laminated mud. It also shows the prominent redox front. Two other sapropels shown in Figs. 6 and 7, however, lack the clear distinction between anoxic and bioturbated, oxic facies. 655A-3H-4, 65–75 cm and 652A-6R-1, 30–60 cm have high organic carbon contents, but appear to have been bioturbated. Possibly, these layers show that anoxia did not last long enough to bury the layer of high organic carbon beneath the zone of bioturbation. They show, however, that productivity in the surface waters must have been extraordinarily high during their deposition.

## 2.1 Organic Matter Character in the Sapropels

The character of organic matter preserved in sapropels and sapropelic sediments of the Mediterranean Sea, whether they be Quaternary or older, has attracted surprisingly little attention in the past. Even though molecular markers and elemental and isotopic investigation are potentially useful tools for resolving questions of origin and depositional environment, published results are sketchy and in part contradictory. Ten Haven (1986) investigated several samples of sapropel S1 by analyzing biomarkers, stable isotopes of organic matter, and petrography of kerogen. He concludes that predominantly marine, but also terrestrial, and microbial sources contribute to total organic carbon in the youngest of the Mediterranean sapropels. Based on saturated hydrocarbons in extracts of sapropels, Sigl and Wenzlow (1977) found a dominantly terrestrial origin in Pleistocene sapropels from the Ionian Sea. Hahn-Weinheimer et al. (1978) found extraordinarily low $^{13}C/^{12}C$ ratios in the organic matter from sapropels at DSDP Site 374 ($\delta^{13}C$ ranged from -56.75‰ to -29.58‰), but did not provide a satisfactory explanation for these extremely negative values.

Pyrolysis results are presented in Table 2 for sapropels from Cores 655A-2H and -3H, and for assorted sapropel layers analyzed from Leg 107. They are plotted in a diagram that relates pyrolysis parameters to origin and maturity of organic matter in Figure 8. The parameters chosen for this type of diagram are the temperature of maximum hydrocarbon release during programmed heating of the sample ($T_{max}$) and the amount of hydrocarbons (mg) released per gram of organic carbon during cracking of sedimentary organic matter (Hydrogen Index, HI). Other parameters listed in Table 2 are S1 (amount of hydrocarbons in mg that are liberated at a temperature of 300° C per g of sediment), and $S_2$ and $S_3$, which are the amount of hydrocarbons and $CO_2$, respectively, that are released during thermal cracking of sedimentary organic matter. In conjunction with the amount of organic carbon in weight % they are used to compute HI and the Oxygen Index (OI), which is a measure of $CO_2$ present in organic molecules (given in mg $CO_2/gC_{org}$). PI is a measure of total hydrocarbons to expected from a given source rock if it reaches peak maturity.

While tradionally OI and HI are used to characterize kerogens of various origins, $T_{max}$ versus HI diagrams are more suitable to characterize marine sedimentary organic matter. Most sedimentary organic matter analyzed by pyrolysis during DSDP and ODP is a mixture of recycled terrestrial material, which constitutes a baseline value of less than 0.2% $C_{org}$. At values higher than this baseline value, preservation and burial of fresh or partly degraded marine organic matter contributed to the sedimentary organic pool. In Figs. 8 and 9 we recognize that this applies to the Tyrrhenian sapropels as well. In Fig. 8, sediments constituting background sediments low in organic carbon invariably plot in the Type III field, which is typical for residual, frequently recycled, undegradable particles mainly of terrestrial origin. At elevated organic carbon concentrations, hydrogen-rich organic matter of marine characteristics raises the HI. Samples from sapropels cluster in the Type II field of Fig. 8, which is typical of marine sedimentary organic matter. The close positive correlation of HI with organic carbon concentrations becomes

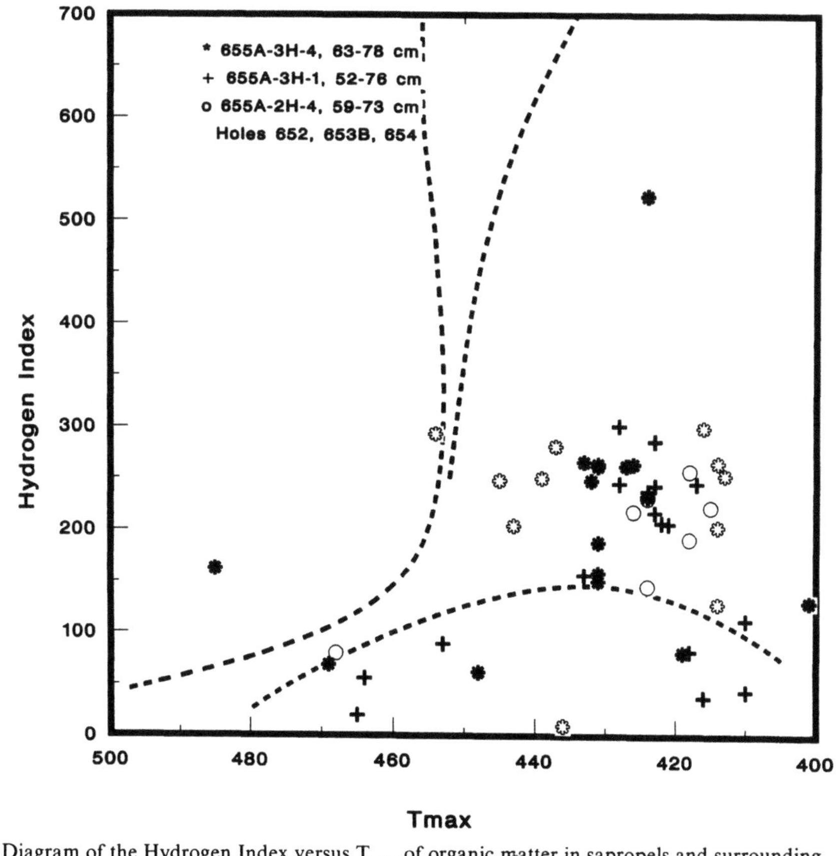


**Fig. 8.** Diagram of the Hydrogen Index versus $T_{max}$ of organic matter in sapropels and surrounding sediment. This type of diagram relates pyrolysis characteristics to maturity and provenance of the organic matter contained in the samples. Three types of organic matter (type I = brackish/bacterial, type II = marine, type III = terrestrial) cluster in specific fields of this diagram. Sapropels plot in the marine field, while organic carbon in the surrounding organic-lean sediments are residual, recycled macerals of terrestrial origin

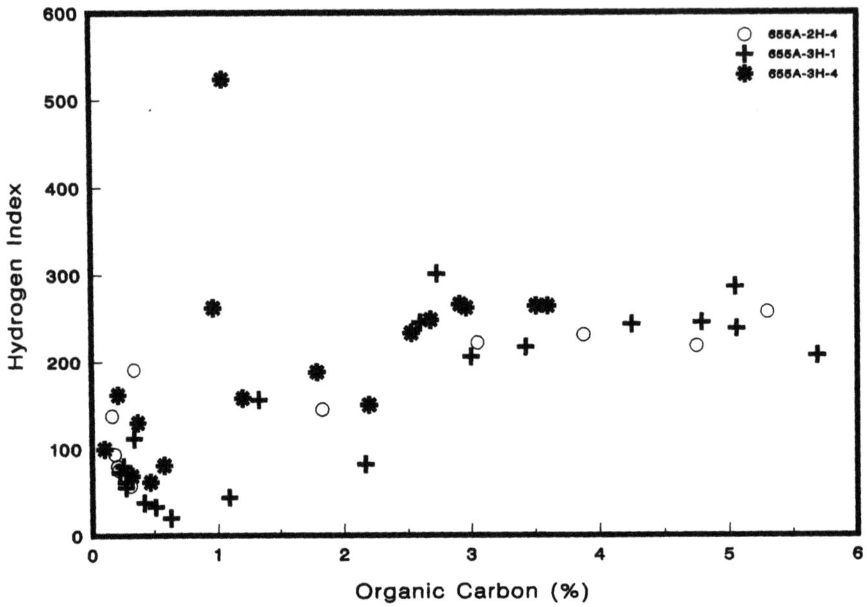


**Fig. 9.** Plot of the Hydrogen Index of sapropel samples and the surrounding nannofossil oozes and marls versus the organic carbon concentrations. At organic carbon values exceeding 0.5%, the contribution of hydrogen-containing organic matter of presumably marine origin becomes significant. This diagram further shows that microbial degradation in the organic-rich samples primarily utilizes hydrogen-rich organic matter, which is more depleted at lower $C_{org}$ concentrations. Symbols are the same as those in Figure 8

apparent in Fig. 9. At organic carbon concentrations $>0.5\%$, a significant correlation of $C_{org}$ and the Hydrogen Index (which is normalized to $C_{org}$ and thus should be uniform for fresh, undegraded organic matter) shows that hydrogen-rich organic matter of marine origin constitutes the bulk of the organic material in sapropels. Furthermore, the decrease of HI towards lower $C_{org}$ abundances implies that hydrogen-containing material is preferentially remineralized during bacterial remineralization of sedimentary organic matter.

## 3 Discussion

Our findings are likely to revive some of the controversy about previously published models of sapropel formation. Problems concern oceanographic and climatic conditions that may have precluded deep water formation and "flushing" of deep basins during deposition of sapropels, if oceanic anoxia is indeed one of the reasons for their deposition. Other aspects worthy of further consideration are the discrepancy between present productivity values of the Mediterranean Sea, which are rather low, and values reconstructed for the sapropel intervals. Even though basinwide anoxia may have occurred and made sapropel deposition possible, changes in primary productivity and organic matter burial must have occurred simultaneously (or as the reason for anoxia) in order to overcome the equilibrated system of organic matter supply and remineralization (Calvert 1987). Productivity spikes of short duration are known to occur in the Mediterranean in historic times in the form of dinoflagellate blooms (red tides), which possibly is the origin of Homer's "wine-red sea". Red tides are common in marginal marine environments elsewhere as results of changing nutrient composition, salinity, or sea levels. Calvert (1987) attributes sapropel formation in the Black Sea to surface productivity spikes of dinoflagellate blooms; total numbers of dinoflagellate cysts reported by Wall and Dale (1974) for the sapropel unit of the Black Sea corroborate this speculation, as do the abundance of molecular marker compounds found by Boon et al. (1979). The boundary conditions for initiation of dinoflagellate blooms are at present not well understood, but salinity fluctuations, transgression and massive resuspension of buried cysts, and changes in the nutrient make-up of waters in the euphotic zone are thought to be among the environmental conditions triggering red tides (E. Cox, pers. comm. 1988).

Environmental changes, such as the massive influx of fresh water from the Black Sea and the Nile, are thought to be prime reasons for bottom water anoxia and sapropel deposition in the eastern Mediterranean (Thunnell et al. 1977) and may thus also be the initial triggering mechanism of red tides. Fresh-water surface layers and higher sea levels after deglaciation may then be common reasons for high surface productivity (by dinoflagellate blooms), for bottom-water anoxia (by supply of increased amounts of organic matter to the deep sea), for isotopic shifts observed in planktonic foraminifers, and ultimately for the deposition of sapropels. Investigation of biological marker compounds and palynology of sapropels from the Tyrrhenian Sea is currently underway, as are isotopic studies of foraminifers and faunal investigations. The results of these studies will certainly shed some light on the significance of the sapropels with respect to changes in the paleo-environment.

Aside from biological factors, environmental processes such as dissolution of Messinian salt during pluvial phases (Rossignol-Strick 1987) and inhibition of deep-water formation (Mangini and Schlosser 1986) may have affected stable stratification in the Meditteranean basins, which would have created a similar situation in the Mediterranean Sea as that postulated for the Kaspian region and the Jurassic of the North Sea by Degens and Paluska (1979). Unbalancing of the ecological system in the wake of transgression and fresh water influx may thus have fostered both increased organic matter production and preservation/burial. These aspects are currently being investigated as well and will be published separately (Emeis and Camerlenghi, in preparation).

## 4 Conclusions

Numerous layers of sapropels and sapropelic sediment have been recovered from Sites 650 to 655 of ODP Leg 107 in the Tyrrhenian Sea. The facies, temporal occurrence, and organic matter character and abundance appear to be similar to the well-studied sapropels of the eastern Mediterranean. According to preliminary age estimates – based on paleomagnetic studies and nannofossil zonations – the organic-rich sediments occurred throughout the Pliocene and Pleistocene as intercalations of few centimeters thickness in hemipelagic calcareous oozes. Maxima in the deposition of sapropels were recognized in all holes, and center around the Pliocene/Pleistocene boundary, the interval from 1.2 to 1.0 Ma, and from 0.4 to 0.2 Ma.

Pyrolysis characteristics of the organic matter are interpreted as typical for immature organic matter from primarily marine sources. From sedimentological observations we infer that bottom water anoxia may indeed have contributed to deposition of the sapropels. We believe, however, that anoxia may have been the result of productivity spikes in the euphotic zone, possibly caused by changes in salinity and nutrient composition. Recent analogues for such enormous, spontaneous and short-lived increases in productivity in response to salinity fluctuation, transgression, and changes in nutrient make-up are blooms of organic-walled algae. Such red tides of dinoflagellates are known to have occurred through historic times in the Mediterranean Sea and in coastal areas of the Atlantic Ocean. The formation of sapropel in the Black Sea coincided with the transition from fresh water to marine conditions; organic marker molecules in the organic fraction of the Black Sea sapropel suggest substantial input from dinoflagellates.

Further study of sedimentological, chemical, and stratigraphic aspects is needed to pinpoint the similarities between sapropels in the eastern and western sub-basins of the Mediterranean Sea. Even at this early stage of probing into the implications of the sapropel occurrences in the western basins we believe that the enigmatic sediment facies and paleoceanographic significance of sapropels warrant further research.

*Acknowledgements.* This work was supported by a grant from USSAC. Technical support by ODP during this research is gratefully acknowledged.

## References

Anastasakis GC, Stanley DJ (1984) Sapropels and organic-rich variants in the Mediterranean: sequence development and classification. In: Stow DAV, Piper DJW (eds) Fine Grained Sediments: Deep-water Processes and Facies. Geol Soc Spec Publ 15:497–510

Boon JJ, Rijpstra WI, de Lange F, de Leeuw JW (1979) Black Sea sterol – a molecular fossil for dinoflagellate blooms. Nature (Lond) 227:125–127

Calvert S (1987) Oceanographic controls on the accumulation of organic matter in marine sediments. In: Brooks J, Fleet AJ (eds) Marine Petroleum Source Rocks. Geol Soc Spec Publ 26:137–151

Channell J (in press) Synthesis of stratigraphic results, ODP Leg 107. In: Kastens K, Mascle J, et al. (eds) Proc Initial Reports (Part B), ODP 107

Cita MB, Grignani D (1982) Nature and origin of late Neogene Mediterranean sapropels. In: Schlanger SO, Cita MB (eds) Nature and Origin of Cretaceous Carbon-rich Facies. Academic Press, New York, pp 165–196

Degens ET, Paluska A (1979) Hypersaline solutions interact with organic detritus to produce oil. Nature (Lond) 281:666–668

Degens ET, Ross DA (1972) Chronology of the Black Sea over the last 25.000 years. Chem Geol 10:1–16

Degens ET, Ross DA (eds) (1974) The Black Sea – Geology, Chemistry, and Biology. AAPG Memoir 20, Tulsa

Hahn-Weinheimer P, Fabricius F, Müller J, and Sigl W (1978) Stable isotopes of oxygen and carbon in carbonates and organic material from Pleistocene to upper Miocene sediments at Site 374 (DSDP Leg 42A). In: Hsü K, Montadert L, et al. (eds) Initial Reports DSDP 42/1: U.S. Govt Printing Office, Washington, pp 483–488

ten Haven HL (1986) Organic and inorganic geochemical aspects of Mediterranean late Quaternary sapropels and Messinian evaporitic deposits. Geol Ultraiectina 46:1–202

Kidd RB, Cita MB, Ryan WBF (1978) Stratigraphy of eastern Mediterranean sapropel sequences recovered during Leg 42A and their paleoenvironmental significance. In: Hsü KJ, Montadert L, et al. Initial Reports DSDP 42/1: U.S. Govt Printing Office, Washington, pp 421–443

Mangini A, Schlosser P (1986) The formation of eastern Mediterranean sapropels. Mar Geol 72:115–124

Müller C (in press) Nannoplankton biostratigraphy and paleoenvironmental results from the Tyrrhenian Sea, ODP Leg 107. In: Kastens K, Mascle J, et al. (eds) Proc Initial Reports (Part B) ODP 107

Olausson E (1961) Studies in deep-sea cores. Rep Swed Deep-Sea Exped 1947–1948 8:337–391

Rossignol-Strick M (1987) Rainy periods and bottom water stagnation initiating brine accumulation and metal concentrations: I. The late Quaternary. Paleoceanography 2:333–360

Ryan WBF (1972) Stratigraphy of Late Quaternary sediments in the eastern Mediterranean. In: Stanley DJ (ed) The Mediterranean Sea: A Natural Sedimentation Laboratory. Hutchinson Ross, Stroudsburg, pp 149–169

Ryan WBF, Hsü K et al. (1973) Initial Reports DSDP 13: US Govt Printing Office, Washington

Selli R, Accorsi CA, Bandini Mazzanti M, et al. (1977) The Vrica Section (Calabria, Italy). A potential Quaternary/Neogene boundary stratotype. Giornale Geol 17:181–204

Sigl W, Wenzlow B (1977) Organic Geochemistry of Mediterranean sapropels and some paleoenvironmental implications. Rapp Comm Int Mer Medit 24(7):195–196

Thunnell RC, Williams DF, Kennett JP (1977) Late Quaternary paleoclimatology, stratigraphy, and sapropel history in the eastern Mediterranean deep-sea sediments. Mar Micropaleontol 2:371–388

Thunnell RC, Williams DA, Belyea PR (1984) Anoxic events in the Mediterranean Sea in relation to the evolution of late Neogene climates. Mar Geol 59:105–134

Wall D, Dale B (1974) Dinoflagellates in late Quaternary deep-water sediments of Black Sea. In: Degens ET, Ross DA (eds) The Black Sea – geology, chemistry and biology. AAPG Memoir 20, Tulsa, pp 364–380

Wilson TRS, Thompson J, Hydes DJ, Colley S, Culkin F, Sorensen J (1986) Oxidation fronts in pelagic sediments: diagenetic formation of metal-rich layers. Science 232:972–975

# Indicators for Holocene Changes in Relative Sea Level

D. NEEV and K.O. EMERY

## 1 Introduction

The Mediterranean coasts of Israel and Sinai have been inhabited by man since at least the Late Pleistocene. This occupation became more intense and almost uninterrupted after Chalcolithic time (through late Holocene). During the past 25 years we have studied the physiography, stratigraphy, and structure of these coasts both offshore and onshore (Neev et al. 1987). Stratigraphic studies included both seismic stratigraphy and sedimentological-environmental aspects. Types of sediments are low energy marine clayey sand, lagoonal evaporites, beach sand, lenticular shell accumulations, mudflat and swamp loams, and coastal dunes. Geological aspects of many archaeological sites along the coasts such as tilt of structures and presence of interbedded or onlapped sediments within them were investigated, in addition to study of relevant archaeological literature. The results indicate environmental changes while the sites were inhabited as well as since then. Available data gathered from each site were integrated and interpreted, including the relevant segment of the eustatic sea level curve. More comprehensive and regional compilations then were made.

Magnitudes of relative sea level changes implied even from preliminary results were so large that an interpretation of tectonic movements across the coastline was preferred rather than just simple additions and withdrawals of water from the ocean due to global climatic changes. Two types of tectonic movements were identified: (1) Differential vertical movement since Late Pleistocene, when the oceanic crust of the eastern Mediterranean Sea (floor of the Levantine Basin) subsided and the continental crust east of it was uplifted. The coastal belt of Israel functioned as a hinge. (2) Oscillatory-type of downward followed by upward movements of the upthrown block across the coastal fault. The latter occurred at least five times during the past 4000 years. During each of these tectonic phases the downthrown side subsided. The cumulative amount of that subsidence was appreciable (varies from a few to tens of meters) whereas the net amount of shift of the upthrown block was small. Such a complicated and unusual type of tectonic movement may be difficult to accept on initial exposure. In fact, our conclusions were countered by several authors (see summaries and discussions of their concepts in Neev et al. 1987, pp 100–102). However, their papers did not contain significant contradiction to our data and descriptions. Two analogous events from other coasts in the world are recorded in the literature. The first is an analysis made by Thatcher (1985) based on 100 years of tide-gauge measurements, indicating the occurrence of two successive oscillatory movements in Japan (each of 2 m amplitudes and 30-year periods). The second is a description by Strabo of an event during Roman times at Mount Casius (northwestern Sinai) (Neev et al. 1987, pp 89 and 99).

## 2 Depositional Environments

The nature of sediments at each coastal locality contains evidence of the mode of deposition. Where the sediments indicate intertidal deposition but are now well above or below the present tide range, there must have been a change of relative sea level. Critical, therefore, are the criteria for identification of intertidal deposition. These include trace fossils, shell beds, lagoonal-hypersaline materials, and sources for Holocene sand dunes.

*Trace fossils within an early Holocene calcarenite bank*. A 4-m-thick cross-stratified calcarenite unit (or bioclastic sandstone) known as the Tel Aviv Kurkar Bed (Horowitz 1979) or the Beit Yanai Kurkar Member (Gavish and Bakler, in press), conformably overlies the Epipaleolithic (17,000 to 12,000 years B.P.) Netanya Hamra (red loam deposited in a mud-flat environment) along the central segment of the coastline of Israel (Figs. 1, 2A). Both units crop out along the coastal zone at elevations of 20 to 50 m above mean sea level. They form lenticular layers, the thickest parts of which correspond with interdune troughs of the underlying Kurkar unit (Late Pleistocene loosely cemented quartz sand dunes). Both the hamra and the calcarenite layers laterally thin and even wedge out toward some of the older dune crests. The prominent white of the calcarenite bank is due to the dominance of bioclastic carbonate components (well-rounded fragments of algae, foraminiferans, and shells). Total carbonate content increases northward, from 60% off Yafo to 90% just south of Mount Carmel (Fig. 1), the remainder (40 to 10%) being well-sorted fine-grained quartz sand. This calcarenitic layer is friable and has a high permeability.

In situ molds of burrowers' tubes are sparsely scattered across this unit, their detrital fill having the same composition as that of the matrix. Locally, near the bottom of the unit, the burrows crowd to form biostromal lenses more than 0.5 m thick and a few meters long (Fig. 2B). The maximum diameter of individual tubes is 2 cm and their length reaches 20 cm. Some secondary borings through the tubes are hollow cylinders (Fig. 2C), that exhibit smooth and perfectly rounded cross-sections; possibly these were made by crabs and pelecypods having long siphons. Occasionally, successions of concave-upward pseudo-septa are recognized across some of the vertical molds, suggesting the animal's adaptation to the accumulation of sediments above it (Fig. 2D). These clusters of tubes are interpreted as trace-fossils of colonies of burrowers that populated the beach or the upper littoral zone. Convolute load structures are locally present within this calcarenite bank. Occasionally, the bank was affected by solution and probably also subjected to vigorous winnowing by waves to form boulder-like tear-apart featuring (Neev et al. 1987, Fig. 6 D,E). These finds corroborate the aquatic-marine origin of the calcarenite bank.

The Calcarenite Bank is younger than 12,000 years B.P. or even 10,000 years B.P. (the youngest age of the Epipaleolithic culture found within the underlying Hamra Unit (A. Ronen pers. commun.) and older than 5,500 years B.P. (the age of Early Bronze settlements within the overlying brownish alluvial loam unit; Gophna 1977). Several radiocarbon dates, ranging between 10,000 and 7000 years B.P. were determined for these calcarenites by Gavish and Friedman (1969).

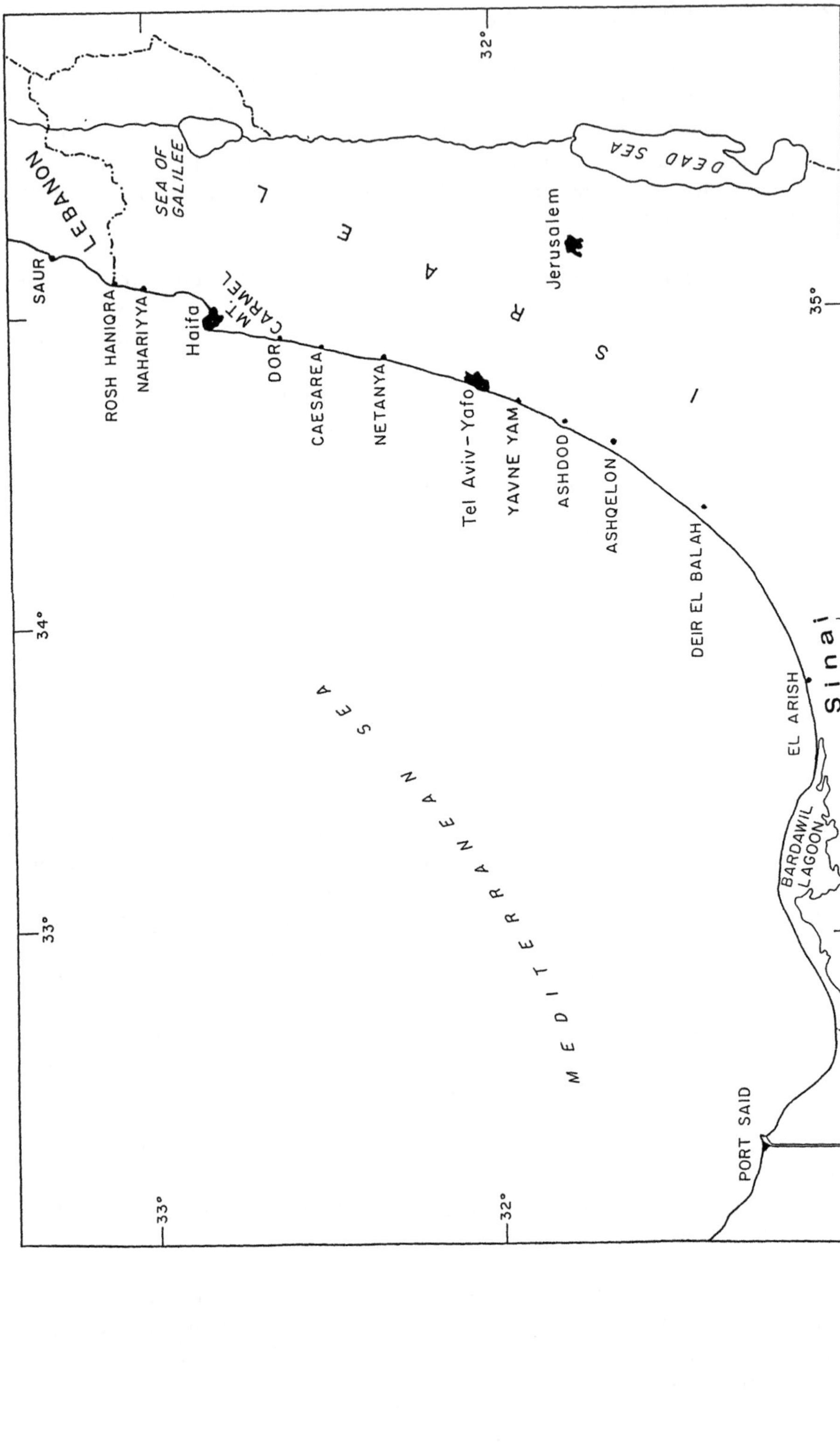


**Fig. 1.** Location map of coastal sites discussed in text

A dense bioclastic limestone unit containing the same faunal assemblage as the friable Calcarenite Bank (H. Frenkel, personal communication, 1970) was found off Natanya (Fig. 1) in an offshore borehole drilled with a chisel and bailer at a water depth of 20 m. It overlies the Late Pleistocene Kurkar unit. A Hamra layer found east of it in a nearby offshore borehole is probably an equivalent of the Epi-paleolithic one (see above), as it is situated in a similar stratigraphic relationship. A unit of loosely cemented calcarenitic bioclastic rock more than 20 m thick and having the same faunal assemblage was found in another drillhole atop a now-submerged kurkar ridge off Yafo (Fig. 1) in a water depth of 30 m. We correlate these two offshore occurrences of calcarenitic units with the Calcarenite Bank of the early Holocene.

We conclude that the Calcarenite Bank was deposited in the intertidal zone during early Holocene times (10,000 to 6000 years B.P.). Maintenance of the same beach environment of deposition through the entire critical period when sea level rose eustatically by several tens of meters suggests that the calcarenite was uplifted simultaneously during that time. However, the uplifting process of this bank was maintained even afterwards when the rate of eustatic change became negligible.

*Shell Beds.* Lenticular layers of loose shells dominated by *Glycymeris violacescens* (Lamarck) and commonly containing imbricated pottery sherds are interbedded in several levels within the late Holocene sedimentary sequence of the coastal cliff. Locally, they occur in terraces at elevations as high as 30 m above m.s.l. to distances of 1 or even 2 km east of the coastline. Texturally, these lenses are similar to the well-cemented shell lenses incorporated within beachrock along the coastline. Pottery sherds also are present within some beachrocks.

The following four criteria were used to determine the origin of the shell-beds: (1) orientation and packing patterns of pelecypod valves; (2) sedimentary structures within interbedded fine-grained sand layers; (3) presence of articulated bivalves; (4) interbedded swamp or lagoonal clay layers.

1. The dominant orientation of the concavity of the *Glycymeris* valves, as well as their pattern of packing: these shells, when transported along the beach by longshore currents and swash, are more stable when they are oriented concave downward, because the friction coefficient is higher than that for the opposite orientation (Fig. 3A,C). In fact, most shells of beaches below high-tide level are oriented that way (Emery 1968). On the other hand, most shells that fall freely into the beach when thrown shoreward by high storm waves or by settling freely to the sea bottom through a relatively thick water column or from a turbidity current (Allen 1984) land in a concave-upward position. This is reasonable, as their center of gravity is near their apex of curvature and their convex side is more streamlined (less friction). Similar results were noted in reproducible experiments that we made by artificially dumping shells from buckets held at about a meter high to form a man-made layer; these shells are oriented dominantly concave upward (Fig. 3B). In both examples the shells are loosely packed.

Another pattern of orientation forms when shells are transported and swept along the sea bottom by strong and persistent currents, such as in tidal channels. When impeded by an obstacle, these shells tend to stabilize with their convex sides facing upcurrent. If a large number of similar shells are involved, they tend to stack

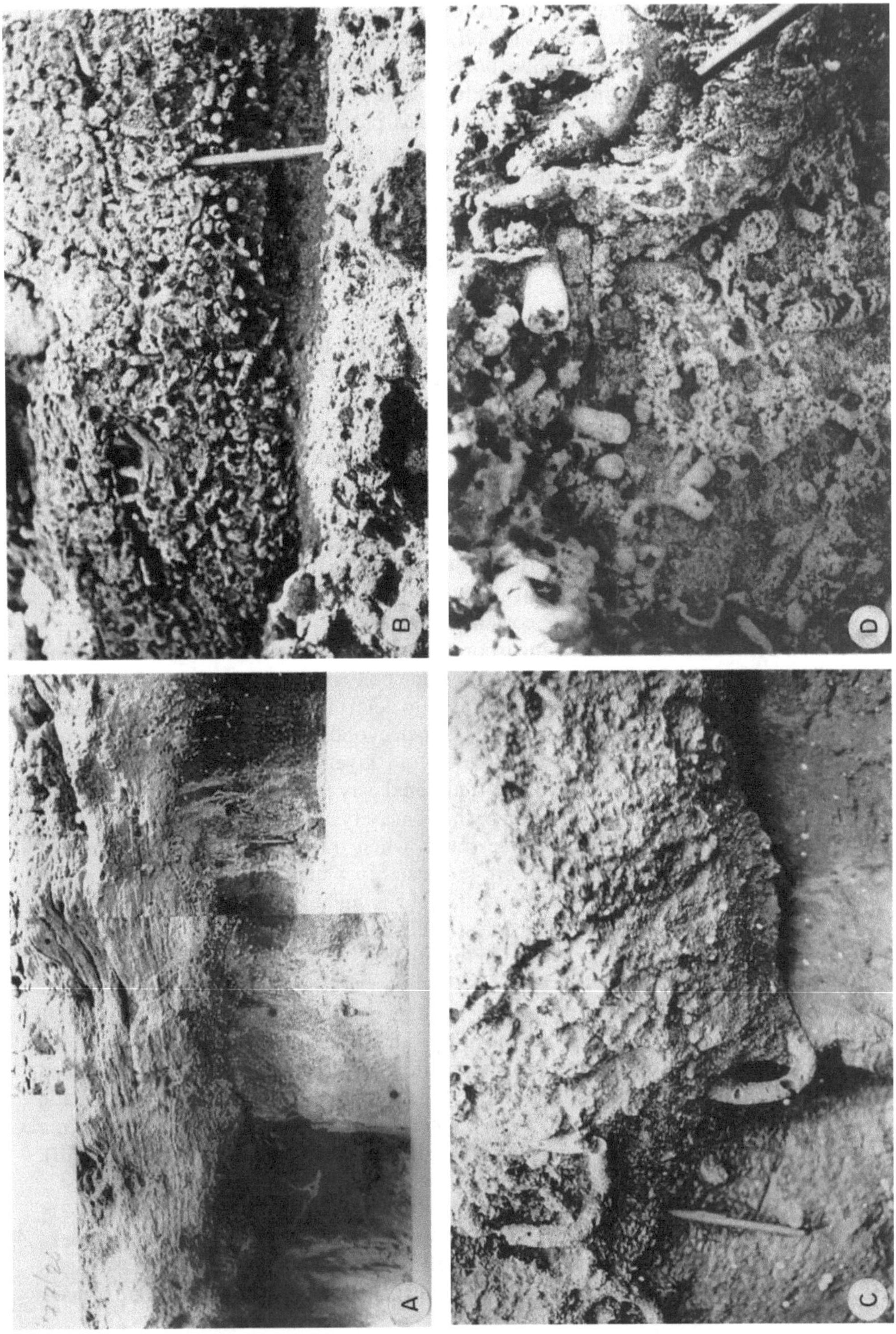
B
D
A
C

one behind the other to form chains in which the shells lean forward and their concave sides point downcurrent. Such a well packed nestling texture forms a systematically imbricated and oriented pattern (Fig. 3E,F,G,H).

To conclude, it appears that shell accumulations exhibiting dominance of either concave downward orientation or the nestling packing attributes indicate natural beach environments of deposition. On the other hand, concave upward accumulations may indicate either an artificial (man-made) origin or a storm-wave deposit.

2. Sedimentary structures in fine-grained sand layers interbedded within the shell beds constitute the second criterion. Good examples are found in the sequences of layered detrital sediments that partly fill remains of medieval structures at Ashqelon or which form the coastal cliff east and northeast of the Yavneh Yam promontory (Figs. 1, 3E,F). These structures include flaser and graded bedding as well as load structures (Fig. 4), thereby indicating an environment of aquatic (probably beach) deposition.

3. The presence of articulated bivalves as individual specimens or in lenses within sand or mud layers constitutes the third criterion. Sediments in which they occur are considered to have been deposited naturally in aquatic environments. If a bivalve did not die in a burrowing position or was not buried by sediments shortly after death, its valves eventually and naturally open and become separated from one another because the muscles and ligament of a bivalve usually decay soon after the animal dies. A sequence of calcarenitic well-rounded loose sand in which lenticular accumulations of articulated bivalves are interbedded was found along a narrow elevated (about + 8 m above m.s.l.) coast-parallel strandline at the low coastal cliff of the ancient Akhziv harbor site (near Rosh Haniqra, Fig. 1; Lewy et al. 1986; Fig. 1). The assemblage of articulated shells within these lenses include both brackish-water species (*Cerastoderma glaucum* Brugiere) and fresh water ones (*Unio sp.*), thereby suggesting an estuarine environment into which the fresh water specimens were swept and rapidly buried. Purely marine gastropod shells also were present, but in much smaller numbers. As this calcarenitic sequence contains Persian to Byzantine pottery sherds, it must have been deposited during late Roman times or somewhat afterward (close to 1500 years BP). About 100 m farther eastward, this sequence onlaps a low coast-parallel kurkar ridge beyond which is the Nahariya Plain. A black sandy clay layer containing fresh water *Unio* sp. shells as well as Roman pottery sherds extensively crops out at the Nahariya Plain (Avnimelech 1943, 42) to an elevation of + 10 m m.s.l. It is highly probable that this late Roman (or Byzantine) fresh water swamp was in hydraulic equilibrium with

**Fig. 2.** The Calcarenite Bank – a bioclastic sandstone unit. Photos are from an outcrop south of Netanya (see also Neev et al. 1987, Fig. 6). **A** The cross-bedded Calcarenite Bank overlying the Epipaleolithic Hamra (red loam) unit. Irregularly bedded biostrome-like lenses of burrowers are developed mostly at the bottom of the bank (see also Fig. 6). **B** A closer view of a densely populated colony of burrowers. **C** Secondary borings manifested by a hollow cylinder drilled by a pelecypod (?) through a long and curved mold of a burrower. **D** Burrowers, to 2 cm diameter, penetrated the underlying Hamra layer. The sequence of concave-upward septa-like features within the vertical tube (*bottom of photo*) suggest upward climbing by the burrower through a succession of bioclastic infilling at its base

Fig. 3A-D

Fig. 3E-H (legend see page 304)

the nearby beach, estuary, and lagoon of the Akhziv Harbor site at the time of its existence.

4. The fourth criterion for intertidal origin of the shell beds is offered by sedimentary sequences where shell lenses are interbedded within clay layers deposited in marine, lagoonal, or swamp environments. It is therefore thought that the shell lenses themselves were deposited in identical environments. Three of many relevant examples are demonstrated in elevated terraces (+ 3 to + 20 m m.s.l.) along the coastline at Dor (Fig. 1, north of the tel), at Mikhmoret (north of Netanya, 33°24′08″N 34°51′54″E), and at Deir el Balah. These terraces consist of dark brown and dark gray clay layers that were deposited in swamp and lagoonal environments and contain shell lenses (Fig. 3C,D and Fig. 5A,B). The time of deposition should be somewhat later than the respective Roman and Iron age sites that they overlie. Considering the nearness of these sites to the coastline and the mean sea level of that time (near that at present) as well as the absence of impermeable barriers between them and the sea, we deduce that these water bodies were in hydraulic equilibrium with sea level at that time.

At present, the loose shell beds are common along the coastal sea cliff at various elevations (to + 30 m m.s.l.). Many onlap archaeological structures (Fig. 3C,D,E,F,G,H), thus their ages of deposition must range from more than 3000 years B.P. to about 500 years B.P.

*Lagoonal hypersaline sediments.* A sequence of lagoonal-evaporitic sediments a few meters thick lies atop the coastal cliff south of the Deir el Balah beach (Fig. 1) at an elevation of about + 20 m (31°24′43″N 34°19′17″E). This sequence onlaps and overlies Iron Age structures, some of whose sun-baked earth-brick walls have been preserved up to a 5 m height (Fig. 5C). The facies of the onlapping sequence

**Fig. 3.** Shell beds – orientation and packing patterns. **A** Concave downward shell laminae (radiocarbon age about 3800 years B.P.) interbedded in laminated calcarenite at Tel Haraz (southern coastal segment – see Neev et al. 1987, p 26–29). **B** Dominantly concave upward-oriented shells, experimentally dumped from waist height. Note the loose packing of the shell assemblage. Both properties result from the free fall mechanism of the shells. **C** About a 2-m-thick sequence of layered shells from + 2 to + 4 m m.s.l. (containing gravels and rounded pottery sherds), dominated by concave downward orientation, alternating with dark brown (swamp environment) clay (seen at *bottom of photo*) as well as clean beach sand crops out at the northern Dor coastline (Fig. 1 and Neev et al. 1987, p 47–49). **D** The sequence of **C** onlapping quarries and structures of Roman age. **E** and **F** A 20-cm-thick lense of laminated shells and sandy marl occasionally interbedded within a layered sequence of Tel-like sediments cropping out along the Yavneh Yam coastline (Fig. 1 and Neev et al. 1987, p 65–67). Note the dense packing (nestling texture of shells) within the two lower shell layers and also within the overlying thin layer of fragmented shells and organogenic pebbles. Such a pattern suggests a grain-size segregation mechanism by relatively strong tidal currents. Also note the wedging out pattern of these layers as well as of the finely laminated underlying sandy marl layer toward a pre-existing wall (*beyond the right hand limit* of the photo). **G** Densely packed shells (nestling texture) in different orientations, suggesting an environment of deposition of turbulent flow within tidal currents. Found at + 6 to + 10 m m.s.l. within a few meters thick sequence of coarse detrital sediments that overlies a Byzantine to Early Moslem suburb at Caesarea. This unit extends from the coastal cliff eastward to a distance of at least 200 m. **H** Wedged-out part of a shell lense deposited within and around remains of Crusaders' buildings at Ashqelon (Fig. 1) at an elevation of about + 8 m m.s.l. Nestling texture, imbrication, and grain-size segregation patterns suggest a mechanism involving high hydraulic energies of tidal currents (the left side of the same lense is described by Neev et al. 1987, Fig. 11D)

**Fig. 4.** Sedimentary structures. **A** and **B** A sequence of alternating finely laminated yellowish quartz sands and dark brown sandy clay, that contains Middle Bronze pottery sherds. This unit unconformably overlies gray, finely laminated Tel-like sandy sediments that contain Early Bronze pottery sherds. Convolute load structures, which occur within the upper unit, suggest that slumping or faulting along the angular unconformity occurred rather soon after deposition of that unit. The outcrop is at the bottom part of the westward-facing coastal cliff at Yavneh-Yam (Fig. 1). See also Neev et al. (1987, Fig. 12)

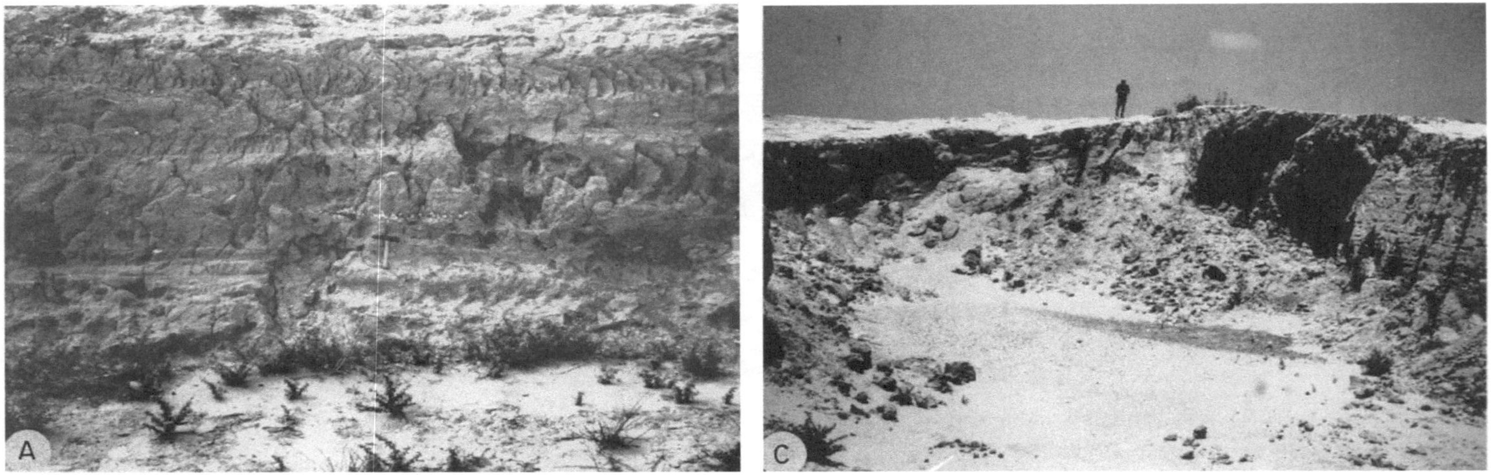

**Fig. 5.** The Deir el Balah evaporites (see also Neev et al. 1987, Fig. 29). **A** A few meters thick dark brown, reddish and greenish loam and marl sequence with some interbedded thin lenses of concave downward oriented shells. That unit overlies Iron Age structures atop the coastal cliff at about + 20 m m.s.l. These marls were once saturated with hypersaline brines, the salts of which are preserved. **B** A closer view of the shell lense shown in A. **C** A more than 5-m-high relict of an Iron Age sun-baked mud-brick structure at about + 20 m m.s.l. close to **A**. This structure is onlapped by a few meters thick, finely laminated, tan colored, and fine-grained sand unit that represents a lateral facies change of the marl and loam unit (**A**)

change laterally from finely laminated fine tan sands to green-gray, reddish and dark-brown sandy marls. Lenses of marine pelecypod shells (dominated by *Glycymeris* sp.) that are oriented concave downward also are interbedded in this sequence (Fig. 5A,B). In some of the green-gray marl layers a dark hue indicated higher moisture content than in most marls, thereby suggesting the presence of hygroscopic salts. Chemical analysis for the main ions of the water-soluble salts were made on samples from these layers. The ionic ratios of Na/Cl, Ca/Mg, $Ca/HCO_3$ and $Ca/SO_4$ were computed and compared with the values of the same ratios in five different water bodies. These are: (1) normal seawater, (2) artificially evaporated Mediterranean seawater, (3) typical marine sabkha brines, (4) calcium-chloride sabkha brines from inland recent sabkhas near Bardawil Lagoon in northwestern Sinai, and (5) drinkable water from wells near Bardawil Lagoon.

The following conclusions concerning original characteristics of the Deir el Balah brines can be drawn from these analyses: (1) The high content of water soluble salts in these marl samples (ranging between 0.5 and 4.7%) from the middle part of the marl sequence suggests that they are dried residues of interstitial pore water of concentrated brines, and not just the residue of seawater spray that fell on the emerged land. (2) The dominance of chlorine in the anions indicates that the original brine was chloridic. (3) The relatively low Na/Cl ratio (0.47 to 0.82) as compared with normal seawater (0.86) suggests that halite (NaCl) was deposited from the original brine because of excess evaporation. Assuming that the source of these brines was seawater, the latter had to become at least 10 times as concentrated to reach saturation for NaCl (Neev and Emery 1967) or to about 30 times (Levy 1977). (4) The very high Ca/Mg ratios (3.7 to 7.5) as compared with normal seawater (0.2) suggest that these brines went through a process of Ca-Mg exchange during dolomitization of previously precipitated $CaCO_3$ minerals. This process is similar to the one described by Levy (1977) for some recent sabkha sediments and brines in the Bardawil area. (5) The relatively high $Ca/(HCO_3 + SO_4)$ ratio corroborates the calcium-chloride nature of the brine. It should be noted that upon dilution of the calcium-chloride brine with meteoric water, the resultant water preserves its original ionic ratios.

These conclusions suggest that the Deir el Balah brines were formed by processes identical with those that formed the calcium-chloride brines of the Bardawil inland sabkhas. The latter also are postulated to be of marine origin. It is assumed that, after a marine regression, they were flushed from the surrounding topographically higher dune areas and concentrated and re-evaporated in the present sabkhas (Levy 1977). Such a stage is an unstable one because, if the flushing process continues, the calcium chloride brines must eventually drain into the Mediterranean Sea. The youth of the Deir el Balah evaporitic sequence (about 2700 years B.P.), the relatively low rainfall in that region, its present topographically isolated and high location, and the impermeable nature of the host sediments (marls) may explain why these soluble salts have been preserved within the sediments until now.

The following reconstruction of tectonic history of the Deir el Balah site was made: (1) During the Iron Age (about 3,000 years B.P.), the structural elevation of the site was near that of the present. (2) It was then downwarped to sea level of that time and covered by a hypersaline lagoon. (3) This site then was tectonically uplifted again to about its original structural elevation.

*Late Holocene coastal dunes atop the coastal cliff.* Two or even three layers of loose quartz dune sand successively accumulated above the Early Bronze swamp and mud-flat deposits all along the coastline from Sinai to Mount Carmel. They represent a phase of rejuvenated extensive sand supply from the Nile Delta to the coast of Israel. In places, a brownish-gray sandy soil a few tens of centimeters thick separates the lower and upper sand layers (first and second generations). Radiocarbon analysis on both charcoal fragments and shells of the terrestrial gastropod *Helicidea* sampled from the intervening soil layer at the same site (Shefayim, at the coastline between Tel Aviv and Netanya) yielded ages of 3690 ± 40 and 3320 ± 120 years BP, respectively. The lower layer (or the older generation) of dune sand is appreciably thinner than the upper one. Barchan dunes to 10 m high with slip-faces that dip to the northeast developed within the upper layer and were active until recently.

The main lobes of these ingressive sand sheets are associated with mouths of Late Pleistocene streams such as the one just north of the Yarkon River (mouth at northern Tel Aviv-Yafo, Fig. 1). Smaller lobes also occur atop the + 30 to + 50 m high coastal cliff, where they form a discontinuous patchy pattern. Long axes of both types of lobe trend northeast (Neev et al. 1987, Figs. 2, 8, and 23). We therefore assume that these sands migrated inland from the nearby beaches and drifted northeastward across the coastline due to the dominant southwesterly storm winds. However, such a process could not occur today across the coastal segments where the cliff is steep (45 to 75 degrees) and high (+ 30 to + 40 m) and where the beach is narrow (10 to 30 m) (Fig. 6). Under present circumstances, massive transport of beach sand to the top of the cliff, even during the strongest storms, seems highly improbable. The youngest settlements covered by the upper dune layer are a Byzantine site atop the cliff (+ 40 m) at Ga'ash (just north of Shefayim) and early Moslem settlements on the Hamra plains near Ashdod (Fig. 1). The Sharon Plain near the coastal zone in the Apollonia-Arsuf and Ga'ash areas (which now are covered by dunes) was rather heavily forested during Crusader times (Lamb 1930, p 152). Therefore, it is assumed that the upper layer of sand was deposited after Roman times and perhaps a third generation of sand-dune ingression occurred during the past few hundred years. These sand ingressions could have occurred only during the subsidence and submergence phases of the oscillatory movements.

## 3 Significance

The intertidal presence of archeological structures built well above the high tide level is widely accepted as indicating submergence of the land. Similarly, the presence of submerged or emerged piscinas, salt ponds, bath houses, or harbor installations that were built within or near the intertidal zone is widely taken as evidence of submergence or emergence of the land or shift of sea level.

Other evidence is provided by naturally deposited sediments, whereby sediments deposited intertidally but now well above high tide or below low tide level must equally mean a past shift of sea level or land level. The coasts of Israel and Sinai contain many such intertidal sediments now uplifted and exposed in

**Fig. 6.** Sand dunes hanging atop the coastal cliff. Northward view of the coastal cliff south of Netanya (Fig. 1) showing narrow beach (10 to 50 m) and the steep high cliff (20 m to 40 m above beach level). The lower two thirds of the cliff consists of kurkar (friably cemented Late Pleistocene quartz dunes). Hamra (red loam) and the Calcarenite Bank (see also Fig. 2A) overlie the kurkar; the Calcarenite Bank forms an extensive caprock that partly protects the coastal cliff from erosion (*foreground* of photo). Post Byzantine sand dunes up to 10 m high have accumulated on the Calcarenite Bank (*background* of photo) and form northeastward facing barchan dunes. (See also Neev et al. 1987, Figs. 2, 8 and 23)

seacliffs well above high tide. Some are associated (and supported) by archeological evidence. Details of the criteria for former intertidal sedimentary environments along the Israel-Sinai coast are discussed above under categories of trace fossils, shell beds, lagoonal-hypersaline properties, and beaches as sources of dune sands. In other regions different evidence for intertidal deposition is applicable, including coral and coralgal reef deposits, intertidal nips, mangrove roots, disruption by ice shove, and others. The abundance and uniqueness of intertidal sediments characteristics mean that any thorough study of either tectonic or eustatic changes of ancient relative sea level must include study of sediments as well as of the less abundant archeological evidences.

*Acknowledgments.* Thanks are due to Mrs. R. Backman, Mr. S. Levy and Mr. J. Levy of the Geological Survey of Israel for the typing, drafting, and photographing for the manuscript, respectively.

# References

Allen JRL (1984) Experiments on their terminal fall of the valves of bivalve molluscs loaded with sand trapped from a dispersion. Sediment Geol 39:197–209

Avnimelech M (1943) Contribution to the geological history of the Palestinian coastal plain; the surroundings of Nahariya. Bull Jewish Expl Soc 10:39–46 (in Hebrew)

Emery KO (1968) Positions of empty pelecypod valves on the continental shelf. J Sediment Petrol 38:1264–1269

Gavish E, Bakler N (in press) The Hof Hasharon coastal zone-geomorphological and sedimentological factors. In: Shmueli A, Grossman B (eds) The Sharon volume. Reshafim (Hebrew)

Gavish E, Friedman GM (1969) Progressive diagenesis in Quaternary to Late Tertiary carbonate sediments, sequence and time scale. J Sediment Petrol 39:980–1006

Gophna R (1977) Archeological survey of the central coastal plain. Tel Aviv 5 (3–4):136–147

Horowitz A (1979) The Quaternary of Israel. Academic Press, London, p 394

Lamb H (1930) The flame of Islam. Doubleday, New York, p 490

Levy Y (1977) Origin and evolution of brines in the coastal sabkhas, northern Sinai. J Sediment Petrol 47:451–462

Lewy Z, Neev D, Prausnitz M (1986) Late Holocene tectonic movements at Akhziv, Mediterranean coastline of northern Israel. Quat Res 25:177–188

Neev D, Emery KO (1967) The Dead Sea, depositional processes and environments of deposition. Geol Surv Israel Bull 41:147 pp

Neev D, Bakler N, Emery KO (1987) Mediterranean coasts of Israel and Sinai, Holocene tectonism from geology, geophysics, and archeology. Taylor & Francis, New York, p 130

Thatcher W (1985) The earthquake deformation cycle at the Nankai Trough, southwest Japan. J Geophys Res 89:3087–3101

Chapter 6

# Organic Geochemistry

# Amino Acids in Marine Aerosol and Rain

A. SPITZY

## 1 Introduction

In 1984 Galloway et al. (1985) stated in a summary of a workshop on the biogeochemical cycling of sulfur and nitrogen in the remote atmosphere that "organic nitrogens are suspected to be major carriers of N in the remote atmosphere" and that "their concentration is unknown, but could be comparable to the concentration of all other species of reactive and reservoir nitrogen". In a few previous works amino acid data from continental and coastal rain samples have been reported by Fonselius (1954), Munczak (1960), Degens et al. (1964), Dean (1963) and Sidle (1965). Williams (1967) found 23 $\mu$g/kg of organic nitrogen in rain sample collected north of Samoa, contributing 40% of the total dissolved nitrogen. Only 20 years later a first detailed account of organic nitrogen species, namely free dissolved amino acids and primary amines in marine rain was published by Mopper and Zika (1987). They found surprisingly high concentrations of amino acids (primary amines were of secondary importance in terms of abundance), comparable to those of the inorganic nitrogen species measured on the same samples. They, however, could not determine whether the amino acid nitrogen was largely recycled over the ocean or had a terrestrial component. Such a component would represent a net transport of new, biologically utilizable nitrogen to the ocean and hence be an important flux in the biogeochemical nitrogen cycle. As an attempt towards resolving this question we started to sample rain and size-fractionated aerosol from the marine atmosphere for amino acid analysis. They are the building blocks of proteins (which together with sugars comprise 40–80% of the organic material of most organisms) and hence a dominant fraction of biogenically derived organic nitrogen in the environment (Degens and Mopper 1976).

In this chapter we report amino acid data obtained from aerosol and rain sampled during a cruise in the northern Indian Ocean (Bay of Bengal) during spring 1987. It turns out that amino acid nitrogen is enriched several-fold in the fine, non-sea salt fraction with radii smaller than 0.5 micron. This indicates a significant terrestrial component in the atmospheric deposition of organic nitrogen to the sea providing a net source of biologically utilizable organic nutrients.

## 2 Materials and Methods

During cruise S059 of RV "SONNE" to the northern Indian Ocean in spring 1987 size-fractionated aerosols were collected on precombusted glass fibre filters in a high volume five-stage cascade impactor pumping at a rate of about 70 $m^3$/h.

Several 1000 m³ air were collected per sample. Pumps were only active as the ship moved and when the wind was within 90° in the direction of steaming. Pumping was interrupted when other ships came nearer than 5 miles and when it rained. After collection the filters were frozen and stored until analysis in the Hamburg laboratory. Rain samples were collected in a plastic beaker after thorough rinsing with the rain water, transferred to a glass bottle, fixed with mercury chloride and stored in the cool dark prior to analysis. For combined amino acid determination samples were hydrolyzed with 6 N HCl under nitrogen at 110°C for 22 h. The hydrolysate was evaporated to dryness and the residue taken up in a sodium citrate buffer of pH 1.8. A subsample was analyzed on a Biotronik Amino Acid Analyzer (Garrasi et al. 1979). The reproducibility of the method is 10–15%. From the individual size fractions we combined the fractions with radii greater than 0.5 $\mu$ (impactor stages 1,2,3,4) and those with radii smaller than 0.5 $\mu$ (impactor stage 5 and backup) in order to obtain two composite samples representing a coarse and a fine fraction. Table 1 gives details on time and location of sampling and on the volumes of air sampled.

**Table 1.** Background data for aerosol and rain samples

| Sample | Date/time | Location lat. N/long. E | Sampled air volume (m³) |
|---|---|---|---|
| Aerosol: | | | |
| AA | 23.3.87/ 5:25 to 24.3.87/22:10 | 13°25′/84°40′ to 17°25′/89°30′ | 3.600 |
| AB | 30.3.87/11:40 to 1.4.87/13:15 | 11°00′/89°15′ to 4°55′/98°50′ | 4.606 |
| Rain: | | | |
| RA | 18.3.87/16:45 | 4°30′/87°25′ | |
| RB | 20.3.87/ 3:45 | 5°50′/87°25′ | |

## 3 Results

Results of analyses are given in Table 2. The aerosol samples show total amino acid concentrations of 0.47 and 1.13 nM/m³, corresponding to 7.4 and 17.4 ngN/m³ respectively. Amino acids are enriched in the fine fraction by a factor of 2.4 and 4.9 respectively versus the coarse fraction. Mopper and Zika (1987) analyzed aerosol by bubbling air through deionized water and then analyzing the water. They obtained 2.2 nM/m³ of dissolved free amino acids. Although analysis of the combined (hydrolyzable) fraction would most probably have shown a higher value (Mopper and Zika found enrichment of the combined over the monomeric fraction in rain samples with up to a factor of 10), we found the agreement with our data within an order of magnitude quite encouraging, considering the minimal amount of samples collected over so different regions of the ocean and sampled and analyzed by differing methods.

**Table 2.** Hydrolyzable amino acids in marine aerosols and rain

| Aerosol | | | | | | | | | |
|---|---|---|---|---|---|---|---|---|---|
| Sample | Fraction with radii $> 0.5\ \mu$ | | | Fraction with radii $< 0.5\ \mu$ | | | Total | | |
| | (ng C) | (ng N) | (nM) | (ng C) | (ng N) $\times m^{-3}$ | (nM) | (ng C) | (ng N) | (nM) |
| AA | 6.3 | 2.2 | 0.14 | 14.3 | 5.2 | 0.33 | 20.6 | 7.4 | 0.47 |
| AB | 8.9 | 3.4 | 0.19 | 44.6 | 14.0 | 0.94 | 53.5 | 17.4 | 1.13 |

| Rain | | |
|---|---|---|
| Sample | N | |
| | ($\mu$g/kg) | ($\mu$M/kg) |
| RA | 42.4 | 2.6 |
| RB | 27.3 | 1.7 |

The amino acid content of the rain samples was 2.6 and 1.7 $\mu$M/kg, or 42.4 and 27.3 $\mu$gN/kg respectively. This is almost identical to the 3.0 $\mu$M/kg obtained by Mopper and Zika (1987) on the only hydrolyzed marine rain sample which they reported (sample I-3). For free amino acids in rain they obtained an average of 6.5 $\mu$M/kg, ranging between 1.1–15.2 $\mu$M/kg. From the respective averages of our rain and aerosol data we obtained a washout ratio of 2.700, roughly meaning that 1 g water scavenges 2 to 3 $m^3$ of air.

## 4 Discussion

About 90% of all $Na^+$ was on the coarse fraction of the samples collected (G. Gravenhorst, personal communication), representing the maritime aerosol associated with sea spray, as found also in earlier work by Hoffman and Duce (1977). Chesselet et al. (1981) found the stable carbon isotope ratio of organic carbon in aerosols to be similar to marine plankton in the coarse fraction and similar to land plants, soil and products of petroleum combustion in the fine fraction. This distinction between marine particulate carbon associated primarily with sea salt and continental particulate carbon attached to the smallest aerosol particles was substantiated by the stable carbon isotope work of Cachier et al. (1986). Based on these data we assume that the coarse and the fine fractions allow the discrimination between marine and terrestrial organic material in the aerosols sampled.

Under this assumption the fact that amino acids are enriched by factors 2.4 and 4.9 in the fine fraction is evidence for a net terrestrial input of biologically utilizable organic nitrogen to the open sea via the atmosphere, being two- to five-fold enriched over the recycled marine component. The higher total content as well as fine fraction enrichment factor in sample AB is reasonable, since this sample was collected as the ship moved from the central Bay of Bengal towards the coast, thus receiving increasing amounts of continental aerosol, raising the total abundance as well as the fine fraction enrichment.

A contribution of gas to particle conversion processes to the content of amino acids in the fine fraction seems very unlikely given the harsh and highly oxidative environment encountered by aerosols in the atmosphere. Moreover, Mopper and Zika (1987) found that the amino acids which they analyzed were present predominantly as their L-optical isomers and therefore most likely of biological origin.

Global extrapolation of our rain amino acid data would result in an estimated annual input of 8.3–12.8 $\times$ $10^{12}$ g amino acid nitrogen to the ocean between 60°S–65°N, based on the estimate of annual precipitation for the world ocean by Elliott and Reed (1984). Since the Bay of Bengal cannot be considered a remote oceanic region, this extrapolation may overestimate the true global flux. The estimate, however, is not far from the value obtained by extrapolating Williams' (1967) organic nitrogen measurement in rain from the Pacific, which would result in 7 $\times$ $10^{12}$ g organic nitrogen deposition by rain annually.

Assuming that the rain got its amino acids by aerosol scavenging and taking 3 as a representative enrichment factor for organic matter in the fine aerosol fraction, the resulting "terrestrial" component would be 6.2–9.6 $\times$ $10^{12}$ gN/year. Wollast (1983) estimates the flux of inorganic nitrogen by rivers to the sea at 2.6 $\times$ $10^{12}$ gN/year under pristine conditions and 24 $\times$ $10^{12}$ gN/year for the present, anthropogenically perturbed state. Ittekkot et al. (1983) estimated the organic nitrogen carried by rivers to the sea at 14 $\times$ $10^{12}$ gN/year, of which 8.2 $\times$ $10^{12}$ gN is in the particulate and the rest in the dissolved form. The atmospheric wet deposition of inorganic nitrogen over the ocean was estimated by Galloway (1985) at 0.07 gN $m^{-2}$ $year^{-1}$, which results in an annual wet deposition flux of 22.9 $\times$ $10^{12}$ gN to the ocean between 60°S–65°N. The dry deposition was estimated at 20% of the wet deposition.

In summary, the data presented indicate a terrestrially derived flux of amino acid nitrogen to the ocean within the order of magnitude of the atmospheric as well as the riverine inputs of inorganic nitrogen and close to the estimate for the river to ocean flux of dissolved organic nitrogen.

*Acknowledgements.* I thank H. Kreilein and M. Müller for taking the samples during the cruise and M. Sternhagen for laboratory assistance.

## References

Cachier H, Buat-Menard P, Fontugne M, Chesselet R (1986) Long-range transport of continentally-derived particulate carbon in the marine atmosphere: evidence from stable carbon isotope studies. Tellus 38B:161–177

Chesselet R, Fontugne M, Buat-Menard P, Ezat U, Lambert CE (1981) The origin of particulate organic carbon in the marine atmosphere as indicated by its stable carbon isotope composition. Geophys Res Letters 8:345–348

Dean GA (1963) The iodine content of some New Zealand drinking waters with a note on the contribution from sea spray to the iodine in rain. New Zealand J Sci 6:208–214

Degens ET, Hunt JM, Reuter H, Reed WE (1964) Data on the distribution of amino acids and oxygen isotopes in petroleum brine waters of various geologic ages. Sedimentology 3:199

Degens ET, Mopper K (1976) Factors controlling the distribution and early diagenesis of organic material in marine sediments. In: Riley JP and Chester R (eds) Chemical oceanography 6, Academic Press, pp 59–113

Duce R (1986) The impact of atmospheric nitrogen, phosphorus, and iron species on marine biological productivity. In: Buat-Menard P (ed) The role of air-sea exchange in geochemical cycling. Reidel, Dordrecht Boston Lancaster Tokyo (NATO ASI Series), pp 497–529
Elliott WP, Reed RK (1984) A climatological estimate of precipitation for the world ocean. J Clim Appl Meteor 23:434–439
Fonselius S (1954) Amino acids in rain water. Tellus 6:90
Galloway JN (1985) The deposition of sulfur and nitrogen from remote atmosphere. In: Galloway JN, Charlson RJ, Andreae MO, Rhode H (eds) The biogeochemical cycling of sulfur and nitrogen in remote atmosphere. Reidel, Dordrecht Boston Lancaster Tokyo (NATO ASI Series), pp 143–175
Galloway JN, Charlson RJ, Andreae MO, Rhode H (1985) Chapter 11 Summary. In: Galloway JN, Charlson RJ, Andreae MO, Rhode H (eds) The biogeochemical cycling of sulfur and nitrogen in the remote atmosphere. Reidel, Dordrecht Boston Lancaster Tokyo (NATO ASI Series), pp 215–224
Garrasi C, Degens ET, Mopper K (1979) Amino acid composition of seawater obtained without desalting. Marine Chemistry 8:71–85
Hoffmann EJ, Duce R (1977) Organic carbon in marine atmospheric particulate matter: concentration and particle size distribution. Geophys Res Lett 4:449–452
Ittekkot V, Martins O, Seifert R (1983) Nitrogenous organic matter transported by the major world rivers. In: Degens ET, Kempe S, Soliman H (eds) Transport of carbon and minerals in major world rivers. SCOPE/UNEP Sonderband 55, Mitt Geol Palöont Inst Univ Hamburg, pp 119–127
Mopper K, Zika R (1987) Free amino acids in marine rains: evidence for oxidation and potential role in nitrogen cycling. Nature 325:246–249
Munczak F (1960) On the appearance of ninhydrin-positive substances in the atmosphere. Tellus 12:282
Sidle AB (1965) Amino acid content of atmospheric precipitation. Tellus 19:128
Williams PM (1967) Sea surface chemistry: organic carbon and organic and inorganic nitrogen and phosphorus in surface films and subsurface waters. Deep-Sea Res 14:791–800
Wollast R (1983) Interactions in estuaries and coastal waters. In: Bolin B, Cook RC (eds) The major biogeochemical cycles and their interactions. Wiley & Sons, Chichester New York Brisbane Toronto Singapore (SCOPE 21), pp 385–407

# The Terrestrial Link in the Removal of Organic Carbon in the Sea

V. Ittekkot and B. Haake

## 1 Introduction

Until recently it has been assumed that most of the organic matter being buried in the sea is of marine origin (Holland 1978). New studies suggest, however, that terrestrial carbon may be an important constituent of organic carbon accumulating in marine sediments. For example, a recent estimate of riverine refractory terrigenous organic carbon almost equals the global organic carbon burial reported earlier (e.g. Ittekkot 1988; Berner 1982). The implication is that most of the organic carbon being buried in modern marine sediments may have a terrestrial origin.

The factors controlling organic carbon preservation in marine sediments are not yet fully known and the subject is still a matter of debate. Previous discussions have mainly centered on the possible role of bottom water oxygen contents (e.g. Emerson 1985) and of sedimentation rates (e.g. Müller and Suess 1979). In this article we discuss the role of lithogenic material in the removal of organic carbon in the sea based on information obtained from recent studies on particle sedimentation in the sea.

## 2 Particle Sedimentation in the Sea

Recent studies on marine sedimentation done with the help of sediment traps moored at various depths in oceanic regions show that planktonic organisms and their metabolic waste products play an important role in the downward flux of particles (e.g. Deuser et al. 1981; Honjo 1982). This participation leads to the formation of large particle aggregates which, though rare, contribute to most of the vertical flux out of the ocean surface (Suess 1980). These particles (e.g. fecal pellets, macroaggregates, marine snow) consist of intact phytoplankton, mineralized and nonmineralized tissues of zooplankton, and lithogenic material (e.g. Fowler and Knauer 1986; Alldredge and Silver 1988).

Filter-feeding zooplankton ingest all sorts of particles from seawater and eject them in the form of fecal pellets. Depending on the feeding pattern of the organisms involved and on the available food supply they contain varying quantities of biogenic and mineral matter (Pilskaln and Honjo 1987; Bienfang 1980). The production of macroaggregates or "marine snow" appears to involve the release of dissolved organic matter by: (i) actively growing phytoplankters, (ii) during cell lysis, and (iii) during grazing by zooplankton (e.g. Alldredge and Silver 1988; for a review). For example, under conditions of nutrient depletion extracellular production and release of polysaccharides is high (Harris and Mitchel 1973; Degens and Ittekkot 1984). The sticky nature of this newly excreted material facilitates the aggregation of fine particles into large aggregates.

Both fecal pellets and macroaggregates can be transported to the deep sea within a short time due to their high sinking rates which can be more than 1000 m $d^{-1}$ for fecal pellets (e.g. Bruland and Silver 1981; Komar et al. 1981) and up to 400 m $d^{-1}$ for marine snow (Shanks and Trent 1980; Asper 1987; Taguchi 1982). Previous studies have shown that sinking particles can contain a significant amount of lithogenic material introduced from various sources (Honjo et al. 1982b; Deuser et al. 1983; Monaco et al. 1987). They include riverine and eolian particles as well as material introduced by horizontal transport processes. The observed seasonality in the downward flux of particles in the ocean in phase with surface biological productivity is a result of this biological mediation over particle sedimentation (Deuser et al. 1981; Honjo 1982).

## 3 Decomposition of Particles

The fate of macroaggregates and fecal pellets during their descent to the deep-sea is still a matter of debate since little is known about the nature of biological cycling in the deep sea. In many sediment trap studies the organic carbon associated with the particle flux shows a decrease with depth (e.g. Suess 1980) which has been interpreted as resulting from organic matter decomposition on particles. Microbial activity on fecal pellets leads to the disintegration of the carbohydrate-rich peritrophic membrane and enhances the breakdown of particles (Honjo 1976). Macroaggregates are also prone to bacterial attack and decomposition (Alldredge and Silver 1988). Their subsequent disintegration might lead to the release of fine particles containing fresh organic matter. Production of new microbial biomass at the expense of this organic matter might in part explain observations of particle maxima in mid-water (Karl and Kanuer 1984) though they might also be a consequence of deep-sea zooplanktonic activity (Walsh et al. 1988).

However, recent studies suggest that sinking particles are poor habitats of microbial activity which is mostly associated with free living bacteria or with those attached to suspended particles (Karl et al. 1988). Decomposition occurs mainly on suspended particles derived from the disintegration of large aggregates during sinking. The role of attached bacteria is mainly to make available dissolved organic matter from particles to be subsequently utilized by free living bacteria (Cho and Azam 1988).

## 4 Role of Mineral Particles: The "Ballast" Effect

In any event, the disintegration of macroaggregates in the water column will depend on how long the aggregates remain suspended there, i.e. on their sinking speeds, which will be directly related to the type of materials involved. It has been suggested that the incorporation of mineral particles such as clays into large aggregates can increase their settling rates (Fowler and Knauer 1986). This facilitates the rapid transfer of both the minerals and the freshly produced organic matter to the deep ocean, or in shallow water environments, to deltaic and shelf sediments. In the deep-sea, bottom-dwelling communities are thus assured of a supply of edible organic matter which is associated with rapidly sinking (hence less degraded) large particle aggregates formed during periods of surface biological productivity (Ittekkot et al. 1984; Tyler 1988). This, and the low sedimentation rates there will entail decomposition of organic matter at the sediment-water interface (Müller and Suess 1979) or in a region close to it – the benthic transition zone of Honjo et al. (1982a) (Fig. 1). Deltaic and shelf environments where an adequate supply of terrigenous mineral matter exists such as the mouths of rivers, and those offshore marine regions influenced by atmospheric dust fallout can be different. For example, in the former, both in situ primary productivity and the inputs of mineral

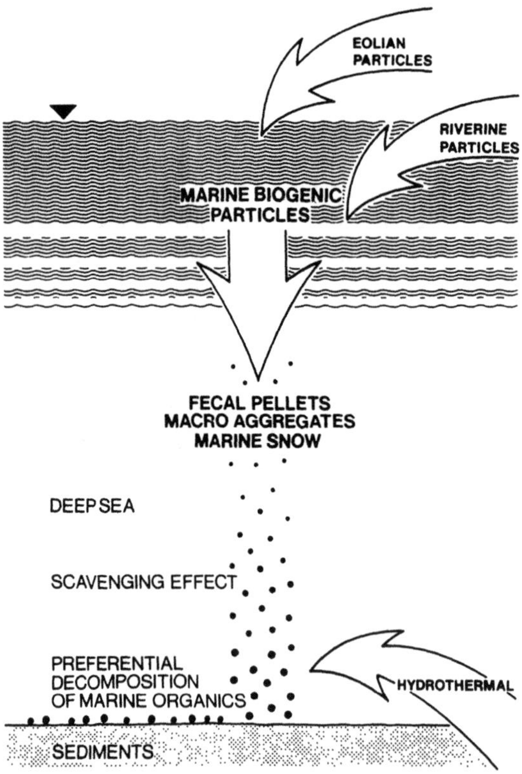


**Fig. 1.** Scheme of particle removal in today's oceans. Particles arriving at the sea surface via dry and wet fallout, and those entering it via the rivers are rapidly removed along with marine biogenic particles to the sea floor. This removal is mediated by biological processes leading to the formation of large particle aggregates such as fecal pellets, macroaggregates, marine snow, which subsequently serve as vehicles for the rapid transport of particles in transit. Once at the seafloor the metabolizable fraction is effectively reworked by bottom-dwelling communities. The allochthonous refractory fraction gets accumulated in sediments along with the biogenic hard parts

and organic matter from terrestrial sources are high. We suggest that the resulting interaction between terrestrial particles and marine organic matter leads to the formation of large particle aggregates which settle rapidly to the sea bottom and facilitate the removal of organic matter derived from both land and sea.

The zones of biological productivity in the coastal seas thus act as biological barriers restricting the dispersal and deposition of fine-grained river sediments within the deltaic and coastal zones. Distinct zonations of wind- and river-derived sediments observed in areas such as off the coasts of Peru and Chile have been suggested to be a result of such barriers (Scheidegger and Krissek 1982). Exceptions are materials derived from rivers such as the Ganges and Brahmaputra which have direct deep-sea connections via canyons. There the zones of deposition of river-derived sediments may be shifted to the lobes of the deep-sea fans away from the landmasses.

## 5 Carbon Input and Burial in the Sea

Distribution of carbon input over the ocean according to water depth and the volume of water over each depth interval shows that the major organic carbon input occurs within a small volume of water over the shelves (Fig. 2; Deuser 1979). These are regions of high primary productivity and they receive large amounts of terrigenous materials discharged by rivers. The burial of organic carbon in the modern ocean distributed according to the major sediment types shows that 83% of burial occurs in deltaic-shelf sediments (Table 1; Berner 1982). Deep-sea envi-

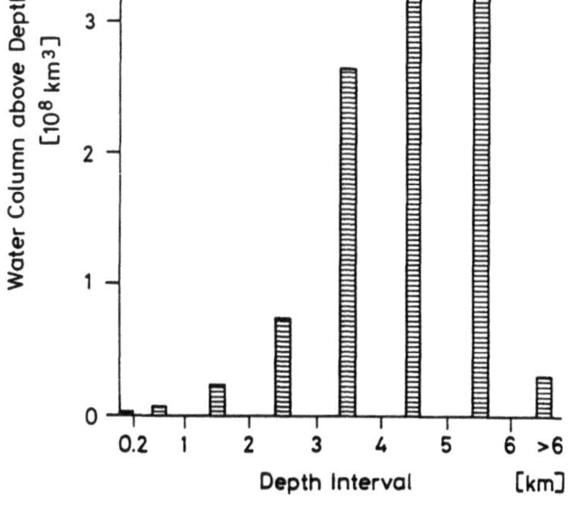


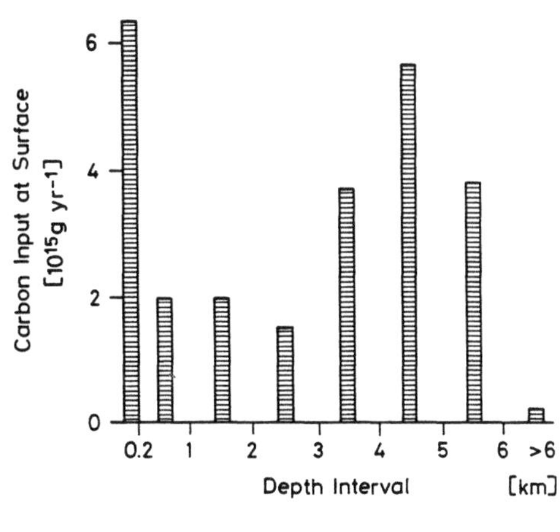


**Fig. 2.** Distribution of volume of water above different depth intervals of the sea floor and the input of organic carbon to the surface waters above the same depths (after Deuser 1979 from Degens and Ittekkot (1987))

**Table 1.** Organic carbon burial in marine sediments (from Berner 1982)

| Sediment type | Organic Carbon Burial ($\times 10^{12}$g) |
|---|---|
| Deltaic-shelf sediments | 130 |
| Biogenous sediments underlying regions of high productivity | 14 |
| Shallow water carbonate sediments | 7 |
| Pelagic sediments (not overlain by regions of high productivity) | 6 |
| Total | 157 |
| Modern day river inputs of refractory organic carbon (Ittekkot 1988) | 150 |

ronments (water depths $> 2000$ m) despite a higher input of organic carbon input at the surface, accounts for no more than 4% of the global annual carbon burial (Deuser 1979; Berner 1982). We interpret such a distribution pattern to be a result of the ballasting effect of lithogenic materials introduced by the rivers, which accelerate the removal from the water column of newly-fixed organic matter to sediments. Additionally, this mineral-organic association leads to an increase in sedimentation rates and subsequently to better preservation of organic matter. In offshore environments with low or little lithogenic inputs degradation of macroaggregates occur within the water column. Exceptions could be offshore areas with lithogenic inputs associated with dust fallout, where newly-produced organic matter can get rapidly transferred to the deep-sea floor. It is interesting in this regard to examine the hypothesis proposed by Tyler (1988) which links the downward flux of organic matter from surface primary productivity and the processes occurring in the deep-sea. It appears that only a limited number of species respond to the vertical flux of organic matter in their physiological processes. It is conceivable that metabolizable organic matter from the surface layers of the open ocean reaches the deep-sea floor only in regions where there is also an adequate supply of lithogenic material.

## 6 Geochemical and Environmental Implications

Thus there appears to be a continental link in the removal of newly fixed carbon to either permanent sinks – deltaic and coastal seas or to transient sinks in the deep-sea. It suggests that terrestrial and recycled organic matter associated with mineral ballasts such as clays will be an ubiquitous component of organic matter accumulating in marine sediments, and may in fact determine even its distribution pattern there.

This continental connection has probably been crucial in the removal of vast quantities of organic carbon to sediments in the recent and ancient geological past. It has been suggested that the low atmospheric $CO_2$ content during the glacials is a result of increased oceanic primary productivity, and Martin and Fritzwater

(1988) have shown the role of iron associated atmospheric dust fallout in stimulating ocean productivity. Transfer of this newly produced organic matter from the surface layers to the deep-sea can be accelerated by the incorporation of mineral particles associated with the dust fallout into macroaggregates. The low sea levels at that time also mean that rivers discharge their materials directly at the shelf edge. In addition, previously deposited sediments on deltaic and shelf environments can be eroded by rivers and wave action and can be mobilized and transported offshore. Thus the glacial oceans might have been different from the modern ones not only in terms of primary productivity but also in terms of the availability of "lithogenic ballasts" which induce the rapid transfer of newly fixed carbon to the deep-sea. The resulting high sedimentation rates will lead to the preservation of large quantities of organic matter derived from sources as varied as in situ primary productivity, land-derived organics and recycled organics derived from eroded deltaic and shelf sediments. The stable carbon isotope ratios of benthic foraminifera from glacial times suggest rapid recycling of part of the organic matter derived from in situ primary productivity at the seafloor. This probably leaves the refractory terrestrial and recycled organic matter as the major sedimentary components. We further suggest that the distribution pattern of organic carbon-rich sediments encountered through the Phanerozoic record (e.g. Berner and Raiswell 1983; Stein et al. 1986) reflects essentially the variations in the availability of lithogenic ballasts through time.

The suggested terrestrial link in organic carbon preservation in the sea implies that man's perturbation of aquatic systems on land with sea-connections might seriously affect the carbon removal rates in the sea. Deforestation in the drainage areas of rivers has increased the sediment load of major rivers. If our hypothesis is correct, then this might lead to higher organic carbon removal rates in seas to which these rivers drain. On the other hand, constructions of dams and barrages in the drainage areas of major rivers, and the diversion of rivers for purposes of irrigation have reduced the river inputs of sediments to the sea (Milliman et al. 1984). This will not only enhance coastal erosion and but also will reduce the input of natural ballasts which are instrumental in carbon removal and preservation. By changing the sediment load of the rivers we are changing biogeochemical cycling of elements in regions where more than 80% of organic carbon is being removed today (Berner 1982).

## References

Alldredge AL, Silver MW (1988) Characteristics, dynamics and significance of marine snow. Prog Oceanogr 20:41–82

Asper VL (1987) Measuring the flux and sinking speed of marine snow aggregates. Deep-Sea Res 34:1–17

Berner RA (1982) Burial of organic carbon and pyrite sulfur in modern ocean: its geochemical and environmental significance. Am J Sci 282:451–473

Berner RA, Raiswell R (1983) Burial of organic carbon and pyrite sulfur in sediments over Phanerozoic time: a new theory. Geochim Cosmochim Acta 47:855–862

Bienfang PK (1980) Herbivore diet affects fecal pellet settling. Can J Fish Aquat Sci 37:1352–1357

Bruland KW, Silver MW (1981) Sinking rates of fecal pellets from gelatinous zooplankton (salps, pteropods, doliolids). Mar Biol 63:295–300

Cho BC, Azam F (1988) Major role of bacteria in biogeochemical fluxes in the ocean's interior. Nature (Lond) 332:441–443

Degens ET, Ittekkot V (1984) A new look at clay-organic interactions. Mitt Geol-Paläont Inst Univ Hamb Festband Georg Knetsch 56:229–248

Degens ET, Ittekkot V (1987) The carbon cycle-tracking the path of organic particles from sea to sediment. In: Brooks J, Fleet AJ (eds) Marine Petroleum Source Rocks. Blackwell, Oxford, pp 121–135 (Geol Soc Sp Publ 26)

Deuser WG (1979) Marine biota, nearshore sediments, and the global carbon balance. Org Geochem 1:243–247

Deuser WG, Ross EH, Anderson RF (1981) Seasonality in the supply of sediment to the deep Sargasso Sea and implications for the rapid transfer of matter to the deep ocean. Deep-Sea Res 28:495–505

Deuser WG, Brewer PG, Jickells TD et al. (1983) Biological control of the removal of abiogenic particles from the surface ocean. Science 214:388–391

Emerson S (1985) Organic carbon preservation in marine sediments. In: Sundquist ET, Broecker WS (eds) The carbon cycle and atmospheric $CO_2$: Natural variations Archean to Present. AGU, Washington, pp 78–87 (Geophys Monogr 32)

Fowler SW, Knauer GA (1986) Role of large particles in the transport of elements and organic compounds through the oceanic water column. Prog Oceanogr 16:147–194

Harris RH, Mitchell R (1973) The role of polymers in microbial aggregation. Ann Rev Microbiol 27:27–50

Holland HD (1978) The chemistry of the atmosphere and the oceans. Wiley, Toronto, 351 pp

Honjo S (1976) Coccoliths: production, transportation and sedimentation. Mar Micropaleontol 1:65–79

Honjo S (1982) Seasonality and interaction of biogenic and lithogenic particulate flux at the Panama Basin. Science 218:883–884

Honjo S, Manganini SJ, Cole JJ (1982a) Sedimentation of biogenic matter in the deep ocean. Deep-Sea Res 29:609–625

Honjo S, Manganini SJ, Poppe LJ (1982b) Sedimentation of lithogenic particles in the deep ocean. Mar Geol 50:199–220

Ittekkot V (1988) Global trends in the nature of organic matter in river suspensions. Nature (Lond) 332:436–438

Ittekkot V, Deuser WG, Degens ET (1984) Seasonality in the fluxes of sugars, amino acids and amino sugars to deep-ocean. Sargasso Sea. Deep-Sea Res 31:1057–1069

Karl DM, Knauer GA (1984) Vertical distribution, transport, and exchange of carbon in the northeast Pacific Ocean: evidence for multiple zones of biological activity. Deep-Sea Res 31:221–243

Karl DA, Knauer GA, Martin JH (1988) Downward flux of particulate organic matter in the ocean: a particle decomposition paradox. Nature (Lond) 332:438–441

Komar PD, Morse AP, Small LF, Fowler SW (1981) Analysis of sinking rates of natural copepod and euphausiid fecal pellets. Limnol Oceanogr 26:172–180

Martin JH, Fritzwater SE (1988) Iron deficiency limits phytoplankton growth in the north-east Pacific subarctic. Nature (Lond) 331:341–343

Milliman JD, Quraishee GS, Beg MAA (1984) Sediment discharge from the Indus River to the ocean: Past, present and future. In: Haq BU, Milliman JD (eds) Marine Geology and Oceanography of the Arabian Sea and Coastal Pakistan. Reinhold, New York, pp 65–70

Monaco A, Heussner S, Courp T et al. (1987) Particle supply nepheloid layers on the north-western mediterranean margin. Mitt Geol-Paläont Inst Univ Hamb SCOPE/UNEP Sonderbd 62:100–109

Müller PJ, Suess E (1979) Productivity, sedimentation rate, and sedimentary organic matter in the oceans – I. Organic carbon preservation. Deep-Sea Res 26:1347–1362

Pilskaln CH, Honjo S (1987) The fecal pellet fraction of biogeochemical particle fluxes to the deep sea. Global Biogeochem Cycles 1:31–48

Scheidegger KF, Krissek LA (1982) Dispersal and deposition of eolian and fluvial sediments off Peru and northern Chile. Geol Soc Am Bull 93:150–162

Shanks AL, Trent JD (1980) Marine snow: Sinking rates and potential role in vertical flux. Deep-Sea Res 27:137–144

Stein R, Rullkötter J, Welte DH (1986) Accumulation of organic-carbon-rich sediments in the Late Jurassic and Cretaceous Atlantic Ocean-a synthesis. Chem Geol 56:1-32
Suess E (1980) Particulate organic carbon flux in the oceans: surface productivity and oxygen utilization. Nature (Lond) 288:260-262
Taguchi S (1982) Seasonal study of fecal pellets and discarded houses of appendiculatia in a subtropical inlet, Kaneoha Bay, Hawaii. Estuar Coast Shelf Sci 14:545-555
Tyler PA (1988) Seasonality in the deep sea. Oceanogr Mar Biol Annu Rev 26:227-258
Walsh JJ, Dymond J, Collier R (1988) Rates of recycling of biogenic compounds of settling particles in the ocean derived from sediment trap experiments. Deep-Sea Res 35:43-58

# Geochemistry and Origin of the Holocene Sapropel in the Black Sea

S.E. Calvert

## 1 Introduction

The Black Sea, the world's largest permanently anoxic marine basin, is a quasi-steady state stratified reservoir with a deep-water replacement time of around 1800 years (Ostlund 1974). The anoxic water mass below the brackish surface layer extends from a depth of 150 to 300 m to the bottom at 2200 m and is stabilized by the supply of Mediterranean water via the 35-m-deep Bosporus sill. The size of the basin and the stability of its hydrographic structure make it an ideal natural laboratory for studies of the chemistry of anoxic marine systems and the geochemistry of permanently anoxic marine sediments. The Black Sea is commonly used as the "type" anoxic basin (see Glenn and Arthur 1985) and as a typical modern analogue for the environment of formation of carbonaceous shales and petroleum source beds (Demaison and Moore 1980). This notion evidently stems from the view that water-column anoxia promotes the preferential preservation of deposited organic matter in marine sediments and that the modern sediments of the Black Sea are indeed enriched in organic matter. The latter view was mentioned by Woolnough (1937) and has been carried into much of the recent literature on this subject.

The nature of the sediments accumulating in the Black Sea have been studied extensively by Soviet workers (see Strakhov 1967, 1969 for a summary), and more recently by other scientists as a result of the cruise of the ATLANTIS II of the Woods Hole Oceanographic Institution (WHOI) in 1969 (see Degens and Ross 1972 for a compilation of results). The early Soviet work has been described by Andrusov (1980), Arkangel'skiy (1927) and Arkangel'skiy and Strakhov (1932), and summaries are given by Caspers (1957) and Strakhov (1967, 1969). The cruise of ATLANTIS II resulted in the collection of a large suite of gravity and piston cores whose lithology and stratigraphy have been described by Ross et al. (1970), Degens and Ross (1972) and Ross and Degens (1974). Using a single core (1974) from the central part of the Black Sea (Fig. 1), Degens and Ross (1972) showed that there are three distinct stratigraphic units in the uppermost 1 m of the deep-water sediments. The youngest is a modern, finely laminated coccolith marl (Recent horizon of Soviet workers; Unit 1 of Ross et al. 1970). This is underlain by a microlaminated sapropel (Old Black Sea horizon of Soviet workers; Unit 2 of Ross et al. 1970). Laminated, carbonate-poor clays (New Euxinian horizon of Soviet workers; Unit 3 of Ross et al. 1970) comprise the oldest unit recovered. These three units can be traced over most of the deep-water area of the Black Sea and reflect the evolution of the hydrographic conditions in the basin following the last glaciation. Thus, Unit 3 is interpreted to be a lacustrine facies that formed during lowered sea level when

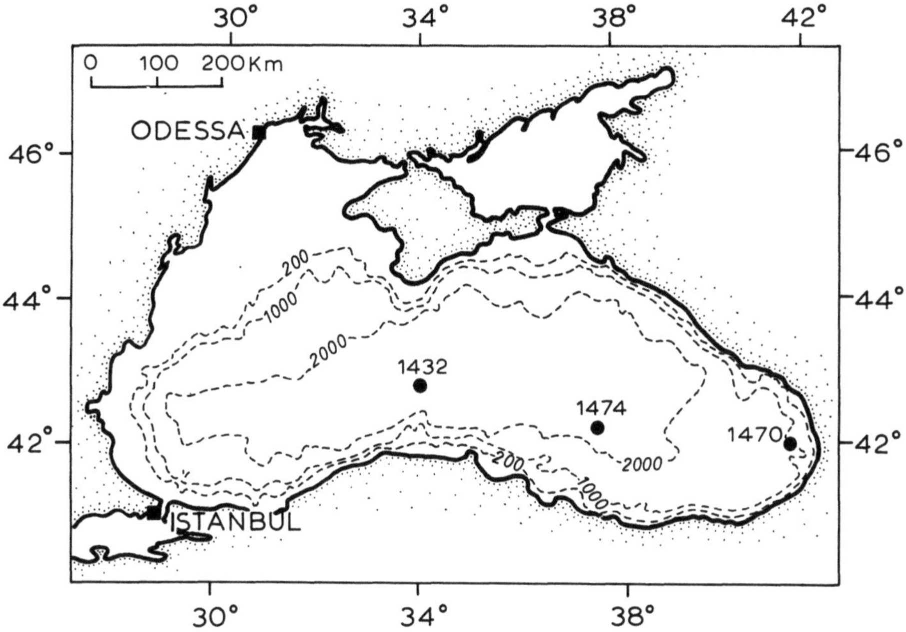


**Fig. 1.** Location of cores discussed in this paper. Core 1474 was used by Degens and Ross (1972) to establish a standard Holocene sediment section in the Black Sea. Cores 1432 and 1470 are examined in this paper

the Black Sea was isolated from the Mediterranean. The sapropel is considered to have formed during the beginning of the anoxic stage of the basin when the connection with the Mediterranean was re-established. And the youngest horizon represents the deposit formed when the salinity of the basin waters reached the present value (22.4‰ at 2000 m) and coccoliths were introduced from the Mediterranean.

Calvert et al. (1987) provided more detail on two of the ATLANTIS II cores, and showed that the pre-modern sapropel is separated from the modern facies by a variably thick mud-flow layer. Such mud-flow layers were originally described from a suite of cores collected by the CHAIN (WHOI) in 1974 and were described by Degens et al. (1978, 1980). Calvert et al. (1987) also showed that the sapropel had age ranges of 1600 to 6600 yr BP, allowing for a correction to the radiocarbon ages by the varve chronology of the sediments established by Degens et al. (1978, 1980).

Glenn and Arthur (1985) and Calvert et al. (1987) have presented estimates of the carbon accumulation rate in the modern facies and the sapropel horizon using data obtained from the available ATLANTIS II cores. They showed that the accumulation rate was higher in the sapropel compared with the modern horizon. Calvert and Fontugne (1987) showed, furthermore, that the organic matter in the sapropel has a predominantly planktonic carbon isotopic signature, whereas the superjacent mud-flow layer and the modern facies contain mixed planktonic and terrestrial organic matter. On the basis of this information, Calvert and Fontugne

(1987) have argued that the sapropel formed as a consequence of increased plankton production during a period of transition from the lake to the modern marine phase of the basin.

The geochemistry of the sediments of the Black Sea has been examined by a number of Soviet workers, notably Baturin et al. (1967), Kochenov et al. (1965), Ostrumov et al. (1961), Pilipchuk and Volkov (1968), Strakhov (1963, 1971), Strakhov et al. (1971) and Volkov and Fomina (1971, 1972). Hirst (1974) reported the major and minor element composition of a suite of the ATLANTIS II cores which showed that there are wide variations in the bulk composition of the various sedimentary units. The purpose of this paper is to re-examine this variability and to examine the extent to which the composition of the sediments is consistent with the formation of the sapropel by increased fertility of the basin and whether the distribution of selected major and minor elements can be used to determine the environment under which the various facies formed, and in particular the degree of oxygenation of the basin during the accumulation of the sapropel.

## 2 Materials and Methods

Samples were obtained from two gravity cores (1432 and 1470) collected by the ATLANTIS II in 1969 (Fig. 1). The samples received from the Woods Hole core repository were dried to constant weight at 70°C and ground to minus 200 mesh powders in tungsten carbide. The bulk mineralogy was determined by standard X-ray diffraction (XRD) techniques using randomly-oriented, pressed-powder mounts. Peak height ratios of the principal minerals were used as a measure of the relative abundances of the phases in the samples. Total carbon and carbonate carbon were determined by dry combustion and by measurement of the $CO_2$ released by treatment with hot 10% HCl, respectively. Organic carbon was obtained by difference. The measurements have a precision of ± 3%.

The major elements (Si, Ti, Al, Fe, Ca, Mg, K and P) were determined by X-ray fluorescence (XRF) using a modification of the fusion-heavy absorber method of Norrish and Hutton (1969). 400 mg sub-samples were mixed with 3.6 g of Spectroflux 105 (Johnson-Matthey Co. Ltd: 47% $Li_2B_4O_7$, 36.7% $LiCO_3$ and 16.3% $La_2O_3$) and fused at 1100°C in a muffle furnace. After cooling, sufficient $Li_2B_4O_7$ (Spectroflux 100) was added to the glass to compensate for the weight-loss on fusion, the sample was remelted and the final glass bead was prepared using the method of Harvey et al. (1973). Calibration of the method was obtained by means of a series of international geochemical reference standards (see Abbey 1980) prepared in the same way.

The minor elements (Ba, Br, Co, Cr, Cu, I, Mn, Mo, Ni, Pb, Rb, Sr, V, Y, Zn and Zr) were also determined by XRF using a method similar to that described by Harvey and Atkin (1982). The samples in this case were prepared by forming a mixture of 4 g sample and 0.5 g finely divided wax (Hoechst Wax C) into 32-mm diameter discs in a hydraulic press. Calibration was again by reference to a set of international standards similar to those used for the major element analyses prepared in the same way. Additional synthetic standards for Br, I and Mo, for which there are few, or no, reference standards, were prepared by diluting appropriate salts with a standard granite (NIM-G). Chlorine and S were also determined by XRF using a third sample preparation technique. In this case 0.2 g of sample was mixed with 0.8 g of Spectroflux 105 and formed into a 32-mm disc. This method was chosen because of the volatility of these two elements under the fusion conditions used for preparing the major element beads and because some matrix modification (by means of the La in the Spectroflux 105) is necessary for these major constituents. Calibration standards were synthetic mixtures of NaCl and $K_2SO_4$ in a standard granite (NIM-G). The Cl values were used to correct all analyses for the diluting effect of salt in the dried samples and for the contribution of Ca, Mg, K, S, Br and Sr to the samples from the sea-salts.

Because the S values obtained by the XRF procedure differed significantly from the total S values published by Ostrumov et al. (1961) and Hirst (1974), separate sub-samples were analyzed by a combustion/amperometric titration procedure described by Guthrie and Lowe (1984). In this case 20 to 50 mg samples were used and the method had a precision of $\pm$ 2.5%. The values obtained by the two methods were in satisfactory agreement.

The major and minor element beads and discs were analyzed using a Philips 1400 fully-automatic X-ray spectrometer under computer control. All calibration curves were linear over the full working range, and analytical precision was better than $\pm$ 3% for the major elements and better than $\pm$ 5% for the minor elements.

# 3 Results and Discussion

## 3.1 General Features

The bulk chemical data (Appendix 1) show that there are substantial variations in the chemical composition of the two cores studied and this reflects the rather complex stratigraphy involved (see Calvert et al. 1987). The distribution of organic C and $CaCO_3$ (derived from the $CO_2$ values assuming no other carbonate-bearing phase is present) can be conveniently used to define this stratigraphy (Fig. 2). In core 1432, the upper 25 cm (Unit A) consists of laminated, coccolith-bearing marls with carbon values ranging from 0.66 to 3.88%, the variability reflecting the laminated nature of the unit. Between 25 and 63 cm (Unit B), the sediment is very homogeneous, with 1.63 $\pm$0.09% C and 13.4 $\pm$0.17% $CaCO_3$. A sapropel (Unit C) lies below 63 cm, the organic C content increasing to a maximum of 14.4% and decreasing rapidly to the base of the core. In core 1470 (Fig. 2), the uppermost 36 cm (Unit A) is again a laminated, coccolith-bearing marl with 1.3 to 2.1% organic C and 12.0 to 24.8% $CaCO_3$. Beneath this unit is another homogeneous unit (Unit B), 90 cm thick, with 1.59 $\pm$0.35% C and 7.5 $\pm$0.94% $CaCO_3$. A sapropel (Unit C) lies between 126 and 165 cm, with a maximum C content of 4.29%. Unit D has the lowest C contents of the entire core and, excluding the sample at 167–169 cm, 15.6 to 26.0% $CaCO_3$.

The various sedimentary units distinguished by the carbon and carbonate profiles have been interpreted by Calvert et al. (1987) to represent (1) the modern

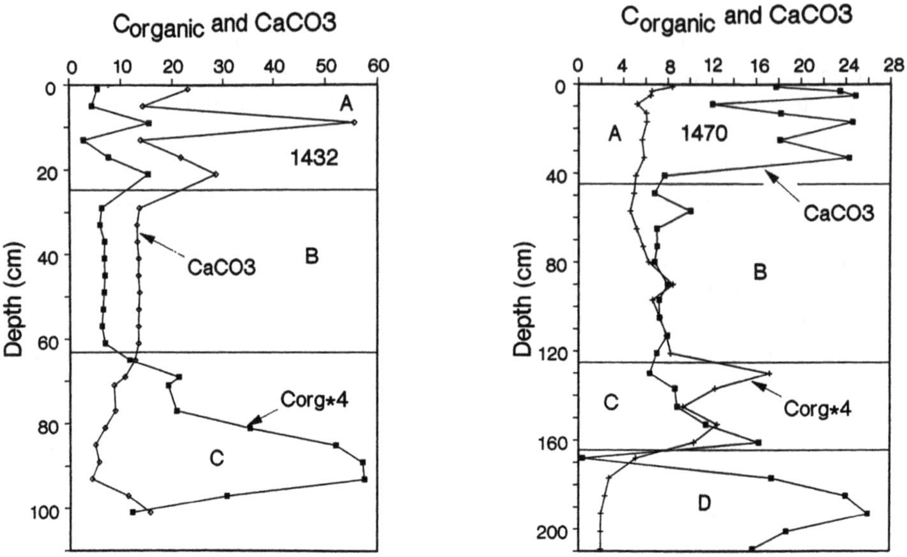


**Fig. 2.** Vertical distribution of organic carbon and calcium carbonate in cores 1432 and 1470. Lithological units *A*-*D* are explained in the text. Note the different depth scales for the two cores. The organic C values have been multiplied by 4

sediment facies of the Black Sea (Unit A) which is equivalent to the Recent horizon of Soviet workers (Arkangel'skiy and Strakhov, 1938) and Unit 1 of Ross et al. (1970), (2) the Recent sapropel (Unit C) which is equivalent to the Old Black Sea Horizon of Soviet workers and Unit 2 of Ross et al. (1970), and (3) laminated, calcareous (lacustrine?) clays (Unit D) which are equivalent to the New Euxinic horizon of Soviet workers and Unit 3 of Ross et al. (1970). The homogeneous unit (Unit B) separating the modern facies and the sapropel in cores 1432 and 1470 is interpreted to be a mud-flow or turbidite horizon; such sediments are evidently very common in cores collected from the deep Black Sea (Degens et al. 1980), although they have not previously been described from the cores collected by the ATLANTIS II in 1969.

### 3.2 Mineralogy

The main features of the mineralogical variation in the two cores are summarized in Fig. 3. In core 1432, there is more chlorite and smectite and less quartz relative to illite in Unit A, and these trends are reversed in Unit B. This is consistent with the presence of coarser-grained sediment in Unit B. In the sapropel, the relative chlorite and quartz contents decrease, and there is a marked quartz minimum at 77 to 82 cm. In core 1470, the mineralogical contrasts between the various units are less clearly marked (Fig. 3). There is slightly more chlorite and smectite relative to illite in Unit A compared with the upper part of Unit B, and they both increase substantially in

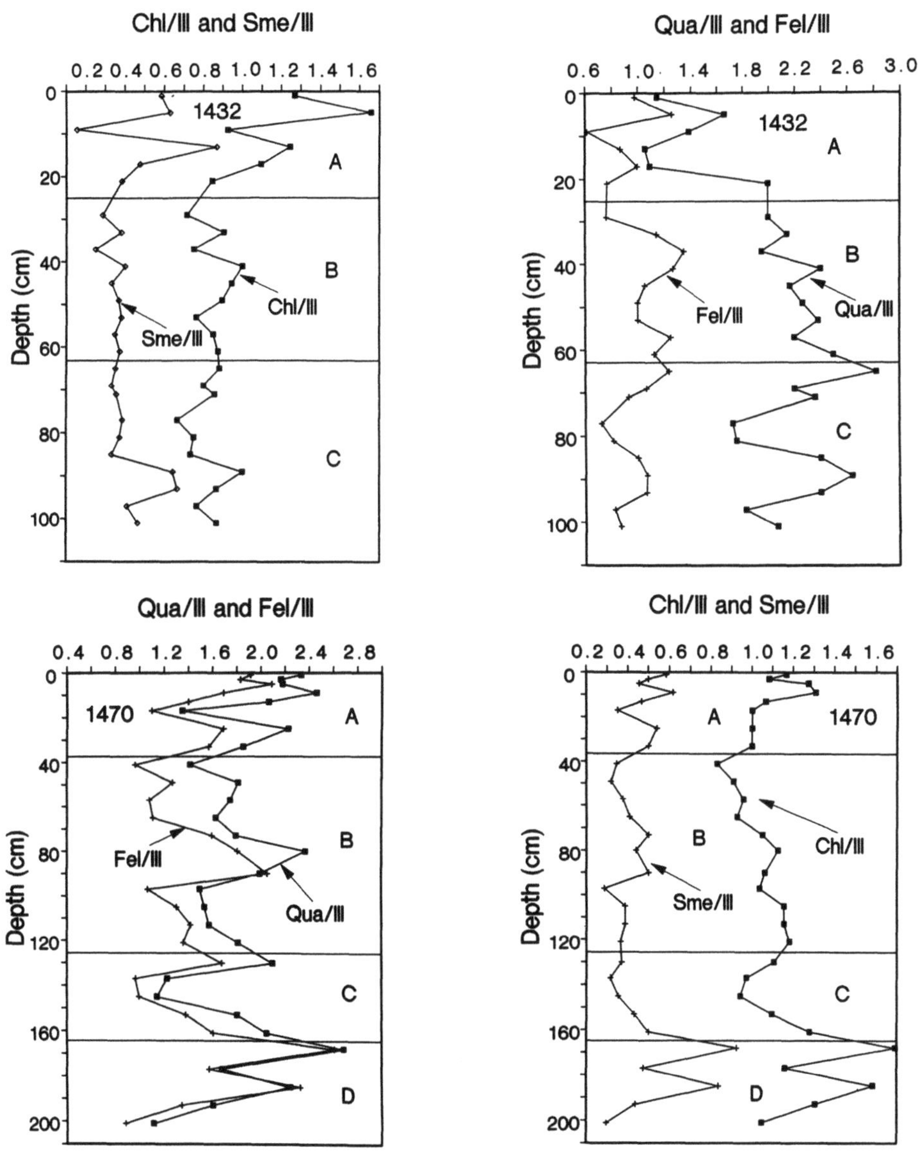


**Fig. 3.** Distribution of the principal aluminosilicate mineral phases in cores 1432 and 1470. The mineral ratios were estimated from XRD peak heights and are therefore only semiquantitative. *Chl/Ill* chlorite/illite; *Sme/Ill* smectite/illite; *Qua/Ill* quartz/illite; *Fel/Ill* quartz/illite

Unit D. The relative quartz content is higher in Unit A, in the centre of Unit B and at the top of Unit D where the $CaCO_3$ content is very low (Fig. 2). The chlorite, smectite and quartz contents appear to co-vary in the lower part of the core.

## 3.3 Chemical Composition

The distribution of organic carbon and $CaCO_3$ (Fig. 2) and the mineralogical variations (Fig. 3) in the two cores show that substantial variations in the bulk chemical composition of the sediments are to be expected. Two methods have been used to unravel the chemical variability observed. Firstly, a factor analysis (Klovan and Imbrie 1971) provides a summary of the variability in terms of a few subjectively identified factors which are thought to control the main inter-element correlations. Secondly, the distribution of individual major and minor elements is examined mainly by means of element/Al ratios. This has been necessary because of the wide variations in the relative proportions of organic matter, $CaCO_3$ and aluminosilicates in the different core units; Al serves as a reference element for the latter fraction of the sediment and the element ratios can be used to identify different aluminosilicate phases and relative element enrichments and depletions in other sediment fractions.

The results of the factor analysis (Fig. 4) show that four factors account for a total of 95.9 and 92.3% of the total sum of squares shown by the data in cores 1432 and 1470, respectively. In core 1432, Factor 1 is a general aluminosilicate factor which accounts for most of the variability of Si, Ti, Al, K, Rb, Y, Zn and Zr. Factor 2 is an organic matter factor, with high scores for C, P, S, Cu, Mo, V and Zn. Factor 3 represents $CaCO_3$, accounting for virtually all the variability in the contents of Ca, $CO_2$ and Sr, together with a very high loading for Co. Factor 4 represents a distinct aluminosilicate factor; the high scores for Fe, Mg, Cr and Ni are interpreted to reflect the distribution of chlorite and smectite.

The identification of the factors in core 1470 (Fig. 4) is more difficult because of the smaller overall variability of the mineralogy and the chemistry in this core. Factor 1 is again a general aluminosilicate factor and Factor 3 probably represents chlorite/smectite. Factor 2 contains most of the variability shown by the carbonate fraction and Factor 4 is probably the organic matter factor. Several of the major and minor elements are loaded differently from those in core 1432 because of the more subtle chemical variations in this core.

### *3.3.1 Major Elements*

The distributions of Si/Al, Fe/Al and Mg/Al in core 1432 (Fig. 5) show that the four units identified by the distribution of carbon and $CaCO_3$ are each easily differentiated by these three major element ratios. In core 1432, Si/Al ratios are highest in Unit B and in the lower part of Unit C. The high ratios in Unit B reflect the higher quartz content already shown by the XRD data (Fig. 3). The high Fe/Al and Mg/Al ratios in Unit A reflect the higher chlorite/ and smectite/illite ratios in this horizon. These ratios also increase in the sapropel (Unit C), but the two peaks

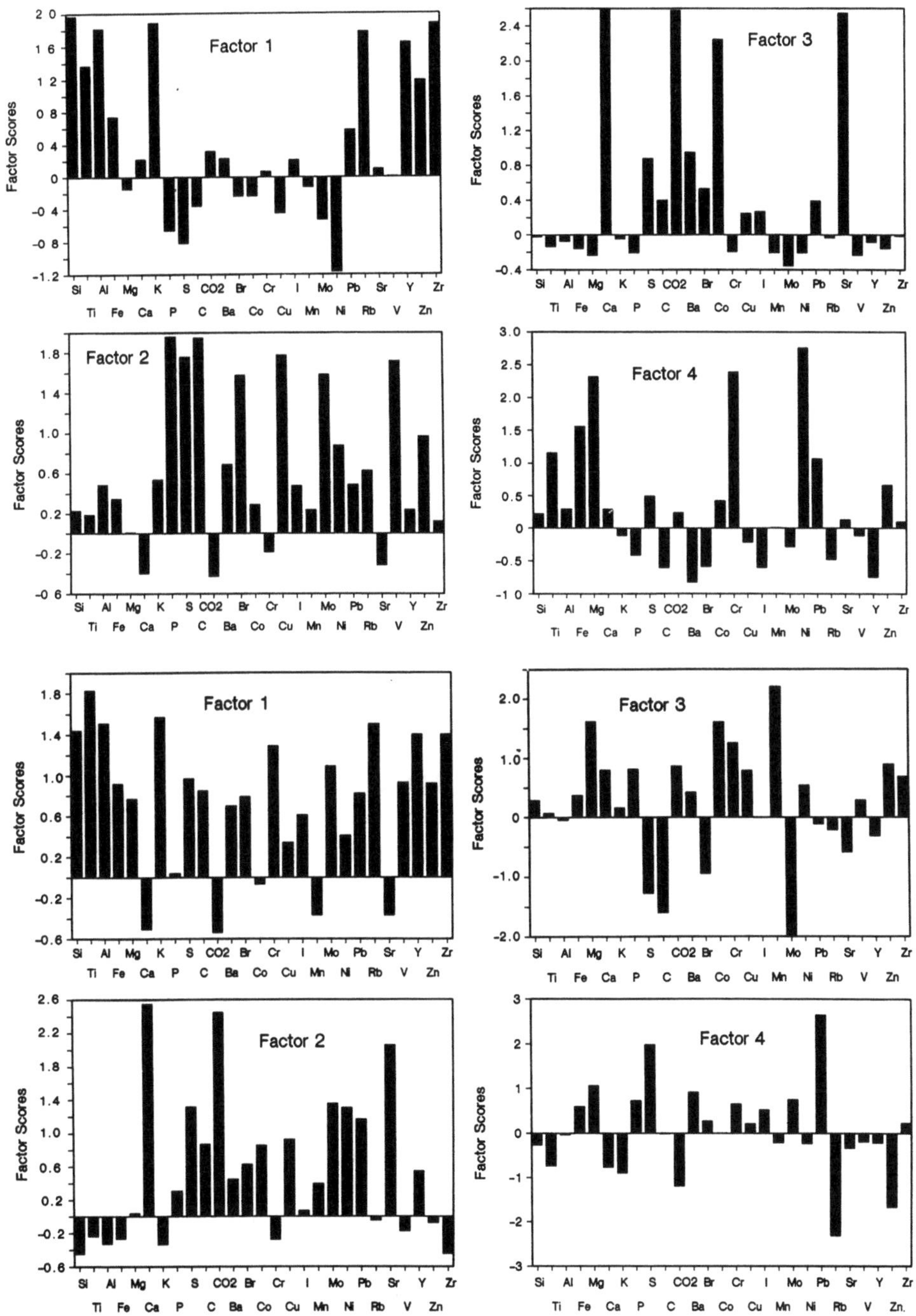


**Fig. 4.** Summary of the results of the Q-mode factor analysis of the major and minor element data. Factor scores greater than unity are significant. Upper four panels core 1432, low four panels core 1470

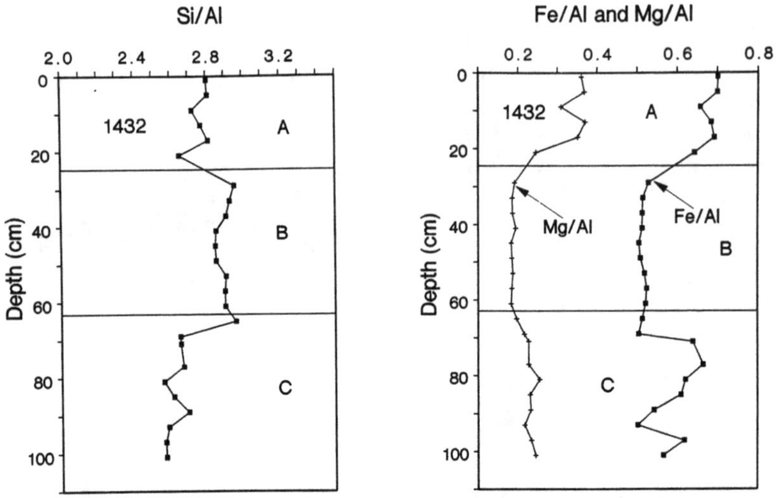


**Fig. 5.** Distribution of Si/Al, Fe/Al and Mg/Al ratios in core 1432

are displaced; the higher Mg/Al ratios occur where there is slightly more chlorite/smectite relative to illite (Fig. 3) and where the organic C peak occurs (Fig. 2), whereas the Fe/Al peak lies above the C peak. The higher Fe/Al ratios occur where total S is also high (see below) so that the distribution of Fe is controlled by the distribution of both chlorite and pyrite.

The differences in the relative Si, Fe and Mg contents between the various units in core 1470 are much smaller (Fig. 6). The Si/Al ratio ranges from 2.58 to 2.93 throughout the core, with only a slight rise at the base. The texture of the sediment, as depicted by the Si/Al ratio, is therefore either more or less constant or other mineralogical variations compensate for variations in the amount of quartz in the core. The Fe/Al ratio is significantly higher in Unit D and the Mg/Al ratio is higher in Units A and D. These variations are consistent with a slightly higher chlorite/illite ratio in the modern facies and a marked increase in chlorite in Unit D (Fig. 3), Mg being a more sensitive reflection of the presence of chlorite than Fe.

The distribution of the P/Al ratio (Fig. 7) in core 1432 largely follows the distribution of organic C. The ratio is higher in Unit A compared with the low and uniform values in Unit B and there is a broad maximum coincident with the C peak (Fig. 2) in the lower part of the sapropel. In core 1470, the ratio is also higher in Unit A and uniformly low in Unit B. A single-point maximum occurs at the C peak at the top of the sapropel and the values increase in Unit D. These distributions show that P is present above a background level (in the aluminosilicate fraction) in the sapropel horizons and in the organic-rich layers in Unit A; this is probably indicative of the presence of organic P.

The distribution of total S in core 1432 (Fig. 8) follows in a general way the distribution of organic C. The values are high in Unit A, lowest in Unit B and there is a broad maximum coincident with the C peak in the sapropel. The high degree of variability of the total S contents, which have been analyzed by two independent

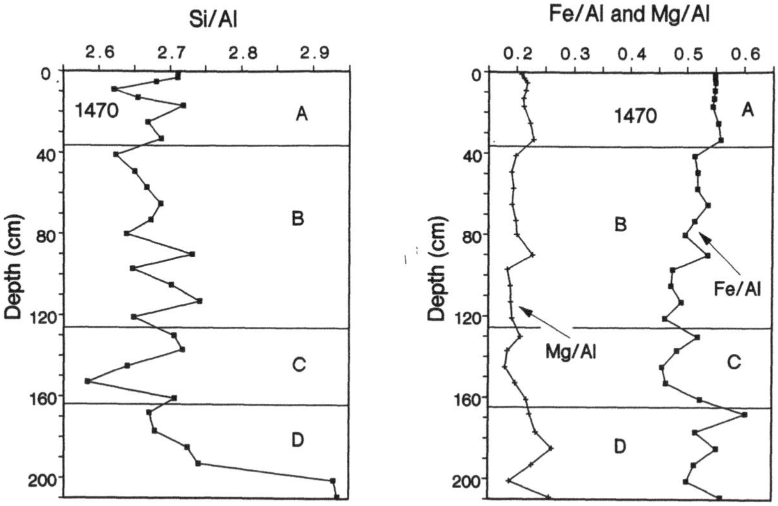


**Fig. 6.** Distribution of Si/Al, Fe/Al and Mg/Al ratios in core 1470

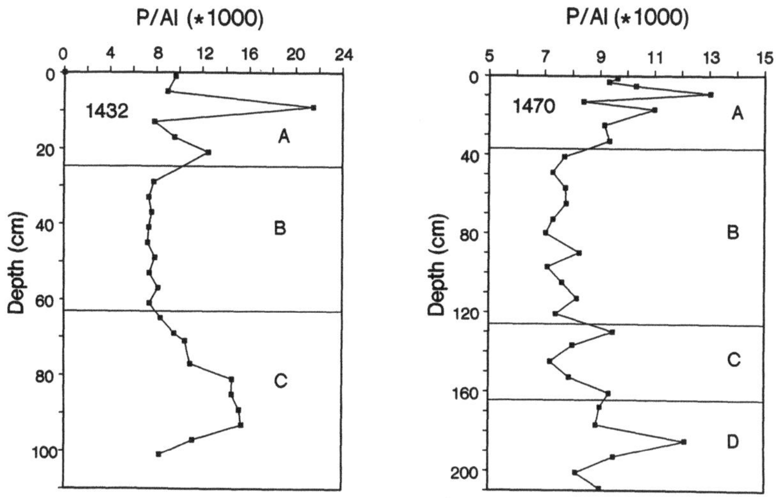


**Fig. 7.** Distribution of P/Al ratios in cores 1432 and 1470

methods (see Appendix 1), is due to the variable amount of gypsum in the samples, as shown by XRD. This has probably formed by desiccation during the storage of the cores in horizontal trays and by the oxidation of authigenic sulphides. Hence, although the distribution of S shown in Fig. 8 reflects in a general way the original distribution of pyrite and FeS, the total S now present in the various samples is not a conservative measure of these phases. In core 1470, the variability of the total S values is much larger, with the values in the sapropel being actually lower than those in Unit B.

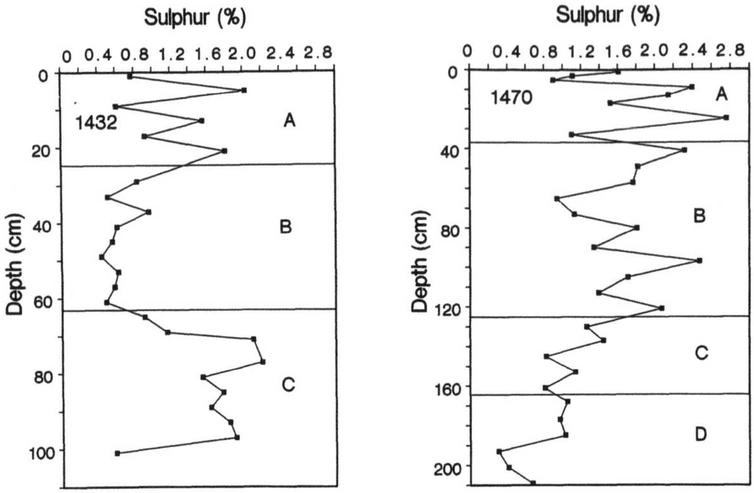


**Fig. 8.** Distribution of total sulphur in cores 1432 and 1470

### *3.3.2 Minor Elements*

The association of the minor elements with the principal sediment components of the two cores is summarized by the results of the factor analysis in Fig. 4. It can be seen that the minor elements are strongly fractionated between the four components and can therefore be used to refine the chemical characterization of the different sedimentary units already revealed by the distribution of the major elements. In addition, the conditions of sedimentation (anoxic vs. oxic) and the effects of diagenesis will also influence the distribution of some of the minor elements.

Apart from Br and I, all of the minor elements are present to some extent in aluminosilicate phases. Hence, the distribution of this group of elements is also examined by means of element/Al ratios in order to allow for the variability caused by the variable mixtures of aluminosilicate and non-aluminosilicate phases at different levels in the cores.

*Rubidium, Zinc and Zirconium.* Rubidium and Zr are present entirely in the alumino-silicate fraction, but they do not have the same distributions in the various components of this fraction. Rubidium is always camouflaged in K-bearing minerals, and its abundance will therefore be largely controlled by the distribution of feldspar and K-mica. The K/Rb ratio in cores 1432 and 1470 ranges between 190 and 270 compared with a ratio of 230 for the continental crust (Heier and Adams 1963). Zirconium, on the other hand, occurs in sediments almost entirely in zircon, which, because of its density, is transported with the coarse silt and the fine sand fractions; hence, the concentration of Zr should vary with the quartz content and, in a general way, the differences in the behaviour of Rb and Zr will reflect the differences in the texture of the sediment. Figure 9 illustrates this contrast; the

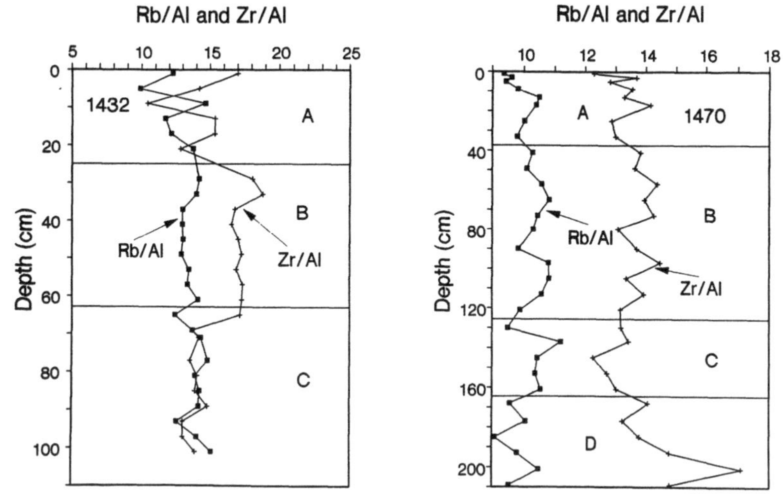


**Fig. 9.** Distribution of Rb/Al and Zr/Al ratios (*100) in cores 1432 and 1470

Rb/Al and Zr/Al ratios are negatively correlated in Unit A because of the laminated nature of this unit. The Zr/Rb ratio is less than unity and therefore the sediment is very fine-grained in the horizons where the organic C and $CaCO_3$ are both high. The Zr/Al ratio is higher in Unit B and decreases in Unit C. The Rb/Al ratio, on the other hand, is distinctly lower than the Zr/Al ratio in Unit B and increases slightly in Unit C. The Zr/Rb is greater than unity in the mud-flow layer because it is coarser-grained, as previously shown by the mineralogy and the Si/Al ratio, than the units above and below it. The Zr/Rb ratio is close to unity in the sapropel, signifying the presence of a finer grained sediment. In core 1470, the Zr/Rb ratio is greater than unity throughout the core, and the differences between the various units are more difficult to discern. The Zr/Al ratio is slightly higher in Unit B compared with the sediments above and below this horizon, and also markedly higher in Unit D. The Rb/Al ratio shows little systematic change apart from a slight decrease in Unit D. Hence, one can conclude that the entire sediment section in core 1470 is coarser-grained than the sediment in core 1432, that Unit B in turn is coarser-grained than Units A and C and that the coarsest sediment is found in Unit D.

The distribution of Zn (not shown here) appears to be similar to the distribution of Rb, but with a slight enrichment in the sapropel in core 1432. The Zn/Al ratio closely follows the variations in the Rb/Al ratio in Unit A, decreases and remains fairly constant in Unit B and increases in Unit C. Although the broad Zn/Al maximum in the sapropel coincides with the modest rise in the Rb/Al ratio, the increase in Zn/Al over the values in Unit B indicate that there is an enrichment in Zn in the sapropel. No such increase in Zn is found in core 1470; here, the Zn/Al ratio decreases slightly from Units A to B, decreases in the upper part of Unit C and finally increases in Unit D. The variations in the Zn/Al ratio can be explained almost entirely by variations in the nature of the aluminosilicate phases in the two cores, although some slight enrichment is observed in the sapropel in Core 1432.

*Chromium and Nickel.* The distribution of Cr and Ni permits a more definitive identification of the mineralogical changes in the two Black Sea cores. In Core 1432, the Cr/Al and Ni/Al ratios co-vary and are highest in Unit A (Fig. 10); both ratios decrease into Unit B where they remain constant, with a Cr/Ni $\gg$ 1. They increase in Unit C, but the Ni/Al ratio increases much more than does Cr/Al so that the Cr/Ni ratio becomes $<$ 1. The contrast between Units A and B is probably brought about solely by the differences in the mineralogy of these two units. Figure 3 shows that Unit A is markedly enriched in chlorite and smectite relative to illite compared with Unit B; Cr and Ni substitute readily for Mg in such mineral phases. On the other hand, there is only a slight relative increase in chlorite and smectite in Unit C, so that the marked increase in Ni/Al and the smaller increase in Cr/Al in Unit C have other causes. Hence, we can identify an enrichment of Cr and Ni, as well as Zn, in the sapropel.

In Core 1470, the distributions of Cr and Ni (Fig. 10) appear to be controlled entirely by the variations in the mineralogy of the sediment. Thus, the Cr/Al ratio is higher in Unit A and slightly higher in Unit D. The Ni/Al ratio, on the other hand, appears to be significantly higher only in Unit D. These variations are consistent with the mineralogical variations shown in Fig. 3; chlorite and smectite are enriched relative to illite in Units A and D. In this core, no metal enrichment in the sapropel horizon can be detected.

*Barium and Strontium.* The distribution of these two minor elements is very similar in Unit A of core 1432 (Fig. 11). The large fluctuations in the Sr/Al ratio are due to the presence of coccolith-bearing marls (Müller and Blaschke 1969; Bukry et al. 1970) intercalated in carbonate- and organic C-poor muds. The carbonate content decreases and is constant in Unit B and the Sr/Al ratio is also constant at around $2.9 \times 10^{-3}$. The ratio decreases further at the base of the sapropel, where the

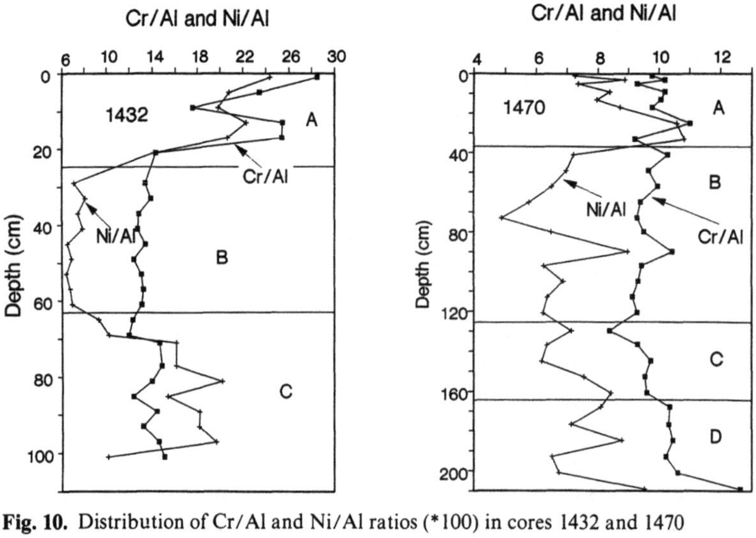


**Fig. 10.** Distribution of Cr/Al and Ni/Al ratios (*100) in cores 1432 and 1470

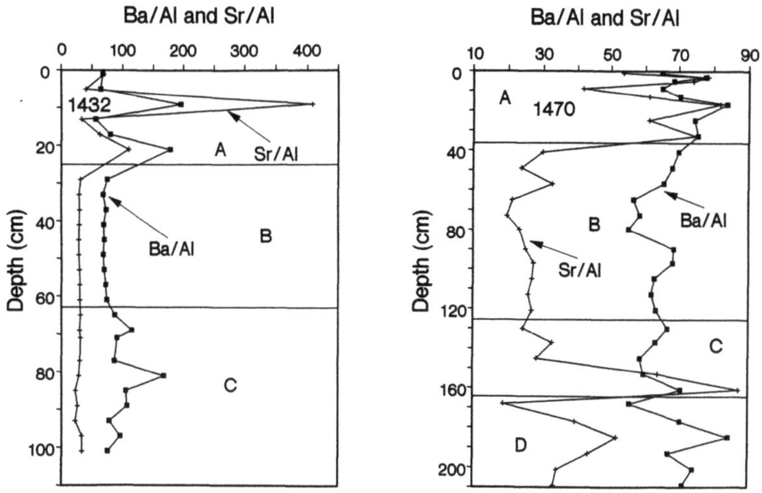


**Fig. 11.** Distribution of Ba/Al and Sr/Al ratios (*100) in cores 1432 and 1470

carbonate content is the lowest for the entire core; the Sr/Al ratio here ($2.2 \times 10^{-3}$) is only slightly higher than that of the aluminosilicate fraction. After the high Ba/Al ratios in Unit A, Ba decreases markedly in Unit B and then increases into Unit C. There are two peaks in the Ba/Al ratio in the sapropel which lie above the organic C maximum (Fig. 2) and which are not accompanied by high Sr/Al ratios. Hence, Ba appears to be present in a phase or phases associated with organic C or carbonate in Unit A, but such phases are not accompanied by carbonate or Ba is contained in a different phase or phases in the sapropel.

In Core 1470, the distribution of Sr again reflects the distribution of carbonate (Fig. 11). The Sr/Al ratio is high, and variably so, in Unit A, it decreases to a constant value of around $2.5 \times 10^{-3}$ in Unit B and increases to a sharp maximum at the base of Unit C coincident with a peak in $CaCO_3$ (Fig. 2). The very similar Sr/Al ratios at the Unit C maximum and in Unit A, although the total carbonate content in the sapropel is only half of that in Unit A (Fig. 2), is due to the presence of aragonitic "rice grains" (Degens and Ross 1972) in the sapropel. This has increased the total Sr content over that associated with calcitic carbonate because of the ease of substitution of Sr in the aragonite structure. The Ba/Al ratio co-varies with the Sr/Al in Unit A and is slightly higher than the values in Unit B. There is a small, single-point maximum in Ba/Al at the base of the sapropel after which there is a slight increase in this ratio in Unit D where Ba/Al and Sr/Al co-vary closely.

The enrichment of Ba in the organic-rich horizons of Unit A and the sapropel in core 1432 can be interpreted as a reflection of the increased fertility of the Black Sea during sapropel formation, as already inferred from other evidence (Calvert and Fontugne 1987). Revelle et al. (1955), Goldberg (1958), Goldberg and Arrhenius (1958) and Gurvich et al. (1978) showed that the Ba contents of pelagic sediments are highest in areas of high productivity, and especially so in siliceous sediments. Goldberg (1958) and Chow and Goldberg (1960) suggested that the Ba was delivered to the sediments in organic-rich particles and subsequently released to the bottom sediments upon decomposition of the organic matrix. The concentration of Ba in sediments has therefore been proposed as a palaeoproductivity indicator (Dymond 1985; Schmitz 1987). Suess (Oregon State University, pers. comm.) has suggested that marine flagellates, some of which are known to secrete barite microcrystals in their storage products (see Fresnel et al. 1979), are responsible for concentrating the Ba in particulate form, while Bishop (1988) has found evidence for the production of barite within empty diatom frustules and the decaying organic matrix of suspended material in the upper water column at several sites in the open ocean. Dehairs et al. (1987) have found a high degree of correlation between the concentration of suspended Ba in the upper part of the water column and the annual average production rate at several sites in the Atlantic. This evidence would explain the original observation of Goldberg (1958) that high Ba contents were correlated with high opal contents in the sediments of the Bering Sea, but this relationship does not appear to exist in coastal sediments, for example the opal-rich sediments of the Namibian Shelf (Calvert and Price 1983). The Black Sea sediment sections examined here do not contain diatoms and the mechanism that concentrates the Ba (active secretion by an organism or secondary production during the decomposition of planktonic organic matter) and the way it is delivered to the sediments are unknown; nevertheless, the pattern of enrichment observed in core 1432 is consistent with the notion that the sapropel and the organic-rich horizons in the modern sediment were formed by increased plankton production.

*Copper, Molybdenum and Vanadium.* These three minor metals appear to behave coherently in Core 1432 (Fig. 12), although there are differences in their detailed behaviours. Cu/Al, Mo/Al and V/Al ratios co-vary in Unit A, the organic C- and carbonate-rich horizons (Fig. 2) being metal-enriched. All three ratios decrease and remain more or less constant in Unit B but then increase dramatically in Unit C to

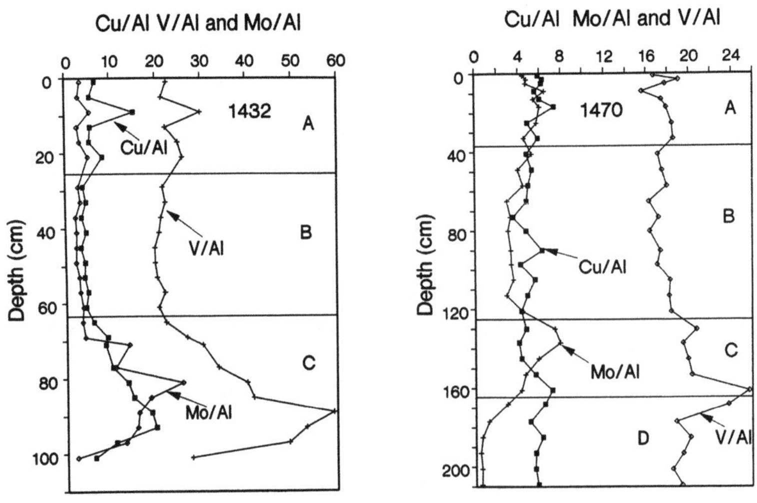


**Fig. 12.** Distribution of Cu/Al, Mo/Al and V/Al (*100) ratios in cores 1432 and 1470

values much higher than those in Unit A. The Cu/Al and V/Al maxima occur at the organic C peak (Fig. 2), whereas the Mo/Al maximum lies above this peak. In Core 1470, the Cu/Al, Mo/Al and V/Al ratios are all slightly higher in Unit A compared with Unit B (Fig. 12). Cu/Al increases slightly at the base of Unit C and remains a little higher in Unit D. Vanadium shows a more marked relative enrichment in Unit C compared with the overlying unit, with a clear maximum at the base of the sapropel. In contrast to the behaviours of Cu and V, Mo is strongly enriched at the top of the sapropel where the organic C content is at a maximum (Fig. 2). The Mo/Al ratio then falls to very low values in Unit D.

Certain minor and trace elements are enriched (i.e. occur at concentrations significantly above their crustal abundances) in many organic-rich sediments (Calvert 1976). Correlations between the abundances of metals and organic C and the presence of high concentrations of metals in extractable organic fractions have suggested that some metals are bound to the organic matter in many marine deposits (see Calvert et al. 1985 for a summary of these results). Metal enrichments have already been observed in the Black Sea sapropel by Pilipchuk and Volkov (1966, 1968) and Volkov and Fomina (1971, 1972) and attributed to the co-precipitation of the metals by Fe sulphides and to the scavenging of dissolved metals from the water column by organic debris. The distributions of Ni in Fig. 10 and Cu, Mo and V in Fig. 12 are similar to those reported by Volkov and Fomina (1971), although Ni is also controlled to a significant extent by the abudance of chlorite in the modern sediment. The very high degree of correlation between Cu and V and organic C supports the main conclusions of Volkov and Fomina (1971). However, the work of Spencer and Brewer (1971) and Jacobs and Emerson (1982) suggests that certain metals can be removed from anoxic waters by the precipitation of their insoluble sulphides. This removal process could also take place in the pore waters of the sediments themselves. Consequently, some of the excess quantities of these metals in the bottom sediments could be present in these phases. The correlation between the abundance of a metal and organic carbon would then be an indirect one, more sulphide being formed where there is more intense sulphate reduction because of a larger burial flux of carbon. The available information does not permit a distinction to be made between these two processes.

The distribution of molybdenum (Fig. 12) suggests that such a mechanism could be involved in the incorporation of this metal in the sediments. The concentration maximum occurs above the C peak and where total S is high (Fig. 8). Korolev (1958) and Bertine (1972) have shown experimentally that Mo is removed very effectively by its coprecipitation with FeS from thio-molybdate solutions and from anoxic sea water, respectively. Hence, Mo may be present to a greater extent in the sulphide fraction compared with Cu and V in the sapropel. Pilipchuk and Volkov (1968, 1974) have shown that the Mo in the Old Black Sea sapropel probably occurs in both the organic and the sulphide fractions. The efficient removal of Mo by coprecipitation and its ready association with organic matter may account for the very high degree of enrichment of this metal relative to its crustal abundance in anoxic sediments compared with other minor metals (Calvert 1976).

*Cobalt and Manganese.* The distribution of Co is somewhat similar to that of Cu, Mo and V (Fig. 13). It is relatively enriched in Unit A in Core 1432, the Co/Al ratio

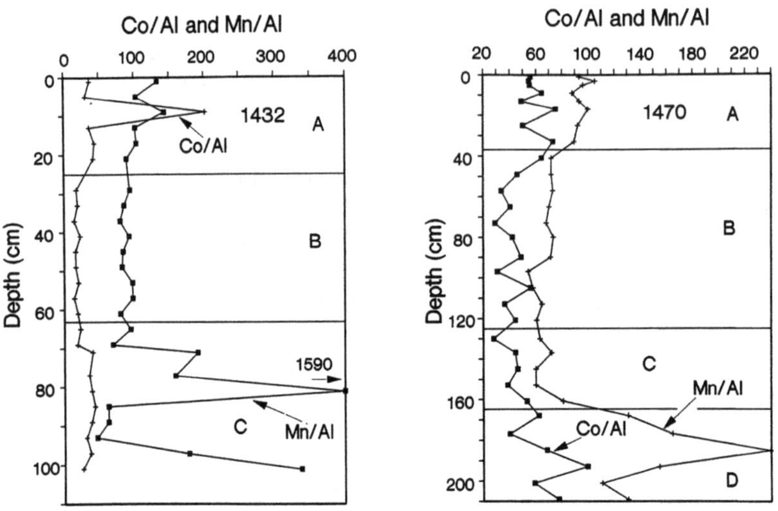


**Fig. 13.** Distribution of Co/Al (*25*104) and Mn/Al (*104) ratios in cores 1432 and 1470. The Co/Al values have been multiplied by 25

co-varying with the organic C and carbonate contents (Fig. 2). The ratio decreases and remains more or less constant in Unit B before increasing in Unit C. In this case, the Co/Al ratio is high both at and above the organic C peak in the sapropel and remains high in Unit D. In core 1470, Co/Al is again higher in Unit A than in Unit B and increases through the sapropel to reach fairly high values in Unit D.

The distribution of Mn in the two cores is unlike that of any other minor element. In Core 1432, the Mn/Al ratio is higher in Unit A compared with the underlying facies; the concentration of Mn in these units reflects the presence of Mn solely in the aluminosilicate phases, as is typical of sediments accumulating in anoxic basins (Calvert 1976). The Mn/Al ratio increases to a very pronounced maximum immediately above the organic C peak in Unit C. The ratio is lowest at the C peak (0.0046) but then increases sharply below this level. In Core 1470, Mn/Al is again higher in Unit A compared with Unit B, the values are low in the sapropel (0.006) and then increase sharply in Unit D.

Concentrations of Mn higher than the levels typical of the aluminosilicate fraction of nearshore sediments were also observed in the sapropel horizons of several of the Black Sea cores studied by Hirst (1974). He explained these enrichments by the precipitation of Mn oxide micronodules, which are known to form in the shallower oxygenated zone of the basin. The present anoxic condition of the deep waters of the Black Sea would, however, preclude the preservation of such oxide phases even if they were originally formed in the sediment. The extreme enrichments observed in core 1432 (Fig. 13) can only be explained by the presence of a mangonoan calcite phase which is known to be formed in anoxic sediments in several nearshore environments and lakes (Manheim 1961; Hartmann 1964; Shterenberg et al. 1966; Calvert and Price 1970; Suess 1979). The waters in such environments are invariably oxygenated, although anoxic conditions prevail below a surface oxygenated horizon. Under these circumstances, Mn is pumped into the subsurface anoxic horizons by the burial of the surface oxides producing high concentrations of pore water Mn. A carbonate phase precipitates because interstitial alkalinities are also high as a consequence of sulphate reduction (see Li et al. 1969). This contrasts with sediments accumulating below anoxic water columns where Mn remains at the crustal abundance level because there is no diagenetic pump which increases the concentration level of Mn within the sediment. The distribution of Mn in core 1432 therefore demonstrates that the surface sediments were oxidized and, therefore, that the basin waters contained free oxygen during the formation of the sapropel.

*Bromine and Iodine.* Br and I are two minor elements whose geochemistries are not influenced by the aluminosilicate fraction in marine sediments (Price et al. 1970). Their distributions in the two cores studied here can therefore be represented by their bulk concentrations. In Core 1432, Br and I co-vary closely in Unit A, with the I/Br ratio close to unity and with the higher values occurring in the organic C- and carbonate-rich horizons (Fig. 14). The concentrations of both halogens are more or less constant in Unit B with an I/Br ratio $> 1$. Iodine increases sharply in the upper part of Unit C and decreases to low values at the organic C peak. Bromine, by contrast, increases steadily through the sapropel to reach a maximum at the C peak, after which it decreases. Hence, the I/Br ratio is much higher than 1 in the upper part of Unit C, but much less than one at the base of this unit. In Core 1470, the concentration levels of the halogens are much lower than they are in Core 1432. They co-vary in Unit A, with an I/Br ratio $> 1$. In Unit B, there are maxima in the centre and at the base of the unit, after which I decreases into Units C and D. Bromine shows a slight increase with depth in Unit B, a peak at the organic C peak in Unit C and a decrease in Unit D. The I/Br ratio is therfore $> 1$ over most of Unit B, $< 1$ in Unit C and $> 1$ in Unit D.

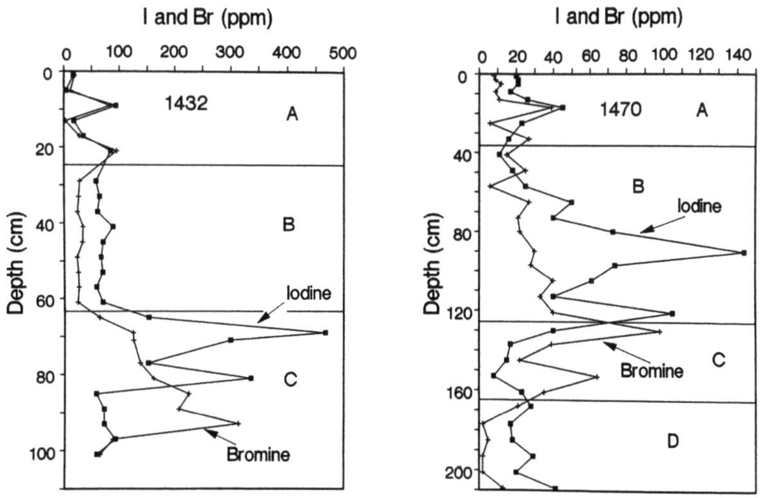


**Fig. 14.** Distribution of I and Br in cores 1432 and 1470

In nearshore sediments, Br and I are closely related to the organic C contents, but the I/$C_{org}$ and I/Br ratios are significantly lower in sediments accumulating under anoxic conditions (Price and Calvert 1977). This is thought to be due to the lack of uptake of iodine by the organic fraction under low redox conditions because iodate, the species which reacts with and is fixed by this fraction (Francois 1987), is reduced to non-reactive iodide at low Eh (Wong and Brewer 1977). The speciation of Br is not affected by the prevailing redox conditions so that there is no difference in the uptake of this element under a wide range of redox conditions. By allowing for some diagenetic loss of iodine with burial (Price et al. 1970; Francois 1987), the distribution of the two halogens can be used to infer whether the sediments originally accumulated under low or high redox potentials. Thus, the relatively high concentrations of I, the presence of I/Br ratios much higher than unity and the presence of I/$C_{org}$ ratios significantly higher than those typical of anoxic sediments (around $20 \times 10^{-4}$) in the upper part of the sapropel in core 1432 all imply that the basin was oxygenated when the horizon formed. This strongly supports the explanation for the presence of high Mn concentrations in the same part of the core, namely that the surface sediment contained Mn oxides when the sapropel formed, and that, consequently, the Black Sea was not anoxic at this time.

## 4 Summary and Conclusions

A detailed examination of the geochemistry of two cores from the Black Sea has amplified the available information on the Holocene stratigraphy of the basin and has provided new evidence for the origin of the sapropel and the hydrographic conditions which existed during its formation. This permits a reevaluation of the environmental conditions leading to the formation of organic-rich marine deposits, which has important implications for the use of the Black Sea as a modern analogue for the origin of carbonaceous sediments and rocks in the modern ocean and in the geological record.

The sapropel horizon occurs below the modern facies and a mud-flow layer in the two cores examined. The age ranges of the sapropel, namely 1600 to 6600 yr BP in deep water and 4000 to 6000 yr BP on the eastern slope (Calvert et al. 1987), show that it formed when sea level had reached its present position and, therefore, after the reconnection of the Black Sea with the Mediterranean. A more precise relationship between the chronology of organic matter deposition and the progressive change in the salinity of the basin cannot be obtained without an even more refined chronology of sedimentation and a more precise determination of the position of the Bosporus sill with respect to changing sea level.

The geochemistry of the sediment section studied here places some important constraints on the oxygen status of the Black Sea during the accumulation of the sapropel. The distribution of Mn, I and Br in the deep water core show conclusively that the surface of the sediment and hence the bottom water was well oxygenated at this time. This evidence, in turn, supports the conclusion, based on other evidence, that the sapropel formed by increased plankton production during the post-glacial evolution of the hydrography of the basin. Thus, the refined chronology of sedimentation and the derived carbon accumulation rates provided by Calvert et al. (1987), the carbon isotope data of Calvert and Fontugne (1987) which show that the sapropel has a fully planktonic signature whereas the associated horizons contain mixed terrestrial and planktonic carbon, and the enrichment of Ba in the sapropel and the organic-rich levels in the modern sediment (Fig. 11) all point to increased fertility as the cause of the accumulation of organic-rich sediments. Although the Black Sea is now intensely anoxic, a sapropel is not forming because production has decreased and the bulk sedimentation rate has increased since sapropel time.

This explanation for the formation of the Black Sea sapropel is broadly similar to that advanced by Strakhov (1971). He has systematically argued that the Old Black Sea horizon was formed by increased plankton production brought about by the upward displacement of deep, nutrient-rich water by the incursion of denser Mediterranean water into the deep basin after sea level rise. He has also pointed out that deep water anoxia probably developed after the deposition of the sapropel because the origin of the $H_2S$ in the basin waters is the bottom sediments themselves, the high rate of accumulation of organic matter in the sediment most probably leading to subsequent sulphide contamination of the deep waters by diffusion.

Corroboration of these conclusions is clearly required. Further work on the nature of the organic matter in the sapropel and associated beds, establishment of a higher resolution chronology of sapropel formation and the differences, if any, in the timing of the beginning of sapropel formation in cores collected from different water depths, and a better chronology of relative sea level change with respect to the depth of the Bosporus sill over the past 10,000 years would all provide important tests of the conclusions drawn here. If confirmed, however, the conclusions would imply that the Black Sea cannot, in fact, be used as a modern analogue for sapropel formation simply because it is presently anoxic. This in turn will require a reappraisal of current theories for the origin of organic-rich horizons in deep ocean drill cores by anoxia (Ryan and Cita 1977; Thiede and van Andel 1977; Arthur 1979; Demaison and Moore 1980) and the majority view of the origin of Pleistocene sapropels in the eastern Mediterranean by the same mechanism (Kullenberg 1952; Olausson 1961; Ryan 1972; McCoy 1974; Ryan and Cita 1977; Rossignol-Strick et al. 1982).

**Appendix 1A.** Core 1432 major and minor element data (major elements in %, minor elements in ppm)

| Sample depth (cm) | Si | Ti | Al | Fe | Mg | Ca | K | P | S (XRF)[a] | S (Comb)[b] | $CO_2$ |
|---|---|---|---|---|---|---|---|---|---|---|---|
| 1 | 18.31 | 0.44 | 6.53 | 4.58 | 2.36 | 9.97 | 1.68 | 0.063 | 0.76 | 0.77 | 10.14 |
| 5 | 18.21 | 0.45 | 6.48 | 4.54 | 2.39 | 8.03 | 1.66 | 0.058 | 1.59 | 2.06 | 6.32 |
| 9 | 8.36 | 0.15 | 3.07 | 2.02 | 0.95 | 24.46 | 0.82 | 0.066 | 1.28 | 0.64 | 24.47 |
| 13 | 20.58 | 0.51 | 7.43 | 5.09 | 2.76 | 6.73 | 1.87 | 0.058 | 1.30 | 1.59 | 6.12 |
| 17 | 18.51 | 0.43 | 6.58 | 4.56 | 2.32 | 9.56 | 1.76 | 0.063 | 1.06 | 0.96 | 9.55 |
| 21 | 14.45 | 0.27 | 5.45 | 3.51 | 1.34 | 13.93 | 1.49 | 0.068 | 2.42 | 1.80 | 12.57 |
| 29 | 22.06 | 0.42 | 7.46 | 3.95 | 1.43 | 6.33 | 2.21 | 0.058 | 0.86 | 0.86 | 6.02 |
| 33 | 23.04 | 0.45 | 7.86 | 4.05 | 1.46 | 5.91 | 2.21 | 0.058 | 0.47 | 0.55 | 5.80 |

**Appendix 1A.** *(Continued)*

| Sample depth (cm) | Si | Ti | Al | Fe | Mg | Ca | K | P | S (XRF)[a] | S (Comb)[b] | $CO_2$ |
|---|---|---|---|---|---|---|---|---|---|---|---|
| 37 | 22.15 | 0.43 | 7.61 | 3.91 | 1.42 | 6.35 | 2.13 | 0.058 | 0.84 | 1.00 | 5.81 |
| 41 | 22.41 | 0.44 | 7.85 | 4.03 | 1.53 | 6.15 | 2.17 | 0.058 | 0.60 | 0.65 | 5.92 |
| 45 | 22.75 | 0.44 | 7.98 | 4.03 | 1.46 | 6.02 | 2.21 | 0.058 | 0.56 | 0.60 | 5.91 |
| 49 | 22.75 | 0.44 | 7.97 | 4.05 | 1.47 | 5.96 | 2.19 | 0.063 | 0.44 | 0.49 | 6.01 |
| 53 | 22.72 | 0.43 | 7.81 | 4.05 | 1.46 | 6.09 | 2.21 | 0.058 | 0.73 | 0.66 | 5.92 |
| 57 | 22.39 | 0.43 | 7.71 | 4.04 | 1.42 | 6.08 | 2.21 | 0.063 | 0.61 | 0.62 | 5.91 |
| 61 | 22.63 | 0.43 | 7.79 | 4.06 | 1.42 | 5.99 | 2.31 | 0.058 | 0.52 | 0.53 | 5.91 |
| 65 | 21.91 | 0.42 | 7.39 | 3.79 | 1.45 | 5.89 | 2.15 | 0.062 | 0.70 | 0.95 | 5.61 |
| 69 | 19.93 | 0.41 | 7.51 | 3.78 | 1.62 | 5.22 | 2.18 | 0.072 | 1.38 | 1.19 | 4.73 |
| 71 | 19.44 | 0.39 | 7.32 | 4.68 | 1.66 | 5.06 | 2.08 | 0.077 | 2.22 | 2.10 | 3.81 |
| 77 | 18.73 | 0.38 | 7.01 | 4.66 | 1.59 | 5.19 | 2.08 | 0.077 | 2.18 | 2.21 | 3.91 |
| 81 | 16.19 | 0.31 | 6.32 | 3.92 | 1.61 | 3.17 | 1.76 | 0.092 | 1.83 | 1.58 | 3.03 |
| 85 | 16.72 | 0.33 | 6.39 | 3.89 | 1.47 | 2.92 | 1.74 | 0.093 | 2.30 | 1.78 | 2.20 |
| 89 | 16.54 | 0.35 | 6.13 | 3.32 | 1.42 | 3.35 | 1.81 | 0.093 | 1.99 | 1.67 | 2.51 |
| 93 | 16.48 | 0.03 | 6.37 | 3.19 | 1.38 | 3.31 | 1.73 | 0.098 | 2.71 | 1.09 | 1.89 |
| 97 | 17.73 | 0.34 | 6.90 | 4.26 | 1.61 | 5.87 | 1.88 | 0.077 | 2.31 | 1.92 | 5.04 |
| 101 | 19.58 | 0.41 | 7.61 | 4.30 | 1.86 | 6.54 | 2.25 | 0.063 | 0.68 | 0.63 | 6.93 |

[a] Sulphur values determined by XRF.
[b] Sulphur values determined by combustion folowed by amperometric titration.

**Appendix 1A.** *(Continued)*

| Sample depth (cm) | $C_{org}$ | Ba | Br | Co | Cr | Cu | I | Mn | Mo | Ni | Pb |
|---|---|---|---|---|---|---|---|---|---|---|---|
| 1 | 1.34 | 440 | 19 | 24 | 186 | 44 | 17 | 873 | 22 | 159 | 19 |
| 5 | 1.07 | 415 | 12 | 20 | 152 | 36 | 4 | 671 | 19 | 135 | 16 |
| 9 | 3.88 | 600 | 83 | 62 | 54 | 47 | 94 | 440 | 17 | 61 | 9 |
| 13 | 0.66 | 415 | 3 | 27 | 189 | 42 | 18 | 761 | 20 | 166 | 23 |
| 17 | 1.89 | 520 | 28 | 29 | 167 | 36 | 35 | 684 | 21 | 136 | 27 |
| 21 | 3.83 | 970 | 95 | 23 | 78 | 46 | 84 | 489 | 28 | 77 | 19 |
| 29 | 1.55 | 555 | 28 | 13 | 100 | 29 | 58 | 704 | 21 | 53 | 21 |
| 33 | 1.46 | 535 | 26 | 15 | 109 | 36 | 64 | 676 | 26 | 64 | 20 |
| 37 | 1.69 | 555 | 24 | 11 | 98 | 27 | 60 | 612 | 17 | 58 | 14 |
| 41 | 1.68 | 540 | 34 | 18 | 100 | 36 | 88 | 732 | 19 | 62 | 16 |
| 45 | 1.71 | 555 | 33 | 13 | 107 | 27 | 70 | 674 | 19 | 52 | 15 |
| 49 | 1.67 | 540 | 23 | 13 | 99 | 34 | 66 | 661 | 18 | 55 | 21 |
| 53 | 1.62 | 540 | 25 | 16 | 102 | 32 | 69 | 766 | 23 | 50 | 18 |
| 57 | 1.57 | 555 | 26 | 11 | 102 | 38 | 58 | 757 | 24 | 52 | 20 |
| 61 | 1.72 | 575 | 24 | 15 | 102 | 34 | 69 | 627 | 29 | 54 | 23 |
| 65 | 2.91 | 640 | 63 | 17 | 91 | 45 | 151 | 703 | 26 | 69 | 5 |
| 69 | 5.33 | 860 | 123 | 14 | 90 | 69 | 467 | 524 | 31 | 77 | 21 |
| 71 | 4.81 | 660 | 124 | 30 | 107 | 64 | 298 | 1398 | 102 | 118 | 21 |
| 77 | 5.22 | 605 | 136 | 25 | 104 | 72 | 150 | 1110 | 77 | 113 | 17 |
| 81 | 8.87 | 1060 | 160 | 25 | 88 | 86 | 335 | 10047 | 162 | 128 | 23 |
| 85 | 13.04 | 670 | 224 | 28 | 79 | 95 | 57 | 403 | 119 | 98 | 17 |
| 89 | 14.31 | 655 | 206 | 24 | 88 | 115 | 71 | 386 | 98 | 112 | 16 |
| 93 | 14.39 | 495 | 312 | 20 | 84 | 126 | 71 | 298 | 100 | 116 | 20 |
| 97 | 7.71 | 660 | 87 | 26 | 100 | 76 | 91 | 1223 | 91 | 136 | 15 |
| 101 | 3.06 | 570 | 64 | 20 | 114 | 48 | 58 | 2579 | 17 | 77 | 22 |

**Appendix 1A.** (*Continued*)

| Sample depth (cm) | Rb | Sr | V | Y | Zn | Zr |
|---|---|---|---|---|---|---|
| 1 | 80 | 451 | 147 | 23 | 86 | 111 |
| 5 | 64 | 262 | 139 | 20 | 71 | 92 |
| 9 | 45 | 1253 | 92 | 18 | 48 | 32 |
| 13 | 87 | 244 | 165 | 22 | 79 | 114 |
| 17 | 80 | 411 | 164 | 20 | 74 | 101 |
| 21 | 75 | 600 | 141 | 20 | 68 | 70 |
| 29 | 106 | 232 | 161 | 25 | 77 | 135 |
| 33 | 110 | 231 | 174 | 29 | 84 | 148 |
| 37 | 99 | 227 | 161 | 25 | 79 | 128 |
| 41 | 102 | 231 | 163 | 23 | 79 | 130 |
| 45 | 104 | 225 | 158 | 31 | 78 | 136 |
| 49 | 103 | 226 | 158 | 27 | 75 | 138 |
| 53 | 105 | 228 | 158 | 27 | 75 | 132 |
| 57 | 103 | 231 | 169 | 26 | 85 | 134 |
| 61 | 110 | 235 | 160 | 28 | 80 | 135 |
| 65 | 92 | 229 | 163 | 26 | 75 | 127 |
| 69 | 103 | 223 | 200 | 22 | 77 | 103 |
| 71 | 105 | 225 | 221 | 25 | 86 | 104 |
| 77 | 104 | 211 | 237 | 25 | 82 | 95 |
| 81 | 88 | 179 | 254 | 26 | 80 | 89 |
| 85 | 91 | 146 | 266 | 24 | 81 | 89 |
| 89 | 87 | 159 | 363 | 21 | 76 | 91 |
| 93 | 80 | 141 | 339 | 21 | 76 | 83 |
| 97 | 97 | 159 | 341 | 24 | 86 | 90 |
| 101 | 115 | 141 | 211 | 7 | 85 | 106 |

**Appendix 1B.** Core 1470 major and minor element data (major elements in %, minor elements in ppm)

| Sample depth (cm) | Si | Ti | Al | Fe | Mg | Ca | K | P | S (XRF)[a] | S (Comb)[b] | $CO_2$ |
|---|---|---|---|---|---|---|---|---|---|---|---|
| 1 | 19.41 | 0.35 | 7.16 | 3.93 | 1.49 | 8.83 | 1.69 | 0.069 | 1.61 | 1.01 | 7.82 |
| 3 | 18.62 | 0.32 | 6.87 | 3.77 | 1.46 | 10.96 | 1.64 | 0.064 | 1.11 | 0.73 | 10.33 |
| 5 | 18.20 | 0.32 | 6.79 | 3.73 | 1.48 | 11.01 | 1.59 | 0.070 | 0.90 | 0.80 | 10.94 |
| 9 | 20.31 | 0.36 | 7.75 | 4.25 | 1.67 | 7.01 | 1.80 | 0.101 | 2.40 | 1.79 | 5.29 |
| 13 | 18.98 | 0.34 | 7.15 | 3.91 | 1.51 | 9.27 | 1.74 | 0.060 | 2.15 | 1.61 | 8.01 |
| 17 | 18.05 | 0.32 | 6.64 | 3.62 | 1.41 | 11.32 | 1.62 | 0.073 | 1.52 | 1.19 | 10.83 |
| 25 | 18.65 | 0.33 | 6.99 | 3.88 | 1.56 | 9.51 | 1.54 | 0.064 | 2.76 | 2.05 | 7.97 |
| 33 | 18.35 | 0.32 | 6.84 | 3.83 | 1.57 | 10.70 | 1.58 | 0.064 | 1.10 | 0.74 | 10.70 |
| 41 | 21.43 | 0.42 | 8.17 | 4.20 | 1.62 | 5.42 | 1.96 | 0.063 | 2.32 | 2.05 | 3.38 |
| 49 | 24.67 | 0.44 | 9.31 | 4.84 | 1.78 | 4.44 | 2.22 | 0.068 | 1.82 | 1.55 | 2.98 |
| 57 | 21.68 | 0.43 | 8.12 | 4.22 | 1.58 | 5.68 | 2.01 | 0.063 | 1.77 | 1.58 | 4.41 |
| 65 | 26.57 | 0.44 | 9.89 | 5.32 | 1.90 | 3.85 | 2.46 | 0.077 | 0.94 | 1.01 | 3.07 |
| 73 | 23.01 | 0.45 | 8.61 | 4.43 | 1.71 | 3.71 | 2.17 | 0.063 | 1.13 | 1.17 | 3.08 |
| 80 | 23.56 | 0.44 | 8.93 | 4.45 | 1.79 | 4.29 | 2.03 | 0.063 | 1.81 | 1.57 | 2.98 |
| 90 | 22.50 | 0.41 | 8.24 | 4.43 | 1.88 | 4.32 | 1.98 | 0.068 | 1.34 | 1.10 | 3.50 |
| 97 | 21.57 | 0.40 | 8.15 | 3.88 | 1.50 | 5.40 | 2.14 | 0.058 | 2.48 | 2.28 | 3.17 |
| 105 | 25.47 | 0.40 | 9.43 | 4.46 | 1.78 | 5.10 | 2.34 | 0.072 | 1.70 | 1.60 | 3.18 |

**Appendix 1B.** (*Continued*)

| Sample depth (cm) | Si | Ti | Al | Fe | Mg | Ca | K | P | S (XRF)[a] | S (Comb)[b] | $CO_2$ |
|---|---|---|---|---|---|---|---|---|---|---|---|
| 113 | 22.81 | 0.42 | 8.32 | 4.09 | 1.58 | 4.51 | 2.24 | 0.068 | 1.39 | 1.30 | 3.49 |
| 121 | 22.51 | 0.40 | 8.50 | 3.93 | 1.63 | 5.06 | 2.06 | 0.063 | 2.08 | 1.90 | 3.07 |
| 130 | 21.94 | 0.41 | 8.11 | 4.22 | 1.68 | 3.57 | 1.97 | 0.077 | 1.26 | 1.16 | 2.79 |
| 137 | 23.04 | 0.41 | 8.48 | 4.10 | 1.56 | 4.74 | 2.10 | 0.068 | 1.44 | 1.42 | 3.79 |
| 145 | 23.02 | 0.42 | 8.72 | 3.99 | 1.57 | 4.03 | 2.24 | 0.063 | 0.82 | 0.90 | 3.88 |
| 153 | 21.91 | 0.38 | 8.48 | 3.94 | 1.68 | 5.57 | 2.01 | 0.067 | 1.14 | 1.03 | 5.01 |
| 161 | 20.54 | 0.37 | 7.59 | 3.98 | 1.65 | 7.31 | 1.94 | 0.071 | 0.81 | 0.85 | 7.12 |
| 168 | 23.71 | 0.46 | 8.88 | 5.36 | 1.98 | 1.51 | 2.11 | 0.080 | 1.05 | 1.16 | 0.15 |
| 177 | 20.22 | 0.35 | 7.55 | 3.90 | 1.77 | 8.63 | 1.82 | 0.067 | 0.97 | 1.11 | 7.62 |
| 185 | 17.98 | 0.30 | 6.60 | 3.65 | 1.73 | 11.69 | 1.62 | 0.080 | 1.03 | 1.10 | 10.56 |
| 193 | 19.29 | 0.34 | 7.04 | 3.62 | 1.60 | 11.05 | 1.74 | 0.067 | 0.30 | 0.42 | 11.45 |
| 201 | 21.23 | 0.40 | 7.25 | 3.63 | 1.36 | 8.49 | 2.04 | 0.059 | 0.41 | 0.72 | 8.21 |
| 209 | 21.83 | 0.39 | 7.44 | 4.17 | 1.92 | 7.36 | 1.87 | 0.067 | 0.67 | 0.30 | 6.89 |

[a] Sulphur values determined by XRF.
[b] Sulphur values determined by combustion followed by amperometric titration.

**Appendix 1B.** (*Continued*)

| Sample depth (cm) | $C_{org}$ | Ba | Br | Co | Cr | Cu | I | Mn | Mo | Ni | Pb |
|---|---|---|---|---|---|---|---|---|---|---|---|
| 1 | 2.10 | 464 | 8 | 16 | 70 | 42 | 20 | 669 | 32 | 52 | 25 |
| 3 | 1.64 | 534 | 9 | 15 | 70 | 43 | 21 | 725 | 33 | 61 | 18 |
| 5 | 1.61 | 464 | 12 | 15 | 63 | 42 | 21 | 654 | 32 | 50 | 29 |
| 9 | 1.30 | 503 | 9 | 20 | 79 | 43 | 17 | 683 | 50 | 65 | 16 |
| 13 | 1.51 | 502 | 11 | 14 | 72 | 43 | 26 | 668 | 39 | 57 | 19 |
| 17 | 1.52 | 556 | 39 | 20 | 65 | 49 | 45 | 665 | 40 | 58 | 19 |
| 25 | 1.41 | 521 | 6 | 14 | 77 | 34 | 23 | 646 | 40 | 74 | 19 |
| 33 | 1.46 | 516 | 27 | 20 | 63 | 40 | 16 | 613 | 31 | 74 | 14 |
| 41 | 1.27 | 570 | 15 | 21 | 84 | 39 | 11 | 586 | 43 | 59 | 29 |
| 49 | 1.23 | 631 | 25 | 17 | 90 | 49 | 18 | 668 | 37 | 65 | 24 |
| 57 | 1.15 | 530 | 6 | 11 | 81 | 40 | 25 | 594 | 36 | 53 | 15 |
| 65 | 1.29 | 558 | 27 | 16 | 93 | 47 | 50 | 694 | 29 | 57 | 23 |
| 73 | 1.43 | 501 | 21 | 10 | 80 | 30 | 40 | 585 | 28 | 42 | 26 |
| 80 | 1.56 | 491 | 22 | 15 | 85 | 42 | 73 | 657 | 27 | 58 | 30 |
| 90 | 2.11 | 562 | 30 | 16 | 86 | 51 | 144 | 587 | 27 | 74 | 24 |
| 97 | 1.65 | 553 | 28 | 10 | 77 | 34 | 74 | 441 | 27 | 51 | 22 |
| 105 | 1.82 | 589 | 40 | 21 | 88 | 52 | 61 | 543 | 33 | 65 | 14 |
| 113 | 1.97 | 513 | 33 | 12 | 76 | 40 | 40 | 538 | 24 | 53 | 23 |
| 121 | 2.06 | 535 | 40 | 15 | 79 | 36 | 105 | 514 | 37 | 53 | 10 |
| 130 | 4.29 | 538 | 98 | 9 | 68 | 38 | 40 | 510 | 60 | 58 | 14 |
| 137 | 3.06 | 533 | 39 | 15 | 79 | 34 | 17 | 661 | 67 | 54 | 22 |
| 145 | 2.34 | 509 | 22 | 16 | 85 | 37 | 15 | 521 | 51 | 54 | 18 |
| 153 | 3.11 | 504 | 64 | 13 | 81 | 47 | 8 | 507 | 39 | 64 | 23 |
| 161 | 2.58 | 534 | 35 | 16 | 73 | 54 | 23 | 614 | 32 | 64 | 20 |
| 168 | 1.27 | 491 | 21 | 22 | 92 | 57 | 28 | 1161 | 26 | 72 | 24 |
| 177 | 0.68 | 530 | 2 | 12 | 78 | 38 | 17 | 1246 | 9 | 54 | 16 |

**Appendix 1B.** *(Continued)*

| Sample depth (cm) | $C_{org}$ | Ba | Br | Co | Cr | Cu | I | Mn | Mo | Ni | Pb |
|---|---|---|---|---|---|---|---|---|---|---|---|
| 185 | 0.59 | 556 | 5 | 18 | 69 | 41 | 18 | 1581 | 4 | 58 | 19 |
| 193 | 0.50 | 469 | 2 | 28 | 72 | 39 | 29 | 1090 | 3 | 46 | 21 |
| 201 | 0.49 | 537 | 2 | 17 | 77 | 40 | 20 | 803 | 4 | 49 | 23 |
| 209 | 0.49 | 529 | 13 | 23 | 94 | 43 | 41 | 972 | 4 | 71 | 13 |

**Appendix 1B.** *(Continued)*

| Sample depth (cm) | Rb | Sr | V | Y | Zn | Zr |
|---|---|---|---|---|---|---|
| 1 | 67 | 383 | 120 | 20 | 70 | 88 |
| 3 | 66 | 539 | 131 | 22 | 75 | 94 |
| 5 | 64 | 504 | 121 | 24 | 65 | 87 |
| 9 | 76 | 323 | 121 | 16 | 70 | 105 |
| 13 | 75 | 437 | 125 | 21 | 71 | 95 |
| 17 | 69 | 543 | 119 | 18 | 71 | 94 |
| 25 | 70 | 426 | 129 | 20 | 74 | 90 |
| 33 | 67 | 513 | 127 | 20 | 71 | 89 |
| 41 | 84 | 244 | 140 | 22 | 78 | 113 |
| 49 | 94 | 221 | 163 | 24 | 90 | 127 |
| 57 | 86 | 265 | 146 | 26 | 80 | 117 |
| 65 | 107 | 210 | 161 | 28 | 98 | 138 |
| 73 | 90 | 170 | 148 | 25 | 70 | 123 |
| 80 | 92 | 208 | 146 | 23 | 89 | 117 |
| 90 | 81 | 206 | 143 | 22 | 87 | 113 |
| 97 | 88 | 222 | 139 | 25 | 77 | 118 |
| 105 | 102 | 253 | 172 | 25 | 90 | 126 |
| 113 | 88 | 215 | 151 | 26 | 81 | 116 |
| 121 | 84 | 228 | 156 | 23 | 76 | 112 |
| 130 | 77 | 197 | 168 | 23 | 71 | 107 |
| 137 | 95 | 277 | 165 | 24 | 77 | 114 |
| 145 | 91 | 246 | 174 | 23 | 75 | 107 |
| 153 | 88 | 539 | 172 | 22 | 80 | 108 |
| 161 | 80 | 662 | 195 | 22 | 82 | 99 |
| 168 | 85 | 167 | 211 | 25 | 89 | 125 |
| 177 | 76 | 297 | 142 | 20 | 73 | 100 |
| 185 | 60 | 340 | 133 | 19 | 71 | 91 |
| 193 | 69 | 304 | 137 | 19 | 74 | 104 |
| 201 | 76 | 247 | 134 | 24 | 78 | 124 |
| 209 | 71 | 246 | 144 | 21 | 84 | 110 |

*Acknowledgments.* The core samples were kindly made available by J. Broda, Woods Hole Oceanographic Institution. Assistance with the analytical work was provided by B. Cousens and M. Soon. The cruise of ATLANTIS II to the Black Sea was funded by the U.S. National Science Foundation (Grant GA 1659) and the Office of Naval Research (Contract N00014-66-CO241-6). The core curation at WHOI is supported by NSF grant OCE85-19889. Laboratory work and the preparation of this paper were supported by the Natural Sciences and Engineering Research Council of Canada.

# References


Abbey S (1983) Studies in "standard samples" of silicate rocks and minerals 1969–1982. Geol Surv Can Pap 83-15, p 114

Andrusov NI (1980) Preliminary account of participation in the Black Sea deep-water expedition of 1890. Isvest Vsesov Geogr Obshch 26:380–409 (in Russian)

Arkangel'skiy AD (1927) On sediments of the Black Sea and their significance in sedimentology. Bvul Mosk Obshch Ispyt Prir 5:199–289 (in Russian)

Arkangel'skiy AD, Strakhov NM (1932) Geological history of the Black Sea. Bvul Mosk Obshch Ispyt Prir 10:3–104 (in Russian)

Arthur MA (1979) North Atlantic Cretaceous black shales: The record at Site 398 and brief comparison with other occurrences. In: Sibuet J-C, Ryan WBF et al. (eds) Initial Reports of the Deep Sea Drilling Project 47, vol 2. Washington DC, pp 719–738

Baturin GN, Kochenov AV, Shimkus KM (1967) Uranium and rare metals in the sediments of the Black and Mediterranean Seas. Geokhim 1:41–50 (in Russian)

Bertine KK (1972) The deposition of molybdenum in anoxic waters. Mar Chem 1:43–53

Bishop JKB (1988) The barite-opal-organic carbon association in oceanic particulate matter. Nature 332:341–343

Bukry D, King SA, Horn MK, Manheim FT (1970) Geological significance of coccoliths in fine-grained carbonate bands of postglacial Black Sea sediments. Nature 226:156–158

Calvert SE (1976) The mineralogy and geochemistry of nearshore sediments. In: Riley JP, Chester R (eds) Chemical Oceanography, 2nd edn, vol 6, Academic Press, London, pp 187–280

Calvert SE, Fontugne MR (1987) Stable carbon isotopic evidence for the marine origin of the organic matter in the Holocene Black Sea sapropel. Isot Geosci 66:315–322

Calvert SE, Price NB (1970) Composition of manganese nodules and manganese carbonates from Loch Fyne, Scotland. Contrib Mineral Petrol 29:215–233

Calvert SE, Price NB (1983) Geochemistry of Namibian Shelf sediments. In: Thiede J, Suess E (eds) Coastal upwelling: its sediment record, vol 1. Plenum, New York, pp 337–375

Calvert SE, Mukherjee S, Morris RJ (1985) Trace metals in fulvic and humic acids from modern organic-rich sediments. Oceanol Acta 8:167–173

Calvert SE, Vogel JS, Southon JR (1987) Carbon accumulation rates and the origin of the Holocene sapropel in the Black Sea. Geology 15:918–921

Caspers H (1957) Black Sea and Sea of Azov. In: Hedgpeth JW (ed) Treatise on marine ecology. Geol Soc Am Mem 67(1):803–890

Chow TJ, Goldberg ED (1960) On the marine geochemistry of barium. Geochim Cosmochim Acta 20:192–198

Degens ET, Ross DA (1972) Chronology of the Black Sea over the last 25,000 years. Chem Geol 10:1–16

Degens ET, Stoffers P, Golubic S, Dickman MD (1978) Varve chronology: Estimated rates of sedimentation in the Black Sea deep-basin. In: Ross DA, Neprochnov YP et al. (eds) Initial Reports of the Deep Sea Drilling Project. Deep Sea Drilling Project, Washington 42B:499–508

Degens ET, Michaelis W, Garrasi C, Mopper K, Kempe S, Ittekkot VA (1980) Warven-Chronologie und Frühdiagenetische Umsetzungen organischer Substanzen Holozäner Sedimente des Schwarzen Meeres. Neues Jahrb Geol Palaeontol Monatsch 1980/2:65–86

Dehairs F, Lambert CE, Chesselet R, Risler N (1987) The biological production of marine barite and the barium cycle in the western Mediterranean Sea. Biogeochem 4:119–139

Demaison GJ, Moore GT (1980) Anoxic environments and oil source bed genesis. Organic Geochem 2:9–31

Dymond J (1985) Particulate barium fluxes in the oceans: An indicator of new productivity. Trans Am Geophys Union 66:1275

François R (1987) The influence of humic substances on the geochemistry of iodine in nearshore and hemipelagic marine sediments. Geochim Cosmochim Acta 51:2417–2427

Fresnel J, Galle P, Gayral P (1979) Résultats de la microanalyse des cristaux vacuolaires chez deux chromophytes unicellulaires marines: *Exanthemachrysis gayraliae*, Pavlova sp. (Prymnesiophycees, Pavlovacees). CR Acad Sci Ser D 288:823–825

Glenn CR, Arthur MA (1985) Sedimentary and geochemical indicators of productivity and oxygen contents in modern and ancient basins: The Holocene Black Sea as the "type" anoxic basin. Chem Geol 48:325–354

Goldberg ED (1958) Determination of opal in marine sediments. J Mar Res 17:178–182
Goldberg ED, Arrhenius G (1958) Chemistry of Pacific pelagic sediments. Geochim Cosmochim Acta 13:153–212
Gurvich YG, Bogdanov YA, Lisitsyn AP (1978) Behaviour of barium in Recent sedimentation in the Pacific. Geokhim 3:359–374 (in Russian)
Guthrie TF, Lowe LE (1984) A comparison of methods for total sulphur analysis of tree foliage. Can J For Res 14:470–473
Hartmann M (1964) Zur Geochemie von Mangan und Eisen in der Ostsee. Meyniana 14:3–21
Harvey PK, Atkin BP (1982) The estimation of mass absorption coefficients by Compton scattering: extensions to the use of RhKa Compton radiation and intensity ratios. Am Mineral 67:534–537
Harvey PK, Taylor DM, Hendry RD, Bancroft F (1973) An accurate fusion method for the analysis of rocks and chemically related materials by X-ray fluorescence. X-ray Spectrom 2:33–44
Heier KS, Adams JAS (1963) The geochemistry of the alkali metals. Phys Chem Earth 5:253–381
Hirst DM (1974) Geochemistry of sediments from eleven Black Sea cores. In: Degens ET, Ross DA (eds) The Black Sea-geology, chemistry and biology. Am Assoc Petrol Geol Mem 20:430–455
Jacobs L, Emerson S (1982) Trace metal solubility in an anoxic fjord. Earth Planet Sci Lett 60:237–252
Klovan JE, Imbrie J (1971) An algorithm and FORTRAN-IV program for large-scale Q-mode factor analysis and calculation of factor scores. J Math Geol 3:61–77
Kochenov AV, Baturin GN, Kovaleva SA, Emel'yanov EM, Shimkus KM (1965) Uranium and organic matter in the sediments of the Black and Mediterranean Seas. Geokhim 3:302–313 (in Russian)
Korolev DF (1958) The role of iron sulphides in the accumulation of molybdenum in sedimentary rocks of the reduced zone. Geokhim 4:452–463 (in Russian)
Kullenberg B (1952) On the salinity of the water contained in marine sediments. Goteb K Vetensk Vitter Hetssamh Handl Sjatte Foljden Ser B 6:3–37
Li Y-H, Bischoff J, Mathieu G (1969) The migration of manganese in the Arctic Basin sediment. Earth Planet Sci Lett 7:265–270
Manheim FT (1961) A geochemical profile in the Baltic Sea. Geochim Cosmochim Acta 25:52–71
McCoy FW (1974) Late Quaternary sedimentation in the eastern Mediterranean Sea. PhD Thesis, Harvard Univ, p 132
Müller G, Blaschke R (1969) Zur Entstehung des Tiefsee-Kalkschlammes im Schwarzen Meer. Naturwissenschaften 56:561–562
Norrish K, Hutton JT (1969) An accurate X-ray spectrographic method for the analysis of a wide range of geological samples. Geochim Cosmochim Acta 33:431–453
Olausson E (1961) Studies of deep-sea cores. Rep Swedish Deep-Sea Expedition 8(6):337–391
Ostlund HG (1974) Expedition "Odysseus 65": Radiocarbon age of Black Sea deep water. In: Degens ET, Ross DA (eds) The Black Sea-geology, chemistry and biology. Am Assoc Petrol Geol Mem 20:127–132
Ostrumov EA, Volkov II, Fomina LS (1961) Distribution of the forms of sulphur compounds in Black Sea bottom deposits (in Russian). Tr Inst Okeanol Akad Nauk SSSR 50:93–129 (in Russian)
Pilipchuk MF, Volkov II (1966) Distribution of molybdenum in Recent sediments of the Black Sea. Dokl Akad Nauk SSSR 167:152–157 (in Russian)
Pilipchuk MF, Volkov II (1968) Geochemistry of molybdenum in the Black Sea. Geokhim 8:977–985 (in Russian)
Pilipchuk MF, Volkov II (1974) Behavior of molybdenum in processes of sediment formation and diagenesis in the Black Sea. In: Degens ET, Ross DA (eds) The Black Sea-geology, chemistry and biology. Am Assoc Petrol Geol Mem 20:542–553
Price NB, Calvert SE (1977) The contrasting geochemical behaviours of iodine and bromine in Recent sediments of the Namibian Shelf. Geochim Cosmochim Acta 41:1769–1775
Price NB, Calvert SE, Jones PGW (1970) The distribution of iodine and bromine in the sediments of the southwestern Barents Sea. J Mar Res 28:22–34
Revelle R, Bramlette M, Arrhenius G, Goldberg ED (1955) Pelagic sediments of the Pacific. In: Poldervaart W (ed) Crust of the Earth. Geol Soc Amer Spec Pap 62:221–125
Ross DA, Degens ET (1974) Recent sediments of the Black Sea. In: Degens ET, Ross DA (eds) The Black Sea-geology, chemistry and biology. Am Assoc Petrol Geol Mem 20:183–199
Ross DA, Degens ET, MacIlvaine J (1970) Black Sea: recent sediment history. Science 170:163–165
Rossignol-Strick M, Nesteroff W, Olive P, Vergnaud-Grazzini C (1982) After the deluge: Mediterranean stagnation and sapropel formation. Nature 295:105–110

Ryan WBF (1972) Stratigraphy of late Quaternary sediments in the eastern Mediterranean. In: Stanley DJ (ed) The Mediterranean Sea. Dowdon Hutchinson & Ross, pp 149–169
Ryan WBF, Cita MB (1977) Ignorance concerning episodes of ocean-wide stagnation. Mar Geol 23:197–215
Schmitz B (1987) Barium, equatorial high productivity, and the northward wandering of the Indian continent. Paleoceanogr 2:63–78
Shterenberg LY, Bazilevskaya YS, Chigivera TA (1966) Manganese and iron carbonates in bottom deposits of Lake Pinnus-Yarvi. Dokl Akad Nauk SSSR 170:205–209 (in Russian)
Spencer DW, Brewer PG (1971) Vertical advection diffusion and redox potentials as controls on the distribution of manganese and other trace metals dissolved in waters of the Black Sea. J Geophys Res 76:5877–5892
Strakhov NM (1963) On some new features of diagenesis in the Black Sea deposits. Litol Poloz Iskop 1:7–27 (in Russian)
Strakhov NM (1967) Principles of lithogenesis, vol 1. Oliver & Boyd, p 245
Strakhov KM (1969) Principles of lithogenesis, vol 2. Oliver & Boyd, p 609
Strakhov NM (1971) Geochemical evolution of the Black Sea in the Holocene. Litol Polez Iskop 3:263–274 (in Russian)
Strakhov NM, Belova IV, Glagoleva MA, Lubchenko IY (1971) Distribution and forms of occurrence of elements in the surface layer of modern Black Sea deposits. Litol Polez Iskop 2:3–31 (in Russian)
Suess E (1979) Mineral phases formed in anoxic sediments by microbial decomposition of organic matter. Geochim Cosmochim Acta 43:339–352
Thiede J, Andel TH van (1977) The paleoenvironment of anaerobic sediments in the late Mesozoic South Atlantic Ocean. Earth Planet Sci Lett 33:301–309
Volkov II, Fomina LS (1971) Dispersed elements in sapropel of the Black Sea and their interrelationship with organic matter. Litol Polez Iskop 6:3–15 (in Russian)
Volkov II, Fomina LS (1972) The role of iron sulphides in the accumulation of minor elements in Black Sea sediments. Litol Polez Iskop 2:18–24 (in Russian)
Wong GTF, Brewer PGG (1977) The marine chemistry of iodine in anoxic basins. Geochim Cosmochim Acta 41:151–159
Woolnough WG (1937) Sedimentation in barred basins and source rocks of oil. Am Assoc Petrol Geol Bull 21:1101–1157

# Early Diagenesis of Organic Matter in Peru Upwelling Area Sediments

J. W. FARRINGTON, M.A. MCCAFFREY, and J. SULANOWSKI

## 1 Introduction

Improved knowledge of the biogeochemistry of organic matter in recently deposited surface sediments and the diagenesis of organic matter buried deeper in sediments is important to a more complete understanding of several important processes in, or issues related to, the marine environment. Among these are: (1) the carbon, nitrogen and sulfur cycles of contemporary and paleoenvironments, (2) nutrient regeneration, (3) interactions of organic matter and organic compounds with trace metals and radionuclides, (4) the influence of organic coatings and organic matrices on resuspension, transport and deposition of particles, (5) the influence of early diagenesis on the eventual composition of organic matter in ancient sediments including the generation, migration and accumulation of oil and gas and (6) the distributions, reactivities and fates in the sea for compounds of anthropogenic origin or mobilization (Degens 1965, 1967; Duursma and Dawson 1981; Berner 1980; Hunt 1979; Tissot and Welte 1978; McCave 1974; Andersen 1977; Farrington and Westall 1986; Brownawell and Farrington 1986; Romankevich 1984; Thurman 1985).

Studies of sediments from highly productive upwelling areas are of interest in regard to the preceding. These areas are protodepositional environments thought to eventually yield conditions conducive to oil and gas generation (Demaison and Moore 1980; Rullkotter et al. 1983). Surface sediments are important links in the nutrient-productivity cycles of these rich fisheries areas. In addition, variations in productivity in the water column related to regional ENSO (El Nino-Southern Oscillation) climatic events influence the C being removed from the contemporary environment by deposition and burial (e.g., Walsh 1981), and accounting for this buried C could be important to the detailed carbon cycle models, as indicated by Deuser (1979). Sediments deposited under conditions of upwelling have been subjected to a variety of sedimentological, paleontological, and geochemical studies (Suess and Thiede 1983; Thiede and Suess 1983).

Relatively few of these studies address the detailed molecular composition of organic matter. This paper compares data on organic carbon, nitrogen, pore water geochemistry, total hydrolyzable amino acids, and interstitial water amino acids for box cores from 15° S off Peru (Henrichs and Farrington 1984; Henrichs et al. 1984) with more recent lipid class biomarker compound data obtained for these same cores (Farrington et al. 1988a) and new data reported in this paper.

## 2 Sampling and Methods (in brief)

Several box cores were obtained on R/V KNORR 73-2 February-March 1978 at 15° S off Peru. The positions of the two Soutar type box cores at 92 and 268 meters water depth from which data for this paper are presented and discussed are given in Fig. 1. These cores, SC4 and SC6, were designated as core stations 4 and 5A respectively in Henrichs and Farrington (1984). Cores were sectioned by sequential removal of screw fastened side plates and use of clean stainless steel spatulas (Henrichs et al. 1984) with subsamples for lipids and amino acids obtained in adjacent portions of the sections of 2, 3, 4, or 6-cm intervals. Samples were placed in cleaned glass jars in a freezer or refrigerator depending on the type of analysis within 1 to 3 h of sectioning.

Details of extraction and analyses are reported in Henrichs et al. (1984) Henrichs and Farrington (1984), and Farrington et al. (1988a). Briefly, amino acids in sediments were analyzed by hydrolysis with 6N HCl under $N_2$ at 10°C for 24 h; cation exchange chromatograph, derivatization to yield (N,O) heptafluoro-butyryl-n-butyl esters of amino acids and analyzed by glass capillary gas chromatography and glass capillary gas chromatography-mass spectrometry. Lipids were extracted from separate subsamples using sonication assisted solvent extraction with isopropanol, methanol-chloroform; activated copper removal of elemental sulfur; column chromatography on silica gel; and glass capillary gas chromatography, glass capillary gas chromatography-mass spectrometry analyses of lipids. Fatty acids were analyzed after base saponification of the lipid extract and

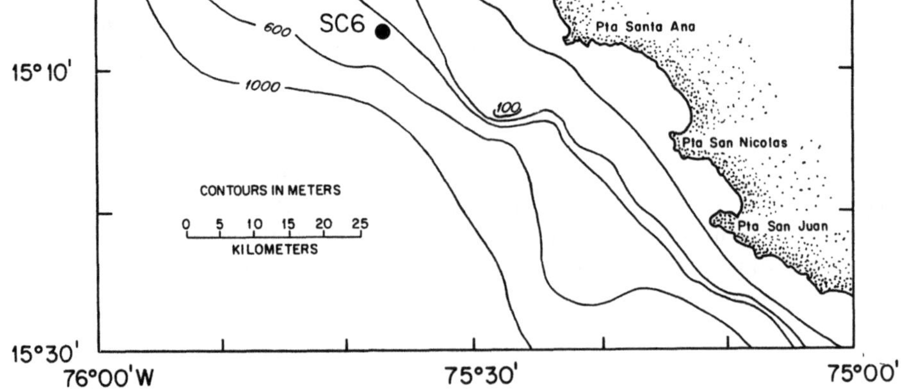


**Fig. 1.** Sampling stations R/V KNORR Cruise 73/2, 1978. SC refers to Soutar Core

methylation of the fatty acids to their methyl esters. Use of internal recovery standards and external standard response curves were the basis of quantification. Data are precise to ± 5–20% depending on the parameter measured as determined by analyses of duplicate subsamples of sediment from several core sections. Procedural blanks were negligible at the concentrations of compounds analyzed and reported. Organic carbon and nitrogen data were obtained by acidification to remove carbonate and measurement by CHN analyzer (Henrichs and Farrington 1984). All other data discussed were reported with methods of measurement in Henrichs and Farrington (1984).

## 3 Results and Discussion

### 3.1 Total Hydrolyzable Amino Acids and Organic Matter

A brief synopsis of Henrichs et al. (1984) and Henrichs and Farrington (1984) is given in this subsection. Total hydrolyzable amino acids (THAA) have depth profiles that show higher concentrations in surface sections compared to deeper sections, and THAA concentrations have some concordance with organic carbon concentrations in maxima and minima in depth profiles for SC4 (Fig. 2). The THAA-N/TN ratio indicates that 40 to 70% of the total nitrogen (TN) in surface sediments of SC4 and SC6 are THAA. The ratio of THAA-N/TN is higher for these Peru surface sediments than for surface sediments at comparable water depths in less productive areas (e.g., Mopper and Degens 1972; Henrichs 1980). Overall, the ratio of THAA-N/TN decreases with depth in SC4 and SC6 (Figs. 2 and 3) when comparing top sections of the cores with bottom sections, probably due to more rapid remineralization of THAA than of other organic nitrogen. Several fluctuations of THAA-N/TN ratio (Figs. 2 and 3) demonstrate that it is likely that some factors are influencing the depth profiles in addition to a simple sequence of events of constant deposition rates followed by early diagenesis-remineralization. Henrichs and Farrington (1984) note that the required C/N ratio of remineralized organic matter calculated from interstitial water total $CO_2$ and $NH_4^+$ measurements and early diagenesis modeling is consistent with amino acids as a source of remineralized organic matter. Despite this remineralization , the THAA composition expressed as mol % of individual amino acids is remarkably uniform, with variability within and between cores of only about twice analytical error (Henrichs 1980; Henrichs et al. 1984). This could be due to near uniformity in amino acid composition of various proteins and other forms of THAA undergoing remineralization, or it may be that relatively little of the total THAA is remineralized thereby having minimal effect on THAA composition. We have calculated that between 70 and 80% of the organic matter accumulating in surface sediments is still present at 50 cm depth (Henrichs and Farrington 1984). Conversely, this means that 20 to 30% of the organic matter is undergoing remineralization.

The amino acids in interstitial waters have distinct compositional differences compared to THAA composition. This is consistent with contributions of amino acids via microbial activity, and the depth profiles of interstitial waters total $CO_2$

and $NH_4^+$ indicate active remineralization of organic matter is taking place in these sediments (Henrichs et al. 1984). Despite this active remineralization, we have stated the hypothesis that the fluctuations in depth profiles of organic carbon for these cores are due mostly to the preservation of an historical record of periods of ENSO (lower water column organic matter productivity) times alternating with periods of non-ENSO (higher productivity) (Henrichs and Farrington 1984).

Since that publication, we have obtained organic carbon data for several more sections of the cores to better define the depth profiles (Figs. 2 and 3), and we have obtained data for several lipid class compounds. These are discussed next.

## 3.2 Selected Lipid Class Compounds

It is important to note that the sectioning intervals for cores for organic carbon and lipid compounds for SC4 and SC6 were not always the same as for THAA (Figs. 2 and 3) because of various subsample size requirements. Results and initial interpretations of n-alkane, lycopane, alkenone and a few fatty acid data have been presented in Farrington et al. (1988a). Table 1 presents the total set of fatty acid data for SC6. The entire set of fatty acid data for both cores and sediment traps in the region are presented and discussed elsewhere (Farrington et al. 1988b). We present here a subset of the data for both cores and compare these data with the amino acid and organic carbon data.

The concentrations of fatty acids in surface section sediments (Table 1, Figs. 2 and 3) are factors of four to six less than concentrations reported for a sample of the sediment/water interface obtained by Smith et al. (1983) at 12°01.8′S, 77°29.3′W further north on the Peru shelf where organic carbon concentrations are higher than at 15° S (Suess et al. 1987).

The depth profiles of concentrations of total fatty acids show a general trend of decreasing concentrations for SC4 and SC6 with some fluctuations (Figs. 2 and 3). The trend of decreasing concentration is expected from earlier reports (e.g., Farrington et al. 1977; Simoneit 1978; DeBaar et al. 1983; Kawamura et al. 1987) and is consistent with early diagenesis-remineralization of fatty acids. A few of the minor fluctuations of maxima and minima of total fatty acids appear to be concordant with maxima and minima in the depth profiles of organic carbon: especially for SC4 (Fig. 2). Data for a selection of individual fatty acids are plotted as depth profiles for SC4 and SC6 in Figs. 2 and 3. We will use the shorthand notation for fatty acids, e.g., 16:1Δ9 refers to the sixteen carbon chain length (16), one double bond (:1) at the 9 carbon position (Δ9). The maxima and minima in depth profiles of concentrations of 14:0, 16:0, 18:0, 16:1Δ9, 18:1Δ9, 18:1Δ11, correspond to the profiles of total fatty acids. The polyunsaturated fatty acids 20:5 and 22:6 also show maxima and minima in depth profiles that correspond with maxima and minima in organic carbon. These fatty acids, 20:5 and 22:6, are

**Fig. 2.** Depth profiles of organic geochemical data R/V KNORR Cruise 73/2, SC Box Core 4 – KNSC4. See text for compound abbreviations. Data are plotted mid-point for each depth section

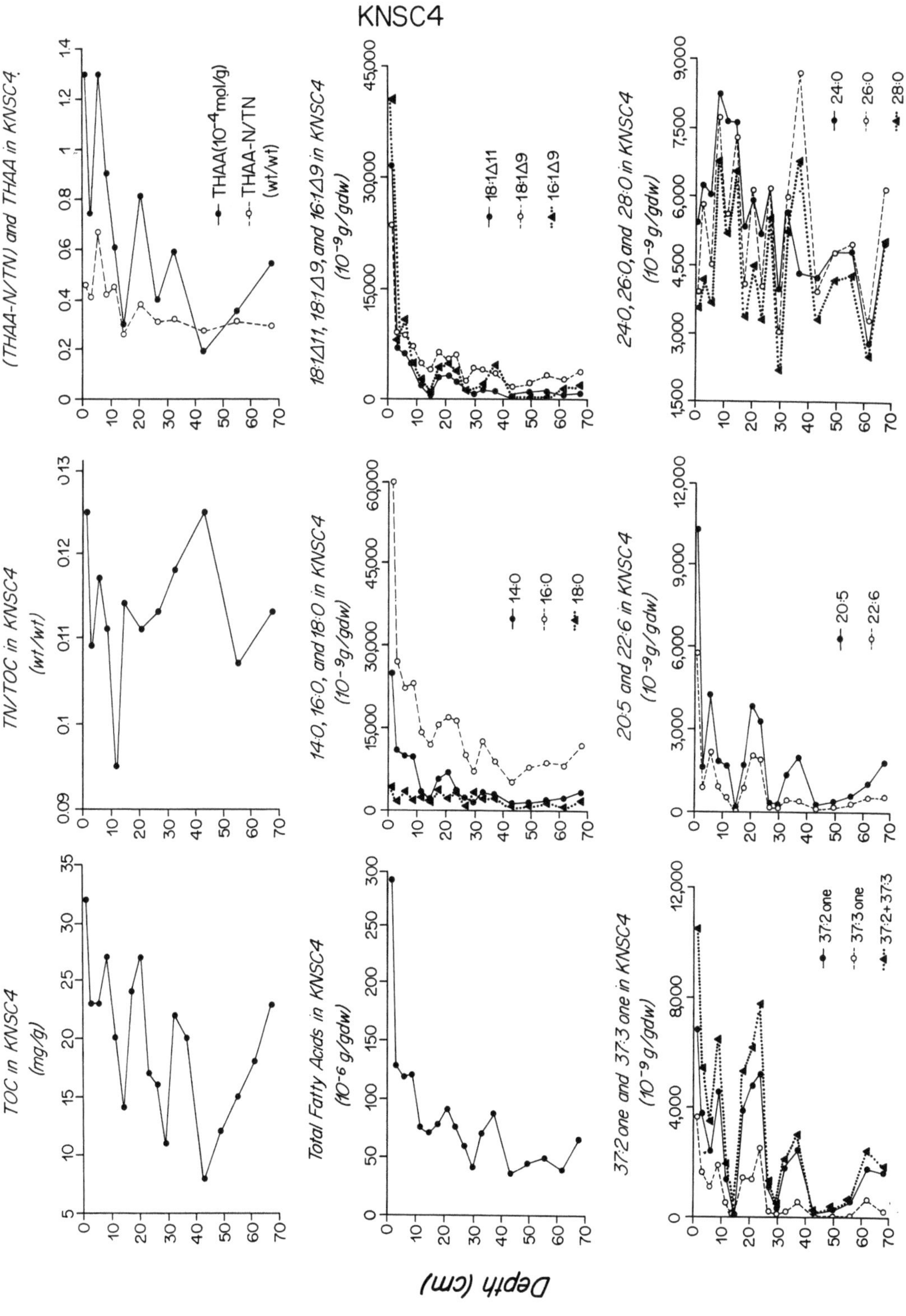
KNSC4
(THAA-N/TN) and THAA in KNSC4
THAA($10^{-4}$mol/g)
THAA-N/TN (wt/wt)
18:1Δ11, 18:1Δ9, and 16:1Δ9 in KNSC4 ($10^{-9}$g/gdw)
18:1Δ11
18:1Δ9
16:1Δ9
24:0, 26:0, and 28:0 in KNSC4 ($10^{-9}$g/gdw)
24:0
26:0
28:0
TN/TOC in KNSC4 (wt/wt)
14:0, 16:0, and 18:0 in KNSC4 ($10^{-9}$g/gdw)
14:0
16:0
18:0
20:5 and 22:6 in KNSC4 ($10^{-9}$g/gdw)
20:5
22:6
TOC in KNSC4 (mg/g)
Total Fatty Acids in KNSC4 ($10^{-6}$ g/gdw)
37:2one and 37:3one in KNSC4 ($10^{-9}$g/gdw)
37:2one
37:3one
37:2+37:3
Depth (cm)

thought to be very susceptible to loss during early diagenesis because of their numerous double bonds (e.g., Farrington and Quinn 1973; Boon et al. 1975) and their presence in high concentrations in surface sediments has been interpreted as indicating relatively freshly deposited planktonic organic matter (Smith et al. 1983).

Depth profiles of concentrations of long chain alkenones, $C_{37:2}$, $C_{37:3}$, and the sum of $C_{37:2}$ + $C_{37:3}$ are plotted in Figs. 2 and 3 for purposes of comparison with organic carbon and fatty acid data. Marlowe (1984), Brassell et al. (1986), and Prahl et al. (1988) have noted the potential utility of these compounds for providing a "molecular stratigraphic" record. The compounds are ubiquitous to *Prymnesiophyceae* and absent in other algal classes analyzed to date (Marlowe et al. 1984). The ratio of $C_{37:2}$ to $C_{37:3}$ calculated as a parameter:

$$U^K_{37} = \frac{[37:2] - [37:4]}{[37:2] + [37:3] + [37:4]}$$

has a strong relationship with temperature of culture medium or with temperature of the euphotic zone from which particles containing 37:2, 37:3 are sampled (Marlowe 1984; Prahl and Wakeham 1987). The $C_{37:4}$ alkenone is present in detectable quantities only at temperatures less than 15 °C (Prahl et al. 1988). A more extensive discussion is beyond the scope of our paper. It suffices to note that measurement of the $U^K_{37}$ in a core from the Kane Gap spanning 120,000 years showed reasonable correlations of $U^K_{37}$ with $\delta^{18}O/^{16}O$ paleoclimatic record (Brassell et al. 1986).

We have shown that the depth profiles of $C_{37}$ alkenone concentrations, $U^K_{37}$, and organic carbon for SC6 correspond in maxima and minima to expected fluctuations of higher productivity, colder temperature conditions and lower productivity, warmer temperature conditions as indicated in Fig. 4 taken from our paper (Farrington et al. 1988a). Data for SC4 do not fit this interpretation. We think that this is a result of SC4 being inshore of the main upwelling zone, and thus material deposited there does not accurately reflect the alternating ENSO low productivity and normal higher productivity events of the main upwelling zone (Farrington et al. 1988a).

Prahl et al. (1988) have shown that the sum of the $C_{37}$, $C_{38}$ and $C_{39}$ alkenones constitute 5 to 13% of the total cellular organic carbon of *E. huxleyi*. The depth profiles of 37:2, 37:3 and their sum for SC4 and SC6 are concordant in their maxima and minima with the maxima and minima for the organic carbon (Figs. 2 and 3). A plot of alkenones versus organic carbon for all data points for SC4 and SC6 shows a significant correlation (Fig. 5). This is intriguing but it is too soon to do more than speculate that the historical record of alkenone concentrations in sediments at 15 °S, and for similar depositional sequences, may provide a clue to prymnesiophyte productivity in the euphotic zone at the time of sediment deposition. We have no information available at this time as to the extent to which *Prymnesiophyceae*

**Fig. 3.** Depth profiles of organic geochemical data R/V KNORR Cruise 73/2, SC Box Core 6 – KNSC6. See text for compound abbreviations. Data are plotted mid-point for each depth section

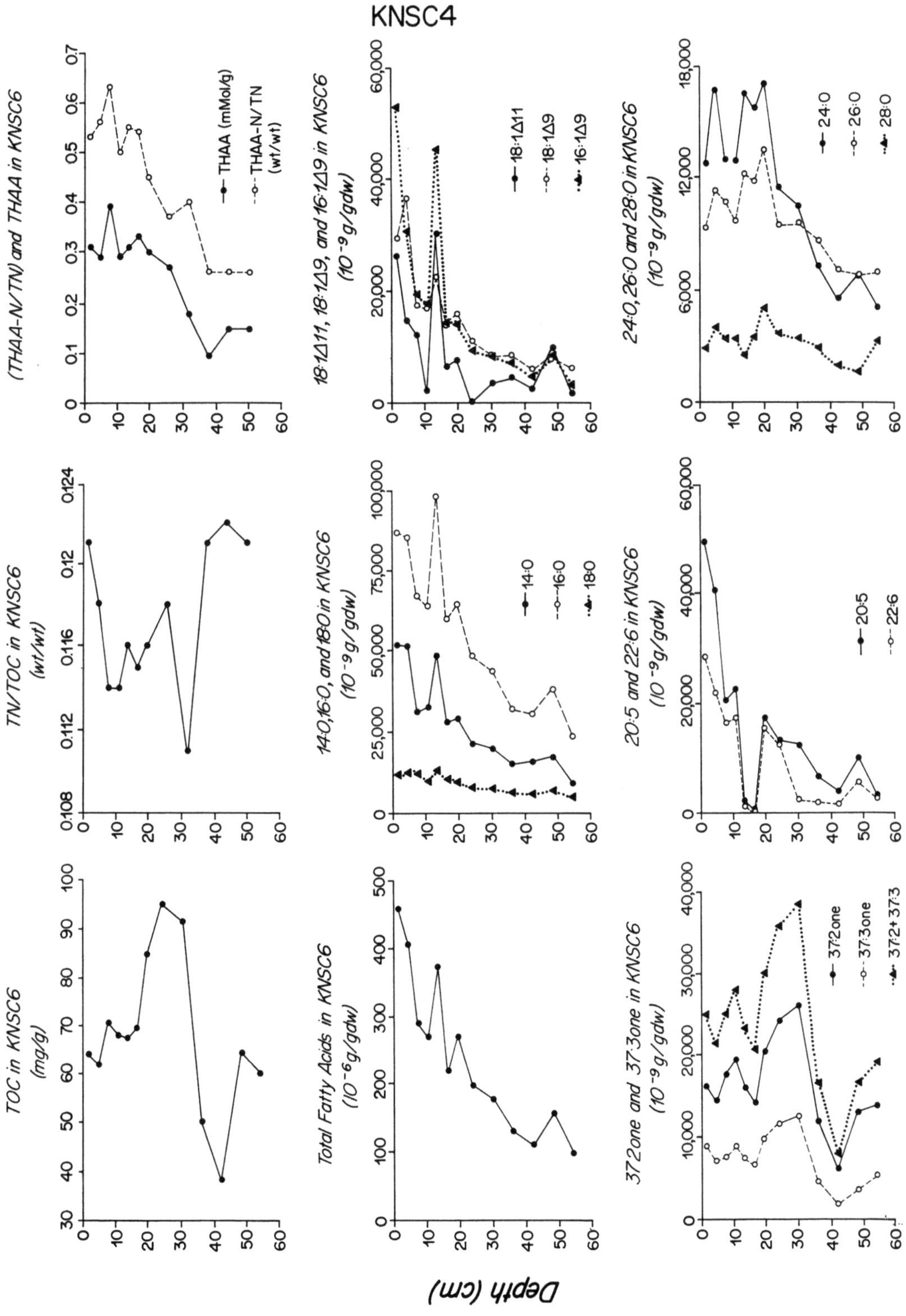
KNSC4
(THAA-N/TN) and THAA in KNSC6
THAA (mMol/g)
THAA-N/TN (wt/wt)
18:1Δ11, 18:1Δ9, and 16:1Δ9 in KNSC6 (10⁻⁹ g/gdw)
18:1Δ11
18:1Δ9
16:1Δ9
24:0, 26:0 and 28:0 in KNSC6 (10⁻⁹ g/gdw)
24:0
26:0
28:0
TN/TOC in KNSC6 (wt/wt)
14:0, 16:0, and 18:0 in KNSC6 (10⁻⁹ g/gdw)
14:0
16:0
18:0
20:5 and 22:6 in KNSC6 (10⁻⁹ g/gdw)
20:5
22:6
TOC in KNSC6 (mg/g)
Total Fatty Acids in KNSC6 (10⁻⁶ g/gdw)
37:2one and 37:3one in KNSC6 (10⁻⁹ g/gdw)
37:2one
37:3one
37:2+37:3
Depth (cm)

contribute to the plankton biomass and productivity of the waters at 15°S at various times of the year.

The depth profiles of the fatty acids 24:0, 26:0, and 28:0 are plotted in Figs. 2 and 3. These profiles are different than the other fatty acid profiles and the alkenone profiles. We attribute this to a combination of at least two sources of input for the 24:0, 26:0, and 28:0. First, there are probably some contributions from marine bacteria and plankton and second, there are probably contributions from land plant detritus delivered to the area by either aeolian or fluvial inputs. The fluvial inputs would be delivered to the nearshore regions by flash flood type conditions that occur during ENSO periods (Farrington et al. 1988a; Volkman et al. 1987). Land plants have significant proportions of 24:0, 26:0, and 28:0 in their lipids (Kawamura et al. 1987; Simoneit 1978). We have shown that, compared to SC6, land plant *n*-alkanes predominate in the nearer shore SC4 (Farrington et al. 1988a) and in a box core even closer to shore (Volkman et al. 1983). The average concentration of the 24:0, 26:0, and 28:0 relative to organic carbon is higher for SC4 compared to SC6 as would be expected if a major proportion of the 24:0, 26:0, and 28:0 are contributed by land detritus being delivered to the nearshore zone (Figs. 2 and 3). We have yet to complete depth profile data interpretation for *n*-alkanols and certain triterpanols present in these sediments (Volkman et al. 1987) that are land plant biomarkers and should allow us to deconvolute the depositional record with respect to land plant detritus input.

Irrespective of the source for the 24:0, 26:0, and 28:0 fatty acids, their depth profiles in both cores are markedly different compared to the other fatty acids plotted in Figs. 2 and 3. Their depth profiles (24:0, 26:0, 28:0) are not consistent with rapid early diagenesis, i.e., there is not a rapid decrease in concentration with increasing depth in the core and the profiles do not have maxima and minima corresponding to periods of presumed high productivity interdispersed with lower productivity ENSO periods as indicated in the profiles of organic carbon and other lipids plotted in Figs. 2 and 3.

## 4 General Discussion

The entire data set presented here, when interpreted within the context of previous studies (Henrichs and Farrington 1984; Henrichs et al. 1984) and our recent studies (Farrington et al. 1988a) demonstrates that early diagenesis of organic matter in surface sediments at 15°S proceeds with utilization of total hydrolyzable amino acids, presumably mainly proteins, and certain lipid class compounds such as the fatty acids 14:0, 16:0, 16:1Δ9, 18:0, 18:1Δ9, 18:1Δ11. During this early diagenesis process, the relative composition of THAA is not changed appreciably while the relative composition of extractable fatty acids is changed significantly. Obviously, this early diagenesis also involves the remineralization of organic compounds we have not measured.

Despite early diagenesis and remineralization of organic matter, it appears that some compounds such as the $C_{37}$ alkenones may be incorporated and preserved in the sediment record in a manner that will allow a reading of the historical record of

**Table 1.** K-73-2, SC6, fatty acids as methyl esters (FAME concentrations ng/g)

| Fatty acid | Depth in core in cm 0–3 | 3–6 | 6–9 | 9–12 | 12–15 | 15–18 | 18–21 | 21–27 | 27–33 | 33–39 | 39–45 | 45–51 | 51–57 |
|---|---|---|---|---|---|---|---|---|---|---|---|---|---|
| 14:1Δ9[a] | 1500 | 1290 | 821 | 785 | 1200 | 820 | 1080 | 784 | 1380 | 955 | 649 | 1080 | 828 |
| 14:0 | 51700 | 51200 | 31500 | 33000 | 48400 | 28300 | 29500 | 21800 | 20100 | 15500 | 16100 | 17600 | 9660 |
| a15:0 | 21500 | 18300 | 15300 | 10600 | 14600 | 8410 | 8460 | 6130 | 6160 | 3400 | 2450 | 3770 | 2610 |
| 15:0 | 10400 | 10700 | 5000 | 4050 | 7570 | 3830 | 3870 | 2950 | 3110 | 2540 | 2400 | 3010 | 1610 |
| i16:0 | 7910 | 6300 | 4810 | 3850 | 5130 | 2230 | 3640 | 2770 | 2590 | 2100 | 1330 | 2350 | 1220 |
| 16:1Δ9 | 52900 | 30500 | 19500 | 18000 | 45300 | 14600 | 14300 | 9610 | 8480 | 7350 | 4830 | 8660 | 3280 |
| 16:0 | 86900 | 85500 | 66900 | 63700 | 98100 | 59600 | 64100 | 48300 | 43800 | 32400 | 31000 | 38500 | 24100 |
| a17:0 | 7700 | 5480 | 4810 | 3360 | 5630 | 2520 | 2970 | 2510 | 2140 | 1520 | 1080 | 1690 | 1330 |
| 17:0 | 4700 | 4060 | 3370 | 3140 | 4790 | 2450 | 3610 | 2470 | 2580 | 1530 | 1540 | 1920 | 1280 |
| 18:4 | 8570 | 3760 | 3040 | 5070 | 1810 | 2270 | 5110 | 3600 | 3120 | 1440 | 1510 | 1560 | 1150 |
| i18:0 | 6650 | 4840 | 4480 | 5530 | 7490 | 4710 | 5090 | 3540 | 4950 | 1460 | 1630 | 2960 | 956 |
| 18:1Δ9 | 29400 | 36500 | 17600 | 17200 | 22500 | 14600 | 16200 | 11200 | 8580 | 8620 | 6160 | 8820 | 6230 |
| 18:1Δ11 | 26200 | 14900 | 12400 | 2280 | 30100 | 6590 | 7860 | 147 | 3580 | 4590 | 2580 | 10100 | 1740 |
| 18:0 | 12300 | 12700 | 12500 | 10300 | 13200 | 10700 | 9860 | 8270 | 7850 | 6630 | 5990 | 7320 | 5320 |
| 20:5 | 49600 | 40600 | 20600 | 22500 | 2220 | 525 | 17300 | 13200 | 12400 | 6660 | 4000 | 10100 | 3260 |
| 20:1Δ11 | 4940 | 3220 | 2390 | 2220 | 3110 | 1020 | 1770 | 1480 | 1490 | 829 | 541 | 918 | 549 |
| 20:0 | 4100 | 3900 | 3720 | 3500 | 5560 | 3190 | 3850 | 3740 | 4040 | 2320 | 2100 | 3160 | 3050 |
| 22:6 | 28400 | 21900 | 16400 | 17400 | 1250 | 229 | 15500 | 12500 | 2360 | 1970 | 1660 | 5560 | 2560 |
| 22:1Δ13 | 1650 | 2120 | 1750 | 789 | 2580 | 1210 | 1760 | 1490 | 924 | 739 | 349 | 768 | 901 |
| 22:0 | 5560 | 6620 | 5330 | 5050 | 5980 | 5230 | 5920 | 4780 | 4660 | 3050 | 2500 | 4110 | 2660 |
| 23:0 | 1180 | 1520 | 1120 | 1180 | 1070 | 1030 | 1490 | 1220 | 1170 | 845 | 629 | 1000 | 753 |
| 24:1Δ15 | 2950 | 2040 | 1800 | 4540 | 7260 | 4790 | 2540 | 2240 | 1880 | 1160 | 896 | 1260 | 826 |
| 24:0 | 12760 | 16730 | 13000 | 12900 | 16600 | 15800 | 17100 | 11500 | 10500 | 7280 | 5630 | 6920 | 5210 |
| 25:0 | 1200 | 1590 | 1260 | 1420 | 1020 | 1380 | 2490 | 1560 | 1440 | 1110 | 863 | 1110 | 1210 |
| 26:0 | 9290 | 11300 | 10700 | 9650 | 12200 | 11800 | 13500 | 9490 | 9570 | 8670 | 7130 | 6880 | 7020 |
| 27:0 | 1660 | 1440 | 1200 | 1280 | 1880 | 1590 | 1540 | 1520 | 1410 | 984 | 656 | 882 | 1300 |
| 28:0 | 2930 | 4010 | 3420 | 3410 | 2550 | 3540 | 5120 | 3740 | 3480 | 2980 | 1970 | 1700 | 3390 |
| 29:0 | 2230 | 526 | 3030 | 997 | 1700 | 3510 | 3030 | 1690 | 1030 | 812 | 661 | 2000 | 1840 |
| 30:0 | 2420 | 2740 | 2560 | 2560 | 2740 | 2670 | 1260 | 3000 | 2270 | 1940 | 1240 | 980 | 2150 |
| Total | $459 \times 10^3$ | $406 \times 10^3$ | $290 \times 10^3$ | $270 \times 10^3$ | $373 \times 10^3$ | $219 \times 10^3$ | $270 \times 10^3$ | $197 \times 10^3$ | $177 \times 10^3$ | $131 \times 10^3$ | $110 \times 10^3$ | $157 \times 10^3$ | $98 \times 10^3$ |

[a]Shorthand notation – 14 refers to carbon number, :1 refers to number of double bonds, Δ9 refers to position of double bond.

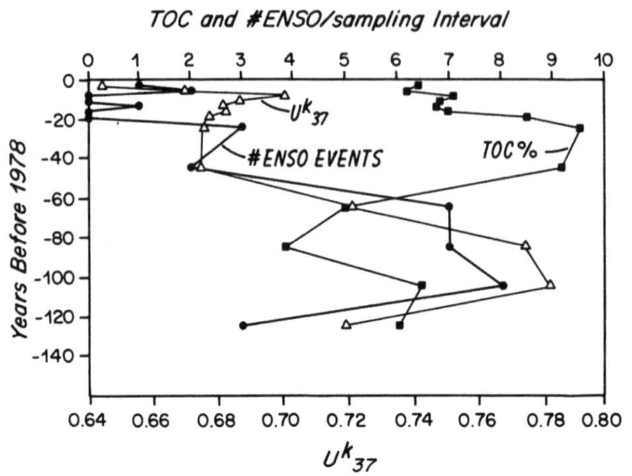


**Fig. 4.** Depth-time profile of organic carbon, numbers of intense ENSO events during the time interval of a section of the core and $U^K_{37}$ taken from Farrington et al. (1988a)

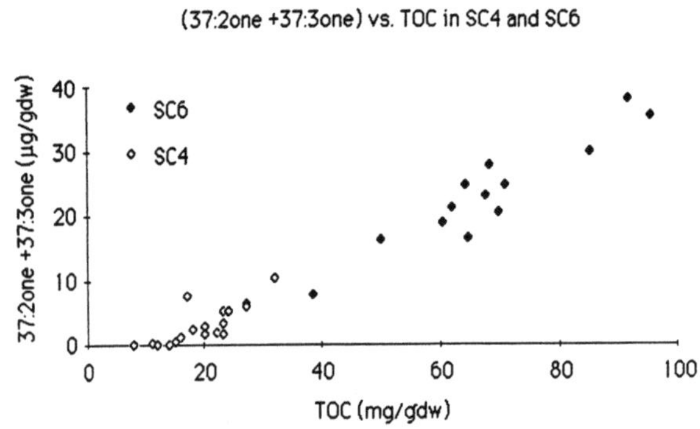


**Fig. 5.** Correlation plot of organic carbon vs. $C_{37}$ alkenones for KNSC4 and KNSC6

fluctuations of years of predominantly normal productivity with lower productivity years of ENSO events. The concentrations of the $C_{37}$ alkenones, perhaps the $C_{38}$ and $C_{39}$ alkenones (Prahl et al. 1988) as well as the $U^K_{37}$ ratio (indicative of temperature) have promise as molecular paleontological biomarkers.

The depth profiles of the polynsaturated fatty acids 20:5 and 22:6 are consistent with the interpretation that the profiles represent an attenuated input signal of marine fatty acid deposition to the sediment. We hypothesize that the depth profile of the 20:5 and 22:6 would have an appearance almost identical to the $C_{37}$ alkenones (Figs. 2 and 3) except for the influence of early diagenesis which remineralizes or transforms some of the 20:5 and 22:6 to the extent that maxima in the depth profiles are progressively attenuated deeper in these cores. We theorize that most marine derived sterols will behave similar to the 20:5 and 22:6 and we are in the process of extending our recent work on sterols (Volkman et al. 1987) to these cores.

Two papers have reported on precise, detailed, important measurements of the regional climate record in the Andean Quelecaya Ice Cap inshore of our study region (Thompson et al. 1984, 1986). The ENSO event periods are recorded from approximately the 1500s to the present (Thompson et al. 1984). Recently we have obtained more box cores and piston cores at 15°S on R/V MOANA WAVE Cruise 87-08 (PUBS I – Peru Upwelling Biogeochemistry Studies, July 1987). We are planning to apply molecular stratigraphy to these cores in an attempt to obtain a regional climate record for the aquatic ecosystem of the shelf at 15°S that we can compare with the Quelecaya Ice Cap record. The box core analyses will provide the corroborative evidence to initiate this study because the intense 1982-83 ENSO event should be recorded in the upper 5 to 10 cm of these box cores (Farrington et al. 1988a).

*Acknowledgements.* J.W.F. wishes to acknowledge Professor Egon Degens for many valuable discussions concerning organic geochemistry during their years together at Woods Hole Oceanographic Institution. We thank the officers, crew and scientific group of R/V KNORR 73/2 for assistance in sampling. A.C. Davis, P.A. Dickinson, N.M. Frew, C. Johnson, C.H. Clifford provided valuable assistance. S.G. Wakeman, J.K. Volkman, S.M. Henrichs, C. Lee, R.B. Gagosian, D.J. Repeta and K.J. Whelan provided valuable discussion of various points in the paper. P. Chandler typed the manuscript. Financial support from the U.S. National Science Foundation grant OCE 85-09859 is gratefully acknowledged. This is Contribution No. 7239 from Woods Hole Oceanographic Institution.

## References

Andersen N (ed) (1977) Concepts in marine organic chemistry. Special Issue Mar Chem 5(4-6):303-638

Brassel SC, Eglinton G, Marlowe IT, Sarnthein H, Pflaumann U (1986) Molecular stratigraphy – a new tool for climate assessment. Nature 320:129-133

Berner RA (1980) Early Diagenesis. Princeton University Press, Princeton NJ

Boon JJ, Leeuw JW De, Schenck PA (1975) Organic geochemistry of Walvis Bay diatomaceous ooze I. Occurrence and significance of fatty acids. Geochim Cosmochim Acta 39:1559-1565

Baar H De, Farrington JW, Wakehan SG (1983) Vertical fluxes of fatty acids in the North Atlantic Ocean. J Mar Res 41:19-41

Brownawell BJ, Farrington JW (1986) Biogeochemistry of PCBs in interstitial waters of a coastal marine sediment. Geochim Cosmochim Acta 50(1):157-169

Degens ET (1965) Geochemistry of Sediments. Prentice-Hall, Englewood Cliffs, New Jersey, p 342

Degens ET (1967) Diagenesis of organic matter. In: Larsen G, Chilingar GV (eds) Diagenesis in Sediments: Developments in Sedimentology 8. Elsevier, Amsterdam, p 343-390

Demaison GJ, Moore GT (1980) Anoxic environments and oil surface bed genesis. Org Geochem 2:9-31

Deuser WG (1979) Marine biota, nearshore sediments and the global carbon balance. Org Geochem 1:243-247

Duursma EK, Dawson R (eds) (1981) Marine Organic Chemistry. Elsevier, Amsterdam

Farrington JW, Quinn JG (1973) Biogeochemistry of fatty acids in Recent sediments from Narragansett Bay, Rhode Island. Geochim Cosmochim Acta 37:259-268

Farrington JW, Westall J (1986) Organic chemical pollutants in the oceans and groundwater: A review of fundamental chemical properties and biogeochemistry. In: Kullenberg G (ed) The Role of the Oceans as Waste Disposal Option. NATO Advanced Study Institute Series. Reidel, Boston, MA, pp 361-425

Farrington JW, Henrichs SM, Anderson RF (1977) Fatty acids and Pb-210 geochronology of a sediment core from Buzzards Bay, Massachusetts. Geochim Cosmochim Acta 41:289-296

Farrington JW, Davis AC, Sulanowski JK et al. (1988a) Biogeochemistry of lipids in surface sediments of the Peru upwelling area at 15°S. Org Geochem 13:607-617

Farrington JW, Sulanowski J, McCaffrey MA, Wakeham SG, Volkman JK (1988b) Fatty acid biogeochemistry of sediments and particulate matter of the Peru upwelling area at 15°S. (In prep)

Henrichs SM (1980) Biogeochemistry of dissolved free amino acids in marine sediments. PhD Thesis, Massachusetts Institute of Technology/Woods Hole Oceanographic Institution Joint Program, WHOI Tech Rpt WHOI-80-39, pp 253

Henrichs SM, Farrington JW (1984) Peru upwelling region sediments near 15°S-I. Remineralization and accumulation of organic matter. Limnol Oceanogr 29(1):1-19

Henrichs SM, Farrington JW, Lee C (1984) Peru upwelling region sediments near 15°S-II. Dissolved free and total hydrolyzable amino acids. Limnol Oceanogr 29(1):20-34

Hunt JM (1979) Petroleum Geochemistry and Geology. Freeman, San Francisco, CA

Kawamura K, Ishiwatari R, Ogura K (1987) Early diagenesis of organic matter in the water column and sediments: microbial degradation and resynthesis of lipids in Lake Hurana. Org Geochem 11:251-264

Marlowe IT (1984) Lipids as paleoclimate indicators. PhD Thesis, Univ Bristol

McCave IN (1974) The Benthic Boundary Layer. Plenum, New York

Mopper K, Degens ET (1972) Aspects of the biogeochemistry of carbohydrates and proteins in aquatic environments. Woods Hole Oceanographic Institution Tech Rpt WHOI-72-68, p 118

Prahl FG, Wakeham SG (1987) Calibration of unsaturation patterns in long-chain ketone compositions for paleotemperature assessment. Nature 330:367-369

Prahl FG, Muehlhausen LA, Zahnle DL (1988) A well preserved geochemical record for long chain unsaturated ketones. Geochim Cosmochim Acta 52:2303-2310

Romankevich EA (1984) Geochemistry of Organic Matter in the Ocean. Springer, Berlin Heidelberg New York, p 334

Rullkotter J, Vuchev V, Hinz K et al. (1983) Potential deep sea petroleum beds related to coastal upwelling. In: Thiede J, Suess E (eds) Coastal Upwelling: Its Sediment Record Part B. Nato Conference Series. Plenum, New York, pp 467-484

Simoneit BRT (1978) The organic chemistry of marine sediments. In: Riley JP and Chester R (eds) Chemical Oceanography. 2nd edn, vol 7, Academic Press London, 233-311

Smith DJ, Eglinton G, Morris RJ (1983) The lipid chemistry of an interfacial sediment from the Peru continental shelf: Fatty acids, alcohols, aliphatic ketones and hydrocarbons. Geochim Cosmochim Acta 47:2225-2232

Suess E, Thiede (eds) (1983) Coastal Upwelling. Its Sediment Record Part A: Responses of the Sedimentary Regime to Present Upwelling. Plenum, New York, p 604

Suess E, Kulm LD, Killingley JS (1987) Coastal upwelling and a history of organic-rich mudstone deposition off Peru. In: Brooks J, Fleet AJ (eds) Marine Petroleum Source Rocks. Geol Soc Spec Pub No 26: 181-197

Thiede J, Suess E (eds) (1983) Coastal Upwelling. Its Sediment Record Part G: Sedimentary Records of Ancient Coastal Upwelling. Plenum, New York London, p 604

Thompson LG, Mosley-Thompson E, Arano BM (1984) El Nino-Southern Oscillation events recorded in the stratigraphy of the tropical Quelecays Ice Cap, Peru. Science 226:50-53

Thompson LG, Mosley-Thompson E, Dansgaard W, Grootes PM (1986) The little ice age as recorded in the stratigraphy of the tropical Quelecaya Ice Cap. Science 234:361-364

Thurman EM (1985) Organic Geochemistry of Natural Waters. Nijhoff, Boston, MA, p 497

Tissot BP, Welte DH (1978) Petroleum Formation and Occurrence. Springer, Berlin Heidelberg New York

Volkman JK, Farrington JW, Gagosian RB, Wakeham SG (1983) Lipid composition of coastal marine sediments from the Peru upwelling region. In: Bjoroy et al. (eds) Advances in Organic Geochemistry 1981. Wiley, New York, pp 228-240

Volkman JK, Farrington JW, Gagosian RB (1987) Marine and terrigenous lipids in coastal sediments from the Peru upwelling region at 15°S. Sterols and triterpene alcohols. Org Geochem 11:463-477

Walsh JJ (1981) A carbon budget for overfishing off Peru. Nature 290:300-304

# Hydrothermal Petroleum Generation from Immature Organic Matter-Implications to the Oceanic Carbon Cycle

B.R.T. Simoneit

## 1 Introduction

Organic matter of marine sedimentary basins is derived from the syngenetic residues of posthumus biogenic debris and is composed of both autochthonous (marine) organic matter and allochthonous residual organic matter from continental sources (e.g. Simoneit 1978; 1982a). The preservation of organic matter in sediments depends on the initial diagenetic processes involving microbial degradation and chemical conversions, coupled with the environmental conditions of acidity and redox potential (e.g. Demaison and Moore 1980; Didyk et al. 1978). Subsequent sediment maturation and lithification causes metamorphosis of the organic matter over geological time periods, yielding petroleum products by the effects of temperature, pressure and petrology (Hunt 1979; Tissot and Welte 1984). However, the action of hydrothermal processes on such sedimentary organic matter was found to generate petroleum-like products in Guaymas Basin essentially "instantaneous" with respect to geological time (Simoneit and Lonsdale 1982). This process and its implications will be reviewed here with the extensively studied Guaymas Basin, Gulf of California and other areas where hydrothermal activity is occurring, in both sedimented and bare-rock regions.

The analytical techniques of organic geochemistry have been used extensively to examine the character of organic matter in the geologic record in terms of its structural and compositional makeup (e.g. Reed 1977; Boon et al. 1977; Deroo et al. 1978; Johns 1986; Philp et al. 1978; Simoneit et al. 1978; 1979; 1980; 1981; Simoneit 1978; 1981; 1982a,b; Stuermer and Simoneit 1978; Stuermer et al. 1978; van de Meent et al. 1980). The sources, the diagenetic and catagenetic histories and the migration mechanisms of this organic matter can then be evaluated from such data, i.e. the organic carbon record is characterized further. It should be noted that in the following discussion organic matter is comprised of gas, lipids (bitumen), humic substances (with fulvic substances, or asphaltenes) and kerogen (with "pseudokerogen") as discussed elsewhere in terms of contemporary (Recent) versus ancient organic matter (e.g. Simoneit 1983; 1986; 1988).

### 1.1 Nature of Organic Matter in Conventional Maturing Basins

Organic matter in immature, recent sediments is comprised of minor amounts (based on total organic carbon content) of biogenic gas ($CH_4$ and $CO_2$, sometimes $H_2S$), significant lipid residues of terrigenous and/or marine origins and a major macromolecular fraction consisting of fulvic and humic acids and particulate detritus (e.g. biopolymer fragments, cell membranes and miscellaneous carbonaceous matter)-the pseudokerogen (Simoneit 1983; Tissot and Welte 1984). The lipids and macromolecular material undergo diagenetic (including microbiological) alteration according to the environmental conditions during transport and in the depositional sinks, i.e. oxidative degradation in high-energy, oxygenated environments and reductive alteration in anaerobic environments (Didyk et al. 1978; Demaison and Moore 1980; Simoneit 1983).

Maturation of organic matter commences after cessation of diagenesis with increasing burial and concomitant rise in the geothermal gradient. This produces some low temperature cracking products from the kerogen, such as natural gas ($CH_4$-$C_{8^+}$) and bitumen ($C_8$-$C_{40^+}$, including a large envelope of an unresolved complex mixture-UCM-of compounds in gas chromatograms), which are superimposed on the endogenous biogenic gas and lipid residues (Simoneit 1983; Tissot and Welte 1984). Maturation of pseudokerogen occurs via molecular rearrangements and addition of geomonomers by copolymerization. Further thermal stress during catagenesis generates additional bitumen and gas, which far exceeds the original concentrations of endogenous lipids and gas, thus erasing their compositional signatures and resulting in the characteristic distributions of petroleum compounds (Simoneit 1983; Tissot and Welte 1984). The spent kerogen remains as amorphous carbon. The late stages of catagenesis and the subsequent high-temperature phase called metagenesis (very deep burial) generate primarily methane from both bitumen and kerogen, and $H_2S$ can also be formed, especially in carbonate sequences.

## 1.2 Nature of Organic Matter in Hydrothermal Systems

Hydrothermal processes can also act on sedimentary organic matter and on entrained ambient organic detritus (e.g. in water), which results in "instantaneous" diagenesis and catagenesis of this recent biogenic detritus and thus produces analogous petroleum products (Simoneit and Londsdale 1982; Simoneit 1983; 1984a,b; 1985a). Gas ($CH_4$-$C_{8^+}$, $CO_2$ and $H_2S$) and bitumen ($C_8$-$C_{40^+}$, with a large UCM) are cracked from the pseudokerogen and biopolymers, becoming superimposed on the endogenous gas and lipids. Additionally, products characteristic of high temperature processes (e.g. olefins, PAH, stabilized molecular markers, etc.) are also found in the bitumen. The spent kerogen remains as amorphous "activated" carbon (Simoneit 1982b).

Pressure, temperature and time effects on the chemistry of the organic matter are all interrelated. Temperature and time primarily effect petroleum generation and are coupled with mineralogical interactions and the elemental compositions of kerogens. Migration processes are understood least. In sedimentary basins migration appears to occur by diffusion and in solution ($CH_4/CO_2/H_2O$ solvent), whereas, in hydrothermal areas migration proceeds by the previous processes and also by thermally driven diffusion, advection and mass transport as oil and/or emulsions (e.g. Whelan et al. 1984; 1988; Kawka and Simoneit 1987). Although the overall result may be the same for both regimes, migration of petroleum in hydrothermal systems appears to be more efficient due to the enhanced fluid flow.

## 1.3 Organic Matter Type

The nature or constitution of the organic matter being altered determines the types of petroleum products that form under both kinds of generation regimes, i.e. normal geothermal versus hydrothermal. The syngenetic sedimentary lipid matter undergoes alteration due to mainly diagenetic and early catagenetic processes thus changing the hydrocarbon signature. For example, in Guaymas Basin and in the Escanaba Trough (Fig. 1) the two end-member signatures can be observed. The lipid hydrocarbons of the seabed sediment in Guaymas Basin (Fig. 2a) exhibit *n*-alkanes primarily $< C_{21}$, and only a minor series of homologs $> C_{23}$ with a strong odd carbon number predominance. This composition is typical of a predominantly microbial origin from marine productivity and a minor influx of terrigenous vascular plant wax ($> C_{25}$) (Simoneit 1975, 1978). The kerogen composition also indicates such an origin. The other example, the lipid hydrocarbons of the seabed sediment in the Escanaba Trough (Fig. 2b), shows *n*-alkanes primarily $> C_{23}$ with a strong odd carbon number predominance and lesser amounts at lower carbon numbers. This signature is derived mainly from terrigenous vascular plant wax, with a minor microbial component (Kvenvolden et al. 1986). The kerogen of similar samples from that region also reflects a primarily terrigenous composition (Simoneit 1978).

However, during oil generation large amounts of additional hydrocarbons are superimposed on the syngenetic lipids, thus diluting such signatures (e.g. loss or reduction of the odd carbon number predominance $> C_{25}$, cf. Fig. 2 vs. Figs. 3 and 4). The major source of petroleum compounds is from the sedimentary macromolecular organic detritus, called kerogen, which generally constitutes the

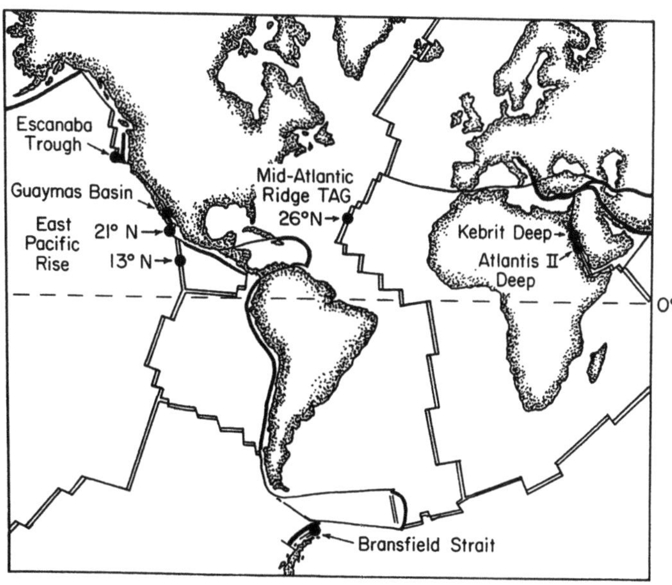


**Fig. 1.** General location map of the hydrothermal vent fields discussed here

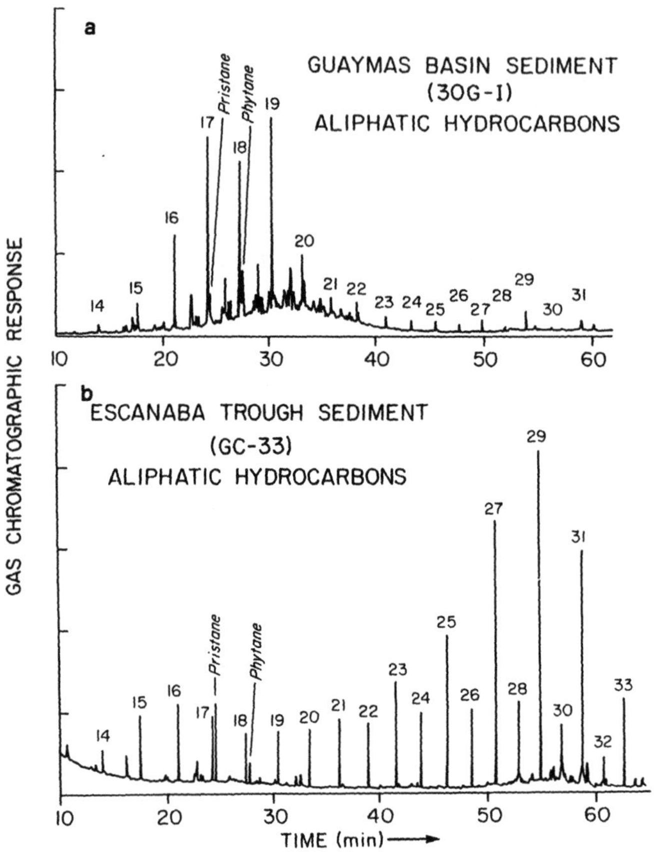


**Fig. 2a,b.** Gas chromatograms of total hydrocarbons in extracts from surface sediments of (**a**) Guaymas Basin, Site 30G, Simoneit et al. (1979), and (**b**) Escanaba Trough, Kvenvolden et al. (1986)

bulk of the total organic carbon content (Tissot and Welte 1984). It is the constitution and source of the kerogen that determines the nature of the petroleum generated. In general, terrestrial organic detritus from mainly vascular plants yields an aromatic kerogen (e.g. coal) which has a natural gas potential, and marine/lacustrine organic matter from primarily microbial residues yields an aliphatic kerogen (e.g. sapropel) which has a paraffinic petroleum potential (Tissot and Welte 1984; Hunt 1979). Kerogens in sedimentary basins are always mixtures of these inferred endmembers. The syngenetic lipids of sediments can usually be utilized to elucidate the various sources of the total organic matter (e.g. Simoneit 1978, 1981, 1982a).

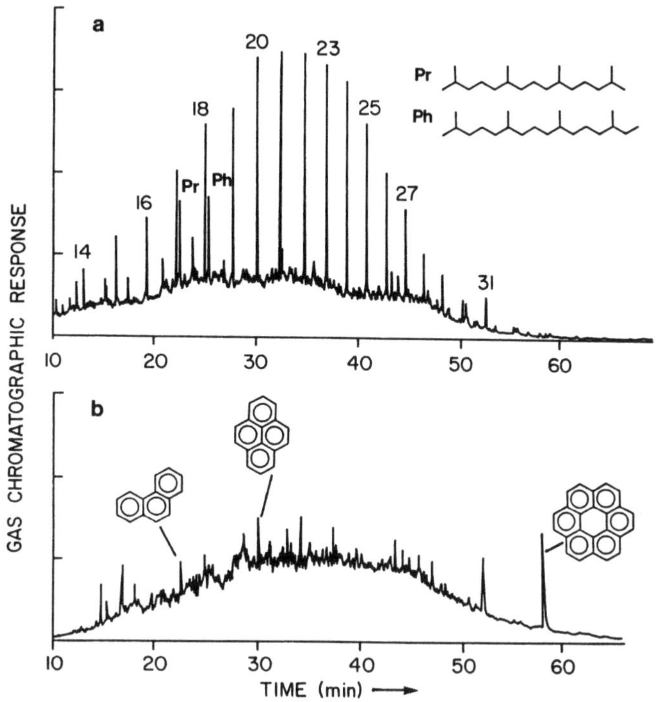


**Fig. 3a,b.** Gas chromatograms of the (**a**) aliphatic and (**b**) aromatic hydrocarbons from dredge sample 7D-2B in Guaymas Basin (Simoneit and Lonsdale 1982)

## 2 Geological Locales and Hydrothermal Petroleum Generation

Although this overview describes data on the occurrence of petroleum products in a few geographic locations of hydrothermal activity (Fig. 1), it is the author's opinion that petroleum generation and migration is a ubiquitous process associated with hydrothermalism and ore formation. The following data illustrate the diverse suite of samples from different environments and wide ranges of organic carbon source material.

### 2.1 Guaymas Basin, Gulf of California

Guaymas Basin is an actively-spreading oceanic basin, which is part of the system of spreading axes and transform faults that extend from the East Pacific Rise to the San Andreas fault (Fig. 1) (Curray et al. 1982; Lonsdale 1985). Ocean plate accretion results in high conductive heat flow with dike and sill intrusions into the unconsolidated sediments (Curray et al. 1982; Einsele 1985; Einsele et al. 1980; Lonsdale and Becker 1985). Sediments accumulate rapidly (> 2 m/1000 yr) and have covered the rift floors to a depth of ~400 m (Curray et al. 1982). The organic matter of these recent sediments is derived primarily from marine microbial detritus and averages about 2% organic carbon. Thermal stress causes rapid maturation with concomitant petroleum generation and the "oil window" appears to migrate upward as the magmatic heat front rises in the sedimentary column (Simoneit 1984a; Simoneit et al. 1984). Petroleum products have been characterized in samples derived from shallow gravity coring (Simoneit et al. 1979), piston coring in the North Rift (Simoneit 1983) and deep coring by the Deep Sea Drilling Project (DSDP), Leg 64 (Curray et al. 1982), and in accumulations at the seabed which were recovered by dredging operations (Simoneit and Lonsdale 1982) and sampling with the submersible *Alvin* during 1982 and 1985 (Simoneit 1984a,b; 1985a; Simoneit and Kawka 1987; Kawka and Simoneit 1987).

Leg 64 of the DSDP encountered intrusives and hydrothermal alteration at depth in all holes drilled in the basin (Curray et al. 1982). Thermogenic hydrocarbon gas $H_2S$ and $CO_2$ were identified for all sites and the lipids (bitumen) were thermally altered close to and below sills, especially for Site 477 in the South Rift (Simoneit et al. 1984; Simoneit and Philp 1982). This alteration is indicated by the loss of the odd carbon number predominance of the *n*-alkanes (the carbon preference index, CPI, approaches 1.0); the appearance of a broad hump of unresolved complex hydrocarbon material (UCM, naphthenes); the isomerization of various biomarkers; the presence of large amounts of primary olefins (alkenes) isoprenoid hydrocarbons and elemental sulfur; and the appearance of polynuclear aromatic hydrocarbons (PAH).

The petroleum products have migrated away from heat sources (e.g. sills or heat flow from magma at depth) by advection, diffusion, distillation and especially by cosolution in hydrothermal fluids (Simoneit et al. 1984; Simoneit 1982b; Simoneit and Philp 1982). The kerogens (i.e., insoluble detrital organic matter in sediments) of the shallow DSDP core samples are typical of unaltered marine organic matter (Simoneit et al. 1984; Simoneit and Philp 1982), whereas those in deeper sections reflect essentially complete expulsion of pyrolysate. The in situ appearance of the kerogen in zones of high thermal stress (at depth of Sites 477 and 478) resembled activated amorphous carbon and was the spent (exhausted) kerogen residue (Simoneit 1982b).

Numerous hydrothermal mounds rise to 20–30 m above the South Rift floor (water depth about 2000 m) and most have active hydrothermal plumes, with water temperatures measured = 315° C at ~200 bars (Lonsdale 1985; Lonsdale and Becker 1985; Merewether et al. 1985). Typical samples from these mounds are cemented and stained with petroleum and have a strong odor similar to diesel fuel (Simoneit and Lonsdale 1982).

Samples recovered with the DSV *Alvin* had very diverse petroleum contents and hydrocarbon distributions (e.g., Figs. 3 and 5) (Simoneit 1984a,b; 1985a; Simoneit and Kawka 1987; Kawka and Simoneit 1987), but they were analogous to those described for bitumens at depth in the DSDP holes (Simoneit 1983, 1984b). The *n*-alkanes range from methane to greater than *n*-$C_{40}$, with usual maxima in the mid-$C_{20}$ region and no carbon number predominance (CPI = 1.0, Table 1) (cf. Simoneit and Kawka 1987; Kawka and Simoneit 1987 for overviews). The generation of these petroleums was by rapid intense heating as indicated by these data and the kinetic parameters of the biomarkers.

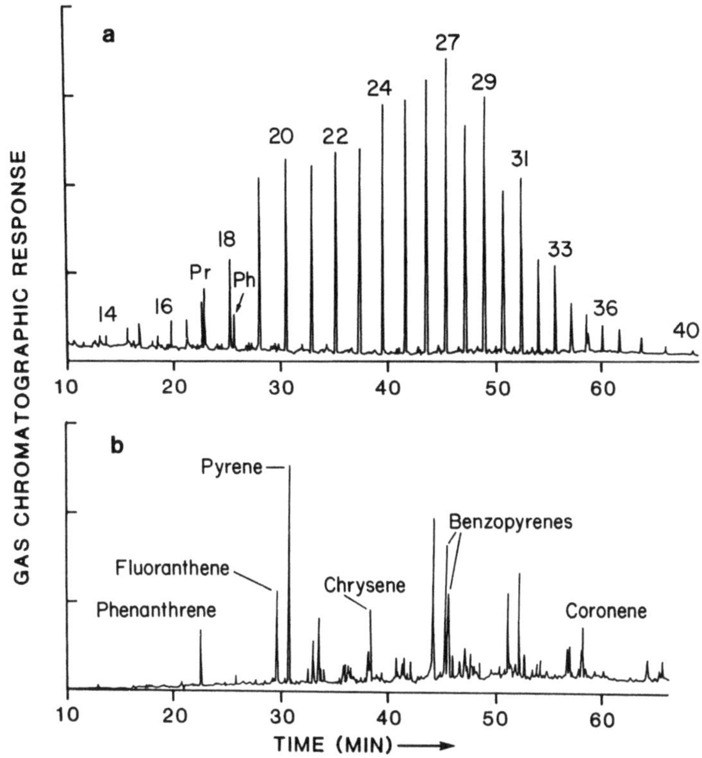


**Fig. 4a,b.** Gas chromatograms of the (**a**) aliphatic and (**b**) aromatic hydrocarbons in the petroleum sample from the Escanaba Trough (Kvenvolden et al. 1986)

An example of a GC trace of an aromatic/naphthenic fraction (F2) of a sample is shown in Fig. 3b. The data indicate that the major resolved peaks are PAH, a group of compounds uncommon in petroleums but ubiquitous in higher temperature pyrolysis residues (Geissman et al. 1967; Hunt 1979; Blumer 1975). The dominant analogs are the pericondensed aromatic series as for example phenanthrene, pyrene, benzopyrenes, perylene, benzoperylene and coronene. A further indication for a pyrolytic origin is the presence of five-membered alicyclic rings (e.g. acenapthene, methylenephenanthrene, fluorene, fluoranthene, etc.), which are found in all pyrolysates from organic matter, since once formed they do not easily revert to pericondensed aromatic hydrocarbons (Blumer 1975; 1976; Scott 1982). It should also be noted that these fractions contain significant amounts of toxic PAH, e.g. the benzopyrenes. Perylene is also present in these fractions and it is the predominant PAH in the unaltered sedimentary lipids of samples deposited in the Gulf from oxygen-minimum environments (Simoneit and Philp 1982; Simoneit 1982a). Thus, the chemical compositions of the aromatic fractions indicate a source from pyrolysis (high temperature) with rapid quenching by hydrothermal removal.

Hydrothermal petroleum migration in Guaymas Basin appears to occur as bulk phase, cosolute fluid and aqueous solution upward to the seabed, where it

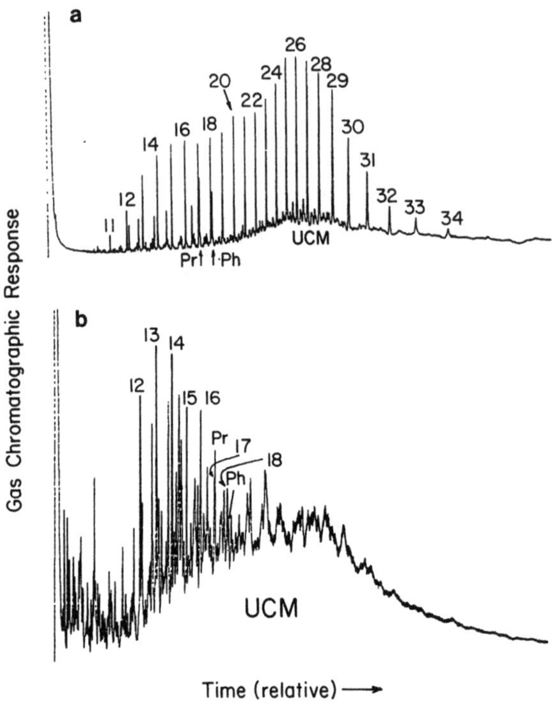


**Fig. 5a,b.** Representative gas chromatograms of total oils extracted from samples recovered by DSV *Alvin* in Guaymas Basin: (**a**) 1170–1 interior (massive); (**b**) 1177–3 oily crust on lithified rock; (carbon chain length of the *n*-alkanes indicated by the *arabic numerals*; *Pr* = pristane, *Ph* = phytane; GC conditions as in Simoneit 1984a)

condenses (solidifies) and collects according to the ambient temperatures in the conduits and vugs of the hydrothermal mineral mounds. PAH and sulfur accumulate in the hot vents; waxes crystallize in intermediate temperature regions (20° to 80° C); and the volatile petroleum partially collects in cold areas (3° C) and emanates into the ambient sea water as plumes (Simoneit 1984a,b; 1985; Merewether et al. 1985).

Both the extensive organic matter maturation to bitumen in the DSDP holes and the significant accumulations of petroliferous exudate at the rift floor, confirm the importance of hydrothermal metamorphism (pyrolysis) of immature organic matter as a feasible mechanism for the formation of petroleum (Simoneit 1983; 1984a,b; 1985a).

## 2.2 Escanaba Trough, Northeastern Pacific

The Escanaba Trough (Fig. 1) is the southern extension of the Gorda Ridge, an active oceanic spreading center about 300 km long and bounded on the north and south by the Blanco and Mendocino fracture zones, respectively (McManus et al.

**Table 1.** Summary of the characteristics of the hydrothermal petroleums described for the various geographic areas

| Location | Total organic carbon of seds (ave. %) | Petroleum Fractions (wt. %) Aliphatic hydrocarbons | Aromatic hydrocarbons | NSO cpds & asphalt | *n*-Alkane range | CPI[1] | Pristane: Phytane | References |
|---|---|---|---|---|---|---|---|---|
| Guaymas Basin, Gulf of Calif. | 2 | 3<br>65 | 23<br>15 | 74<br>20 | 13–35<br>1–40 | 1.03<br>1.02 | 1.11<br>0.3–2.5 | Simoneit & Lonsdale, 1982; Simoneit and Kawka, 1987 |
| Escanaba Trough, Gorda Ridge | 0.4 | 2 | 44 | 54 | 14–40 | 1.25 | 1.7 | Kvenvolden et al. 1986 |
| Bransfield Strait, Antartica | 0.8 | n.d. (0.5–1.2μg/g)[2] | n.d. | n.d. | 15–33 | 1.6–2.0 | 0.3–5 | Brault & Simoneit, 1988a; Whiticar et al. 1985 |
| Atlantis II Deep, Red Sea | 0.14 | n.d. (230 ng/g) | n.d. | n.d. | 15–40 | 1.1 | 0.8 | Simoneit et al. 1987a |
| East Pacific Rise, 13°N | 0.4 | n.d. (1.0μg/g) | n.d. | n.d. | 15–34 | 1.1 | 1.2 | Brault et al. 1985 |
| East Pacific Rise, 21°N | n.d. | n.d. (0.2–6 ng/g) | n.d. | n.d. | 14–40$^+$ | 0.9–1.03 | 0.5–1.0 | Brault et al. 1988b |
| Mid-Atlantic Ridge, TAG Area 26°N | n.d. | n.d. | n.d. | n.d. | 10–25 | 1.01 | | Brault & Simoneit, 1988b |

[1] CPI – carbon preference index, calculated here over the range $C_{24}$ to $C_{34}$.
[2] Values in parentheses are total yield.

1970). The Escanaba Trough is filled with up to 500 m of Quaternary turbidite sediment (McManus et al. 1970).

Petroleum cements the sediments/ores that blanket the ridge axis and is derived from hydrothermal alteration of sedimentary organic matter (Kvenvolden et al. 1986). An example GC trace of each a saturated and aromatic hydrocarbon fraction separated from a total extract is shown in Fig. 4. The organic source material appears to be terrigenous for these petroleums, based on the CPI, carbon number range (Table 1) and biomarker composition, coupled with the sedimentological considerations (Kvenvolden and Simoneit 1987).

In the aliphatic hydrocarbon fraction the *n*-alkanes range from $C_{14}$ to $C_{40}$, with a carbon number maximum at *n*-$C_{27}$ and still a significant odd carbon number predominance $>$ *n*-$C_{25}$ (CPI = 1.25, Fig. 4a), typical of a terrestrial higher plant origin. Homologs of a marine origin ($<$ *n*-$C_{25}$) are less concentrated. The generation of this petroleum was probably by intense heating of short duration, as indicated by kinetic parameters of the biomarkers and by the high concentrations of unsubstituted PAH (Fig. 4b) (Kvenvolden and Simoneit 1987).

## 2.3 Bransfield Strait, Antarctica

The Bransfield Strait, Antarctica is a classical back-arc rift, which is tectonically active with extensional features such as dip-slip faults and intrusives, and is also heavily sedimented (Whiticar et al. 1985; Suess et al. 1988). Gravity coring was carried out in areas where the sub-bottom signal of the 3.5 kHz survey was lost, generally in the eastern portion of the basin (Fig. 1) (Suess et al. 1988). This signal loss was interpreted to be due to hydrothermal activity which had generated interstitial natural gas (Whiticar et al. 1985). All the cores had a petroliferous odor in the hydrothermally altered zones, although weaker in intensity than was the case for such samples from Guaymas Basin. The lipid/bitumen compositions of two piston core examples, PC 1341-1 and PC 1347-1, will be discussed here (Brault and Simoneit 1988a).

The compositions of the hydrocarbon fractions separated from the lipid/bitumen extracts of the cores confirm the interpretations of the thermal regimes based on fluorescence data (Suess et al. 1988) and some examples of GC traces are shown in Fig. 6. The two extreme GC patterns for core 1341-1 are shown for samples from the 5 and 7.5 m horizons, respectively. The thermally unaltered sample exhibits a compound distribution that can be correlated with its marine biogenic origin (Fig. 6a). The same is also the case for the shallow sample of core 1347-1 (Fig. 6c) and Venkatesan and Kaplan (1987) observed analogous GC patterns for samples from an earlier core in the Bransfield Strait. The *n*-alkanes range from about $C_{15}$ to $C_{33}$, with only a minor carbon number predominance ($CPI_{25\text{-}33}$ = 1.6–2.0) and maximum at *n*-$C_{29}$ (Table 1). The *n*-alkanes $>$ $C_{25}$ with the stronger carbon number preference may have originated from vascular plant wax in aerosols derived from the southern hemisphere continents. Preliminary analyses of aerosols indicate such waxes in samples north of the Drake Passage (Simoneit, unpublished data). However, the bulk of the *n*-alkanes, the additional biomarkers as for example diploptene, $C_{28}$-steradienes and $C_{25}$-polyalkenes, are

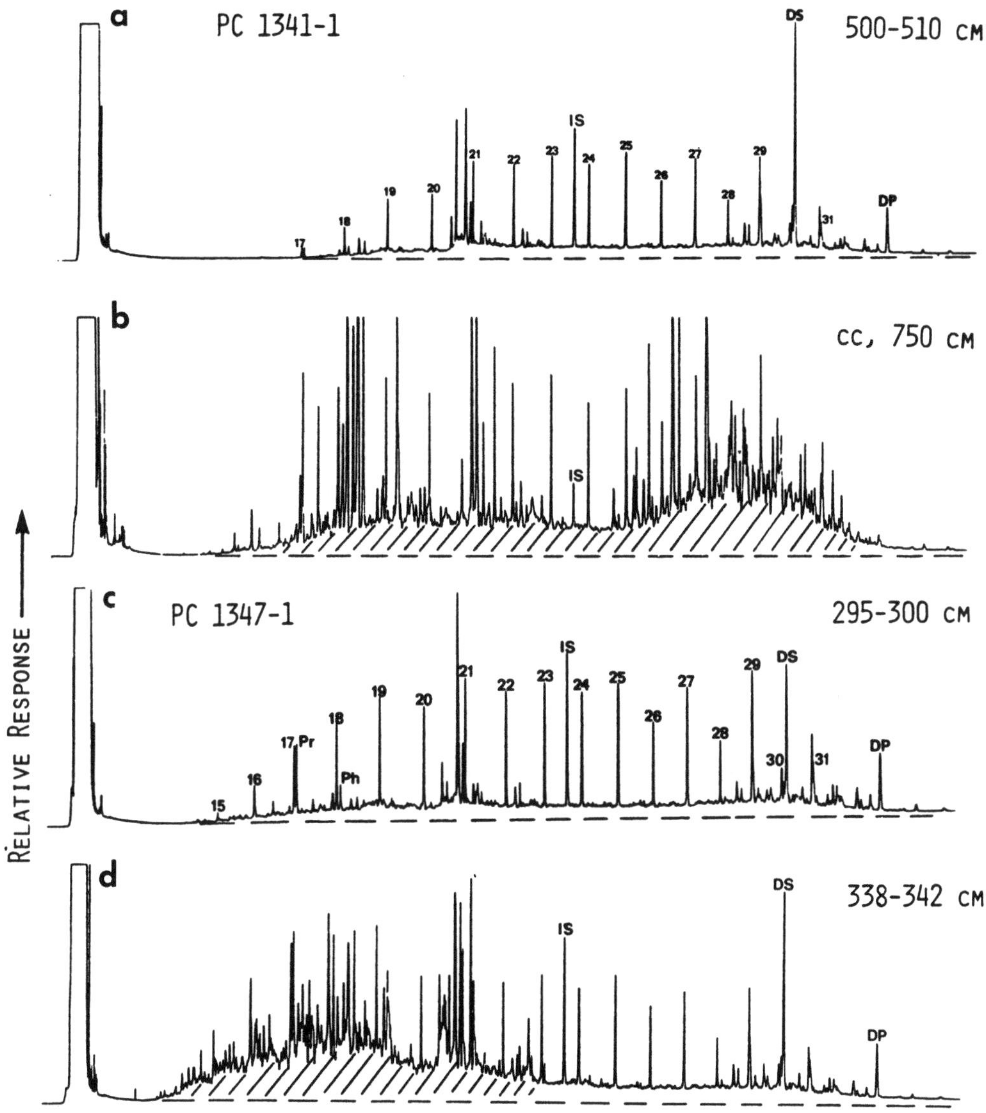


**Fig. 6a-d.** Gas chromatograms of total hydrocarbon fractions from various core samples in the Bransfield Strait (Suess et al. 1988; Brault and Simoneit 1988a) (*Pr* = pristane, *Ph* = phytane, *IS* = internal standard, *DS* = $C_{28}$ steradiene, *DP* = diploptene, *arabic numerals* refer to the carbon chain length of the *n*-alkanes, *hachured areas* are the UCM of thermogenic origin): **a** Core 1341-1, 500-510 cm; **b** Core 1341-1, cc 750 cm; **c** Core 1347-1, 295-300 cm; **d** Core 1347-1, 338-342 cm derived from autochthonous marine microbial sources. This has also been reported by Venkatesan and Kaplan (1987).

The hydrocarbon patterns in the hydrothermally-altered zones are dramatically different (cf. Fig. 6b and d, Brault and Simoneit 1988a). The *n*-alkane distributions do not change significantly in range or CPI, only some samples have the carbon number maximum at $C_{27}$. However, in addition there is a superposition on the *n*-alkane pattern of complex thermal products, both resolved and unresolved (UCM, the hachured areas in Fig. 6b and d). In the case of core 1341–1 these thermogenic products comprise the light to heavy molecular weight range ($C_{15}$-$C_{33}$), whereas the example from core 1347–1 contained only light (i.e. volatile) thermogenic compounds. The former pattern is indicative of hydrothermal alteration versus depth. The latter example supports the interpretation of a migrated condensate located at the 3.5 m horizon of core 1347–1, since the hydrocarbon distribution > *n*-$C_{23}$ is basically unaltered (cf. Fig. 6d vs. a or c, Table 1). The endogenous *n*-alkanes are still discernable and generally unaltered in the thermogenic hydrocarbon mixture, which is ulike the process of hydrothermal petroleum generation in Guaymas Basin, where the *n*-alkane yield is also enhanced and the carbon number predominance is lost (i.e. CPI = 1.0, cf. Fig. 3 or 5). Thus, the hydrocarbon patterns for the hydrothermally-altered samples from Bransfield Strait indicate localized heating and only limited migration.

## 2.4 Atlantis II Deep, Red Sea

The Atlantis II Deep (Fig. 1) is a low temperature hydrothermal system where the present ambient water (brine) temperature is 62°C (Hartmann 1980; 1985). Hydrocarbons and bulk organic matter of two sediment cores (No. 84 and 126, CHAIN 61 cruise) located within the Deep have been analyzed (Simoneit et al. 1987). Although the brine overlying the coring areas was reported to be sterile, autochthonous marine microbial inputs and minor terrestrial sources represent the major sedimentary organic material. This input is derived from the upper water column above the brine and down to/at the brine-marine water interface where microorganisms can metabolize the suspended organic matter. The reworked compounds may then be incorporated into the sediments under the brine by association with sinking particles of metallic oxide precipitates.

In the sediments low temperature maturation results in petroleum generation from the low amounts of organic matter (ave. 0.05%). Both steroid and triterpenoid hydrocarbons (biomarkers) show that extensive acid-catalyzed reactions are occurring in the sediments. In comparison with other hydrothermal (e.g. Guaymas Basin) or intrusive systems (e.g. Cape Verde Rise, Simoneit et al. 1981), the Atlantis II Deep exhibits a lower degree of thermal maturation. This is easily deduced from the elemental composition of the kerogens and the absence of PAH of a pyrolytic origin in the bitumen.

The lack of carbon number preference among the *n*-alkanes (CPI = 1.0) suggests, especially in the case of the long chain homologs (e.g. Fig. 7, Table 1), that the organic matter of Atlantis II Deep sediments has undergone some degree of catagenesis. However, the yields of hydrocarbons are much lower than those

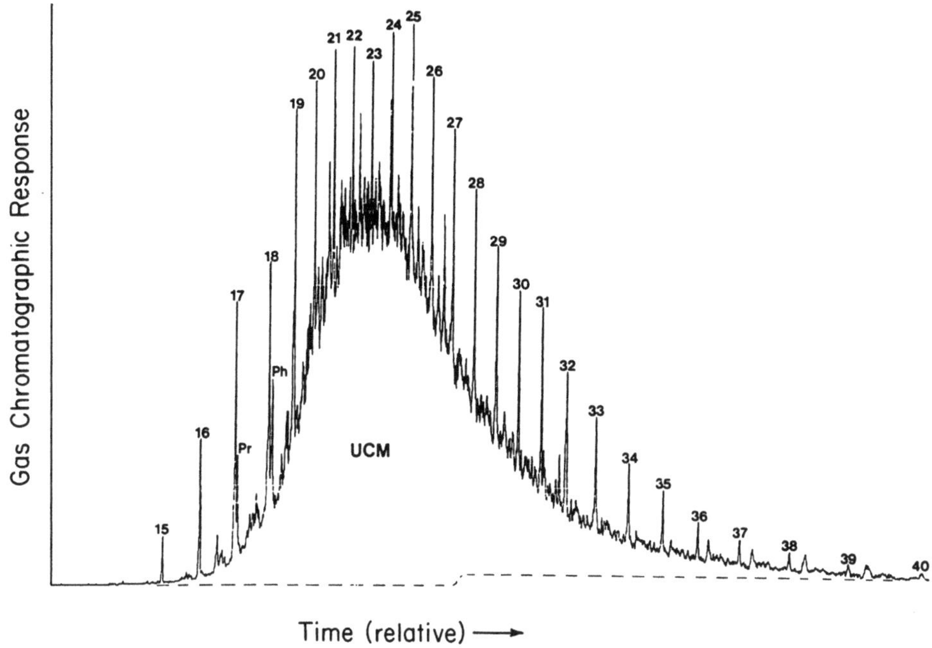


**Fig. 7.** Gas chromatogram of the total hydrocarbon fraction extracted from the Atlantis II Deep core sample 84, 443–453 cm (Simoneit et al. 1987)

observed in other hydrothermal areas. The effect of lower temperature and poor source-rock characteristics (i.e. low organic carbon content of the sediments) appear to be responsible for this difference.

Petroliferous material has also been recovered from the Kebrit Deep in the northern Red Sea. This material appears to be derived from higher temperature hydrothermal activity (Michaelis et al. 1988).

## 2.5 East Pacific Rise, 13° N and 21° N

Active hydrothermalism occurs on the unsedimented axis of the East Pacific Rise (EPR) in the region of 13° N (Fig. 1) with abundant faunal communities (Hekinian et al. 1983). Aliphatic hydrocarbons have been analyzed in hydrothermal metalliferous sediments sampled at the base of an inactive chimney, close to active vents and waters collected in the maximum of hydrothermal plumes (Brault et al. 1985; 1988a). Hydrocarbons from hydrothermal sediments exhibit characteristics of immature organic matter, i.e. recently biosynthesized and microbiologically degraded, as indicated by the importance of low molecular weight ($> C_{25}$) *n*-alkanes and phytane, and a contribution of ubiquitous continental higher plant wax input shown by the dominant high molecular weight *n*-alkanes with an odd carbon number predominance (Fig. 8a, Table 1). The immature character of organic matter is also indicated by the presence of coprostane and cholestane, and the

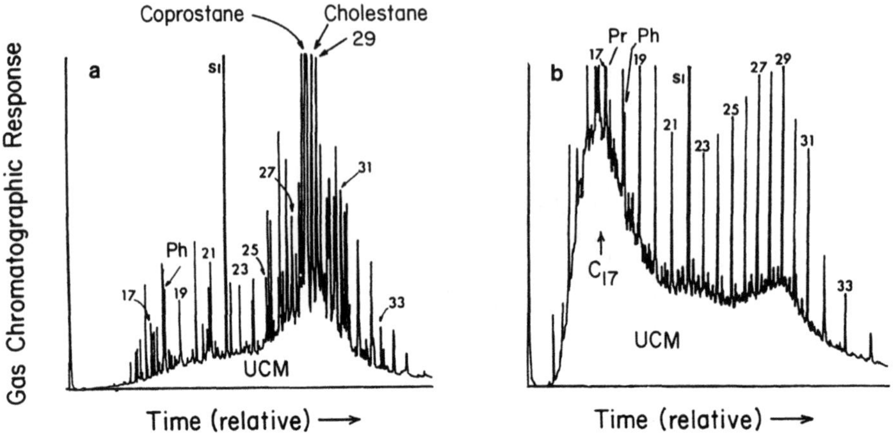


**Fig. 8a,b.** Gas chromatograms of aliphatic hydrocarbons from extracts of: (**a**) hydrothermal metalliferous sediment and (**b**) surrounding ambient water from the EPR at 13°N (Brault et al. 1985). The *n*-alkanes are indicated by their number of carbon atoms. *SI* = internal standard (*n*-$C_{22}$); *Pr* = pristane; *Ph* = phytane

predominance of 17β(H),21β(H)-hopanes over the 17α(H)-configuration. These biomarker distributions are the result of mild thermal stress as would be expected in the talus an extinct vent system. The contents of a sediment trap in the area are characterized by the same type of biologically-derived material and also by the presence of compounds such as 17α(H)-hopanes and intermediates of thermal alteration, which confirm the importance of higher temperature degradation of entrained organic detritus near hydrothermal systems (Brault et al. 1985). Thermally matured compounds and intermediates are also present at trace levels in waters collected near the discharges of hydrothermal plumes (e.g. Fig. 8b), again supporting their origin from higher thermal stress. The hydrocarbon pattern of the ambient waters is indicative in many cases of a bacterial origin, i.e. from their pyrolysis in entrained ocean water during vent cooling (Brault et al. 1988a).

Extensive hydrothermal activity has also been described for the EPR in the region of 21°N (Fig. 1), occurring also on unsedimented oceanic crust with associated abundant faunal commmunities (Spiess et al. 1980; Ballard et al. 1981). Various samples of massive sulfides from vent chimneys have been analyzed for hydrocarbon contents, which are extremely low but of a definite thermogenic origin (Table 1). The *n*-alkanes range from $C_{14}$ to greater than $C_{40}$, with no carbon number predominance for samples of massive sulfides and a slight odd carbon number predominance for a sample of pyritized tube worm from a chimney (Brault et al. 1988b). All samples contain PAH, supporting evidence for an origin of the hydrocarbons from hydrothermal activity. This coupled with the carbon number maxima at *n*-$C_{27}$ or higher indicates that these hydrocarbons were entrapped/condensed in a high temperature regime such as an active chimney. It should be pointed out that the sample with the pyritized tube worm residues also contains hydrothermally altered derivatives of biomarkers (e.g. cholestenes, hopenes) from the vent biota, i.e. mainly tube worms and bacteria.

## 2.6 Mid-Atlantic Ridge, TAG Area 26°N

The Trans-Atlantic Geotraverse (TAG) hydrothermal field on the Mid-Atlantic Ridge crest at 26°N (Fig. 1) is to date the only known active vent system on a slow-spreading mid-oceanic ridge (Rona et al. 1984). Various hydrothermal ores deposited directly on oceanic crust have been dredged from the area (TAG 1985–1) and four types of samples have been examined for lipid/bitumen content (Brault and Simoneit 1988b).

A sample consisting of predominantly ferric oxide contained no hydrocarbons derived from hydrothermal alteration of associated organic detritus. However, three other samples (consisting of mainly anhydrite, sphalerite and chalcopyrite, respectively) did contain minor amounts of more volatile ($C_{10}$-$C_{22}$) hydrothermal petroleums. The saturated and aromatic hydrocarbon fractions separated from the extracts of the sphalerite sample are shown as an example in Fig. 9. The *n*-alkanes range from $C_{11}$ to $C_{22}$ with a CPI = 1.0, pristane and phytane are present and the UCM maximizes at the GC retention time for *n*-$C_{17}$ (Table 1). This pattern is analogous as observed for the samples from the EPR at 13°N and from the Atlantis II Deep, Red Sea. The supporting evidence for a hydrothermal origin is found in the aromatic fraction which contains naphthalene, phenanthrene, their alkyl homologs and sulfur aromatic compounds.

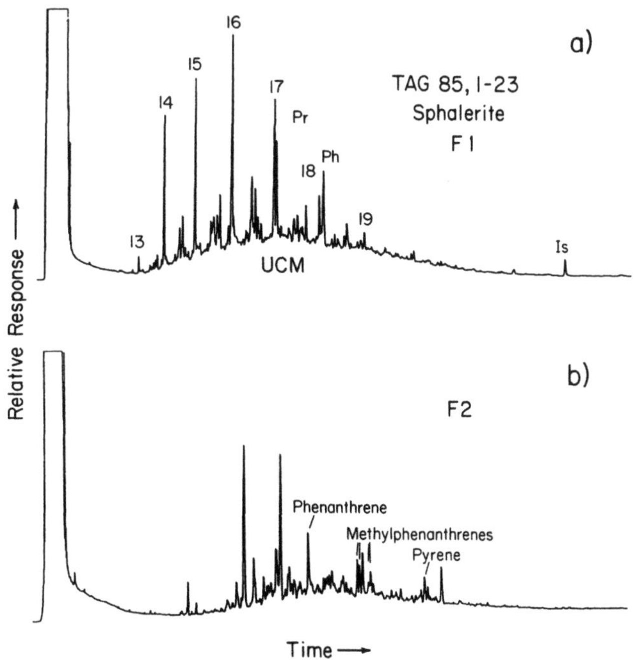


**Fig. 9a,b.** Gas Chromatograms of (**a**) aliphatic and (**b**) aromatic hydrocarbons from the extract of a massive sphalerite sample from the Mid-Atlantic Ridge, TAG '85, 1-23 (Brault and Simoneit 1988b)

# 3 Implications

Since pressure, temperature and time all affect the nature of sedimentary organic matter, the end results of these processes yield various grades of petroleums and cause its migration and interactions with the inorganic surroundings. In addition, hydrothermal activity provides a major input of organic carbon to the deep ocean, which is utilized in part by chemosynthetic biota.

## 3.1 Petroleum Generation

The principal zone of petroleum formation in sedimentary sequences under normal geothermal gradients commences at about 1 km to as low as 3 km (e.g. Tissot and Welte 1984; Hunt 1979). This corresponds to a temperature range of 50°–120° C and is dependent on the geologic age of the sediments (duration of heating). The effect of pressure on this process is significant, but less important and needs more supportive data (Tissot and Welte 1984). The cracking of organic matter to natural gas is believed to take place at elevated temperatures of 150°–250° C (e.g. Hunt 1979; Vassoevich et al. 1974; Kartsev et al. 1972). These proposed temperature regimes for the oil and gas "windows" may need some adjustment in consideration of more recent data on hydrothermal systems and ultra-deep wells.

The "instantaneous" petroleum generation in hydrothermal systems is a facile process, occurring at temperatures approaching a maximum of 400° C. In this case, the lack of extensive organic matter destruction can be interpreted to be due to the rapid removal of the thermogenic products from the hot zone. The formation of this hydrothermal petroleum appears to commence in low temperature areas, first generating products from weaker bonds (e.g. ether, sulfide, carbonyl, tertiary carbon linkages), and later as the temperature regime rises, products from more refractory organic matter and even "resynthesized" compounds (e.g. PAH). The major similarities and differences between hydrothermal petroleums and reservoir petroleums are summarized in Table 2. Most of the hydrocarbon products are the

**Table 2.** Hydrothermal petroleum compared to reservoir petroleum

| Similarities to Reservoir Petroleum- |
|---|
| 1. Natural gas and gasoline-range hydrocarbons |
| 2. Full range of *n*-alkanes, no carbon number predominance (CPI = 1.0–1.2) |
| 3. Naphthenic components (major-hump, UCM) |
| 4. Isoprenoid hydrocarbons (including significant pristane and phytane) |
| 5. Biomarkers (e.g. mature 17α(H)-hopanes and steranes) |
| 6. Alkylaromatic hydrocarbons and asphaltenes |
| **Differences from Reservoir Petroleum-** |
| 1. Polynuclear aromatic hydrocarbons (PAH) > alkyl aromatic hydrocarbons |
| 2. Residual immature biomarkers and intermediates (e.g. 17β(H)-hopanes, hopenes, sterenes) |
| 3. Significant aromatic sulfur hetero compounds |
| 4. High sulfur content |
| 5. Alkene content near "source rock" |

same for both, the major difference is the enhanced content of PAH and sulfur in the hydrothermal products.

The organic matter associated with deeper hydrothermal systems (e.g. epithermal ores in volcanic terranes) is usually more asphaltic with a high PAH content. Such organic matter is widely distributed and for example has been studied from the California mercury deposits (idrialite, Blumer 1975; Geissman et al. 1967) and other hydrothermal sulfide deposits (Germanov and Bannikova 1972). The advent of deep well drilling (> 7000 m) has yielded core materials (Cretaceous shales) which were at in situ temperatures of about 260°–300° C (e.g. Price 1982; Price et al. 1981). These samples had high concentrations of bitumen components and the kerogens still had a significant hydrocarbon generation potential. This indicates that in situ petroleum is stable at much higher temperatures and pressures as discussed above and over long geologic time periods. Metagenesis also appears to occur over much wider and higher temperature conditions than normally believed.

### 3.2 Migration Processes

Migration of petroleum in sedimentary sequences proceeds in solution and in gas/fluid phases (possibly also supercritical) from the source rocks to the traps (Hunt 1979). The aqueous solubility of petroleums and various hydrocarbon fractions has been determined experimentally (Price 1976). It was found that petroleum solubility increased exponentially above 100° to 180° C and these solubilities were high enough to account for the formation of petroleum reservoirs by the primary migration mechanism of molecular solution. Salinities of 150‰ NaCl caused drastic exsolution of the petroleum and at 350‰ essentially total "salt-out" was observed (Price 1976). This finding supports the requirement for the exsolution of the petroleum from the migration solution in the salty waters of reservoir sands. In addition, it has been demonstrated that methane in the presence of water is an even better carrier for petroleum than each alone (Price et al. 1983). Both increases in pressure (to about 1800 atmospheres) and temperature (to 250° C) raised the solubility of petroleum. Cosolubility was found at rather mild conditions (e.g. 100° C at 1000 atm., 200° C at 500 atm.). The addition of other gases (e.g. $CO_2$, ethane) to this mixture also has a positive effect. Thus, primary migration appears to proceed as gas/fluid and aqueous solution phases.

In the case of hydrothermal systems, e.g. Guaymas Basin, the petroleum products appear to migrate by advection, diffusion and hydrothermal circulation as fluids away from the heat sources upward to the seabed. There the petroleum condenses (solidifies) according to the ambient temperatures in the conduits and vugs of the hydrothermal mineral mounds. PAH and sulfur condense in the hot vents; waxes crystallize in intermediate temperature regions (~20°–80° C); and the volatile petroleum partially collects in cold areas (~0° C) and emanates mainly into the ambient sea water as plumes (Simoneit 1984a; 1985a; Merewether et al. 1985).

### 3.3 Inorganic Interactions

Processes involving organic matter contribute to the formation of a variety of ore deposits (Saxby 1976). The deposition of metal carbonates and phosphates is a process with a definite biogenic origin, whereas the deposition of sulfides represents a mineralization at the oxidative expense of organic matter. Metal-organic complexing has been invoked to concentrate metals and such entities may be derived from biogenic precursors (e.g. porphyrins) or be generated de novo.

During mineral diagenesis and metamorphism under non-oxidizing conditions the organic matter composition changes progressively to more aromatic and asphaltic residues by the expulsion of volatile components (e.g. $CO_2$, $CH_4$, $H_2O$, etc.). The inferred residuum is graphite and in the Guaymas Basin hydrothermal system the spent kerogen remaining in the altered sediments (at about 300°C) consisted of amorphous, activated carbon (Simoneit 1982b). This higher temperature aromatic and asphaltic organic matter is often associated with heavy metal enrichments as for example uranium (e.g. Schidlowski 1981) or Carlin-type gold and silver ores (e.g. Radtke and Scheiner 1970). Heavy aromatic hydrocarbons (PAH), present in all the hydrothermal sites described here, are a product of high temperature alteration and thus may be good indicators for such processes in the periphery of sulfide ore bodies (e.g. Blumer 1975; Germanov and Bannikova 1972).

### 3.4 Biodegradation and Chemosynthesis

Where the condensed hydrothermal petroleum is exposed, as on exterior surfaces of mounds, and accessible, as in unconsolidated surface sediments, it is rapidly degraded to a residue of an unresolved complex mixture of hydrocarbons, the UCM or "hump" (e.g. Simoneit 1985b). These residues do not result from water leaching alone, because that would have left the heavy components ($> C_{20}$) unaltered. They appear to be a result of microbial alteration, however, some water leaching cannot be excluded (Simoneit 1985b).

Many of the Guaymas Basin samples show evidence of biodegradation. Microbial degradation of petroleum, as studied both in the laboratory and in reservoirs, proceeds by initial attack on the light normal paraffins (*n*-alkanes $< C_{25}$), then on the heavy normal paraffins ($> C_{25}$), followed by the isoprenoids and, lastly, on the alkyl naphthenes (polycycloalkanes) and alkyl aromatics (Winters and Williams 1969; Bailey et al. 1937a,b; Connan et al. 1980). This process leaves behind a nondegradable residue of branched and cyclic hydrocarbons, the UCM or "hump". The same types of hydrocarbon distributions were found for many of the Guaymas Basin samples. Microbial degradation of these samples is further confirmed by their molecular markers, analogous to those described for petroleums microbially degraded in the laboratory (Connan et al. 1980).

Thus microbiological alteration and utilization of the organic carbon from the petroliferous matter accessible in hydrothermal vent systems constitute a major biogenic carbon source in the ocean. It may be equivalent in magnitude to $CO_2$ fixation by vent fauna in sedimented rift basins. Therefore, a carbon source from hydrothermal petroleum must be incorporated in the ecological studies of the biota of such areas.

## Conclusions

Recent immature sediments are the receptacles of posthumus biogenic detritus, which upon deposition undergoes diagenetic and additional microbial alteration. Increasing burial in sedimentary basins results in the onset of organic matter maturation, which generates some volatile products from the kerogen (easily cracked moieties) that become superimposed on the endogenous lipid residues. This is the beginning of petroleum formation. As the depth of burial (i.e. temperature) keeps increasing catagenesis commences and here major petroleum generation takes place. At still greater depths of burial the metagenetic stage is envisaged, where extensive cracking, disproportionation and reforming of the organic matter, both petroleum and kerogen residues, occur to yield primarily gases and amorphous carbon.

In the case of hydrothermal systems the previous processes are compressed into an "instantaneous" geological time frame. Hydrothermal systems operative below a sediment blanket (e.g. Guaymaş Basin, Escanaba Trough, Bransfield Strait, and Atlantis II Deep) generate petroleum from that sedimentary organic matter (generally immature), which migrates upward or laterally and leaves behind a spent carbonaceous residue. Hydrothermal systems operating in unsedimented rift areas (e.g. East Pacific Rise at 13°N and 21°N, Mid Atlantic Ridge) generate trace amounts of petroleum and emit mainly methane from deeper sourced carbon. The petroleum is generated by pyrolysis of suspended and dissolved biogenic organic detritus (including bacteria and algae) entrained during the turbulent cooling of the vents. Also, low level maturation is observed in the surrounding area, probably due to intermittent warming of ambient detritus. This same process of organic matter pyrolysis and product movement also appears to occur in hydrothermal regions where ore deposits form and the organic matter content of the country rocks is low. Organic matter associated with such minerals can reflect the temperature of their formation.

Accessible and exposed petroleums on the exterior regions of hydrothermal mounds or in unconsolidated surface sediments are microbially degraded and leached, whereas interior samples are generally unaltered. Since primary productivity of the faunal ecosystems at hydrothermal vent sites is based on chemosynthetic pathways, the oxidative degradation of these petroleums represents a major carbon source for bio-assimilation in addition to $CO_2$-fixation.

In conclusion, gases, bitumen (lipids) and kerogen are ideal carbonaceous fractions that complement each other in providing information about the sources and thermal history of sedimentary organic matter. Kerogen is a sensitive in situ indicator for thermal stress, and bitumen (petroleum-asphalt with gas) represents the product mixture of that stress-products which may have remained in situ or migrated. These carbonaceous fractions are amenable for study in hydrothermal systems to elucidate temperature regimes, migration, bio-assimilation and resource potential.

*Acknowledgements.* I thank the Deep Sea Drilling Project and the National Science Foundation for access to DSDP samples, the NSF for my participation on the D.S.V. *Alvin* cruises; Dr. P. Lonsdale, Dr. R.P. Philp, Dr. P. Jenden, Dr. M.A. Mazurek, Mr. E. Ruth, Mr. O.E. Kawka, Dr. M. Brault, Dr. J. Baross, Dr. K.A. Kvenvolden, Dr. P.A. Rona and Ms. A. Lorre for samples, data and assistance. Funding from the National Science Foundation, Division of Ocean Sciences (Grants OCE81-18897, OCE-8312036, OCE-8512832 and OCE-8601316) is gratefuly acknowledged.

## References

Bailey NJL, Jobson AM, Rogers MA (1973a) Bacterial degradation of crude oil: Comparison of field and experimental data. Chem Geol 11:203-221

Bailey NJL, Krouse HR, Evans CR, Rogers MA (1973b) Alteration of crude oil by waters and bacteria – evidence from geochemical and isotope studies. Am Assoc Petrol Geol Bull 57:1276-1290

Ballard RD, Francheteau J, Juteau T, Rangan C, Normark W (1981) East Pacific Rise at 21°N: the volcanic, tectonic and hydrothermal processes of the central axis. Earth Planet Sci Lett 55:1-10

Blumer M (1975) Curtisite, idrialite and pendletonite, polycyclic aromatic hydrocarbon minerals: Their composition and origin. Chem Geol 16:245-256

Blumer M (1976) Polycyclic aromatic compounds in nature. Sci Am 234(3):34-45

Boon JJ, deLange R, Schuyl PJW, deLeeuw JW, Schenck PA (1977) Organic geochemistry of Walvis Bay diatomaceous ooze-II. Occurrence and significance of the hydroxy fatty acids. In: Campos R, Goni J (eds) Advances in Organic Geochemistry 1975. ENADIMSA, Madrid, pp 255-272

Brault M, Simoneit BRT (1988a) Steroid and triterpenoid distributions in Bransfield Strait sediments: Hydrothermally-enhanced diagenetic transformations. In: Advances in Organic Geochemistry 1987. Org Geochem, in press

Brault M, Simoneit BRT (1988b) Trace petroliferous organic matter associated with hydrothermal minerals from the Mid-Atlantic Ridge at the TAG 26°N Site, J Geophys Res, in press

Brault M, Simoneit BRT, Marty JC, Saliot A (1985) Les hydrocarbures dans le systeme hydrothermal de la ride Est-Pacifique, a 13°N. CR Acad Sci Paris 301, II:807-812

Brault M, Simoneit BRT, Marty JC, Saliot A (1988a) Hydrocarbons in waters and particulate material from hydrothermal environments at the East Pacific Rise, 13°N. Org Geochem, 12:209-219

Brault M, Simoneit BRT, Saliot A (1988b) Trace petroliferous organic matter associated with massive hydrothermal sulfides from the East Pacific Rise at 21°N and 13°N, Oceanol Acta, in press

Connan J, Restle A, Albrecht P (1980) Biodegradation of crude oil in the Aquitaine basin. In: Douglas AG, Maxwell JR (eds) Advances in Organic Geochemistry 1979. Pergamon Press, Oxford, pp 1-17

Curray JR, Moore DG, Aguayo JE, Aubry MP, Einsele G, Fornari DJ, Gieskes J, Guerrero JC, Kastner M, Kelts K, Lyle M, Matoba Y, Molina-Cruz A, Niemitz J, Rueda J, Saunders AD, Schrader H, Simoneit BRT, Vacquier V (1982) Initial Reports of the Deep Sea Drilling Project, Vol 64, Parts I and II, US Govt Printing Office, Washington DC, 1314 pp

Demaison GJ, Moore GT (1980) Anoxic environments and oil source bed genesis. Org Geochem 2:9-31

Deroo G, Herbin JP, Roucache JR, Tissot B, Albrecht P, Dastillung M (1978) Organic geochemistry of some Cretaceous claystones from site 391, Leg 44, Western North Atlantic. In: Benson WE, Sheridan RE et al. (eds) Initial Reports of the Deep Sea Drilling Project, Vol 44, US Govt Printing Office, Washington DC, pp 593-598

Didyk BM, Simoneit BRT, Brassell SC, Eglinton G (1978) Organic geochemical indicators of paleoenvironmental conditions of sedimentation. Nature (Lond) 272:216-222

Einsele G (1985) Basaltic sill-sediment complexes in young spreading centers: Genesis and significance. Geology 13:249-252

Einsele G, Gieskes J, Curray J, Moore D, Aguayo E, Aubry MP, Fornari DJ, Guerrero JC, Kastner M, Kelts K, Lyle M, Matoba Y, Molina-Cruz A, Niemitz J, Rueda J, Saunderss A, Schrader H, Simoneit BRT, Vacquier V (1980) Intrusion of basaltic sills into highly porous sediments and resulting hydrothermal activity. Nature (Lond) 283:441–445
Geissman TA, Sim KY, Murdoch J (1967) Organic minerals. Picene and chrysene as constituents of the mineral curtisite (idrialite). Experientia 23:793–794
Germanov AI, Bannikova LA (1972) Alteration of organic matter of sedimentary rocks during hydrothermal sulfide concentration. Dokl Akad Nauk SSSR 203:1180–1182
Hartmann M (1980) Atlantis II Deep geothermal brine system. Hydrographic situation in 1977 and changes since 1965. Deep-Sea Res 27:161–171
Hartmann M (1985) Atlantis II Deep geothermal brine system. Chemical processes between hydrothermal brines and Red Sea deep water. Mar Geol 64:157–177
Hekinian R, Fevrier M, Avedik F, Cambon P, Charlou JL, Needham HD, Raillard J, Boulegue J, Merlivat L, Moinet A, Manganini S, Lange J (1983) East Pacific Rise near 13°N: geology of new hydrothermal fields. Science 219:1321–1324
Hunt JM (1979) Petroleum Geochemistry and Geology, Freeman, San Francisco, 617 pp
Johns RB (1986) Biological Markers in the Sedimentary Record, Methods in Geochem and Geophys. 24, Elsevier, Amsterdam, 364 pp
Kartsev AA, Vassoevich NB, Geodekian AA, Neruchev SG, Sokolov VA (1972) The principal stage in formation of petroleum. Proc 8th World Petrol Congr 2:3–11
Kawka OE, Simoneit BRT (1987) Survey of hydrothermally-generated petroleums from the Guaymas Basin spreading center: Org Geochem, vol 11, pp 311–328
Kvenvolden KA, Simoneit BRT (1987) Petroleum from Northeast Pacific Ocean hydrothermal systems in Escanaba Trough and Guaymas Basin. Am Meet Am Assoc Petrol Geol Abstr, Los Angeles, June 7–10
Kvenvolden KA, Rapp JB, Hostettler FD, Morton JL, King JD, Claypool GE (1986) Petroleum associated with polymetallic sulfide in sediment from Gorda Ridge. Science 234: 1231–1234
Londsdale P (1985) A transform continental margin rich in hydrocarbons, Gulf of California. Am Assoc Petrol Geol Bull 69:1160–1180
Lonsdale P, Becker K (1985) Hydrothermal plumes, hot springs, and conductive heat flow in the Southern Trough of Guaymas Basin. Earth and Planet Sci Lett 73:211–225
McManus DA, Burns RE, Weser O, Vallier T, von der Borch RK, Goll RM, Milow ED (1970) Initial Reports of the Deep Sea drilling Project, Vol 5, Washington, DC, US Government Printing Office, pp 165–172
Merewether R, Olsson MS, Lonsdale P (1985) Acoustically detected hydrocarbon plumes rising from 2-km depths in Guaymas Basin, Gulf of California. J Geophys Res 90:3075–3085
Michaelis W, Jenisch A, Richnow HH (1988) Biomarker composition related to Red Sea hydrothermal systems. Third Chem Congr North Am, Toronto, June 5–11
Philp RP, Calvin M, Brown S, Yang E (1978) Organic geochemical studies on kerogen precursors in recently-deposited algal mats and oozes. Chem Geol 22:207–231
Price LC (1976) Aqueous solubility of petroleum as applied to its origin and primary migration. Am Assoc Petrol Geol Bull 60:213–244
Price LC (1982) Organic geochemistry of core samples from an ultra-deep hot well (300°C, 7 km). Chem Geol 37:215–228
Price LC, Clayton JS, Rumen LL (1981) Organic geochemistry of the 9.6 km Bertha Rogers No 1 well, Oklahoma. Org Geochem 3:59–77
Price LC, Wenger LM, Ging T, Blount CW (1983) Solubility of crude oil in methane as a function of pressure and temperature. Org Geochem 4:201–221
Radtke AS, Scheiner BJ (1970) Studies in hydrothermal gold deposition (I). Carlin gold deposit, Nevada: The role of carbonaceous materials in gold deposition. Econ Geol 65:87–102
Reed WE (1977) Molecular compositions of weathered petroleum and comparison with its possible source. Geochim Cosmochim Acta 41:237–247
Rona PA, Thompson G, Mottl MJ, Karson JA, Jenkins WJ, Graham D, Mallette M, von Damm K, Edmond JM (1984) Hydrothermal activity at the Trans-Atlantic Geotraverse hydrothermal field, Mid-Atlantic Ridge Crest at 26°N. J Geophys Res 89:11365–11377

Saxby JD (1976) The significance of organic matter in ore genesis. In: Handbook of Stratabound and Stratiform Ore Deposits: I, Principles and General Studies, Vol 2, Geochemical Studies, Wolf KH (ed), Elsevier, Amsterdam, pp 111–133
Schidlowski M (1981) Uraniferous constituents of the Witwatersrand conglomerates: Ore-microscopic observations and implications for the Witwatersrand metallogeny. US Geol Survey Prof Paper 1161-N, 29 pp
Scott LT (1982) Thermal rearrangements of aromatic compounds. Acc Chem Res 15:52–58
Simoneit BRT (1975) Sources of Organic Matter in Oceanic Sediments, Ph D Thesis, University of Bristol, England, December 1975, 300 pp
Simoneit BRT (1978) The organic chemistry of marine sediments. In: Riley JP, Chester R (eds) Chemical Oceanography, Vol 7, Academic Press, London, pp 233–311
Simoneit BRT (1981) Utility of molecular markers and stable isotope compositions in the evaluation of sources and diagenesis of organic matter in the geosphere. The Impact of the Treibs' Porphyrin Concept on the Modern Organic Geochemistry, A Prashnowsky (ed), Bayerische Julius Maximilian Universität, Würzburg, pp 133–158
Simoneit BRT (1982a) The composition, sources and transport of organic matter to marine sediments – the organic geochemical approach. In: Thompson JAJ, Jamieson WD (eds) Proc Symp Mar Chem into the Eighties, Nat Res Counc Can, pp 82–112
Simoneit BRT (1982b) Shipboard organic geochemistry and safety monitoring, Leg 64, Gulf of California. In: Curray JR, Moore DG et al. (eds) Initial Reports of the Deep Sea Drilling Project, Vol 64, US Govt Printing Office, Washington DC, pp 723–728
Simoneit BRT (1983) Organic matter maturation and petroleum genesis: Geothermal versus hydrothermal. In: Proc Symp The Role of Heat in the Development of Energy and Mineral Resources in the Northern Basin and Range Province, Geotherm Res Counc, Spec Rep No 13, Davis, California, pp 215–241
Simoneit BRT (1984a) Hydrothermal effects on organic matter high versus low temperature components. In:Schenck PA, de Leeuw DW, Lijmbach GWM (eds) Advances in Organic Geochemistry 1983, Org Geochem 6, 857–864
Simoneit BRT (1984b) Effects of hydrothermal activity on sedimentary organic matter: Guaymas Basin, Gulf of California petroleum genesis and protokerogen degradation. In: Rona PA et al. (eds) Hydrothermal Processes at Seafloor Spreading Centers, NATO-ARI Series, Plenum Press, New York, pp 453–474
Simoneit BRT (1985a) Hydrothermal petroleum: Genesis, migration and deposition in Guaymas Basin, Gulf of California. Can J Earth Sci 22:1919–1929
Simoneit BRT (1985b) Hydrothermal Petroleum: Composition and Utility as a Biogenic Carbon Source. In: Jones ML (ed) Hydrothermal Vents of the Eastern Pacific: An Overview, Bull Biol Soc Wash 6:49–56
Simoneit BRT (1986) Organic geochemistry of black shales from the Deep Sea Drilling Project, a summary of occurrences from the Pleistocene to the Jurassic. In: Degens ET, Meyers PA, Brassell SC (eds) Biogeochemistry of Black Shales, Mitt Geol-Paläontol Inst Univ Hamburg 60:275–309
Simoneit BRT (1988) Petroleum generation in submarine hydrothermal systems: An update. Can Mineral 26:827–840
Simoneit BRT, Kawka OE (1987) Hydrothermal petroleum from diatomites in the Gulf of California. In: Brooks J, Fleet AJ (eds) Marine Petroleum Source Rocks, Geol Soc Lond Special Publication No.26, pp 217–228
Simoneit BRT, Lonsdale PF (1982) Hydrothermal petroleum in mineralized mounds at the seabed of Guaymas Basin. Nature (Lond) 295:198–202
Simoneit BRT, Philp RP (1982) Organic geochemistry of lipids and kerogens and the effects of basalt intrusions on unconsolidated oceanic sediments: Sites 477, 478 and 481, Guaymas Basin, Gulf of California. In: Curray JR, Moore DG et al. (eds) Initial Reports of the Deep Sea Drilling Project, Vol 64, US Govt Printing Office, Washington DC, pp 883–904
Simoneit BRT, Brenner S, Peters KE, Kaplan IR (1978) Thermal alteration of Cretaceous black shale by basaltic intrusions in the Eastern Atlantic. Nature (Lond) 273:501–504
Simoneit BRT, Mazurek MA, Brenner S, Crisp PT, Kaplan IR (1979) Organic geochemistry of recent sediments from Guaymas Basin, Gulf of California. Deep-Sea Res 26A:879–891
Simoneit BRT, Halpern HI, Didyk BM (1980) Lipid productivity of a high Andean lake. In: Trudinger PA, Walter MR, Ralph BJ (eds) Biogeochemistry of Ancient and Modern Environments, Aust Acad Sci Canberra, pp 201–210

Simoneit BRT, Brenner S, Peters KE, Kaplan IR (1981) Thermal alteration of Cretaceous black shale by basaltic intrusions in the eastern Atlantic. II: Effects on bitumen and kerogen. Geochim Cosmochim Acta 45:1581–1602

Simoneit BRT, Philp RP, Jenden PD, Galimov EM (1984) Organic geochemistry of Deep Sea Drilling Project sediments from the Gulf of California hydrothermal effects on unconsolidated diatom ooze. Org Geochem 7:173–205

Simoneit BRT, Grimalt JO, Hayes JM, Hartman H (1987) Low temperature hydrothermal maturation of organic matter in sediments from the Atlantis II Deep, Red Sea. Geochim Cosmochim Acta 51:879–894

Spiess FN, Macdonald KC, Atwater T, Ballard R, Carranza A, Cordoba D, Cox C, DiazGarcia VM, Francheteau J, Guerrero J, Hawkins J, Haymon R, Hessler R, Juteau T, Kastner M, Larson R, Luyendyke B, Macdougall JD, Miller S, Normark W, Orcutt J, Rangin C (1980) East Pacific Rise; hot springs and geophysical experiments. Science 207:1421–1433

Stuermer DH, Simoneit BRT (1978) Varying sources for the lipids and humic substances at Site 391, Blake-Bahama Basin, DSDP Leg 44. In: Initial Reports of the Deep Sea Drilling Project, Vol 44, Benson WE, Sheridan RE et al. US Govt Printing Office, Washington DC, pp 587–591

Stuermer DH, Peters KE, Kaplan IR (1978) Source indicators of humic substances and protokerogen: Stable isotope ratios, elemental compostions and electron spin resonance spectra. Geochim Cosmochim Acta 42:989–997

Suess E, Fisk M, Whiticar MJ, Wefer G, Wittstock R, Theilen F, Schreiber R, Simoneit BRT, Laban C, Kadko D, Schlosser P, Top Z (1989) Hydrothermalism in the Bransfield Strait, Antarctica. Earth Planet Sci Lett in preparation

Tissot BP, Welte DH (1984) Petroleum Formation and Occurrence: A New Approach to Oil and Gas Exploration. Springer, Berlin Heidelberg New York Tokyo, 2nd edn 699 pp

van de Meent D, Brown SC, Philp RP, Simoneit BRT (1980) Pyrolysis-high resolution gas chromatography and pyrolysis gas chromatography-mass spectrometry of kerogens and kerogen precursors. Geochim Cosmochim Acta 44:999–1013

Venkatesan MI, Kaplan IR (1987) Organic geochemistry of Antarctic marine sediments, Part I: Bransfield Strait. Mar Chem 21:347–375

Vassoevich NB, Akramkhodzhaev AM, Geodekyan AA (1974) Principal zone of oil formation. In: Tissot B, Bienner F (eds) Advances in Organic Geochemistry 1973, Technip, Paris, pp 309–314

Whelan JK, Hunt JM, Jasper J, Huc A (1984) Migration of $C_1$-$C_8$ hydrocarbons in marine sediments. In: Schenck PA et al. (eds) Advances in Organic Geochemistry 1983, Org Geochem 6:683–694

Whelan JK, Simoneit BRT, Tarafa M (1988) Composition and Geochemical Implications of $C_1$-$C_8$ Hydrocarbons in Sediments from Guaymas Basin, Gulf of California. Org Geochem 12:171–194

Whiticar MJ, Suess E, Wehner H (1985) Thermogenic hydrocarbons in surface sediments of the Bransfield Strait, Antarctic Peninsula. Nature (Lond) 314:87–90

Winters JC, Williams JA (1969) Microbiological alteration of crude oil in the reservoir. In: Symp on Petroleum Transformation in Geologic Environments. Am Chem Soc Div Petrol Chem Preprints NY Mtg Vol 14(4), pp E22–E31

# Structural Inferences from Organic Geochemical Coal Studies

W. MICHAELIS, H. H. RICHNOW, A. JENISCH, T. SCHULZE, and B. MYCKE

## 1 Introduction

Coal results from the combination of many chemical reactions comprising biological or thermal degradations on organic constituents like plants, algae, bacteria, spores, etc. (Degens 1965; Tissot and Welte 1984). As a result the organic matter of coal is rather heterogeneous in nature and precludes an exact knowledge of its macromolecular chemical structure. However, structural information at the molecular level may provide useful implications for various coal technological aspects such as gasification, liquefaction, and coking.

In general, the macromolecular chemical structure of coal is described as a three-dimensional, cross-linked network (Larsen 1981; Lucht and Peppas 1981; Dong and Ouchi 1988). Basic data on elemental composition, extractability, molecular weight determinations, or spectroscopical studies only give gross structural information and often differ between several investigations. Some aspects of the basic molecular skeleton that comprises the coal structure may not be clarified easily by applying these techniques. In particular, a great deal of importance has been attached to the debate describing coal as a two-component system (Given et al. 1986). There are indications that coals contain a considerable fraction of low-molecular-weight compounds of limited accessibility to solvent extraction within the three-dimensional network (Jurkiewicz et al. 1982; Green and Larsen 1984; Monthioux and Landais 1987). A better knowledge of this important fraction should be obtained by degradation studies.

Many elucidations of coal organic structure are based on degradative approaches comprising the thermal or chemical breakdown of the macromolecular network. Pyrolysis coupled with gas chromatography-mass spectrometry can reveal useful information about the degraded entities on a molecular level (Philp and Saxby 1980; Nip et al. 1986; Bertrand et al. 1986). For chemical degradation of coals different techniques have been used. They include oxidative as well as reductive attempts like permanganate oxidation (Ishiwatari et al. 1985), peroxytrifluoroacetic acid oxidation (Deno et al. 1978; Verheyen et al. 1985; Choudhury et al. 1988), alkylation (Stock 1982), and O-alkylation (Shaw et al. 1988). These reactions yield variable amounts of soluble low-molecular-weight material from the coal. But they widely differ in reactivity and therefore specificity. Oxidative reactions often provide considerable fractions degraded from the coal organic matter but the achieved molecular fragments sometimes give little information about the original polycondensed matrix of the coal. Therefore, in our coal structural studies we apply selective non-oxidative chemical degradation procedures which provide information on the type of bonds between the released fragments and the coal matrix. We have chosen a low-temperature, low-pressure hydrogenation with rhodium on charcoal as a catalyst. By this degradation we could cleave ester and ether bonds selectively in macromolecular organic material of geological origin like humic substances and kerogens (Mycke and Michaelis 1986a,b). This study presents part of our structural studies on coals of different rank and origin. The low-molecular-weight compounds obtained from hydrogenolyses of the coals are discussed in terms of their origin, maturity of the organic matter, and their structural relevance as building blocks of the coal matrix.

## 2 Samples and Methods

Hydrogenolyses were performed with two humic coal samples: a lignite of Tertiary age (Miocene, Lower Rhine Basin, FRG) and a high volatile C bituminous coal of Westphalian D, Upper Carboniferous age (Herrin Coal; IBC-101; Illinois, USA). Basic parameters are listed in Table 1. The samples were dried and finely ground. The ground coals were extensively extracted by ultrasonic treatment using chloroform/methanol as solvent. A fractionation of the extracts by column chromatography ($SiO_2$/n-hexane) revealed the hydrocarbons which were further purified by thin layer chromatography and separation of n-alkanes by molecular sieve treatment.

The hydrogenolysis experiments were run in a stainless steel autoclave. Dioxane/distilled water (1/1; v/v) were used as solvents and preextracted rhodium on charcoal as catalyst. The initial pressure was 50 kg/cm² $H_2$. The reaction temperature of 200 °C was held for a period of 5 h. Experimental details are given by Mycke (1985). The obtained hydrogenated substances were extracted with chloroform and soluble compounds subsequently separated into acidic, phenolic, and neutral fractions by liquid/liquid extraction at various pH values. The neutral compounds were further purified as described for extracted compounds.

**Table 1.** Basic analytical parameters of coal samples. Results of the elemental analyses are given on a dry ash-free basis

| Sample | $R_m$ (%) | C (%) | H (%) | O (%) | N (%) | S (%) | E[a] | H[b] |
|---|---|---|---|---|---|---|---|---|
| Lignite | 0.30 | 67.87 | 5.59 | 25.37 | 0.85 | n.d. | 27.40 | 200.4 |
| Bituminous Coal | 0.46 | 75.43 | 5.42 | 12.97[c] | 1.32 | 4.75 | 7.56 | 40.7 |

[a] Yields of solvent extracts in mg/g $C_{org}$.
[b] Yields of products after hydrogenolysis in mg/g $C_{org}$.
[c] Value calculated by difference.

## 3 Analyses

The hydrocarbon fractions were analyzed by high resolution gas chromatography on a 25 m or 50 m × 0.3 mm SE 54 fused silica capillary column; temperature program: 80°C 5 min isothermal; 80–300°C, 3°C/min; 300°C, 15 min isothermal; carrier gas: $H_2$.

GC/MS investigations were performed on a gas chromatograph coupled to a Varian CH7A mass spectrometer; ionizing energy: 70 eV; source temperature: 250°C; carrier gas: He; GC conditions and temperature program as above. The compounds were identified by comparison of their mass spectra and retention times with standards.

## 4 Results and Discussion

The amounts of solvent extractable material from the lignite and the high volatile bituminous coal are presented in Table 1. Hydrogenolysis of the preextracted coal samples afforded a substantial release of soluble organic material. Total amounts of solubilized material for the lignite and sub-bituminous coal after degradation were 200.4 and 40.7 mg/g $C_{org}$, respectively. The hydrogenation reaction is known to cleave ether and ester bonds (Burwell 1954). Major cleavage products are acids, alcohols, and hydrocarbons. The molecular composition of the coal-derived degradation products is compared with the coal extracts to obtain more information on the entire coal structure. When deuterium instead of hydrogen was used in the degradation experiments bonding sites between the coal matrix and the released compounds could be described more precisely.

## 5 Acyclic Alkanes

The aliphatic hydrocarbon fraction extracted from the lignite shows a rather simple compound distribution pattern which is well known from many other coal samples of this rank (Van Dorsselaer et al. 1977). Normal alkanes in the carbon range of $C_{25}$-$C_{33}$ are the major products with a clear odd/even predominance and a maximum chain length at n-$C^{21}$ (Fig. 1A). Neither diterpenes nor acyclic isoprenoids such as pristane or phytane are present in considerable concentrations. Terrestrial higher plant material can account for a source of these hydrocarbons (Bray and Evans 1961; Eglinton and Hamilton 1967). A completely different alkane pattern is released from the lignite after the catalytic hydrogenation of the preextracted sample. The short-chain alkanes are dominated by pristane and phytane, long-chain n-alkanes from $C_{21}$ to $C_{33}$ show their distribution maximum at $C_{29}$ with a lower carbon preference index than the extract. These differences between the lignite extract and the chemical degradation products should be expected. They have been observed after thermal and chemical treatments of coals (Radke et al. 1982; Mudamburi and Given 1985; Shaw et al. 1988). However, some

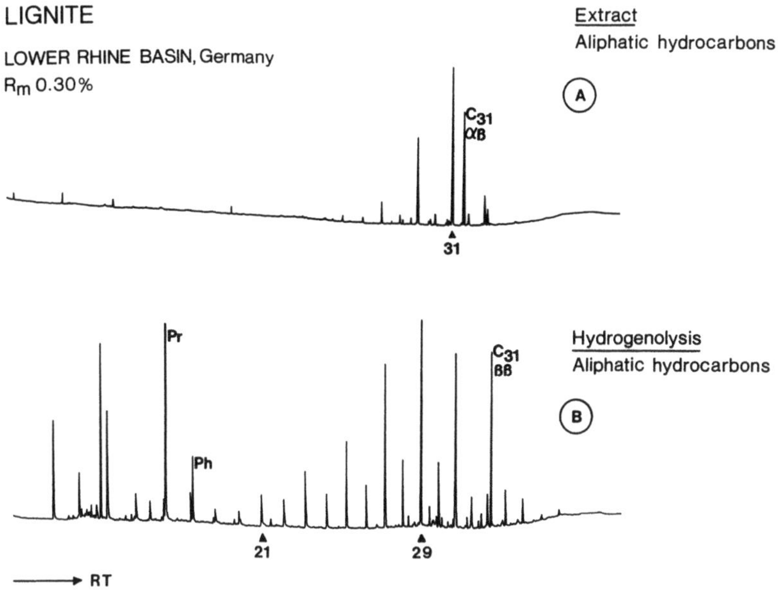


**Fig. 1A,B.** Gas chromatograms of the alipathic hydrocarbons obtained from **A** the solvent extracts and **B** the hydrogenolytically derived compounds. *Black triangles* numbers of carbon atoms of n-alkanes; *Pr* pristane; *Ph* phytane; $C_{31}\alpha\beta$ 17α(H),21β(H)-22R-homohopane; $C_{31}\beta\beta$ 17β(H),21β(H)-homohopane. For analytical conditions, see Section 2

of the differences in the alkane distribution of our chemical degradation reactions are difficult to explain. If we assume a dominating input of plant material to the lignite organic matter, there should be present a considerable amount of even-numbered carbon chains ester- or ether-linked to the polymeric matrix. Their hydrogenolytic degradation should release components with mainly even-over-odd predominance. This trend is observed (Fig. 1), but odd-numbered n-alkanes are also present in the reaction products as major compounds. These results could mean that parts of the alkane fraction are physically entrapped in the macromolecular structure of the coal and have been liberated during the reaction procedure.

The application of deuterium instead of hydrogen in our degradation reactions could give an estimate of the amount of hydrocarbons occluded in the coal network. The degraded alkanes showed incorporation of 0–9 deuterium atoms. An incorporation of deuterium only takes place during bond cleavage, which is coupled with a transfer of deuterium along the carbon chain. Added saturated alkanes did not undergo deuterium substitution under the reaction conditions applied. Peaks of non-deuterated and deuterated alkanes split off in gas chromatographic analyses when a slow heating rate is used. From these gas chromatograms it already became obvious that the major part of the alkanes incorporated deuterium. A precise analysis by mass spectrometry for each alkane peak confirmed these results. In the range of n-$C_{16}$ to n-$C_{29}$ the amount of non-deuterated saturated hydrocarbons varied between 3 and 14%. Maximum deuterium content centered around seven

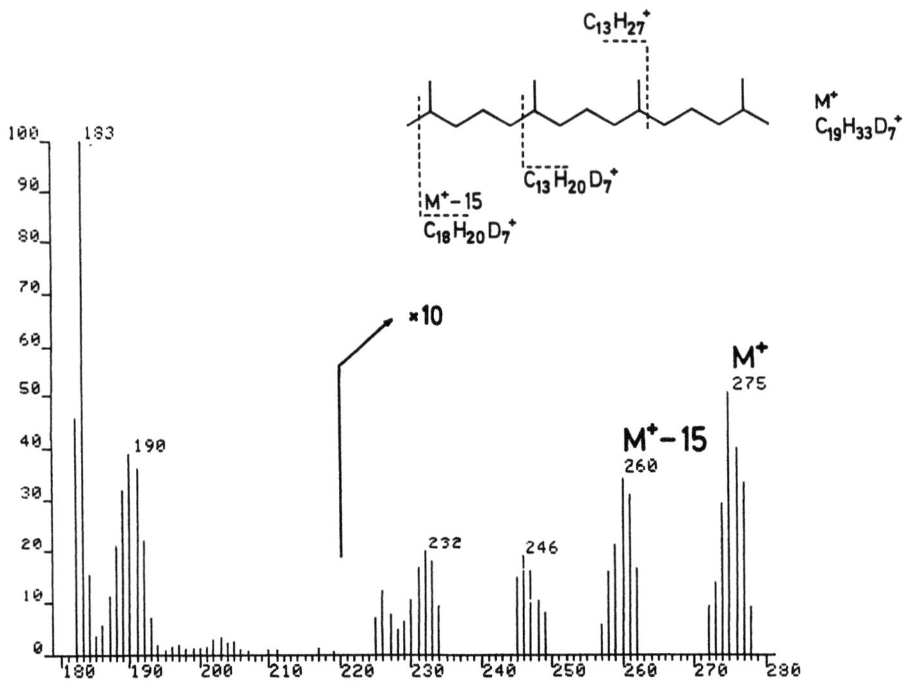


**Fig. 2.** Mass fragmentation pattern of pristane after hydrogenolysis performed with deuterium. Characteristic pristane fragments like $C_{31}$ show a bimodal distribution. Splitting of the carbon chain including the non-deuterated end of pristane yielded in the fragment m/z 183 ($C_{13}H_{27}^+$). Splitting with the deuterated end included gives several fragments revealing a maximum at m/z 190 ($C_{13}H_{20}D_7^+$)

deuterium atoms per molecule. No particular preference for even or odd carbon chain numbers was observed. In many cases the mass fragmentation pattern of the deuterated alkanes allowed a differentiation between deuterium enrichment at one or two ends of the molecule. Comparable deuteration results have been observed for isoprenoid alkanes. In Fig. 2 a partial mass spectrum of deuterated pristane is presented. Molecular ion and fragments with incorporation of seven deuterium atoms show the highest relative abundances ($M^+$ 275 and $M^+$–15). A bimodal structure is observed for the $C_{13}^+$ fragment. Deuterium enrichment at one end of the molecule should afford the two fragments $C_{13}H_{27}^+$ (m/z 183) and $C_{13}H_{20}D_7^+$ (m/z 190). This can be explained by a linkage of the degraded pristane to the macromolecular matter of the coal via one ester or ether bond. Probable natural precursors for pristane, which is found in many ancient sediments, crude oils, and coal extracts, are believed to be chlorophyll a and b (Dean and Whitehead 1961) and tocopherols (Goossens et al. 1984). They contain the pristane skeleton as ether or ester bonded moieties. Our degradation experiments suggest that molecular entities of these biologically produced precursor molecules are incorporated into the macromolecular network of the coal.

## 6 Cyclic Alkanes

Major cyclic alkanes in the lignite extract and in the hydrogenation products of the lignite residue are triterpenoids of the hopane type (Figs. 1 and 3). The extract contains mainly 17$\alpha$(H),21$\beta$(H)-22R-hopanoids (hereafter $\alpha,\beta$) in the carbon number range $C_{27}$-$C_{32}$ with homohopane predominating. Instead, 17$\beta$(H), 12$\beta$(H)-isomers (hereafter $\beta,\beta$) with carbon numbers up to $C_{35}$ (maximum homohopane) become more abundant in the products released by hydrogenolysis.

Assuming these hopanoids were released during hydrogenolysis from pores of the polymeric coal network they should show distributions and isomerization degrees similar to the extractable ones. For example, both organic fractions differ significantly in the ratios of $C_{32}\alpha\beta/C_{31}\beta\beta$ hopanes which are found to be 12 for the extractable and 0.5 for the degraded compounds. Because of these large differences we believe that saturated compounds of the hopane series are highly extractable from the coal (Dong et al. 1987). The $\beta,\beta$-hopanes in the hydrogenolysis reaction products should have been cleaved from the coal matrix.

Cyclic terpenoids of the hopane type are abundant and widespread geofossils (e.g. Ensminger et al. 1974). Hopanoids with less than 31 carbon atoms may be derived from sources like ferns, lichens, and higher plants. Extended hopanoids ($> C_{31}$) are produced by numerous bacteria which mostly use bacteriohopanepolyols as membrane rigidifiers (Ourisson et al. 1979; Rohmer et al. 1980).

The hopane fraction found in the lignite sample exhibits mainly higher homologues ($> C_{31}$). Therefore, besides higher plants, bacteria seem to be an important contributor to the lignite organic matter. It is striking that $C_{31}$ $\alpha,\beta$-22R-hopane is by far the most abundant cyclic compound (22S/22S + 22R = 0.03) in the lignite extract. Corresponding hopane distributions are reported from peat environments (Quirk et al. 1984). These authors could show that the 17$\alpha$(H),21$\beta$(H)-22R-homohopane is formed at the very early stages of diagenesis – despite the fact that $\alpha,\beta$-hopanes are usually regarded as thermodynamically more stable isomers occurring at higher maturity degrees. This early diagenetic

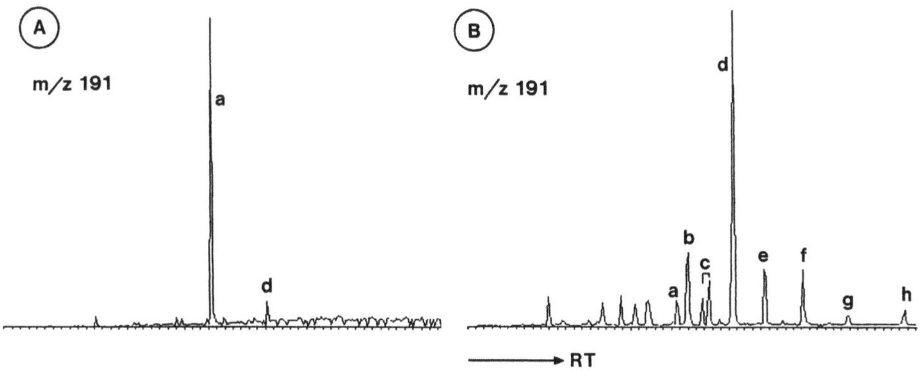


**Fig. 3A,B.** Hopane distributions shown by mass fragmentogram m/z 191 in lignite solvent extracts (**A**) and in lignite hydrogenolysis products (**B**). *a* $\alpha,\beta$-22R-$C_{31}$; *b* $\beta,\beta$-$C_{30}$; *c* $\alpha,\beta$-$C_{32}$ (22R and 22S); *d* $\beta,\beta$-$C_{31}$; *e* $\beta,\beta$-$C_{32}$; *f* $\beta,\beta$-$C_{33}$; *g* $\beta,\beta$-$C_{34}$; *h* $\beta,\beta$-$C_{35}$

$\alpha,\beta$-hopane formation probably associated with bacterial decay of plant matter might be responsible for the hopane composition found in the lignite extract (Quirk et al. 1984; Rohmer et al. 1980).

In contrast, $\beta,\beta$-homohopane dominates the composition of the cyclic hydrocarbons derived from hydrogenolysis. This compound may originate from $\beta,\beta$-homohopane acid or alcohol entities linked to the coal macromolecule. There are two hypotheses for the incorporation of these compounds into the macromolecular organic material of the lignite: either these entities are directly derived from a biological precursor or diagenetically induced reactions of bacteriohopane polyols ($C_{35}$) lead to $C_{31}$ alcohols or $C_{31}$ acids which are subsequently bound to the coal matrix.

The formation of hopanoic alcohols ($C_{32}$ only) and acids ($C_{32}$ mainly) is reported to appear at the earliest stages of diagenesis in peats. The origin of these compounds is probably a result of the degradation of bacteriohopane polyols and is presumably associated with bacterial activity in these acidic, nutrient-deficient environments (Rohmer et al. 1980; Quirk et al. 1984).

Therefore, we favour a diagenetically induced reaction sequence to cause the high concentrations of $C_{31}$ hopane in the lignite sample. A cleavage of the side chain of bacteriohopanepolyols may occur prior to incorporation of these compounds to the macromolecular coal network.

Furthermore, integral parts of bacterial membranes containing $C_{35}$ hopane precursor polyols seem to be incorporated into the coal network retaining their unaltered biological configuration. This is indicated by the presence of $C_{35}$ hopanes in the hydrogenolysis products (Fig. 3).

In support of this hypothesis the catalytic hydrogenation was performed with deuterium instead of hydrogen to obtain the site of bonding of the degraded compounds to the coal matrix.

Mass spectra of the $C_{32}$ $\alpha,\beta$-hopane and the $C_{31}$ $\beta,\beta$-hopane (Fig. 4) demonstrate a selective uptake of deuterium during the reaction. Fragment I and III remain non-deuterated. However, $M^+$, $M^+$-15, and the fragment II reveal the incorporation of three additional deuterium atoms. Therefore, the site of linkage should be the side chains of the hopanoids. Interestingly, not only the $C_{31}$ $\beta,\beta$-isomers but even the $\alpha,\beta$-bishomohopane took up deuterium. This may be explained by binding of diagenetically derived free $\alpha,\beta$-hopane acids or alcohols to the coal matrix via ester or ether bonds.

Chemical degradations of the lignite and the high volatile bituminous coal afforded variable amounts of steranes. In particular from the bituminous coal relatively high amounts of steranes of the 5$\alpha$(H),14$\alpha$(H),17$\alpha$(H)-configuration (hereafter $\alpha,\alpha,\alpha$) with 20R-epimers dominating over the 20S could be released. Sterane distributions differ significantly from those obtained from the extracts (Fig. 5), in which $\alpha,\alpha,\alpha$-20S-isomers predominate over 20R-isomers. Diasteranes were also present in the extractable fraction but not in the hydrogenolysis products.

Biogenic precursors of steroids found in the geologic environment are sterols synthesized by various eukaryotic organisms (Mackenzie et al. 1982). Most of the steroidal lipids found in geologic materials have undergone several alterations. The steranes formed during early diagenesis retain the biological configuration

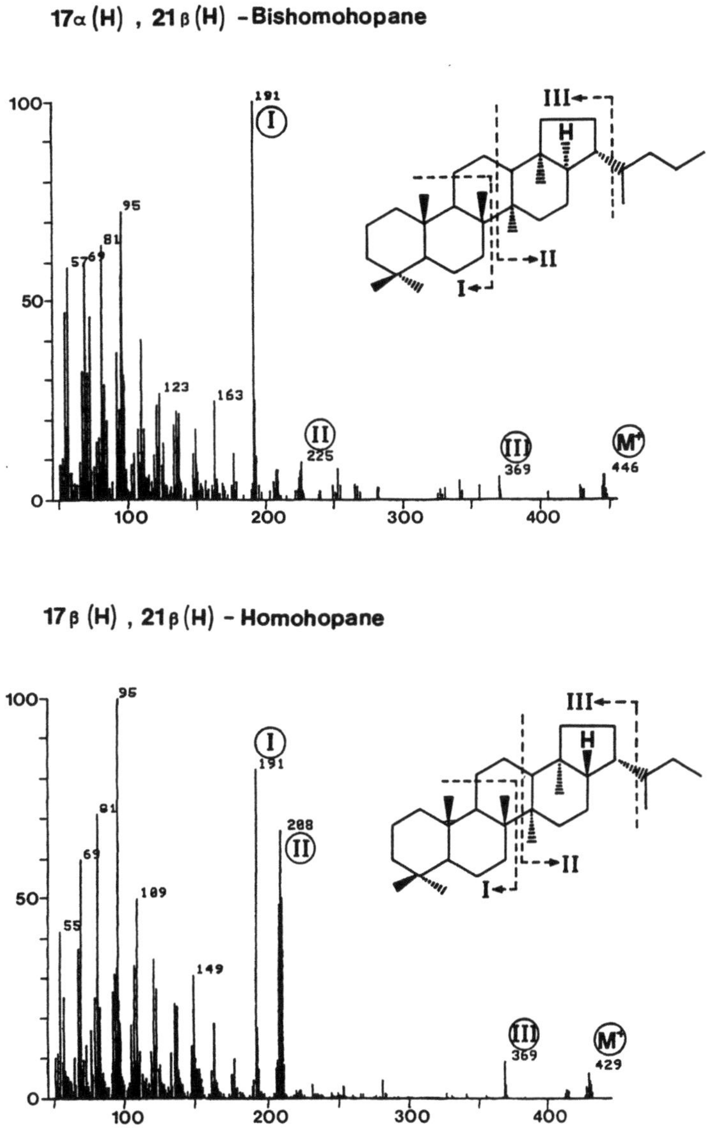


**Fig. 4.** Fragmentation pattern for deuterated hopanes, lignite degradation

($\alpha$,$\alpha$,$\alpha$-20R) of their precursor sterols. With increasing maturity of the samples isomerization to the thermodynamically more stable 20S-configuration and at catagenetic conditions to the 14$\beta$(H),17$\beta$(H)-steranes is observed. The isomerization ratios 20S/20S+20R differ between 0 in immature samples and 0.55 reflecting mature conditions. The 20S/20S+20R-value (0.51) calculated from extractable steranes is in accordance with the maturity of this sample ($R_m$ 0.46%). Steranes linked to the polymer and released by hydrogenolysis show a lower value

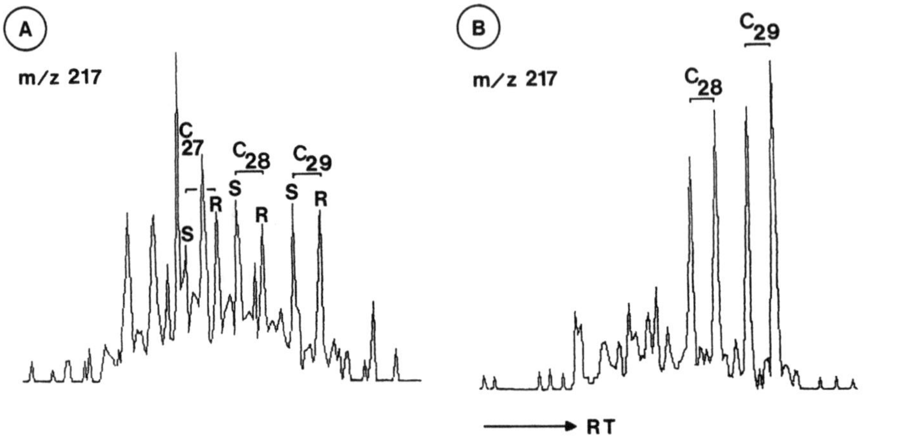


**Fig. 5A,B.** Mass fragmentogram m/z 217 displaying the sterane distributions in the solvent extract (**A**) and the degradation products (**B**) of the Herrin coal. Indicated are the 14α(H),17α(H),21α(H)-steranes. *S* and *R* refer to the 20S- and 20R-isomers

(0.46). Comparable observations have been made for hopanes in extracts and thermal or chemical degradation products of kerogens and asphaltenes (Seifert and Moldowan 1980; Jones and Douglas 1987). The reason for this is not clear, but it may be caused by restricted sensitivity to thermal alteration for compounds bound into the macromolecular network.

A series of tetracyclic compounds, which consists of *ent*-beyerane, 16α(H)- and 16β(H)-phyllocladane, 16α(H)- and 16β(H)-*ent*-kaurane, could be identified in the hydrogenolysis products of the coal samples. The extract of the lignite did not contain these hydrocarbons.

Tetracyclic diterpenoids have been found not only in oils and sediments (Simoneit 1977; Philp et al. 1981; Livsey et al. 1984; Noble et al. 1985a), but also in coals (Hollerbach 1979; Hagemann and Hollerbach 1980; Noble et al. 1985b; Alexander et al. 1987).

Their precursor compounds are particularly abundant in leaf resins of conifers (Thomas 1969). Compounds of the beyerane skeleton have only been recognized in a small number of higher plant species, whereas the kaurane skeleton is widely distributed (Noble et al. 1985a). Kauranes occur in high amounts in waxes of the Araucariaceae family (Coniferales). Instead, the phyllocladanes are proposed as markers for the Podocarpaceae family (Noble et al. 1985b). Both families are distributed all over the world since the Triassic but in Recent times they are restricted to the southern hemisphere (Ehrendorfer 1983). It is interesting that the hydrogenolytic degradation of the Herrin coal sample (Carboniferous) revealed tetracyclic terpanes of the phyllocladane type, because until now these compounds have been reported to appear for the first time in samples of Permian age (Noble et al. 1985b).

The partial mass fragmentograms (m/z 123) in Fig. 6 show the distribution of tetracyclic terpanes obtained after chemical degradation of the lignite (A) and the Herrin coal sample (B). Araucariaceae as well as Coniferales may have contributed

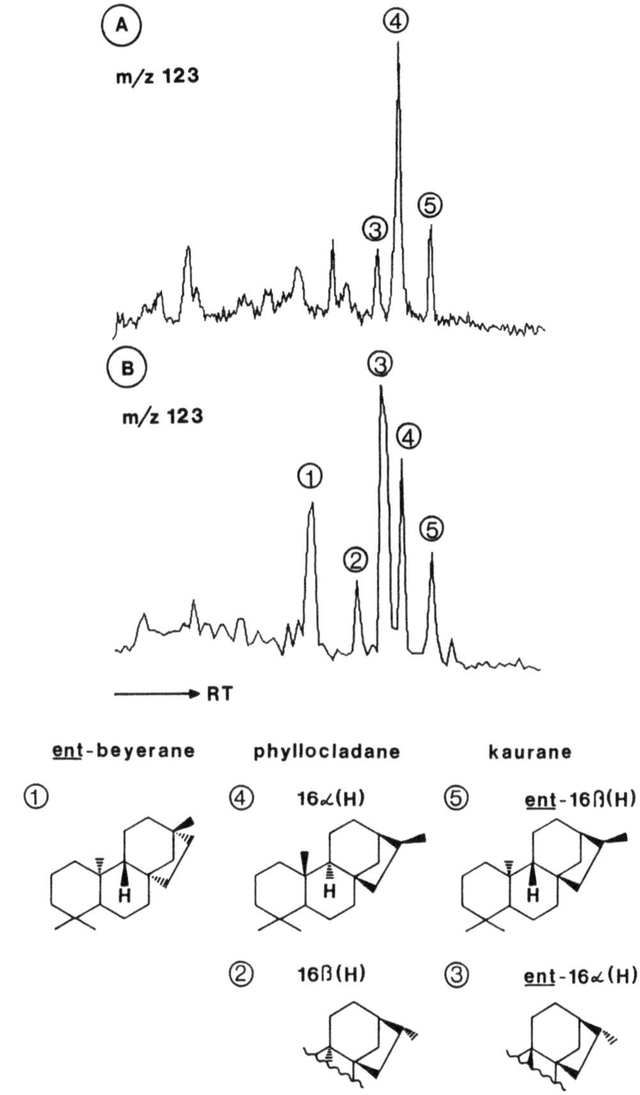


**Fig. 6A,B.** Distributions of tetracyclic compounds (mass fragment m/z 123) obtained from the aliphatic hydrocarbon fractions of degradation products: **A** lignite and **B**Illinois No. 6 coal. *Numbers in circles refer to the compound structures depicted below*

to the organic matter of the two coals. The differences in terpane distributions between the two samples are not only induced by varying inputs of organic matter but also by different degrees of maturation. During maturation the amount of thermodynamically more stable 16β(H)-phyllocladane increases, while 16α(H)-compounds nearly disappear at a rank of $R_m$ 0.7%. For the epimers of kauranes this maturation effect is reverse.

The lignite reveals after degradation 16α(H)-phyllocladane as the dominating compound, whereas in the degradation products of the sub-bituminous coal

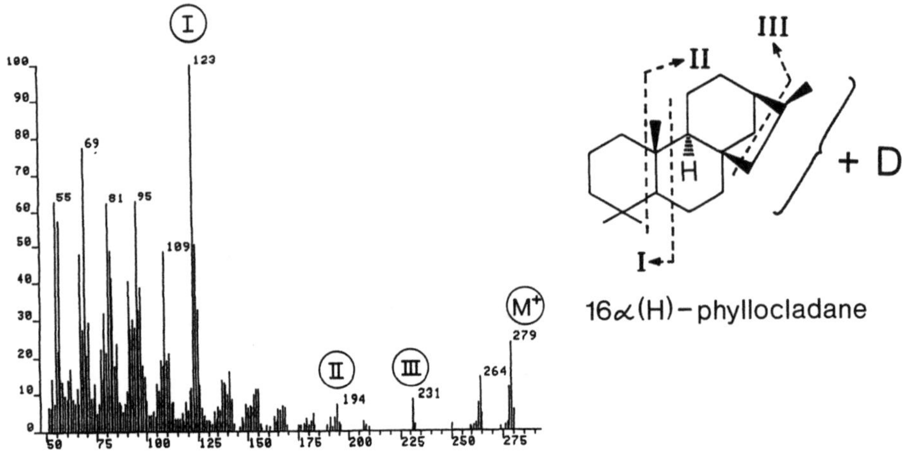


**Fig. 7.** Mass fragmentation pattern of deuterated 16α(H)-phyllocladane yielded from lignite degradation with deuterium

16β(H)-epimers are more abundant. The ratios of 16α(H)- to 16β(H)-epimers are infinite for the phyllocladanes and 0.5 for the kauranes in the lignite degradation products. The values of 2.6 and 0.5 for the respective ratios were obtained for the high volatile bituminous coal. This trend of decreasing ratios with increasing rank has also been observed in extracts of coals (Noble et al. 1985b).

These results on tetracyclic terpane distribution in degraded low-molecular-weight fractions of coal residues again provide evidence that parts of the biological precursor material is incorporated into the coal matrix. The bonding sites of the tetracyclic compounds to the macromolecular coal organic matter was determined by catalytic hydrogenolysis of the lignite sample with deuterium. As a typical example the mass fragmentation pattern of 16α(H)-phyllocladane is shown in Fig. 7. No deuterium uptake occurs in fragments I and III, which represent the A-ring and the ABC-ring system. Instead, fragment II is shifted by 5 amu. Identical deuterium enrichment is observed for the molecular ion $M^+$ and the $M^+$-15 fragment. This suggests a possible linkage from ring D to the coal matrix by ester or ether bonds. The probable type and site of linkage is in accordance with the known biological precursors of phyllocladanes, which contain an alcohol or acid group in the D-ring.

## 7 Conclusion

Selective chemical degradation yielded considerable amounts of soluble low-molecular-weight material from macromolecular coal organic matter. Catalytic hydrogenolyses of coal residues afforded product mixtures which show distributions significantly different from solvent extractable compounds. Analytical emphasis was put on the study of n-alkanes, acyclic and cyclic terpenoids. These products are not simply released from a closed pore system of the coal. The major part of these compounds has shown characteristic deuterium incorporations when the degradation was performed with deuterium instead of hydrogen. This suggests that part of the macromolecular coal network has been attacked by cleavage of ester or ether bonds using catalytic hydrogenolysis.

Mass spectrometric investigation of the deuterated coal degradation products provided further insight into the site of bonding by which they are attached to the coal matrix. $\beta,\beta$-and $\alpha,\beta$-hopanes clearly show deuterium incorporation in the side chain indicating a linkage of this part of the molecule to the macromolecular coal organic matter. Deuteration of the $\beta,\beta$-hopanes and the thermodynamically more stable $\alpha$-$\beta$-hopanes degraded from the lignite residue suggests an early diagenetic formation of $\beta,\beta$- as well as $\alpha,\beta$-hopanoic acids or alcohols which are subsequently bound to the coal network. Selective deuterium enrichment at one end of the molecule was also found in the released pristane. This is in accordance with the chemical structure of the natural sources for this abundant isoprenoid.

The first discovery of tetracyclic terpanes of the kaurane and phyllocladane type in coal degradation products opens interesting possibilities of palaeoenvironmental discussions. The structural information at a molecular level obtained from selective chemical degradations may further enhance our knowledge on the complex basic skeleton of coals.

*Acknowledgements.* The authors want to thank Prof. Dr. E.T. Degens for encouraging our work. We would like to acknowledge Dr. B. Wutzler, Rheinbraun AG (Hambach) and Dr. R.D. Harvey, Illinois State Geological Survey, for providing us the samples. The technical assistance of Mrs. U. Kruse is appreciated.

## References

Alexander G, Hazai I, Grimalt J, Albaigés J (1987) Occurrence and transformation of phyllocladanes in brown coals from Nograd Basin, Hungary. Geochim Cosmochim Acta 51:2065–2073

Bertrand P, Behar F, Durand B (1986) Composition of potential oil from humic coals in relation to their petrographic nature. In: Leythaeuser D, Rullkötter J (eds) Advances in organic geochemistry 1985. Pergamon Journals, Oxford, pp 601–508

Bray EE, Evans ED (1961) Distribution of n-paraffins as a clue to recognition of source beds. Geochim Cosmochim Acta 22:2–15

Burwell BL (1954) The cleavage of ethers. Chem Rev 54:615–685

Choudhury D, Sanyal PK, Banerjee AK (1988) Stepwise fragmentation of coal by $H_2O_2$-trifluoroacetic acid oxidation. Fuel 67:177–181

Dean RA, Whitehead EV (1961) The occurrence of phytane in petroleum. Tetrahedron Lett 21:768–770

Degens ET (1965) Geochemistry of sediments a brief survey. Prentice-Hall, Englewood Cliffs, New Jersey

Deno NC, Greigger BA, Stroud SG (1978) New method for elucidating the structures of coal. Fuel 57:455–459

Dong J-Z, Ouchi K (1988) Structure in Wandoan coal from analysis of mild hydrogenation products: 2. Analysis of first- to third-stage hydrogenolysis products. Fuel 67:541–551

Dong J-Z, Katoh T, Itoh H, Ouchi K (1987) Origin of alkanes in coal extracts and liquefaction products. Fuel 66:1336–1346

Eglinton G, Hamilton RJ (1967) Leaf epicuticular waxes. Science 155:1322–1335

Ehrendorfer F (1983) Übersicht des Pflanzenreiches. In: von Denffer D, Ziegler H, Ehrendorfer F, Bresinsky A (eds) Strasburgers Lehrbuch der Botanik. Gustav Fischer, Jena, pp 758–915

Ensminger A, Van Dorsselaer A, Spyckerelle C, Albrecht P, Ourisson G (1974) Pentacyclic triterpanes of the hopane type as ubiquitous geochemical markers: origin and significance. In: Tissot B, Bienner F (eds) Advances in organic geochemistry 1973. Editions Technip, Paris, pp 245–260
Given PH, Marzec A, Barton WA, Lynch LJ, Gerstein BC (1986) The concept of a mobile or molecular phase within the macromolecular network of coals: A debate. Fuel 65:155–163
Goossens H, de Leeuw JW, Schenk PA, Brassell SC (1984) Tocopherols as likely precursors of pristane in ancient sediments and crude oils. Nature 312:440–442
Green TK, Larsen JW (1984) Coal swelling in binary solvent mixtures: pyridine-chlorobenzene and N,N-dimethylaniline alcohol. Fuel 63:1538–1547
Hagemann HW, Hollerbach A (1980) Relationship between the macropetrographic and organic geochemical compositions of lignites. In: Douglas AG, Maxwell JR (eds) Advances in organic geochemistry 1979. Pergamon Press, Oxford, pp 631–638
Hollerbach A (1979) Vorkommen und Bedeutung von terpenoiden Chemofossilien in Erdölen und Sedimenten. Thesis habil, Universität Aachen
Ishiwatari R, Morinaga S, Simoneit BRT (1985) Alkaline permanganate oxidation of kerogens from Cretaceous black shales thermally altered by diabase intrusions and laboratory simulations. Geochim Cosmochim Acta 49:1825–1835
Jones DM, Douglas AG (1987) Hydrocarbon distributions in crude oil asphaltene pyrolysates. 1. Aliphatic compounds. J Ener Fuels 1:468–476
Jurkiewicz A, Marzec A, Pislewski N (1982) Molecular structure of bituminous coal studied with pulse nuclear magnetic resonance. Fuel 61:647–650
Larsen JW (1981) Coal structure. In: Cooper BR, Petrakis L (eds) Chemistry and physics of coal utilization – 1980. Am Inst Phys, New York, pp 1–27
Livsey A, Douglas AG, Connan J (1984) Diterpenoid hydrocarbons in sediments from an offshore (Labrador) well. Org Geochem 6:73–81
Lucht LM, Peppas NA (1981) Cross-linked macromolecular structure in bituminous coals: theoretical and experimental considerations. In: Cooper BR, Petrakis L (eds) Chemistry and physics in coal utilization. Am Inst Phys, New York, pp 28–48
Mackenzie AS, Brassell SC, Eglinton G, Maxwell JR (1982) Chemical fossils: the geological fate of steroids. Science 217, No. 4559:491–504
Monthioux M, Landais P (1987) Evidence of free but trapped hydrocarbons in coals. Fuel 66:1703–1708
Mudamburi Z, Given PH (1985) Multifacetted study of a Cretaceous coal with algae affinities – II. Composition of liquefacation products. Org Geochem 8:221–231
Mycke B (1985) Chemofissilien aus selektivem Abbau organischer Geopolymere. Thesis, Universität Hamburg
Mycke B, Michaelis W (1986a) Lignin-derived molecular fossils from geological material. Naturwiss 73:731–734
Mycke B, Michaelis W (1986b) Molecular fossils from chemical degradation of macromolecular organic matter. In: Leythaeuser D, Rullkötter J (eds) Advances in organic geochemistry 1985. Pergamon Journals, Oxford, pp 847–858
Nip M, Tegelaar EW, Brinkhuis H, de Leeuw JW, Schenk PA, Holloway PJ (1986) Analysis of recent and fossil plant cuticles by Curie point pyrolysis-gas chromatography and Curie point pyrolysis-gas chromatography-mass spectrometry: recognition of a new, highly aliphatic and resistant biopolymer. In: Leythaeuser D, Rullkötter J (eds) Advances in organic geochemistry 1985. Pergamon Journals, Oxford, pp 769–778
Noble R, Knox J, Alexander R, Kagi R (1985a) Identification of tetracyclic diterpene hydrocarbons in Australian crude oils and sediments. J Chem Soc, Chem Commun:32–33
Noble RA, Alexander R, Kagi RI, Knox J (1985b) Tetracyclic diterpenoid hydrocarbons in some Australian coals, sediments and crude oils. Geochim Cosmochim Acta 49:2141–2147
Ourisson G, Albrecht P, Rohmer M (1979) The hopanoids – the palaeochemistry and biochemistry of a group of natural products. Pure Appl Chem 51:709–729
Philp RP, Saxby JD (1980) Organic geochemistry of coal macerals from the Sydney Basin (Australia). In: Douglas AG, Maxwell JR (eds) Advances in organic geochemistry 1979. Pergamon Press, Oxford, pp 639–651
Philp RP, Gilbert TD, Friedrich J (1981) Bicyclic sesquiterpenoids and diterpenoids in Australian crude oils. Geochim Cosmochim Acta 45:1173–1180
Quirk MM, Wardroper AMK, Wheatley RE, Maxwell JR (1984) Extended hopanoids in peat environments. Chem Geology 42:25–43

Radke M, Willsch H, Leythaeuser D, Teichmüller M (1982) Aromatic components of coal: relation of distribution pattern to rank. Geochim Cosmochim Acta 46:1831–1848
Rohmer M, Dastillung M, Ourisson G (1980) Hopanoids from $C_{30}$ to $C_{35}$ in recent muds – chemical markers for bacterial activity. Naturwiss 67:456–458
Seifert W, Moldowan JM (1980) The effect of thermal stress on source-rock quality as measured by hopane stereochemistry. In: Douglas AG, Maxwell JR (eds) Advances in organic geochemistry 1979. Pergamon Press, Oxford, pp 229–237
Shaw PM, Brassell SC, Assinder DJ, Eglinton G (1988) Stepwise chemical degradations of a UK bituminous coal. Fuel 67:557–564
Simoneit BRT (1977) Diterpenoid compounds and other lipids in deep sea sediments and their geochemical significance. Geochim Cosmochim Acta 41:463–476
Stock LM (1982) The reductive alkylation reaction. Coal Science 1:161–279
Tissot BP, Welte DH (eds) (1984) Petroleum formation and occurrence 2nd ed. Springer, Berlin Heidelberg New York
Thomas BR (1969) Kauri resins – modern and fossil. In: Eglinton G, Murphy MTJ (eds) Organic geochemistry – methods and results. Springer, Berlin Heidelberg New York, pp 599–618
Van Dorsselaer A, Albrecht P, Connan J (1977) Changes in composition of polycyclic alkanes by thermal maturation (Yallourn lignite, Australia). In: Campos R, Goni J (eds) Advances in organic geochemistry 1977. Enadisma, Madrid, pp 53–59
Verheyen TV, Pandolfo AG, Johns RB, Mackay GH (1985) Structural investigations of Australian coals – VI. The effect of rank as elucidated by per trifluoroacetic acid oxidation. Geochim Cosmochim Acta 49:1603–1614

# Catalytic Versus Noncatalytic Degradation of Organic Matter Related to Its Gas Productivity

Y.G. Zhang and X.Z. Feng

## 1 Introduction

In order to understand better the chemical degradation of organic matter during diagenesis, catagenesis and metagenesis, many simulation experiments have been done. Since the seventies, laboratory pyrolysis has been widely used (Barker 1974; Harwood 1977; Peters et al. 1981; Durand et al. 1982; Rohrback et al. 1984), in which the heating temperature was elevated to compensate for the short duration of the laboratory pyrolysis in comparison with geological time. The results of the experiments do resemble some aspects of the natural process, but the discrepancy with the latter is obvious.

An effort has been made to prolong the heatıng time in laboratory to six years (Saxby et al. 1986). Unfortunately this kind of experiment can rarely be duplicated. Winters et al. (1983) stated that hydrous pyrolysis gave better result than anhydrous pyrolysis in respect of its resemblance to the natural process. On the contrary, Comet et al. (1986) and Tannerbaum et al. (1986) found no advantage of hydrous pyrolysis over anhydrous pyrolysis.

More and more scientists emphasize the role of mineral matrix as catalyst in the degradation of geopolymers (Espitalle et al. 1980; Goldstein 1983; Horsfield 1984). Chung and Sackett (1979) discussed the difference in the mechanism of cracking organic matter between catalytic and noncatalytic pyrolysis. Durand et al. (1982) pyrolysed kerogens with zeolite and ferric oxide to produce methane. Tannerbaum et al. (1986) conducted pyrolysis experiments with and without calcite, illite or montmorillonite in which the montmorillonite was found to be the most effective catalyst. However the productivity of methane is considerably low in these experiments, far less than that of butane at 300°C for 1000 hours.

It implies that, for the laboratory experiment, catalysts much more effective than clay minerals are needed to intensively breakdown organic matter in a period much shorter than geological times. Frenkel and Hellre-Kallai (1979) pointed out that clay minerals act like acidic catalysts in geological bodies. In fact, Lee et al. (1981) used aluminum chloride to gasify Green River oil shale kerogen. Metal halides have been widely applied as catalysts for the study of coal gasification, the literature of which was reviewed by Gavin (1982). The catalytic mechanism of metal halides, e.x. zinc chloride and aluminium chloride, was discussed by Taylor and Bell (1980). These compounds act something like a Lewis acid catalyst.

In this paper, the results of pyrolyses of both kerogen and whole rocks with and without metal halides is reported, placing emphasis on hydrocarbon gas productivity to explore its possible application to natural gas geochemistry. Additionally, it provides a tool for estimating the flux of gaseous carbon, especially methane and carbon dioxide, from the lithosphere to the atmosphere. It is believed that only a very small part of the hydrocarbon gas could be retained in the geological bodies. A substantial part of the carbon dioxide would also escape to the atmosphere.

## 2 Materials and Methods

Six immature to low-mature samples whose vitrinite reflectance ranges from 0.45 to 0.71% were collected. Recent sediments were not considered in view of the fact that biogeochemical processes during early diagenesis can not be simulated by laboratory pyrolysis. Among these six samples, two of them were of sapropelic nature, including a Cretaceous shale from the southern Songliao Basin and a Lower-Tertiary oil shale from Maoming Opencut. Four of them were of humic nature, including two lignites, one fusite and one shale. Thus, these samples were widely representative from Type I to Type IV as well as from DOM to concentrated organic matter as coal. The characteristics of the samples are shown in Table 1.

The kerogens (50 mg each) and the pulverized whole rocks (500 mg each) with and without metal halides, were vacuum-dried and sealed in pyrex tubes. The tubes were heated in a precisely controlled furnace at temperatures of 270°C, 300°C, 330°C, 360°C, 390°C, 420°C, 450°C and 550°C for 48 hours. After cooling to ambient temperature, the tubes were put into a breaker which was directly connected with a Hewlett Packard 5880A Gas Chromatograph and **vacuumed** beforehand. Then, the tube was broken and the gas was conducted into the GC via a sample loop. The column used was 3m × 2mm, packed with Porapak Q. the gaseous hydrocarbons and carbon dioxide were detected by both TCD and FID. The solid residues were washed by dilute HCl and distilled water to PH = 7 for elemental analysis.

Additional series of catalytic and noncatalytic pyrolysis experiments were carried out for the carbon isotopic analysis of the produced gaseous constituents which were seperated into $C_1$, $C_2$, $C_3$, $C_4$, $C_5$ and $CO_2$ by GC, and combusted to form $CO_2$ for measurements with a Finnigen MAT 250 mass spectrometer.

**Table 1.** Characteristics of samples

| Sample | Location | Age | TOC (%) | H/C | O/C | $R_o$ (%) | Type |
|---|---|---|---|---|---|---|---|
| (1) Shale | Southern Songliao | K | 5.30 | 1.567 | 0.068 | 0.71 | Sapr. |
| (2) Oil Shale | Maoming | Tr | 12.85 | 1.384 | 0.076 | 0.52 | Sapr. |
| (3) Lignite | Huang County | Tr | 57.60 | 0.898 | 0.196 | 0.52 | Humic |
| (4) Lignite | Pingzhuang | J | 56.09 | 0.794 | 0.276 | 0.45 | Humic |
| (5) Fusite | Pingzhuang | J | 66.68 | 0.624 | 0.164 | 0.48 | Humic |
| (6) Shale | Huang County | Tr | 0.53 | 0.659 | 0.198 | 0.54 | Humic |

# 3 Results and Discussions

## 3.1 Hydrocarbon Gas Productivity

The hydrocarbon gas productivity is expressed in terms of the yield of $C_1$-$C_5$ hydrocarbons in cubic meter normalized against one ton of total organic carbon (TOC). The three seperated series of experimental results are shown in Fig. 1. The result of noncatalytic pyrolysis of the kerogens (on the left) is compared with that of the catalytic pyrolysis of the same kerogens (the center part), as well as the result of the catalytic pyrolysis of the whole rocks (on the right). For the humic samples,

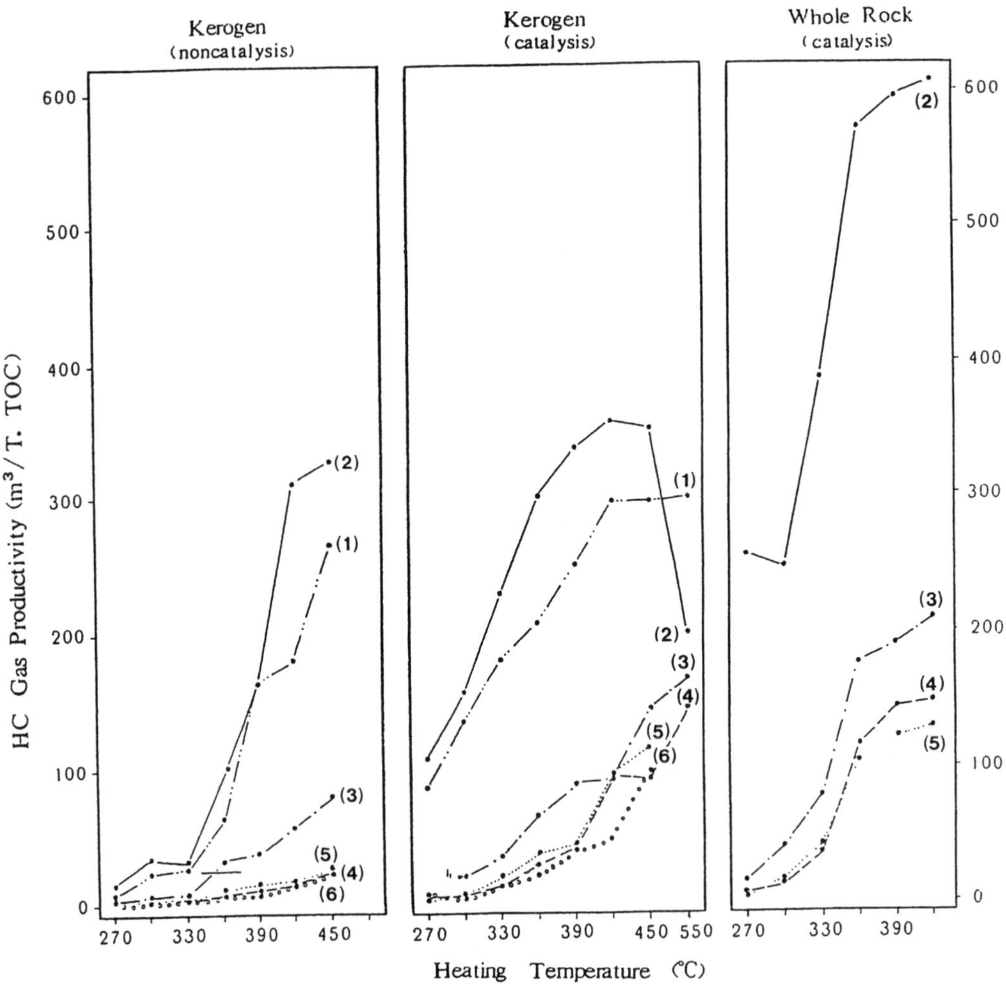


**Fig. 1.** Hydrocarbon gas productivities at various heating temperatures in noncatalytic pyrolyses of the kerogens (left), in catalytic pyrolyses of the kerogens (center) and in catalytic pyrolyses of the whole rocks (right). (*1*) Southern Songliao Shale; (*2*) Maoming Oil Shale; (*3*) Huang County Lignite; (*4*) Pingzhuang Lignite; (*5*) Pingzhuang Fusite; (*6*) Huang County Shale

the HC gas productivities of the catalytic pyrolysis of the whole rocks are much higher than those of the kerogens which in turn are much higher than those of the noncatalytic pyrolysis of the kerogens. Obviously, the catalytic effect played an important role in the degradation of organic matter, causing the gas productivities to be raised by two to five times. As to the sapropelic samples, their HC gas productivities are as high as 605 $m^3$/T-TOC, three or more times higher than those of the humic samples. In this case, catalytic pyrolysis did cause the gas production to occur at lower temperatures than noncatalytic pyrolysis, but the ultimate gas productivities (i.e. at 450°C for 48 hrs.) remain essentially the same for the catalytic and noncatalytic experiments.

For the conversion of liquid hydrocarbons and hydrocarbon side chains of geopolymers to gaseous hydrocarbons, additional hydrogen split off from geopolymers and from substances of higher molecular weight would be needed. In sapropelic samples there are more hydrocarbon chains and less hydrogen available from the kerogen, therefore the ultimate HC gas productivity is limited by the availability of the hydrogen flux from the kerogen (Zhang 1981), for which catalysis can do nothing. What catalysis can do is to accelerate the hydrogenation and breakdown of longer hydrocarbon chain to form shorter ones. Whereas, in humic samples there are fewer hydrocarbon chains and more hydrogen available from the kerogen, so the HC gas productivity is controlled primarily by the availability of hydrocarbon chains and secondly by the catalytic efficiency of the conversion of the latter to gas.

When the Maoming oil shale was heated to 550°C, the gas productivity was much lower than that at 450°C. Such a phenomenon was also observed by Harwood (1977) when a Type I kerogen was heated at temperatures higher than 450°C. It seems that structural condensation of hydrocarbons may take place above 450°C. This is one of the reasons why pyrolysis experiments at temperatures higher than 450°C are not recommended.

All the HC gas productivities of the whole rocks are higher than those of the kerogens in terms of per unit weight of organic carbon. Considerable amounts of acid-soluble substances which possess high gas-generative potential might be lost during the isolation of kerogen with HCl-HF acid treatment. The mineral matrix remaining in the rocks may further promote the catalysis. So, it is preferable to work on whole rock rather than on kerogen.

The increments of methane productivity against temperature intervals are shown on Fig. 2. The disadvantage of noncatalytic pyrolysis can be readily discerned, as the increment grows higher and higher even when the organic matter approached the stage of semianthracite (420°C) or even entered the stage of anthracite. This is not the case in the natural geological process. The distribution of methane increments are more reasonable in catalytic pyrolysis of the whole rocks. The methane increment peaks at the temperature interval from 330°C to 360°C when the H/C atomic ratio decreases from 0.45 to 0.35 for the Pingzhuang lignite and from 0.58 to 0.46 for the Maoming oil shale. It corresponds to the evolutionary stage of just exceeding "oil window". Therefore the result of catalytic pyrolysis is much closer to the actual process of catagenesis and metagenesis which in turn might involve the catalytic effect of mineral matrix.

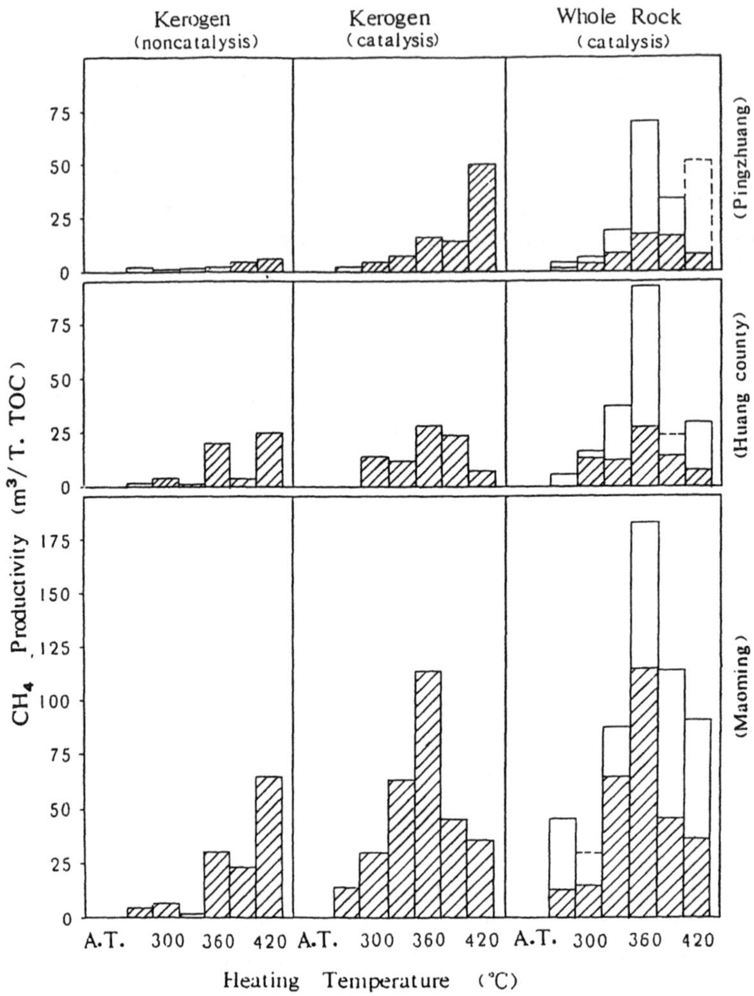


**Fig. 2.** Distributions of the increments of hydrocarbon gas productivity among various heating temperature intervals in noncatalytic pyrolyses of the kerogens (left), in catalytic pyrolyses of the kerogens (center) and in catalytic pyrolyses of the whole rocks (right). The upper two are lignites and the lower one is an oil shale. A.T. denotes ambient temperature

The ultimate HC gas productivities of the humic samples are 130 to 208 $m^3$/T-TOC in catalytic pyrolysis. This figure may be obtained by noncatalytic pyrolysis too at higher heating temperatures (550°C to 600°C), when organic matter already becomes anthracite, but the result is likely to be false as much of the water derived from kerogen is dissociated and reacts with the carbon residue to form methane and carbon dioxide (so-called town gas reaction). This is another reason why pyrolysis experiments at temperatures higher than 450°C are not recommended. It is believed that the temperature for the formation of thermogenic natural gas does not exceed 300°C in geological bodies.

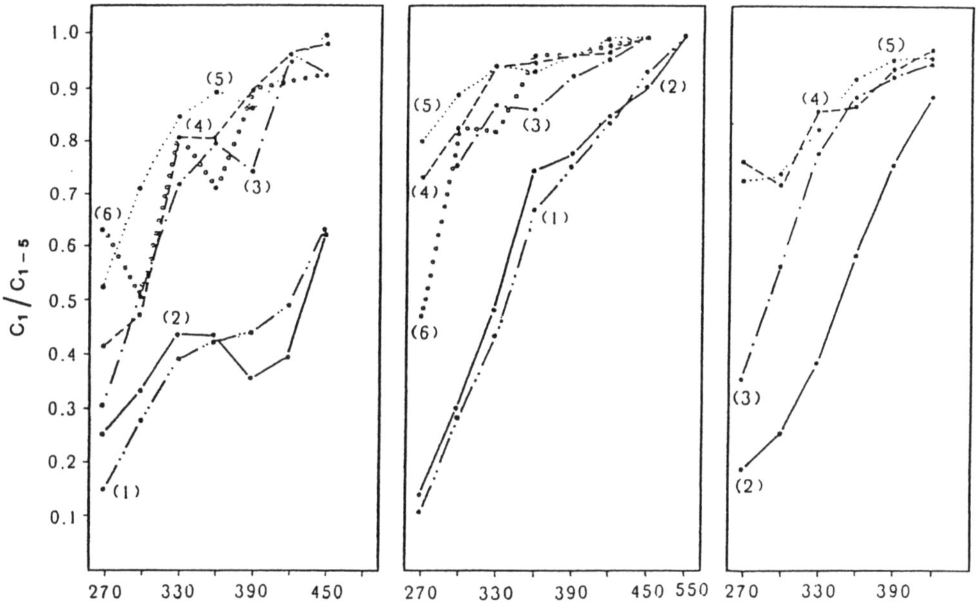


**Fig. 3.** Variation of the wetness of the gas produced in noncatalytic pyrolyses of the kerogens (left), in catalytic pyrolyses of the kerogens (center) and in catalytic pyrolyses of the whole rocks (right). The numbers stand for the same samples as Fig. 1

The wetness indices of the gases expressed in terms of $C_1/C_{1\text{-}5}$ are shown in Fig. 3.

The gases produced in noncatalytic pyrolysis are too wet as compared with natural gases, especially for the humic samples. The situation of catalytic pyrolysis is more plausible although the wetness of gases is still higher than that of natural gases. This could be so interpreted – heavier hydrocarbons ($C_2$–$C_5$) could be retained and left behind in rocks during gas migration, resulting in lower wetness of natural gases.

The hydrocarbon and carbon dioxide gas productivities of the whole rocks during catalytic pyrolysis and the H/C atomic ratios of the residues after heating are shown in Table 2.

It is noticeable that the ultimate HC gas productivity of the fusite pertaining to Type IV amounts to 125.32 $m^3$/T-TOC. It is actually not "Dead Carbon" as Tissot (1984) put it.

The release of large amounts of carbon dioxide occurred even at a heating temperature of 270°C for all the four samples and essentially flattened out at 360°C for the fusite. The fluctuation of the carbon dioxide productivity is larger than those of gaseous hydrocarbons. In spite of the larger analytical error, the ultimate carbon dioxide productivities are around 150 $m^3$/T-TOC for all the samples regardless of their humic or sapropelic nature.

**Table 2.** Gas Productivities of the whole rocks in catalytic pyrolysis and the elemental compositions of the residues after heating for 48 hours

| Samples | Heating Temperature (°C) | Atomic Ratio | | Gas Productivity ($m^3$/T-TOC) | | | | | | |
|---|---|---|---|---|---|---|---|---|---|---|
| | | H/C | O/C | $C_1$ | $C_2$ | $C_3$ | $C_4$ | $C_5$ | Total HC | $CO_2$ |
| Maoming Oil Shale | 270 | 0.80 | 0.17 | 48.09 | 19.38 | 41.25 | 90.43 | 58.44 | 257.59 | 69.96 |
| | 300 | 0.62 | 0.09 | 62.18 | 26.07 | 36.96 | 75.02 | 43.97 | 243.74 | 51.21 |
| | 330 | 0.58 | 0.077 | 149.88 | 47.39 | 57.90 | 90.43 | 40.47 | 386.07 | 50.19 |
| | 360 | 0.46 | 0.072 | 333.46 | 75.33 | 84.75 | 67.32 | 10.35 | 571.21 | 64.05 |
| | 390 | 0.47 | 0.05 | 449.73 | 86.30 | 47.08 | 9.42 | 0.03 | 592.53 | 87.16 |
| | 420 | 0.39 | 0.047 | 534.55 | 65.91 | 4.20 | 0.06 | ---- | 604.75 | 150.43 |
| Huang County Lignite | 270 | 0.54 | 0.12 | 5.02 | 1.98 | 2.50 | 3.44 | 1.15 | 14.08 | 53.40 |
| | 300 | 0.50 | 0.11 | 21.82 | 7.22 | 4.71 | 4.17 | 0.94 | 38.86 | 76.79 |
| | 330 | 0.45 | 0.11 | 59.24 | 9.13 | 5.02 | 2.45 | 0.24 | 76.08 | 70.94 |
| | 360 | 0.43 | 0.11 | 154.67 | 17.05 | 3.19 | 0.14 | --- | 175.05 | 105.61 |
| | 390 | 0.41 | 0.098 | 168.85 | 11.67 | 0.61 | 0.02 | --- | 181.15 | 123.98 |
| | 420 | 0.32 | 0.06 | 199.25 | 9.18 | --- | --- | --- | 208.44 | 135.83 |
| Pingzhuang Lignite | 270 | 0.50 | 0.14 | 3.44 | 0.23 | 0.50 | 0.24 | 0.06 | 4.48 | 113.85 |
| | 300 | 0.46 | 0.11 | 9.57 | 1.78 | 1.25 | 0.49 | 0.08 | 13.18 | 125.83 |
| | 330 | 0.45 | 0.089 | 29.74 | 2.82 | 1.68 | 0.29 | 0.23 | 34.75 | 115.01 |
| | 360 | 0.35 | 0.09 | 105.65 | 7.49 | 1.73 | 0.07 | 0.02 | 114.96 | 148.44 |
| | 390 | 0.33 | 0.062 | 134.28 | 8.00 | 0.62 | 0.02 | 0.01 | 143.02 | 159.64 |
| | 420 | 0.28 | 0.08 | 141.99 | 4.40 | 0.10 | --- | --- | 146.50 | 174.08 |
| Pingzhuang Fusite | 270 | 0.51 | 0.11 | 2.70 | 0.18 | 0.49 | 0.28 | 0.05 | 3.70 | 120.39 |
| | 300 | 0.44 | 0.096 | 10.25 | 1.47 | 1.56 | 0.74 | 0.09 | 14.13 | 160.93 |
| | 330 | 0.45 | 0.091 | 33.04 | 3.52 | 2.59 | 0.90 | 0.01 | 40.10 | 135.99 |
| | 360 | 0.38 | 0.065 | 96.42 | 7.27 | 1.09 | 0.03 | --- | 104.81 | 189.47 |
| | 390 | 0.36 | 0.055 | 115.45 | 5.37 | 0.24 | 0.01 | --- | 121.06 | 207.17 |
| | 420 | 0.33 | 0.032 | 125.32 | 4.24 | 0.10 | --- | --- | 129.66 | 184.49 |

## 3.2 Maturation Stages

It is quite surprising that the maturation of the kerogens in pyrolysis was essentially not affected by the catalyst. The elemental compositions, i.e. H/C and O/C atomic ratios, of the kerogen residues of the sapropelic shale from the southern Songliao Basin and the Pingzhuang lignite after heating with and without catalyst are shown in Fig. 4. The catalysis can promote the hydrogen transfer from one soluble constituent to another soluble constituent and the degradation of the latter, but cannot promote the disintegration of kerogen and even the hydrogen transfer from kerogen to soluble constituents. The kerogen must have a more rigid framework than expected.

It has been widely accepted that sapropelic substances tend to generate wetter hydrocarbon gas and humic substances tend to generate drier hydrocarbon gas. It is therefore quite surprising to find through the catalytic experiments that the wetness of hydrocarbon gas derived from organic matter is controlled by the H/C atomic ratio of organic matter, rather than by their sapropelic or humic nature. This point could be readily discerned from Fig. 5, as all the data points follow the same trend regardless of sapropelic or humic samples.

In Fig. 5, there is a turning point near H/C 0.35. Below this point the gases produced are wet, with $C_1/C_{1-5} < 0.93$. Therefore H/C 0.35 could be taken as a

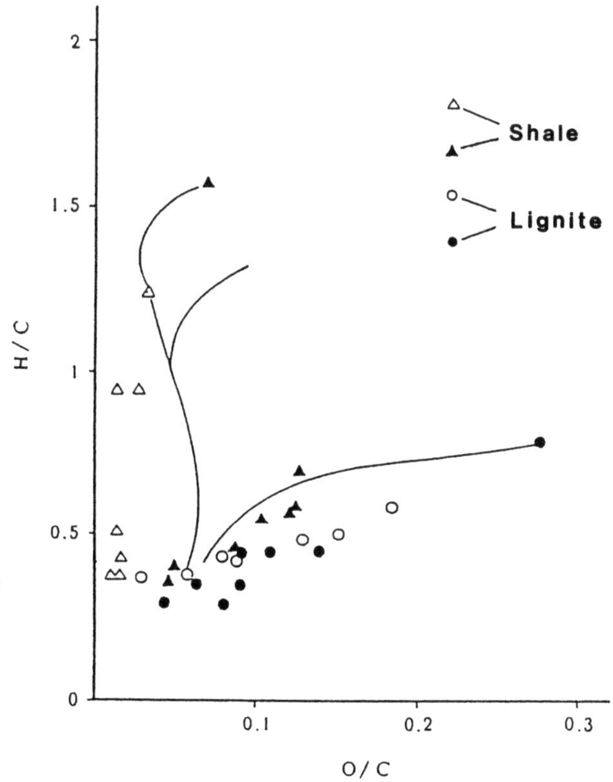


**Fig. 4.** Variations of the *H/C* and *O/C* atomic ratios of the kerogens from the southern Songliao sapropelic shale and the Pingzhuang lignite in noncatalytic pyrolyses (*open symbols*) and in catalytic pyrolyses (*solid symbols*)

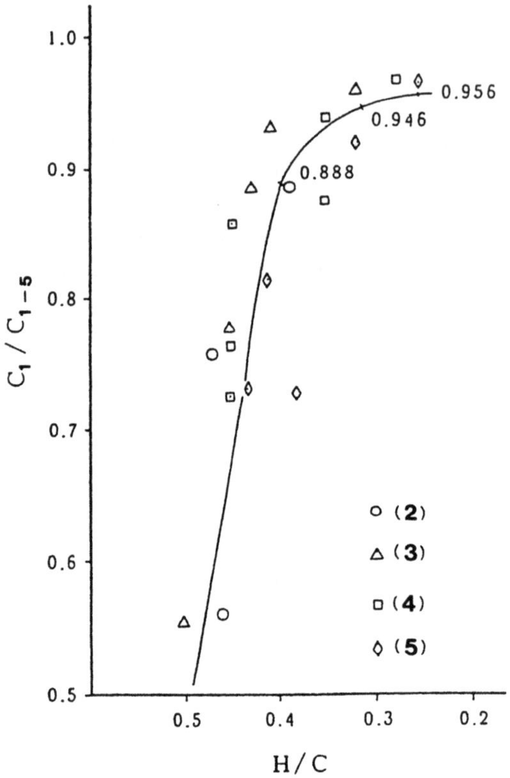


**Fig. 5.** Relationship between the wetness index of the pyrolytic gases and the H/C atomic ratios of the residues in catalytic experiments of the whole rocks. The numbers stand for the same samples as Fig. 1

maturity interface between the wet gas stage and the dry gas stage. It should be noted that natural gas would be somewhat drier than the gas generated in vitro owing to migration. It is found that, when H/C> 0.52, the gases are very wet with $C_1/C_{1\text{-}5} < 0.5$ and the pyrolysis products are virtually light oil and condensate. When H/C< 0.52, the pyrolysis products can be regarded as mainly wet gas. Thus, H/C 0.52 could be taken as a maturity interface between the light oil-condensate stage and the wet gas stage, as well as the interface between catagenesis and metagenesis beyond which the chemical difference between sapropelic and humic substances becomes hardly recognizable.

The HC gas productivity curves of the whole rocks tend to be flattened out at heating temperature of 420°C (Fig. 1). And the wetness curve on Fig. 5 tends to be flattened out near H/C 0.25. Taking these two facts into consideration, it is reasonable to assume that the generation of hydrocarbon would be ceased at H/C 0.25. This would be the lower limit of hydrocarbon gas generation beyond which anthracite becomes superanthracite. This concept coincides with the knowledge of the aromaticity of organic matter which reaches 100% when H/C approaches 0.25 (Fig. 6, after Van Krevelen, 1981). The carbon atom of polyaromatic structures can normally contribute little if any to the generation of hydrocarbons except in the case of the presence of an external hydrogen source and intensive hydrogenation at elevated temperatures.

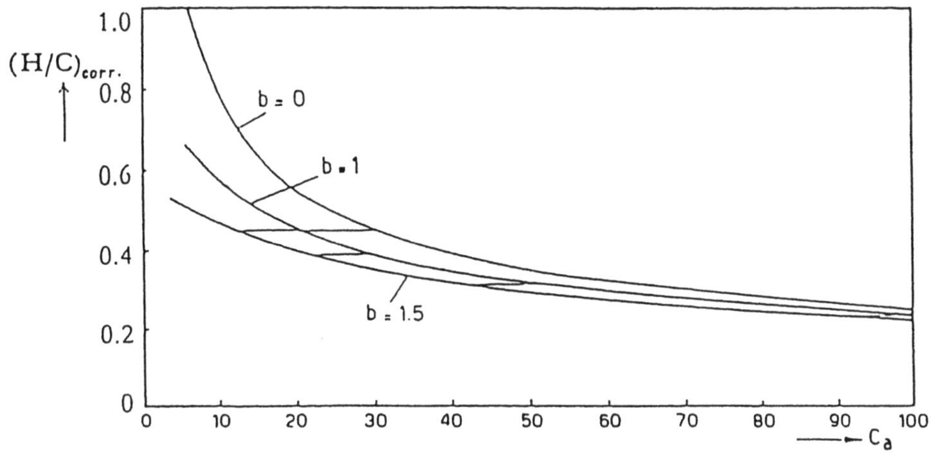


**Fig. 6.** Relationship between the aromaticity(Ca) and the H/C atomic ratio of organic matter(after Van Krevelen, 1981)

### 3.3 Carbon Isotopes

The carbon isotopic composition of the kerogen samples expressed in per mil PDB ranges widely, $\delta^{13}C = -30.33‰$ for the sapropelic shale of southern Songliao Basin, -26.40‰ for the lignite of Huang County, -24.20‰ for the Pingzhuang lignite and -20.18‰ for the Maoming oil shale. They became only slightly heavier (1‰ to 2‰) when heated at various temperatures with a catalyst. These data agree with the statement of Galimov (1980), but conflict with those of the noncatalytic pyrolysis of recent sediments (Peters et al. 1981).

No regular change of carbon isotopic composition of the $C_1$-$C_5$ gaseous hydrocarbons could be observed in the noncatalytic experiments reported here, and that of McCarty and Felbeck (1986). On the other hand, carbon isotopic composition did change regularily in catalytic experiments. This is another important advantage of the catalytic experiment. The isotopic data of the latter are listed in Table 3.

It is interesting that the differences in carbon isotopic composition between the residual kerogen ($\delta^{13}C_0$) and the individual gaseous hydrocarbons produced ($\delta^{13}C_1$, $\delta^{13}C_2$, $\delta^{13}C_3$, $\delta^{13}C_4$) are dependent on the H/C atomic ratio of the residual kerogen (Fig. 7), except during the early maturation stage of the coals when methoxyl group rich in $^{13}C$ was preferentially liberated (Smith et al. 1985). The relationships between $\delta^{13}C_1$, $\delta^{13}C_2$, $\delta^{13}C_3$, $\delta^{13}C_4$ and $\delta^{13}C_0$ ($\delta^{13}C_1$-$\delta^{13}C_0$ . . .) were employed to calculate the relationships between $\delta^{13}C_2$, $\delta^{13}C_3$, $\delta^{13}C_4$ and $\delta^{13}C_1$ ($\delta^{13}C_2$-$\delta^{13}C_1$, . . .). Based on these data, a maturation diagram for natural gas is established (Fig. 8). In fact, the carbon isotopic differences between ethane, propane, butane and methane were used by James (1983) to estimate LOM (Level of Metamorphism) based on field data and by Sundberg and Bonnet (1983) to calculate the paleotemperature of the formation of natural gas based on theoretical grounds. But this was the first time that these relationships are confirmed by experiment, i.e. catalytic pyrolysis.

**Table 3.** Carbon isotopic compositions of the gas constituents and the kerogens in the catalytic pyrolyses for 48 hours

| Samples | Heating Temper-ature °C | H/C after Heating | $\delta^{13}C$ (‰ PDB) Kerogen | $C_1$ | $C_2$ | $C_3$ | $C_4$ | $C_5$ | $CO_2$ |
|---|---|---|---|---|---|---|---|---|---|
| | 270 | 0.70 | -28.85 | -57.83 | -41.50 | -34.37 | -33.38 | -32.59 | -20.63 |
| Sapropelic | 300 | 0.58 | -29.19 | -55.33 | -36.60 | -33.01 | -32.31 | -31.59 | -22.57 |
| Shale | 330 | 0.59 | -28.88 | -47.04 | -33.01 | -32.82 | -30.98 | -30.46 | -22.00 |
| (Songliao) | 360 | 0.55 | -29.44 | -41.05 | -33.02 | -30.86 | -28.53 | -29.14 | -23.44 |
| Lignite | 270 | 0.54 | -26.28 | -38.47 | -30.21 | -28.21 | -25.21 | – | -20.03 |
| (Huang | 300 | 0.50 | -25.68 | -45.55 | -33.53 | -25.42 | -23.81 | – | -22.18 |
| County) | 330 | 0.45 | -25.44 | -37.79 | -28.61 | -24.01 | -21.01 | – | -23.28 |
| | 360 | 0.43 | -26.35 | -34.80 | -25.60 | -18.78 | – | – | -23.31 |
| Lignite | 270 | 0.50 | -23.70 | -28.11 | -30.26 | -24.85 | – | – | -17.15 |
| (Pingzhuang) | 300 | 0.46 | -23.66 | -30.28 | n.d. | -21.89 | -21.85 | – | -18.27 |
| | 330 | 0.45 | -23.20 | -34.79 | -26.19 | -23.47 | – | – | -20.33 |
| | 360 | 0.35 | -23.54 | -31.42 | -23.72 | -19.85 | – | – | -22.05 |
| Oil Shale | 270 | 0.80 | -19.83 | -45.90 | -31.32 | -24.66 | -20.19 | -20.83 | -13.89 |
| (Maoming) | 300 | 0.62 | -18.80 | -43.79 | -25.88 | -22.03 | -23.09 | -19.77 | -16.15 |
| | 330 | 0.58 | -18.49 | -38.26 | -24.03 | -21.99 | -20.27 | -19.71 | -16.72 |
| | 360 | 0.46 | -18.49 | -31.28 | -22.58 | -20.28 | -18.41 | – | -16.47 |

The maturity of a gas on the maturation diagram suggested here (the left side of Fig. 8) is expressed in terms of the H/C atomic ratio of its source material when it was formed. For comparison, that proposed by Sundberg and Bonnet (1983) is shown on the right side of Fig. 8. Through practice, it is found to be more reasonable to estimate in terms of H/C ratio instead of paleotemperature which is sometimes too high to be acceptable. The application of this maturation diagram of natural gas will be presented in another paper.

The carbon isotopic composition of the pyrolytical carbon dioxide became a bit lighter as the heating temperature went up, in the opposite direction of that of the kerogen. The range of variation of the $\delta^{13}C_{CO_2}$ is 1 to 5‰, comparable to that of the $\delta^{13}C_{kerogen}$ which is 1 to 2‰. The $\delta^{13}C_{CO_2}$ is constantly heavier than the $\delta^{13}C_{kerogen}$, but not necessarily heavier than that of propane or butane. It seems that the fractionation of carbon isotopes among individual hydrocarbons bears no definite relation to $\delta^{13}C_{CO_2}$.

It is noteworthy that when the pyrolysis temperature was raised above 360°C, i.e. 390°C, 420°C, 450°C, the carbon isotopic compositions of ethane and propane became heavier than what was expected, even sometimes much heavier than those of the kerogens and carbon dioxide. No regular change of the differences between $\delta^{13}C_2$, $\delta^{13}C_3$ and $\delta^{13}C_1$ could be observed any more. The reason is unclear, probably due to the speedup of the breakdown of ethane and propane.

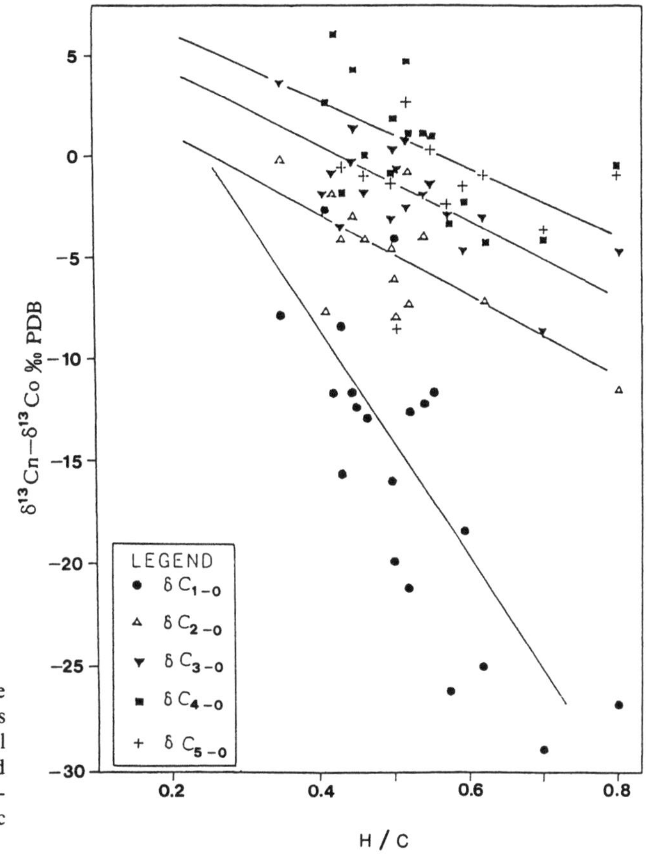


**Fig. 7.** Depedence of the carbon isotopic differences between various pyrolytical gaseous hydrocarbons and the source kerogens ($\delta^{13}C_n$-$\delta^{13}C_0$) on the H/C atomic ratio of the latter

# 4 Conclusions

1. The results of the catalytic pyrolysis simulation experiment resemble the natural geological process much closer than the noncatalytic experiment in respect to the productivity and wetness of the hydrocarbon gas produced and their distributions among various maturation stages, as well as the carbon isotopic fractionation among various gaseous hydrocarbons.
2. Catalysis could promote the hydrogenation and breakdown of longer hydrocarbon chains to form shorter hydrocarbon chains, as well as the split off of oxygen-containing functional groups to form lower molecular weight substances, but it can not promote the disintegration of the kerogen and hydrogen transfer from kerogen to soluble constituents.
3. The wetness of hydrocarbon gas is dependent on the H/C atomic ratio of its source material rather than the sapropelic or humic nature of the latter.
4. The interfaces between various maturation stages can be set at H/C 0.52 for that between light oil-condensate stage and wet gas stage or catagenesis stage and

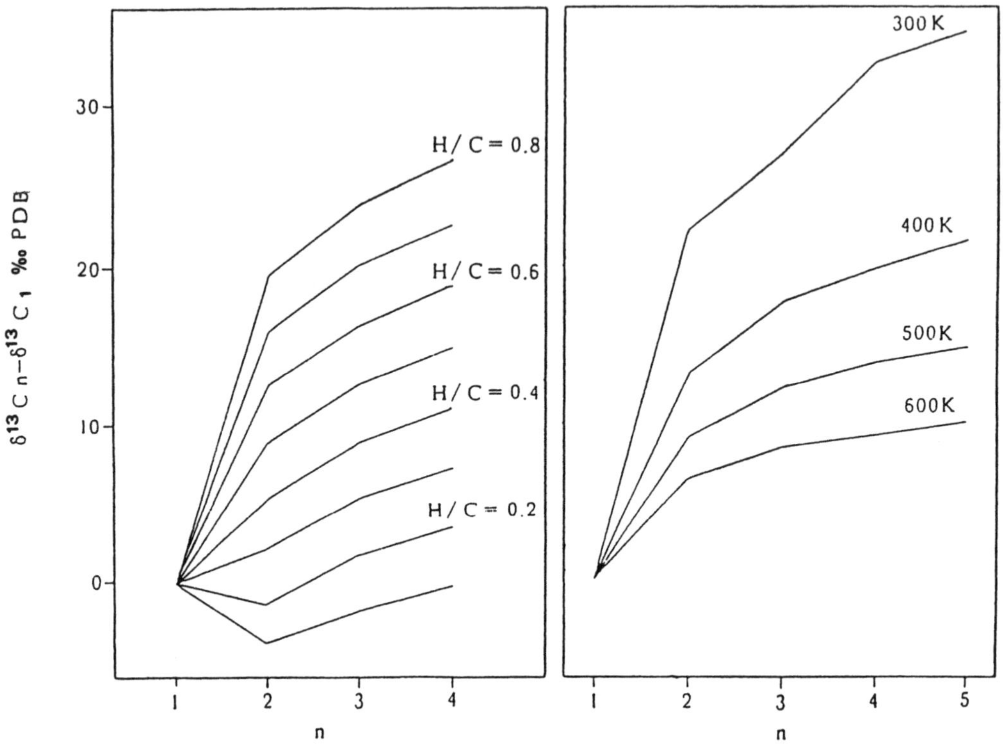


**Fig. 8.** Maturation diagram of natural gas proposed by this paper (left) in comparison with the paleothermometric diagram advanced by Sundberg and Bonnet, 1983 (right)

metagenesis stage, at H/C 0.35 for that between wet gas stage and dry gas stage, and at H/C 0.25 for the lower limit of hydrocarbon gas generation.

5. The carbon isotopic fractionation between various gaseous hydrocarbons during catalytic pyrolysis allows a maturation diagram of natural gases to be established in which their maturities are expressed in terms of H/C atomic ratio of their source material when they are formed.
6. The ultimate thermogenic hydrocarbon gas productivity was around 600 $m^3$/T-TOC for sapropelic organic matter and 130 to 208 $m^3$/T-TOC for humic organic matter in catalytic pyrolysis, while the ultimate thermogenic carbon dioxide productivity is roughly around 150 $m^3$/T-TOC for both sapropelic and humic samples. Thus, as the subsidence of sedimentary basins proceeds, a significant part of sedimentary organic carbon is going back to the atmosphere and hydrosphere as thermogenic gases which could account for 10 to 40% of the original organic carbon.

## References

Barker C (1974) Pyrolysis Techniques For Source Rock Evaluation. AAPG Bull 58:2349-2361
Chung HM, Sackett WM (1979) Use of stable carbon isotope composition of pyrolytically derived methane as maturity indices for carbonaceous materials. Geoch et Cosmoch Acta 43:1979-1988
Comet PA, McEvoy J, Giger W, Douglas AG (1986) Hydrous and Anhydrous Pyrolysis of DSDP Leg 75 Kerogens – A Comparative Study Using a Biological Marker Approach. Org Geochem 9:171-182
Davis JB, Stanley JP (1982) Catalytic Effect of Smectite Clays in Hydrocarbon Generation Revealed by Pyrolysis-gas Chromatography. J Anal Appl Pyrolysis 4:227-240
Durand SC, Boulet R, Durand B (1982) Formation of Methane and Hydrocarbons by Pyrolysis of Immature Kerogens. Geochim Cosmochim Acta, 46:1193-1202
Espitalie J, Madea M, Tissot B (1980) Role of Mineral Matrix in Kerogen Pyrolysis: Influence on Petroleum Generation and Migration. AAPG Bull 64:59-66
Frenkel K, Hellre-Kallai L (1979) Aromatization of Ilmonene – A Geochemical Model. Org Geochem 1:3-6
Galimov EM (1980) $^{13}C/^{12}C$ in Kerogen. In: Kerogen, Insoluble Organic Matter From Sedimentary Rocks. Technip, Paris, pp 271-299
Gavin DG (1982) Coal Hydrogenation Catalysis. In: Durend B (ed) Catalysis 5 (ed Bond GC, Webb G), R Soc Chem, London, pp 220-272
Goldstein TP (1983) Geocatalytic Reactions in Formation and Maturation of Petroleum. AAPG Bull 67:152-159
Harwood RJ (1977) Oil and Gas Generation by Laboratory Pyrolysis of Kerogen. AAPG Bull 61:2082-2102
Horsfield B (1984) Pyrolysis Studies and Petroleum Exploration. In: (ed Brooks J, Welet D) Advances in Petroleum Geochemistry, 1 pp 247-198
James AT (1983) Correlation of Natural Gas By Use of Carbon Isotopic Distribution Between Hydrocarbon Components. AAPG Bull 67:1176-1191
Lee C, Starbuck B, Olsen G, Osteryoung R, Bugle RC (1981) Gasification of Green River Oil Shale Kerogen. Fuel 60:967-970
McCarty HB, Felback Jr GT (1986) High Temperature Simulation of Petroleum Formation-IV Stable Carbon Isotope Studies of Gaseous Hydrocarbons. Org Geochem 9:183-192
Peters KE, Rohrback BG, Kaplan IR (1981) Geochemistry of Artificially Heated Humic and Sapropelic Sediments-I: Protokerogen. AAPG Bull 65:688-705
Rohrback BG, Peters KE, Kaplan IR (1984) Geochemistry of Artificially Heated Humic and Sapropelic Sediment-II: Oil and Gas Generation. AAPG Bull 68:961-970
Saxby JD, Bennett AJR, Corcoran JF, Lambert DE, Riley KW (1986) Petroleum Generation: Simulation Over Six Years of Hydrocarbon Formation From Torbanite and Brown Coal in A Subsiding Basin. Org Geochem 9:69-82
Smith JW, Rigby D, Gould KW, Hart G (1985) An Isotopic Study of Hydrocarbon Generation Processes. Org Geochem 5:341-347
Sundberg KR, Bonnett CR (1983) Carbon Isotopic Paleothermometry of Natural Gas. In: (ed Bjory M et al.) "Advances in Organic Geochemistry", 1981, Wiley, pp 769-774
Tannerbaum E, Huizinga BJ, Kaplan IR (1986) Role of Minerals in Thermal Alteration of Organic Matter-II: A Material Balance. AAPG Bull 70:1156-1165
Taylor ND, Bell AT (1980) Effects of Lewis Acid Catalysts on the Cleavage of Aliphatic and Aryl-aryl Linkages in Coal-related Structures. Fuel 59:499-510
Tissot BP (1984) Recent Advances in Petroleum Geochemistry, Applied To Hydrocarbon Exploration. AAPG Bull 68:545-563
Van Krevelen (1981) Coal, Typology-Chemistry-Physics-Constitation. Elsevier, Amsterdam, 514 pp
Winters JC, Williams JA, Lewan MD (1983) A Laboratory Study of Petroleum Generation By Hydrous Pyrolysis. In: (eds Bjoroy et al.) "Advances in Organic Geochemistry", 1981, Wiley, Chichester, pp 524-533
Zhang Yi Gang (1981) An Approach To Estimate the Evolution of Organic Matter and Its Application To Resource Evaluation. Chem Geol 34:165-177

# Hydrofluoric Acid Induced Alterations of Sedimentary Humic Acids

R.I. HADDAD, B.G. ROHRBACH, and I.R. KAPLAN

## 1 Introduction

For more than a decade geochemists have been attempting to understand the mechanisms and processes by which biochemically produced organic matter is degraded, transported and ultimately incorporated into the geologic record. It is now evident that the large complex organic macro-molecules known as fulvic acids, humic acids and kerogens are major pieces in this geochemical puzzle. Characterization of these pieces has progressed slowly but steadily during the past three decade. Unfortunately, these compounds are operationally defined and standard concentration and separation procedures are necessary to ensure reliable and comparable results. Recently Thurman and Malcolm (1981) proposed a standardized methodology for the isolation of aquatic humic matter using non-ionic macroporous sorbents (e.g. amberlite XAD-2). However, there still does not appear to be a consensus reached on methodology for dealing with sedimentary humic substances (cf. Hayes and Swift 1978; Aiken et al. 1985).

One of the obvious problems in dealing with aquatic and sedimentary humic substances is the need to isolate the humic matter from the inorganic (carbonate/silicate/oxide) matrix. A method used at UCLA in laboratory simulated geothermal maturation studies on recent sediments (Peters 1978; Rohrback 1979) involved HCI/HF digestion of the carbonate/silicate matrix prior to the extraction of humic acid, fulvic acid and kerogen. An important consideration in these digestions was the effect on the organic matter present in the sample of increased temperatures resulting from the exothermic reaction between the silicate matrix and the 60% hydrofluoric acid used to digest the unconsolidated sediment matrix. Such a procedurally-induced effect, depending on its magnitude, could have a significant impact on any interpretations and/or conclusions derived from the data.

Since the chemical composition of the source material (i.e., bulk organic composition of the sediment) is important in determining the types of diagenetic and catagenetic products (Tissot et al. 1974; Tissot and Welte 1978; Ishiwatari et al. 1978; Peters 1978; Rohrback 1979), it was decided to evaluate the potential for procedurally inducing humic alterations on sediment collected from two different environments on which studies had been conducted at UCLA. These are:

1. Tanner Basin – an outer continental shelf basin, characterized by a high marine organic content (4% TOC), derived primarily from zooplankton, and
2. Staten Island Peat – a fresh water swamp, characterized by a high terrigenous organic content (20% TOC) derived primarily from tules and reeds (Stuermer et al. 1978).

## 2 Methods

Humic and fulvic acids were batch extracted with 0.2 N NaOH (prepared from sodium metal) from whole freeze-dried sediments under nitrogen for a minimum of two days until the extract was straw yellow in color. The humic acid was separated from the fulvic fraction by precipitation at pH $< 1$, purified, rinsed with slightly acidic water to remove salt, and freeze dried (Fig. 1). The isolated humic acid from

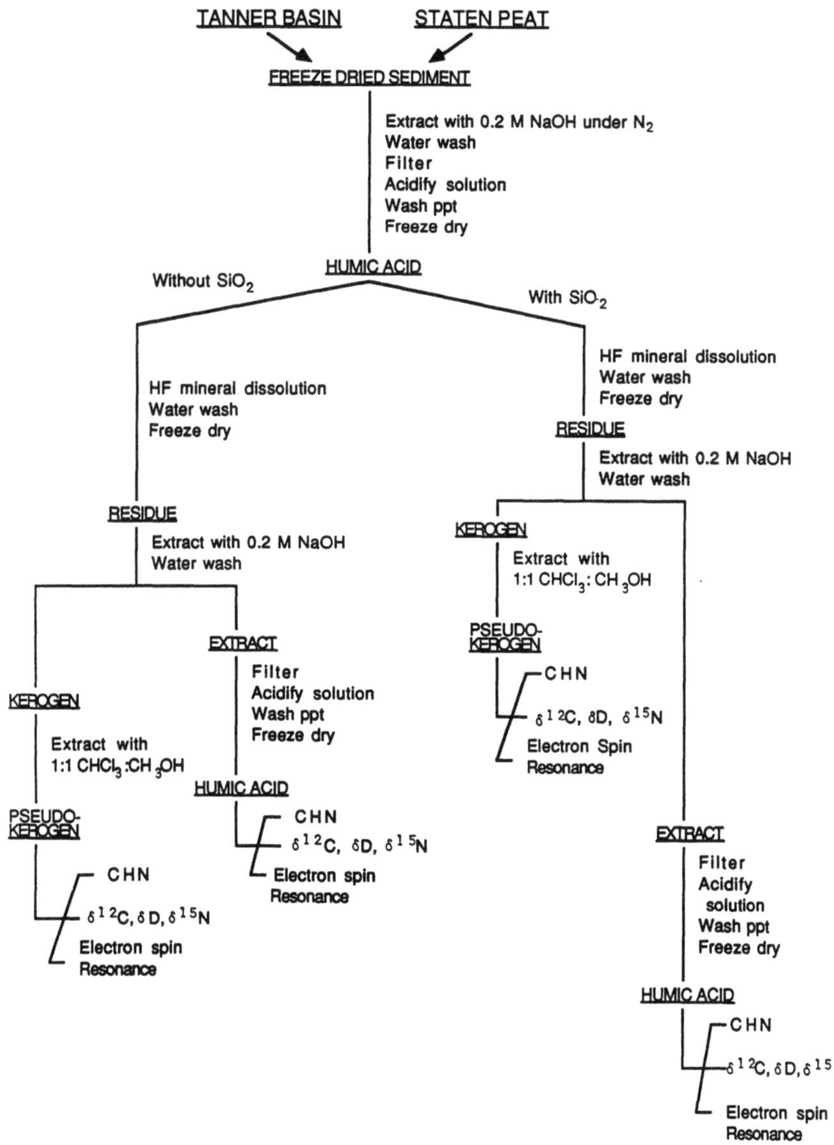


**Fig. 1.** Flow chart of methodology used

each type environment was next statistically separated into two sub-samples (Fig. 1). To one of each pair of the subsamples (TBHA/S, SPHA/S), 40–140 mesh silica gel (Baker Analyzed Reagent) activated at 235°C for 16 hours was added in a 5:1 (silica gel:humic acid) ratio. This was done to test the possibility of procedurally induced humic acid alteration as a result of the exothermic HF-Silica reaction. Nothing was added to the remaining subsample of each pair (TBHA, SPHA).

Sixty percent HF (Baker Analyzed Reagent) was slowly added (dropwise) to nalgene centrifuge bottles containing the samples. Each mixture was then agitated for 20 minutes, separated by centrifugation and the supernatant removed. This process was repeated until the addition of HF produced no further reaction. The samples were then washed 3 to 5 times with distilled water and re-extracted with 0.2 N NaOH. The extract (TBHA-HF, TBHA/S-HF, SPHA-HF, SPHA/S-HF) was again decanted off, acidified, washed, purified, freeze dried and set aside for subsequent characterization. The residue in each sample was extracted with 1:1 $CHCl_3$:$CH_3OH$ to qualitatively examine the possibility of labile organic matter generation. The remaining residue (TBHA-KF, TBHA/S-KF, SPHA-KF, SPHA/S-KF) operationally defined for this study as pseudo-kerogen after Stuermer et al. (1978) was freeze dried.

The elemental carbon, hydrogen and nitrogen contents of humic and kerogen substances were measured (Tables 1, 2) prior to (TBHA-Std, SPHA-Std) and following HF treatments (TBHA-HF, KF; TBHA/S-HF, KF; SPHA-HF,KF; SPHA/S-HF,KF) using a procedure described in Stuermer et al. (1978) (combustion of the samples at 900°C in the presence of CuO and Ag metal). Water vapor produced by the combustion was converted to $H_2$ gas by reaction with uranium turnings at 800°C. $CO_2$, $N_2$ and $H_2$ were purified, measured volumetrically and sealed in Pyrex tubes for subsequent isotopic analysis.

Stable isotope analyses were carried out by methods described in Kaplan et al. (1970). Nitrogen isotope analyses were carried out following methods described by Cline and Kaplan (1975). All isotope ratio values are given in the standard "$\delta$" notation (Tables 3, 4). The isotope standards used for carbon, nitrogen and hydrogen were Pee Dee Belemnite (PDB), atmospheric nitrogen and standard mean ocean water (SMOW) respectively.

Electron spin resonance (ESR) spectroscopy measurements were made (Table 3, 4) using a Varian model E-12 spectrometer, employing a modulation frequency of 3370 KHz and a range of 100 G. Weighed samples were sealed in evacuated quartz tubes for analysis. Strong pitch (Varian Part No. 121573, 0.1% KCl) was used as the external standard.

## 3 Results and Discussion

Examination of the results in Tables 1–4 and Figure 2 show that there are virtually no differences between samples acidified with $SiO_2$ present (TBHA/S, SPHA/S) and samples acidified with $SiO_2$ (TBHA, SPHA). We therefore conclude that the starting humic acid undergoes no significant chemical and/or thermal reactions resulting from the presence of $SiO_2$ during the HF treatment.

**Table 1.** Results of gravimetric and elemental analyses of humic acid and pseudo-kerogen from Tanner Basin

| Sample | Sample Weights | | Ash Free | | | | | | | | Atomic Ratios | | |
|---|---|---|---|---|---|---|---|---|---|---|---|---|---|
| | Initial (g) | Final (g) | Sample Weight | | % Yield | Elemental Composition | | | | % Ash | H/C | N/C | O/C |
| | | | Initial (g) | Final (g) | | %C | %H | %N | %O[a] | | | | |
| TBHA Std | 1.5000 | – | 1.1700 | – | – | 51.5 | 4.6 | 6.5 | 37.4 | 22.0 | 1.072 ±0.011 | 0.108 ±0.013 | 0.545 ±0.110 |
| TBHA HF | – | 0.0638 | – | 0.0602 | 5.2 | 52.0 ±1.5[b] | 4.3 ±0.2 | 6.3 ±0.2 | 34.7 ±4.2 | 5.6 | 0.992 ±0.004 | 0.104 ±0.005 | 0.500 ±0.084 |
| TBHA KF | – | 0.9604 | – | 0.8452 | 72.2 | 50.8 ±4.2 | 5.0 ±0.4 | 5.9 ±0.9 | 34.4 ±6.2 | 12.0 | 1.181 ±0.011 | 0.099 ±0.020 | 0.508 ±0.134 |
| TBHA/S Std | 1.5010 | – | 1.1708 | – | – | 51.5 | 4.6 | 6.5 | 37.4 | 22.0 | 1.072 | 0.108 | 0.545 |
| TBHA/S HF | – | 0.0698 | – | 0.0649 | 5.5 | 53.6 ±1.6 | 4.4 ±0.3 | 6.6 ±0.4 | 33.4 ±3.5 | 7.0 | 0.985 ±0.006 | 0.105 ±0.008 | 0.467 ±0.068 |
| TBHA/S KF | – | 1.0211 | – | 0.8812 | 75.3 | 55.2 | 5.1 | 6.4 | 33.3 | 13.7 | 1.109 | 0.099 | 0.452 |

[a] % Oxygen determined by difference.
[b] Represents 1σ standard error.

**Table 2.** Results of gravimetric and elemental analyses of humic acid and pseudo-kerogen from Staten Island Peat

| Sample | Sample Weights | | Ash Free | | | | | | | | Atomic Ratios | | |
|---|---|---|---|---|---|---|---|---|---|---|---|---|---|
| | | | Sample Weights | | | Elemental Composition | | | | | | | |
| | Initial (g) | Final (g) | Initial (g) | Final (g) | % Yield | %C | %H | %N | %O[a] | % Ash | H/C | N/C | O/C |
| SPHA Std | 2.5000 | - | 2.2275 | - | - | 51.0 | 3.6 | 3.8 | 41.6 | 10.9 | 0.847 | 0.064 | 0.612 |
| SPHA HF | - | 1.4430 | - | 1.4300 | 64.20 | 54.0 | 4.5 | 3.3 | 38.2 | 0.9 | 1.000 | 0.052 | 0.530 |
| SPHA KF | - | 0.0018 | - | 0.0013 | 0.06 | [b] | [b] | [b] | | 27.7 | | | |
| SPHA/S Std | 2.5032 | - | 2.2304 | - | - | 51.0 | 3.6 | 3.8 | 41.6 | 10.9 | 0.847 | 0.064 | 0.612 |
| SPHA/S HF | - | 1.2847 | - | 1.2307 | 55.20 | 54.5 | 4.5 | 3.2 | 37.8 | 4.2 | 0.991 | 0.050 | 0.520 |
| SPHA/S KF | - | 0.0010 | - | 0.0009 | 0.04 | [b] | [b] | [b] | | | | | |

[a] % Oxygen determined by difference.
[b] Insufficient sample for analysis.

**Table 3.** Isotope and ESR data from Tanner Basin samples

| Sample | Mass (mg) C | H | N | O | Isotope (‰) $\delta^{13}C$ | $\delta D$ | $\delta^{15}N$ | ESR Spin density ($\times 10^{16}$) ($g^{-1}$) | Line width (gauss) | g-value |
|---|---|---|---|---|---|---|---|---|---|---|
| TBHA Std | 602.5 | 53.8 | 76.0 | 437.6 | -21.10 | -115.4 | +7.24 | 1.52 | 7.82 | 2.0008 |
| TBHA HF | 31.3 | 2.6 | 3.8 | 22.5 | -20.86 | -106.5 | +6.78 | 1.62 | 4.76 | 2.0040 |
| TBHA KF | 429.4 | 42.3 | 49.9 | 290.7 | -21.22 | -120.6 | +6.83 | 0.672 | 7.48 | 2.0030 |
| TBHA/S Std | 603.0 | 53.9 | 76.1 | 437.9 | -21.10 | -115.4 | +7.24 | 1.52 | 7.82 | 2.0008 |
| TBHA/S HF | 34.8 | 2.8 | 4.3 | 21.7 | -20.74 | -108.6 | +6.00 | 0.667 | 6.46 | 2.0018 |
| TBHA/S KF | 486.4 | 44.9 | 56.4 | 293.4 | -21.56 | -125.5 | +2.37 | 0.505 | 7.48 | 2.0034 |

**Table 4.** Isotope and ESR data from Staten Peat samples

| Sample | Mass (mg) C | H | N | O | Isotope (‰) $\delta^{13}C$ | $\delta D$ | $\delta^{15}N$ | ESR Spin density ($\times 10^{16}$) ($g^{-1}$) | Line width (gauss) | g-value |
|---|---|---|---|---|---|---|---|---|---|---|
| SPHA Std | 1136.0 | 80.2 | 84.6 | 926.6 | -26.41 | -87.1 | +2.70 | 0.147 | 9.01 | 2.0024 |
| SPHA HF | 772.2 | 64.4 | 47.2 | 546.3 | -26.49 | -99.0 | +1.09 | 5.98 | 6.80 | 1.9990 |
| SPHA/S Std | 1137.5 | 100.4 | 84.8 | 927.8 | -26.41 | -87.1 | +2.70 | 0.147 | 9.01 | 2.0024 |
| SPHA/S HF | 670.7 | 35 4 | 39.4 | 465.2 | -26.49 | -98.7 | -0.74 | 5.99 | 6.46 | 2.0013 |

The most significant and surprising results appear to be related to the acidolysis, and more importantly the type of organic matter (i.e. the "source effect"). This can be most clearly seen in the gravimetric analyses (Fig. 2). Acidification of the terrestrial Staten Peat humic acid yielded a 55–65 wt.% recovery of the initial humic acid as operationally defined. Less than 1 wt.% of the initial humic acid was recovered as operationally defined non-soluble pseudo-kerogen. However, acidification of the marine Tanner Basin humic acid extract yielded only a 5–6% recovery as humic acid while the bulk of the material (70–75%) was transformed into operationally-defined non-soluble pseudo-kerogen.

## Tanner Basin

The lost fraction estimated from mass balance calculations accounts for ~20–25% of the initial material (Table 3). These calculations suggest that this fraction may represent acid catalyzed decarboxylation (loss of $CO_2$), loss of other volatile species ($N_2$, $NO_2$, $H_2O$, etc.), and low molecular weight polar compounds such as carboxylic acids, carbohydrates, amino acids, etc. Due to the oxidative environment of

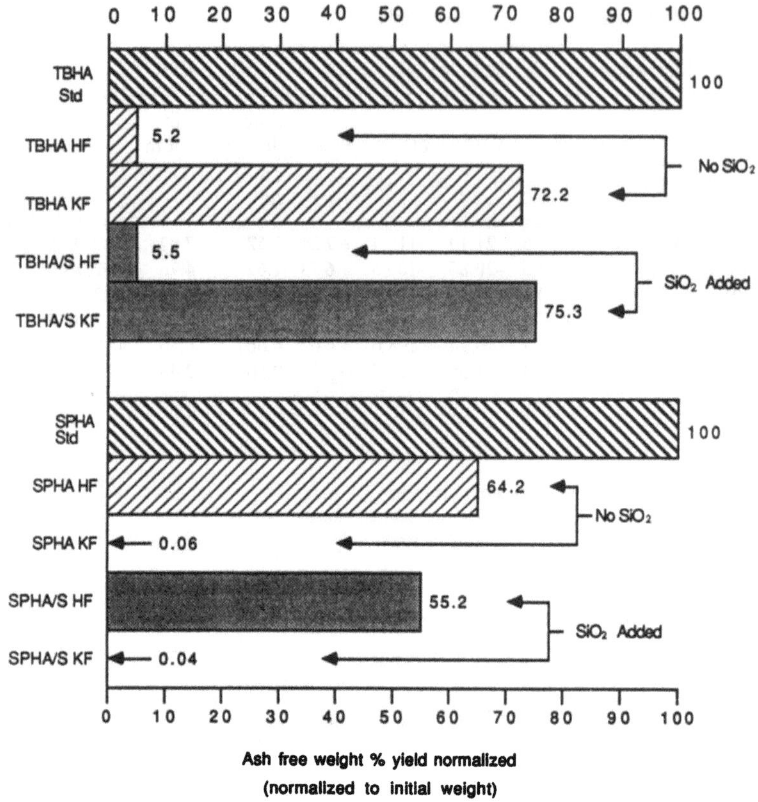


**Fig. 2.** Gravimetric analysis

the reaction it is believed that reduced volatile species (i.e. $CH_4$, $NH_3$) were not evolved in significant quantities. Unfortunately the uncertainty in the elemental oxygen determination makes the lost fraction difficult to quantify. Isotope mass balance calculations, indicating a preferential incorporation of $^{13}C$ into the lost fraction, also support a decarboxylation mechanism (see Volger and Hayes 1980; Abelson and Hoering 1961) or loss of oxidized polar groups.

Elemental comparison of the Tanner Basin humic acid standard, recovered humic acid and pseudo-kerogen have been interrelated as atomic H/C, N/C and O/C ratios (Table 1). The three fractions exhibit similar N/C and O/C atomic ratios, whereas the H/C ratio increased in the pseudo-kerogen fraction (TBHA/S-HF) relative to the initial humic acid standard (TBHA/S-Std). The H/C shifts are statistically significant and suggest an increase in hydrogen in the TBHA/S-KF fractions and a decrease in hydrogen in the TBHA/S-HF fractions. The magnitude of these changes is small and therefore their significance is unclear.

Carbon and hydrogen isotope data suggest a small amount of fractionation during the acidification (Table 3). Lighter values (more depleted in $^{13}C$) are found in the pseudo-kerogen and heavier values in the acid treated humic acid. $\delta D$ values for the TBHA/HF samples become isotopically heavier, whereas the values for TBHA/KF were isotopically depleted in D by 5–10‰. $\delta^{15}N$ shows only small changes (~0.5‰ lighter) except for sample TBHA/S KF which is isotopically lighter by almost 5‰.

The results from ESR spectroscopy (Table 3) show that the g-values range from the value of the starting humic acid (2.0008) to the values of the pseudo-kerogen (2.0030–2.0034). The acidified humic acid shows an intermediate g-value of 2.0011–2.0018. These data suggests an overall increase in the oxygen free-radical concentration.

The line width, indicative of free radical diversity, remains relatively constant between the starting material and the pseudo-kerogen. However, a large decrease in line width is seen in the HF treated humic acids. This is interpreted as a decrease in the free-radical diversity. The spin concentration shows a general decrease from the starting sediment to the products, indicating more unpaired electrons are stabilized in the products.

## Staten Peat

Acidification of humic acid standard (SPHA-Std, SPHA/S-Std) from this terrestrial environmental resulted in a 36–45% loss of the starting material. This loss is on the order of 40% greater than the loss associated with the Tanner Basin samples. The composition of this lost fraction is only approximate, however mass balance calculations do indicate that ~16–20% of the initial C, ~17–21% of the initial O, ~0.2–0.7% of the initial H, and ~1.7–2.0% of the initial N are evolved in this fraction (Table 4). As stated above, reduced volatile species are probably not evolved in significant quantities. Unlike the Tanner Basin samples, pseudo-kerogen was not generated in yields large enough for isotopic analysis. Atomic ratios for the standard and for the acidified humic acids (SPHA-HF, SPHA/-HF) (Table 2) indicate small but significant changes in the elemental composition. Carbon isotopic values are essentially identical between the standard and HF treated humic acids, whereas the hydrogen isotopes are ~10‰ lighter and $\delta^{15}N$ values are 1.7 to 2.0‰ lighter in the HF treated humic acid (Table 4).

As in the Tanner Basin samples, ESR analyses indicate a decrease in the line width in the HF treated humic acid while the spin concentration increases. This is interpreted as a decrease in the free radical diversity and an increase in the free radical concentration in the products. Overall, the Staten Peat humic acid, while not altered as much as the Tanner Basin sample, does appear to be affected by the HF acidification.

## Comparison with Sedimentary-Derived Proto-Kerogen

Comparison of Stuermer et al. (1978), proto-kerogen values obtained from Tanner Basin with values from the generated pseudo-kerogen show certain trends. The pseudo-kerogen shows higher carbon, and lower oxygen values relative to the proto-kerogen. The nitrogen content remains fairly consistent between the two sets of data. The $\delta^{13}C$ values of the proto-kerogen and pseudo-kerogen are very similar suggesting either natural fractionation in the proto-kerogen or induced fractionation due to the HF treatment (see Stuermer et al. 1978).

Electron Spin Resonance spectroscopy shows that Stuermer's proto-kerogen has a spin value of $1.88 \times 10^{17}$ spins/gm as compared with $6.72 \times 10^{15}$ spins/gm for Tanner Basin pseudo-kerogen. These data indicate that the generated pseudo-kerogen has less free radicals than naturally formed kerogen. The difference in line widths between proto- and pseudo-kerogen suggests that the generated pseudo-kerogen contains a greater diversity of free radicals. Comparison of g values indicate a higher oxygen free radical concentration in the pseudo-kerogen. The important point to note here is the susceptibility to change of natural proto-kerogen parameters due to contamination by pseudo-kerogens produced during the extraction process. Also to be noted is the % yield of the pseudo-kerogen (72–75%). This could lead to erroneous yields for both humic acids and proto-kerogen fractions isolated from recent sediments. Furthermore, artificial changes in the production of pseudo-kerogen may also be responsible for certain non-consistent results obtained with kinetic pyrolysis experiments in the presence of minerals (Halpern 1981).

A mechanistic understanding of the above results is beyond the scope of this paper. It is apparent that acidolysis is occurring and to some extent altering the starting organic material. It is also evident that the degree of alteration is dependent upon the type of organic matter. It may be that the less aromatic marine sample undergoes condensation and to some extent aromatization. However, even this relatively simple mechanistic process cannot be conclusively evaluted.

## 4 Conclusions

The hydrofluoric acidification of humic material resulted in (1) procedurally induced humic acid composition change and (2) the generation of an operationally defined pseudo-kerogen product which appears to be dependent on the chemical nature of initial organic matter. It is this artificial alteration due to procedural techniques which must be taken into account when working with recent humic acids and kerogen. The correction due to artificial alteration becomes all the more critical because the results are not constant, but vary according to the environment of deposition at the character of the deposited organic matter. It is therefore suggested that humic acids should be extracted prior to HCl or HF treatments. Additional future work should be done in determining if artificial alteration of natural kerogen occurs during HF or HCl treatment.

*Acknowledgement.* This study was supported by NASA Grant NGR 05-007-221.

## References

Abelson PH, Hoering TC (1961) Carbon isotope fractionation in formation of amino acids by photosynthetic organisms. Proc Nat Acad Sci 47:623-632

Aiken GR, McKnight DM, Wershaw RL, MacCarthy P (1985) Humic substances in soil, sediment and water. Wiley, New York 692 pp

Cline JD, Kaplan IR (1975) Isotopic fractionation of dissolved nitrate during denitrification in the eastern tropical North Pacific Ocean. Mar Chem 3:271-299

Halpern HI (1981) An investigation of mineral – kerogen interactions and their relation to petroleum genesis. Dissertation, Univ of California, Los Angeles

Hayes MHB, Swift RS (1978) The chemistry of colloids. In: (eds. Greeland DJ, Hayes MHB) The Chemistry of Soil Constituents Wiley, New York, pp 179-230

Ishiwatari R, Rohrback BG, Kaplan IR (1978) Hydrocarbon generation by thermal alteration of kerogen from different sediments. Am Assoc Pet Geol Bull 62:687-692

Kaplan IR, Smith JW, Ruth E (1970) Carbon and sulfur concentration and isotopic composition in Apollo II lunar samples. Proc Apollo Lunar Sci Conf Geochim Cosmochim Acta Suppl 1, 2:1317-1329

Peters KE (1978) Effects on sapropelic and humic proto-kerogen during laboratory-simulated geothermal maturation experiments. PhD Dissertation, Univ of California, Los Angeles

Rohrback BG (1979) Analysis of low molecular weight products generated by thermal decomposition of organic matter in recent sedimentary environments. PhD Dissertation, Univ of California, Los Angeles

Stuermer DH, Peters KE, Kaplan IR (1978) Source indicators of humic substances and proto-kerogen. Stable isotope ratios, elemental comparisons and electron spin resonance spectra. Geochim Cosmochim Acta 42:989-997

Thurman EM, Malcolm RL (1981) Preparative isolation of aquatic humic substances. Environ Sci Tech 15:463-466

Tissot BP, Welte DH (1978) Petroleum Formation and Occurrence: A New Approach To Oil and Gas Exploration. Springer, Berlin Heidelberg New York, 538 pp

Tissot B, Durand B, Espitalie J, Combas A (1974) Influence of nature and diagenesis of organic matter in formation of petroleum. Am Assoc Pet Geol Bull 58:499-506

Volger EA, Hayes JM (1980) Carbon isotopic compositions of carboxyl groups of biosynthetized acids. In: (eds. Douglas AG, Maxwell JR) Advances in Organic Geochemistry, 1979, Pergamon Press, Oxford, pp 697-704

# Subject Index

MIX
Papier aus verantwortungsvollen Quellen
Paper from responsible sources
FSC® C105338

If you have any concerns about our products,
you can contact us on
**ProductSafety@springernature.com**

In case Publisher is established outside the EU,
the EU authorized representative is:
**Springer Nature Customer Service Center GmbH**
**Europaplatz 3, 69115 Heidelberg, Germany**

Printed by Libri Plureos GmbH
in Hamburg, Germany